S0-BIS-840

Statistical Reasoning in Sociology

THIRD EDITION

Statistical Reasoning in Sociology

Third Edition

John H. Mueller

Karl F. Schuessler
Indiana University

Herbert L. Costner
University of Washington

HOUGHTON MIFFLIN COMPANY
Boston
Atlanta · Dallas · Geneva, Illinois · Hopewell, New Jersey · Palo Alto · London

Printed in the U.S.A.

Library of Congress Catalog Card Number: 76-13097

ISBN: 0-395-24417-X

Contents

Preface

Among books of its kind, *Statistical Reasoning in Sociology* offers a distinctive blend of substantive sociology and statistical method. In each chapter, our intent has been to present statistical reasoning as a means of discovering and verifying sociological knowledge. The book is thus as much an appreciation of the statistical method in sociology as an account of the technical elements of that method. Designed for use in a first course on statistics for undergraduates in sociology and related fields, this volume is intended to foster a statistical conception of social knowledge as well as to guide the student toward proficiency in statistical manipulations. It thus belongs as much to the liberal arts as to technical sociology.

As a textbook, this volume may be used without prerequisites. However, it will be more accessible to students with some background in sociology, since here and there we have used standard sociological terms without explanation. Similarly, it will present fewer obstacles to the student with college algebra, since a few of our discussions are algebraic in character. These mathematical arguments may be omitted, since their main threads are usually given in verbal form. All in all, the book is within easy reach of college students competent in arithmetic and seriously interested in the study of sociology.

The first chapter of *Statistical Reasoning in Sociology* is a brief essay on the history of statistics. In it, we highlight the social circumstances surrounding important statistical developments and emphasize the practical concerns for which statistics have been collected and analyzed. Statistics have seldom been collected so that statisticians would have something to do. Rather, statistical data and methods for analyzing them have been developed in response to the requirements of pressing problems and in the attempt to enhance our understanding of social and physical events.

Except for the first chapter, the book falls naturally into two parts: *descriptive statistics* and *inferential statistics.* Under descriptive statistics, we discuss first the nature of statistical data and factors affecting their accuracy. The ramifications of the latter topic permit us to stress the theme that statistical conclusions can be no better than the data on which they rest. Beyond this point, our coverage of descriptive statistics is fairly conventional: frequency distributions, measures of central tendency, measures of variation, and measures of correlation. We do give special consideration to ratios and rates, since these measures find so many uses in sociology, particularly in ecological and demographic studies.

Our presentation of inferential statistics begins with a discussion of probability statements as a method of coping with chance and uncertainty. Under sampling, we consider the uncertain relationship between sample and parent populations and the problems that arise because of this uncertainty; under statistical estimation, we consider inferences drawn from sample data when no suppositions about the population parameters have been made; under hypothesis testing, we consider inferences from sample data about the validity of a priori suppositions about the population. In all of these discussions, our object is no less to develop an appreciation for the general logic of statistical inference than to promote skill in the application of statistical formulas.

The third edition differs from the second edition in several minor respects and at least one major one. Where possible, we have replaced population statistics for the 1960s by comparable statistics for the 1970s, on the assumption that students will be more interested in the most current events than in those that are only relatively recent. We have also trimmed and polished the narrative in a number of places, both to enhance readability and to make room for topics that have become prominent in sociology in the last ten years or so. Topics appearing in the third but not the second edition include component analysis, multiple correlation and regression, path analysis, analysis of variance, and multivariate contingency table analysis.

In broadening the book's coverage, we have in effect created a longer book around a shorter one (or a shorter one within a longer one). The shorter book (corresponding to the second edition) covers only the most elementary topics, while the longer book covers several topics that usually appear for the first time in the intermediate course in social statistics. The shorter book requires competence in arithmetic alone, whereas the longer book presumes some knowledge of high school algebra; the shorter book will do for a one-semester course, but the longer book may be stretched to cover two semesters. The shorter book will do for the nonprofessional major and liberal arts student; the longer book will be more suitable for the professionally oriented major with an eye on graduate work in the social sciences.

As in previous editions, we offer problems and questions at the end of each section in every chapter. We have attempted to retain the best from previous editions and to add some new ones of equal quality. We anticipate that each instructor will rely partly, if not heavily, on bodies of quantitative data in which she or he has some special interest and competence. Instructors may also avail themselves of rather large tabulations scattered throughout the text that lend themselves to analysis by a variety of techniques. For example, tables appearing early in the text may be used later to illustrate the method of contingency table analysis; similarly, frequency tables appearing in the chapter on sampling may be used to illustrate the calculation of an average, a topic taken up early in the book. The listing of population sizes, suicide rates, and homicide rates for 229 Standard Metropolitan Statistical Areas in 1974 (Appendix, Table IX) can be used in a number of ways to considerable didactic advantage. These series, or fractions of them, may be used to create many new problems and questions, particularly those having to do with sampling.

We gratefully acknowledge the contributions of those who have assisted, directly or indirectly, in the creation of the third edition. Our own former students have suggested, by their questions and difficulties, textual modifications that purport to make this edition a more effective teaching aid. We also extend our appreciation to instructors who have generously shared with us their experience in teaching from the second edition. It is our pleasure to acknowledge the constructive suggestions in overall critiques supplied by the following professional colleagues: Nicholas L. Danigelis, University of Vermont; Phyllis Ewer, University of Illinois (Chicago); Richard J. Hill, University of Oregon; Louis Rowitz, Illinois State Pediatric Institute; Robert W. Suchner, Northern Illinois University; Judith M. Tanur, State University of New York at Stony Brook; and Robert F. Winch, Northwestern University.

Those who have assisted most tangibly in the third edition include Janice Kennedy and Beulah Reddaway, who have transformed rough copy into typed manuscript, and Patrik Madaras, who assisted us in locating appropriate data sources and in making numerous tabulations and computations. For their efforts in expediting our work, we are warmly grateful.

We are also indebted to the following publishers, journals, and organizations who gave permission to use copyrighted materials: *The American Journal of Sociology*; *The American Sociological Review*; Barnes and Noble, Inc.; *Bulletin of the Institute of Educational Research*; Bureau of Applied Social Research; Chapman and Hall, Ltd.; Columbia University Press; Gallup International; Iowa State College Press; Jossey-Bass, Inc.; *The Journal of Research in Crime and Delinquency*; *The Journal of the American Statistical Association*; The Macmillan Company; McGraw-Hill, Inc.; *The Pacific Sociological Review*; *Population Studies*; Prentice-Hall, Inc.; Princeton University Press; *Social Forces*; *Social Problems*; *Sociometry*; University of Chicago Press; John Wiley & Sons, Inc.; and Yale University Press.

We also wish to express our appreciation to the numerous authors whose scholarly work was published in these various sources and whose materials have enriched our volume. Their names are cited along with references to their work. We are grateful to the Literary Executor of the late Sir Ronald A. Fisher, F.R.S., to Dr. Frank Yates, F.R.S., and to Longman Group, Ltd., London, for permission to reprint Tables III, IV, and V from their book *Statistical Tables for Biological, Agricultural and Medical Research* (6th edition, 1974).

Finally, we note with profound gratitude the assistance, and especially the endurance, of our wives, Lucille and Virginia. Their direct contributions to this volume are noteworthy; their indirect contributions are enormous.

Herbert Costner
Karl Schuessler

An Introduction to Statistics 1

Twentieth-century statistics is a blend of methods and concepts from diverse sources. A brief summary of the rather curious history of these sources will be useful in understanding the nature of modern statistics and its social science applications.

The practice of compiling records of population and resources is very ancient, dating to the Biblical "censuses" and continuing intermittently through the Middle Ages and the Renaissance period in Europe. Although these sporadic and scattered records constitute statistics in a primitive sense, they resemble an inventory of families and possessions more closely than the compact tabulations and summary conclusions characteristic of statistical analysis. Statistical studies, as distinct from sheer compilations, are commonly presumed to have originated in the seventeenth century with the work of a studious London shopkeeper named John Graunt (1620–1674). His ingenious book, entitled *Natural and Political Observations . . . Made Upon the Bills of Mortality*, was first published in 1661 and constituted an analysis of the weekly reports of deaths and christenings from the 122 parishes (plus one "pesthouse" or hospital) of London from 1604 to 1661. The distinctive accomplishment of Graunt's *Observations* was not to make public the facts contained in these "bills of mortality," which had been published each Thursday for several decades, but to quote Graunt himself ". . . to have reduced several great confused *Volumes* into a few perspicuous *Tables*, and abridged such *Observations* as naturally flowed from them, into a few succinct *Paragraphs*. . . ." A distinguishing feature of statistical analysis, now as then, is to reduce masses of data into a compact and readily comprehensible form that will facilitate drawing conclusions. But it did not suffice for Graunt to give his observations without presenting also the foundation for them: "... I have taken the pains, and been at the charge, of setting out those *Tables*, whereby all men may both correct my *Positions*, and raise others of their own." Here again, contemporary statistical standards follow the lead of this early practitioner.

The publication of John Graunt's *Observations* impressed seventeenth-century officials and scholars with the potential utility of such analyses, both theoretical and practical, and led to improved records both in England and on the European continent. In 1693, such improved records permitted Edmund Halley (1656–1742), a distinguished English mathematician and astronomer, to develop the "life table" in essentially the same form as that employed by demographers and actuaries today. Halley described his table by saying that it "... shews the Chances of *Mortality* at all *Ages*, and likewise how to make a certain Estimate of the value of *Annuities* for *Lives*, which hitherto has been only done by an imaginary *Valuation*; Also the *Chances* that there are that a *Person* of any *Age* proposed does live to any other *Age* given. . . ." As this passage suggests, Halley did not regard his conclusions as a mere description of the particular population on which they were based (the city of Breslau in Silesia, now in Poland but then a German province); rather Halley interpreted his table as indicating the "chances" of survival that could be applied in computing life annuities for persons in different populations (for example, in London at a later time). Although contemporary demographers are somewhat more restrained in generalizing than Halley, this seventeenth-century document illustrates a persistent concern in statistics to extend conclusions beyond the immediate events described. Halley's use of observed proportions to infer chances also illustrates an early use of what has become a fundamental statistical concept—*probability*.

The theory of probability emerged as a serious scholarly effort in the latter half of the seventeenth century, even though it was centered on games of cards and dice. Prior attention to problems of probability had been unsystematic and often led to erroneous results, although we have no record of whether the wealthy gamblers who sought the advice of mathematicians in these matters suffered heavy losses as a consequence. Several leading mathematicians worked on problems of probability in the seventeenth century, including Blaise Pascal (1623–1662) and Pierre de Fermat (1608–1665), whose letters to each other laid out the fundamentals of probability theory before their appearance in published sources. The most notable of these early contributions to probability theory, however, was made by the Swiss mathematician, Jacques Bernoulli (1654–1705).

Bernoulli's Theorem, which was subsequently named the "Law of Large Numbers," asserted, in effect, that probabilities other than those known a priori could be determined by observation, and that the accuracy of such a probability estimate would increase with an increase in the number of independent observations. Bernoulli himself had no passion for observation, but he accorded it high importance in the development of knowledge, including knowledge of human affairs, in which he assumed an underlying order: "If all events from now through eternity were continually observed (whereby probability would ultimately become certainty), it would be found that everything in the world occurs for definite reasons and in definite conformity with law, and that hence we are constrained, even for things that may seem quite accidental, to assume a certain necessity and, as it were, fatefulness." In his mathematical proof, Bernoulli necessarily made assumptions about the conditions under which a set of observations would yield an accurate estimate of the true probability, but there is no reason to believe that he foresaw the practical difficulties that were to impede the realization of those conditions in the sampling of complex human events. In contemporary statistics, the nature of the sampling is accorded no less importance than the sample size, in recognition of the assumptions necessary for Bernoulli's mathematical proof.

The major attention of mathematicians during the latter half of the seventeenth century was concentrated on Newton's "celestial mechanics" and the calculus, and interest was thus deflected from probability theory. Another possible reason for the initially slow development of work on probability was the exceedingly tedious calculation required for some of the most interesting problems, especially those entailing what is termed the *binomial expansion.* This difficulty was overcome by Abraham de Moivre (1667–1754), a noted French mathematician who published a seven-page paper in 1733 entitled *Approximatio ad Summam Terminorum Binomii* $(a + b)^n$ *in Seriem expansi* [Approximation to the sum of the terms in the expansion of the binomial $(a + b)^n$]. This seemingly obscure development, initially used as a computational shortcut for gamblers and their mathematical consultants, constituted the discovery of the *normal curve* (see Chapters 7, 11, 13, and 14). De Moivre could not have foreseen all of the varied applications and far-reaching implications of his invention. But, like Newton, whose influence dominated scientific thinking at that time, de Moivre saw in his discoveries evidence for an underlying order in the universe: "And thus in all cases it will be found, that although Chance produces irregularities, still the Odds will be infinitely great, that in process of Time, those Irregularities will bear no proportion to the recurrency of that Order which naturally results from Original Design ... it is thus demonstrable that there are, in the constitution of things, certain Laws according to which Events happen. . . ." Although contemporary statisticians would express it in other terms, and claim mundane rather than supernatural significance for the order they discern, statistics remains a search for order in events, in spite of the "irregularities" of chance.

Many of the fundamentals of modern statistics had been laid down by the middle of the eighteenth century, but strangely, not by persons who would have referred to their work as "statistics." In Elizabethan (seventeenth-century) and early Georgian (early eighteenth-century) England, a "statist" was a functionary of the state, or as we might say, a politician—and with the somewhat shady reputation that is sometimes associated with that term in our own century. Thus Hamlet avers that

> I once did hold it, as our statists do,
> A baseness to write fair . . .

Statistical originally referred to things political, or having to do with statecraft, as its Latin root suggests. The earliest known university course with a statistical title (*Statisticum Collegium*, at the University of Jena, Germany, in the early eighteenth century) was a course of lectures on the constitutions, resources, and policies of the various states of the world. *Statistics* in this political sense was especially well developed in eighteenth-century Germany; a British gentleman in 1798 wrote, ". . . I found that in Germany they were engaged in a species of political inquiry, to which they had given the name of statistics . . . by Statistical is meant in Germany an inquiry for the purpose of ascertaining the political strength of a country, or questions respecting matters of state. . . ." Eighteenth-century "statisticians" were thus the forerunners of contemporary students of comparative politics; since their purposes commonly required some attention to population and resources (for example, to determine military potential or the potentially exploitable tax base), tabular statistics (*Die Tabellen Statistik*) had a close affinity with the demographic researches initiated by John Graunt. Graunt's close friend, Sir William Petty (1623–1687), was the author of an early book in this vein, which was published in 1690 under the title of *Political Arithmetick, or a Discourse concerning the extent and value of Lands, People, Buildings; Husbandry, Manufacture, Commerce, Fishery, Artizans, Seamen, Soldiers; Public Revenues, Interest, Taxes*. As this lengthy title suggests, economics, politics, and demography were closely related concerns even before the eighteenth-century German practitioners of *Statistik* created their distinctive meaning of the term and gave the modern field of statistics its name.

In the late eighteenth and early nineteenth centuries, the normal curve was rediscovered and became known as the *curve of error*, because of its utility in describing errors of observation in the position of heavenly bodies. Although contributors to these developments in mathematics and astronomy were scattered throughout the western world from Russia to the United States, the most notable among them were Pierre-Simon, Marquis de Laplace (1749–1827) in France and Carl Friedrich Gauss (1777–1855) in Germany. This new application of the normal curve in astronomy marked the beginning of a broader conception of the generality of the curve, demonstrating its usefulness in describing phenomena other than those associated with games of chance. This generalization and the development of associated concepts, for example, the *principle of least squares* (Chapter 7) and the concept of *probable error* (Chapters 13 and 14), eventually led to still other applications in biological, psychological, and social statistics and contributed to the fusion of the several streams that make up statistics in the contemporary sense.

The fusion of the several traditions heretofore mentioned—descriptive statistics in demography and economics, probability theory, *Statistik* in the political sense, and the theory of error—was personified in the remarkable nineteenth-century Belgian scholar, Adolphe Quételet (1796–1874). An astronomer and mathematician by training, Quételet became the Astronomer Royal of Belgium; but he was also the organizer and supervisor of official statistics in Belgium, and as president of the *Commission Centrale de Statistique* he organized a system that was a model for other nations. Quételet conceived of statistics as a method of analysis applicable to all the sciences, and hence foresaw the very general applicability of statistical reasoning as it subsequently developed. In accord with his conception of statistics as a broadly applicable method of inquiry, Quételet initiated statistical studies on topics to which statistics was then generally considered inapplicable. In a pioneering study in what was called *moral statistics*, he examined the effects of such factors as sex, age, education, climate, and seasons on criminal behavior (*Recherches sur la penchant au crime aux differens ages*, 1831). While such studies subsequently became commonplace, Quételet's endeavor was not only novel but highly controversial at the time. Some perceived a contradiction between the traditional "free will" conception of human nature and the regularities that Quételet reported: "Thus we pass from one year to another with the sad perspective of seeing the same crimes reproduced in the same order and calling down the same punishments in the same proportions. Sad condition of humanity! . . . We might enumerate in advance how many individuals will stain their hands in the blood of their fellows, how many will be forgers, how many will be poisoners. . . ." But for Quételet, as for twentieth-century practitioners of statistics in the social and behavioral sciences, the discovery of regularities in human behavior is not a denial of humanity, but a means of understanding it.

Following Quételet the term *statistics* was widely used in a sense approximating its contemporary meaning. But although the current meaning of the term had been established by the middle of the nineteenth century, some fundamental concepts of statistics remained to be developed. Intellectual life in the latter half of the nineteenth century was heavily influenced by Darwin's theory of evolution, and the arena of the most intensive statistical development shifted to the field of genetics. Francis Galton (1822–1911), a cousin of Charles Darwin, discovered that the children of very tall parents tend to be shorter than their parents, while the children of very short parents tend to exceed their parents in height. This phenomenon, which he demonstrated with graphs and tables in an 1885 paper, he termed *reversion* or the *law of regression* toward "mediocrity in hereditary stature" (that is, reversion toward the average height of the species). Subsequently he saw in his graphs and tables a more general feature which he named the *co-relation* between variables. He wrote: "Two variable organs are said to be co-related when the variation of one is accompanied on the average by more or less variation of the other. . . . The statures of kinsmen are co-related variables; thus the stature of the father is correlated to that of the adult son, and the stature of the adult son to that of the father. . . ." Galton devised a way to measure the degree of co-relation, but it was Galton's associate, Karl Pearson (1857–1936) who drew on his familiarity with prior work on the "curve of error" (which Pearson renamed the "normal curve") to create the *coefficient of correlation* in its present widely applied form (Chapter 9).

Pearson was an exceptional scholar—creative, energetic, disputatious, skeptical. Trained as a lawyer, he chose instead to pursue a career in mathematics. As a professor of mathematics, he devoted most of his time to the study of heredity. Passionately enthusiastic about the study of heredity, evolution, and biometrics, he is now remembered less for his contributions to biology than for his remarkable contributions to statistics. After deriving the coefficient of correlation (1895) he published a veritable flood of papers in which he developed a variety of statistical methods, primarily as tools for his analysis of heredity, including partial and multiple correlation (Chapter 10) and the "chi-square test" (Chapter 15). Pearson was unquestionably the leading figure in statistics during the early twentieth century and his influence gave a new cast to statistics, focusing attention not simply on the observation of events, but on the connections between events, that is, on the correlation between variables. This focus had been anticipated by Quételet, but it was Pearson who truly made statistics *bivariate* and *multivariate*. The statistics of correlation quickly found application, not only in the study of heredity, but also in psychology and education and subsequently in the analysis of social, economic, and demographic statistics. Pearson's influence on statistics was so great that some features of his philosophy of science (*The Grammar of Science*, 1892) came to be identified with statistics itself, for example, his uncompromising rejection of the concept of causation. It is thus ironic that multiple regression, which Pearson invented, has since become a statistical technique underlying *causal models* (Chapter 10). It is testimony to Pearson's influence, however, that the concept of causation entailed in such models is a multifactor conception that has been heavily influenced by Pearson's own statistical work.

Inferring cause and effect is usually—and quite properly—associated with experimental studies; hence, at first glance, experimental and statistical methods might appear to be antithetical. Such is not the case, however, and contemporary treatises on experimental design are commonly statistical treatises. Statistical treatments of experimental design can be traced primarily to the work of Sir Ronald Fisher (1890–1962), who, like Pearson, was a mathematician interested in the study of genetics. As a statistician at the Rothamsted Agricultural Station, Fisher found the statistical techniques previously developed of limited utility primarily because they were based on the assumption of very large samples and were thus imprecise for the small samples which were more typical of agricultural experiments. Fisher therefore devised new statistical techniques for experiments on small samples, and, in the process, clarified the meaning of the term *experiment* and contributed to more effective and efficient experimental designs. His work is represented primarily in *Statistical Methods for Research Workers* (1925) and *The Design of Experiments* (1935). His most notable contribution is known as the *analysis of variance* (Chapter 15), which is the basis for contemporary statistics of experimental design.

The twentieth century has also witnessed the development of statistical techniques for the analysis of the association of attributes, which find current application especially in the analysis of survey research. Quételet apparently devised the first measure of this kind in 1832, but it was Pearson's student, G. Udny Yule (1871–1951) who first made an intensive study of statistical devices for this purpose. Yule's explorations have been extended, especially by American statisticians and sociologists, to encompass the simultaneous analysis of several qualitative variables, and these techniques are now commonly employed in social research (Chapters 8 and 15). Twentieth-century statistics continues to develop in this and other ways far too numerous to describe here. As in the past, statistics develops as the research enterprise encounters new kinds of problems and hence requires new kinds of analytic tools.

As the foregoing brief sketch of the history of statistics suggests, statistics has many roots and multiple applications. The student interested in a more detailed history of the field should consult the Selected References at the end of this chapter, although the recent history of statistics remains yet to be written. As the references listed will indicate, statistics has drawn from a wider variety of scientific fields than the summary above would suggest. As for the many current applications of statistics, this abbreviated account barely hints at the great variety. The interested student may peruse the volume edited by Tanur (1972) for a more adequate sampling of readily comprehended current applications—from the analysis of historical documents to the population of whales, from the safety of anesthetics to the projections of winning candidates on election night. There are common threads running through statistical analysis, in spite of the great diversity of substantive applications. Statistical analysis, whatever the field of application, entails quantifying observations, ordering and analyzing those quantities to discern patterns and associations, and applying the logic of probability to distinguish random, unsystematic fluctuations from patterns and associations of substantive interest.

These general features of statistical analysis assume somewhat different forms in each branch of applied statistics. In social statistics, the quantification of observations entails not only counting people and events such as marriages, births, and crimes, but also the development of quantitative indicators for more elusive social phenomena such as value consensus in groups, role conflict, interpersonal attraction, and the integration of minorities (see Chapter 2). The distinctive subject matter and substantive questions of sociology have also resulted in an emphasis on certain analytic techniques that gives social statistics its own separate character, even while it shares common features with applied statistics in other fields. Social statistics has been influenced by demographic analysis and hence gives special emphasis to demographic rates and their proper norming and appropriate standardization to facilitate meaningful comparisons between populations. Social statistics has also been influenced by the analytic needs of survey research and hence special emphasis is given to the analysis of qualitative data and the association of attributes, for example, the association of social position (race, occupation, sex, and so forth) with expressed opinions and reported behavior. In social statistics, the statistics of experimental design has been adapted for the analysis of social psychological experiments and quasi experiments. And the complexity of events in social life has given rise to an emphasis on the simultaneous analysis of multiple factors which influence social phenomena, as in multiple regression analysis and path analysis. While none of these techniques is unique to social statistics, and these techniques share certain abstract properties with the generic focus of statistical analysis, their combination and application in the analysis of social events makes social statistics a field of application that is distinctive in its focus and emphasis.

Certain aspects of social statistics have become a familiar part of twentieth-century urban life, and in many instances statistical findings have become an undifferentiated part of the common culture. John Graunt dispelled the supposition, popular in his time, that there were in England three women for every man; the approximately even division between men and women in national populations is now a part of folk knowledge in industrial societies. Although we may be charmed by Tennyson's lines asserting:

> Every moment dies a man,
> Every moment one is born.

the conventional wisdom now recognizes that the population of the world is not as stable as these lines propose. Similarly, people now recognize intuitively (that is, as a part of the common culture) that education and income are correlated, that different income groupings express somewhat different political opinions, that the prevalence of violent crime varies systematically with distance from the center of the city, and that smoking cigarettes is a health hazard. Not only have statistical findings been incorporated into the common culture, but elementary statistical concepts have also become a part of a widely shared intellectual heritage. For the present-day college student, the concepts of percentage and average are part of a familiar vocabulary, and most contemporary college students probably do not need to consult a dictionary to attach some meaning, however imprecise, to such terms as *birth rate, correlation, normal curve, percentile, probability, variation,* and perhaps even *statistical significance*.

But the requirements of citizenship and of professional responsibility often demand more than a cursory acquaintance with statistical concepts and findings. Decisions with potentially great ramifications often rest on the proper use and interpretation of relatively complex statistical results. Social scientists and those in related professions find the current literature relevant to their work increasingly presented in the form of statistical analyses. An informed citizen must be able to absorb and understand statistical findings to know what is happening in the world. H. G. Wells (*Mankind in the Making*, 1904) anticipated a new educational requirement:

> The time may not be very remote when it will be understood that for complete initiation as an efficient citizen of one of the new great complex world wide states that are now developing, it is as necessary to be able to compute, to think in averages and maxima and minima, as it is now to be able to read and to write.

Wells' prophecy already seems outdated, for it hardly suffices "to be able to think in averages"; the "efficient citizen" now needs to be able to employ, and especially to understand and critically evaluate, even more intricate statistical reasoning in sociology, economics, political science, and related disciplines. The statistics already absorbed into the common culture do not suffice, and some additional educational aids must be provided. That is the purpose of the chapters which follow.

SELECTED REFERENCES

Fisher, R. A.
1950 Contributions to Mathematical Statistics. New York: Wiley. (Includes a biography of Fisher by P. C. Mahalanobis.)

Newman, James R. (ed.)
1956 "Statistics and the design of experiments." In The World of Mathematics, Vol. III, Part VIII. New York: Simon and Schuster.

Oxford English Dictionary
1933 Oxford University Press. See *statist, statistic, statistical*, and *statistics*.

Pearson, E. S.
1938 Karl Pearson: An Appreciation of Some Aspects of His Life and Work. Cambridge University Press.

Tanur, Judith M., et al. (eds.)
1972 Statistics: A Guide to the Unknown. San Francisco: Holden-Day.

Walker, Helen M.
1929 Studies in the History of Statistical Method. Baltimore: Williams and Wilkins.

Westergaard, Harald L.
1969 Contributions to the History of Statistics. New York: A. M. Kelley. (Originally published in 1932.)

2 Social Variables and Their Measurement

1 FUNDAMENTALS OF MEASUREMENT

Objects or events of a given kind may differ from each other in a number of ways. Persons vary in age, income, political preference, and place of residence. Small groups differ in size, cohesion, length of acquaintance among members, and frequency of interaction. Complex organizations vary in purpose, number of levels in the hierarchy of authority, and capital assets. Communities vary in size of population, crime rates, and attractiveness to migrants. Nation-states differ from one another in birth rates, standard of living, and degree of industrialization. The aim of social science is to analyze such variations in order to make them understandable. Such an analysis requires that objects or events be classified or measured with respect to the characteristics of interest.

In classifying and measuring, it will be useful to distinguish between the case, the dimension, and the description. The *case* is one of the population of objects or events whose characteristics the analyst seeks to describe. A *dimension* is a characteristic that may vary from case to case. The *description* is the attribute or measure for a particular case on a given dimension, and it presumes that there is some procedure for determining the proper class or measure for each case. In referring to a "tall, blond man," for example, the case is a person, the dimensions are height, hair color, and sex, and the descriptions are "tall," "blond," and "man." Similarly, in the statement, "France is more highly industrialized than Kenya but less highly industrialized than Japan," the cases are nation-states, the dimension is degree of industrialization, and the description is a simple ordering of the three nation-states along this dimension.

Analysis requires that abstractly conceived dimensions be translated into actual measures and classifications by some operation that results in giving a class label or measurement number to each case. We might *operationalize* the measurement of personal income, for example, by asking persons to tell us their income after taxes during the last calendar year, by asking them to indicate their total income last month, by tabulating the income reported on their most recent income tax form, or by asking them to check which of a series of income intervals includes their approximate total income during the last twelve months. Each would provide a similar but somewhat different way of describing the income of a given respondent. Persons might be classified into racial categories by asking them to indicate their race or by having an interviewer classify them on the basis of their appearance. The results of these two procedures for classifying populations by race would undoubtedly be in agreement in most instances, but not in all. Again, then, we have more than one way to operationalize the same underlying dimension. Even such a seemingly straightforward dimension as the population of cities may be determined in more than one way: the population of Indianapolis in 1975 might be (a) the number of people residing within the city limits as reported in the 1970 census; (b) the number of people within the Standard Metropolitan Statistical Area of Indianapolis as reported in the 1970 census; (c) the number of people in 1970—either within the city limits or within the Standard Metropolitan Statistical Area—adjusted in any one of a number of possible ways for growth from 1970 to 1975. The pattern of variation in American city sizes will differ somewhat depending on which of the several possible ways of describing city size the analyst adopts. When an abstractly conceived dimension has been operationalized in some specified way, we are justified in referring to it as a *variable*. Thus each of the ways of determining city size results in a different variable, all of which pertain to the dimension of city size.

The transition from conceptual dimension to operationalized variable constitutes the process of *measurement* or *classification*, and the results of this process (that is, cases classified or measured on selected dimensions) constitute *data*, which are the raw material for statistical analysis.[1] An explicit description of the procedure for generating data (the *operationalization* of the variables) helps clarify the meaning of the variables and permits other investigators to replicate the analysis at another time or with another set of cases. Such an explicit description of the operationalizing procedure also permits social analysts to discern potential shortcomings or peculiarities in the data that might otherwise remain hidden, and a critic may propose that the analyst has drawn an improper conclusion because of shortcomings in the variables themselves. For example, if an analyst measures the "quality

[1] Note that the term *data* (pronounced: day'·tah) is plural. Hence we say that data *are* the raw material for statistical analysis. The singular is *datum*—a term only rarely used in statistics because a single bit of information does not lend itself to statistical analysis.

of education" in high schools by the grades of students attending them, we may reasonably question whether the analysis pertains to the quality of education, as claimed, or to the grading practices of the teaching staff. Operational procedures do not necessarily measure well what the analyst claims they measure.

Quantitative and Qualitative Variables Cases may differ in quantity or quality, and variables are accordingly classified as quantitative or qualitative. A *quantitative* variable is one on which cases may differ in magnitude, that is, cases may be high or low, greater or smaller, more or less. Examples of quantitative variables are age, height, income, size of population, length of prison term, favorableness of opinion, and gross national product. The operationalization of a quantitative dimension locates each case along a scale of magnitude, and the magnitude for a given case is called a *variate*.

A *qualitative* variable is one on which cases may differ in kind; no scale or magnitude is implied, and cases differ in their *attributes* rather than in amount. Thus, sex will vary according to the attributes of male and female; a person's marital status may be single, married, separated, widowed, or divorced; nationality may be French, Norwegian, or Chinese. The attributes constituting the categories of a given qualitative variable cannot be arranged in order of magnitude, and one case is not "higher" or "greater" than another on such variables.[2]

The classification of variables as quantitative and qualitative will turn up apparent ambiguities. Sex is a qualitative variable when the case is the person; for collectivities, however, the percentage male or the percentage female is a quantitative variable. Occupation is a qualitative variable into which members of the labor force may be classified; on the other hand, occupational prestige is a quantitative variable since it ranges from low to high. Similarly, criminal offense is a qualitative variable, but the seriousness of criminal offenses is a quantitative variable. These examples illustrate the distinction between qualitative attributes and the social values (for example, prestige, seriousness, attitudes, attractiveness, and so forth) that often adhere to such attributes and which constitute a dimension distinct from the attribute itself.

[2] Some authors apply the concept *variation* only to variables that can vary in quantity. Therefore qualitative data are sometimes labeled as *nonvariable* or *enumerative*, emphasizing thereby that qualitative categories cannot be measured as magnitudes, but only enumerated as qualitative traits. Magnitude and quality therefore would be antithetical concepts. This text shares the position of those who define variation as covering both quantitative and qualitative characteristics.

This disputation in terminology can hardly be dignified as a semantic problem, since there is never a question of essential meaning if terms are consistently employed. It is simply a problem in labeling. The nomenclature employed by statisticians is by no means uniform, not only as applied to concepts here under discussion, but to other statistical concepts as well. This circumstance leads to confusion in the minds of students who obtain their statistical education from a variety of academic sources. In subsequent instances, this lack of uniformity will not always call for comment, but the student should constantly be alert to that possibility.

Quantitative variables are sometimes described by terms that suggest attributes rather than magnitudes. Various age groupings may be qualitatively designated, in order, as "infants," "children," "adolescents," "adults," and "senior citizens." Letter grades may be substituted for quantitative scores. The United States Census classifies sparsely settled areas with fewer than 2,500 inhabitants as "rural" and more densely settled areas as "urban." But no one should be misled by these labels into believing that the dimensions in question have thereby been transmuted from quantitative to qualitative; we have merely attached verbal symbols to arbitrary segments of a continuum. Anyone familiar with the symbolism still will interpret the terms quantitatively.

Measuring Quantitative Variables *Measurement* entails the assignment of numbers to represent the position of cases along a dimension of interest. Measurement presumes that the assignment of numbers for this purpose proceeds in accord with an explicit rule which can be applied with reasonably consistent results when the procedure is repeated. Measurement is thus a specialized form of description. Like other kinds of description, measurement is not an end in itself but is useful as an aid in analysis and understanding.

We measure someone's height by counting the number of inches (or other units of length, for example, centimeters) in a straight line from the top of that person's head to the floor on which he or she stands. In so doing, we assign a number (the count of inches or centimeters or other units) to represent the position of the case (the person) along a dimension (height), according to a rule of procedure (counting the number of units of measure along a straight line from head to foot). In like manner, we measure the size of small groups by counting the number of members. The age of persons is measured by the number of years or months since birth. We measure the income of households by totaling the number of dollars received by the members of each. These examples are relatively simple illustrations of measurement which do not indicate the full range of complexities and difficulties frequently encountered in measuring social variables. These difficulties arise in (1) devising appropriate units of measure, (2) specifying useful indicators for dimensions not amenable to direct measurement, and (3) gaining access to the requisite information.

The Unit of Measure Units of measure are human inventions, just as the inch and the centimeter are inventions and, in that sense, arbitrary. Although some units of measure may seem natural (for example, the count of members as a measure of group size), such units of measure are human devices for specific purposes. The size of a student body in a college or university is commonly measured, not by the number of different persons who are students, but by the number of "full-time equivalents" so that a student enrolled one-half time is counted as one-half a unit. A simple count of heads is thus not the only way to measure the size of an aggregate; the unit is designed to measure what the purposes of analysis require.

The basic requirement of a unit of measure is that any one unit should be equal to any other unit of the same kind, at least for the purposes of the measurement in question. Thus in measuring height, some inches are "leg inches" and some are "head inches," but these differences are disregarded, and any inch is treated as the equivalent of any other; even though we recognize that they are not equivalent in *every* respect, we are willing to treat each inch as making an equivalent contribution to the height of a person. Similarly, in measuring the duration of a marriage, the first month is treated as equal to the hundredth month, even though the partners may not find them experientially the same. In measuring income, a dollar added to the income of a poor family is given the same measure as a dollar added to the income of a millionaire, even though a dollar may be subjectively larger to the poor than to the rich. In measuring the popularity of political candidates by the percentage of the total vote they receive, each percentage point is treated as the equivalent of any other—for the purpose of this measurement—even though one percentage point may be gained in the ghetto and another may come from a "silk-stocking" district. In each of these instances, a given unit of measure is treated as equal to another of the same kind, even though the units may differ on dimensions other than the one in question. This is analogous to treating one ton of steel as equal to another ton of steel even though the first may be sold as steel plating and the second as steel wire. For other purposes (that is, in respect to other dimensions), the units are not identical; but measurement is designed to measure one dimension at a time and other characteristics are temporarily disregarded for the purposes of measuring the dimension of immediate interest. Measurement necessarily entails disregarding variations other than those upon which the measure is focused. This is not a flaw of measurement, but one of its strengths. However, a given analyst may be criticized for failing to select the right dimensions to measure or for failing to measure enough different dimensions to provide a description adequate for his or her purposes.

Units of measure may be infinitely divisible (for example, years, months, weeks, days, hours, minutes, seconds, milliseconds . . .) or the units may be indivisible (as in the number of children in a family, the number of marriages in a year, and so on). When the units of measure are infinitely divisible, the variable thus measured is called a *continuous* variable. No matter how fine the measurement, a continuous variable can always (theoretically) be measured in finer units; the micrometer which measures to one thousandth of an inch is superseded by a more precise instrument that will measure to one ten-thousandth of an inch; for many purposes such precision may be superfluous and unused, but the variable is continuous if infinitely small gradations along the scale are conceivable. Recognizing that a continuous variable is infinitely divisible implies that actual measures for continuous variables are always approximate measures, that is, they are *rounded* measures. The lack of precision entailed in rounding need not be a source of concern so long as the precision thus lost is inconsequential for our purposes.

When units of measure are indivisible, or when the unit of measure is some function of indivisible units (for example, the ratio of births to women of childbearing age), the variable thus measured is called a *discrete* variable. Even though the values of a discrete variable may cover a wide range, the actual values of the variable are limited to isolated values in that range. Household size may range from 1 to 25, but no household includes 3.4 persons; only certain values within the range are possible for real cases. If the number of women of childbearing age is 100, the ratio of births to the number of such women may be .22 or .23, but not .226 (implying 22.6 births for the 100 women, which is impossible). We note in passing that a discrete variable may give rise to an average that is never realized in an actual case; thus the average number of persons per household may be 3.4 even though no given household will include such a fractional number of persons.

A major issue in measurement is how to devise appropriate units of measure for dimensions of substantive concern. As the discussion above suggests, devising appropriate units of measure is largely a matter of identifying units which are always alike with respect to the dimension to be measured, however much such units might differ in other respects. Thus, devising units of measure necessarily requires an exceptionally clear specification of what the dimension to be measured *is* (and what it is not), expressed in terms that imply or suggest the units by which change or variation might be measured. For many social variables, appropriate units of measure have not been devised—and may never be devised. For example, although job satisfaction has been analyzed quantitatively, no one has defined a unit of measure for job satisfaction. Nor is there a standard unit of measure for the "social distance" between two or more ethnic groups. Stratification systems are recognized as being open or closed to varying degrees, but this conception has not been translated into a unit of measure for "openness."

The lack of direct measures for dimensions of interest to social analysts does not mean that quantitative analysis of those dimensions cannot be undertaken; lacking a unit of measure, social analysts use presumed correlates of such dimensions called *indicators* as surrogates for direct measurement. A substantial part of current social analysis is based on such indicators because usable units of measure have not been developed for many dimensions of interest. Variables with well-defined units of measure may, indeed, be used as indicators for other dimensions for which no appropriate units of measure have been devised. Thus, the percentage of eligible voters who fail to vote may be treated as a measure (indicator) of political apathy.

Indicators The height of a column of mercury is a familiar indicator of temperature. No one presumes that the height of the mercury column constitutes temperature in quite the same sense that length constitutes the number of inches from end to end, but the height of the column of mercury is a dependable correlate of temperature and thus serves as a useful measure of it. An indicator is an accessible and dependable correlate of a dimension of interest, which correlate is used as a measure of that dimension because direct measurement of the dimension is not possible or practical. Thus, attitudes are assessed by responses to carefully selected questions and statements, and the responses are summarized to constitute an indicator (an accessible correlate) of the attitude dimension. Job satisfaction may be measured by the rate of absenteeism, by the amount of "turnover," or by self-reports of satisfaction; all are reasonable indicators of job satisfaction because we assume that each is a correlate of satisfaction, for which no units of measure have been devised to permit direct measurement. Discrepancies between the incomes of similarly qualified men and women, or between the occupational distribution of males and females in the labor force, serve as indicators of sex discrimination. Public opinion polls provide indicators of public sentiment about a variety of current issues. The research achievement of scientists is measured by their number of publications, or by the citations to their work in the publications of others. An investigator may depend on the rankings (or ratings) of knowledgeable informants or "judges" to provide indicators of performance skills, community leadership, or the political integration of societies. Even social variables that are, in principle, directly measurable are commonly measured by indicators in practice. The age of persons is frequently determined simply by asking people to tell their age, on the reasonable assumption that this response will be a dependable correlate (indicator) of the number of years since their birth. Census enumerators do not actually "count heads" but ask respondents in each household to report on the number and characteristics of persons living there. Social and social psychological research consists largely of the analysis of indicators of one kind or another.

Research scholars in the social sciences have exercised remarkable ingenuity in devising indicators for the measurement of social dimensions. But, ingenious or not, indicators are always imperfect measures of the dimensions they represent, and they do not covary exactly with the dimension we seek to measure for a variety of reasons. Some respondents may misinterpret the questions asked. Factors other than job satisfaction may affect absentee rates. In measuring sex discrimination by the discrepancy between the incomes of "similarly qualified" men and women, the qualifications are never exactly identical. The work of a scientist may be cited or not cited for reasons other than the research achievement that work represents. In summary, there is slippage between the dimension of interest and the indicators devised to measure it indirectly. Since any given indicator may be "off its mark," several indicators for the same dimension may be used. In this way, one indicator serves as a check on another, and notable discrepancies in the conclusions suggested by an analysis based on different indicators that are supposed to measure the same thing serve as warning that one or more may be heavily flawed. Although indicators always provide imperfect measures, they commonly provide the only available measures, and our understanding of social life depends heavily on the use in research of these surrogates for the dimensions whose variation we seek to understand.

The term *social indicator* has been given a more specialized meaning than the more generic term employed in the discussion above. Following the lead of economists who have devised a number of periodically reported "economic indicators" designed to reflect the health of the economy and to explore the reasons for its fluctuations (for example, in the unemployment rate, the cost of living index, the balance of payments in international trade, inventories of durable goods, the money supply, and so on), social scientists have perceived the need for analogous "social indicators." Such indicators are designed to reflect the degree to which the society is successfully achieving major social goals such as adequate health care, opportunities for education and leisure, safety and security in one's home and community, and general well-being. Although interest in social indicators has motivated the exploration of some new ways of measuring the achievement of social values, the major accomplishment of researchers focusing on social indicators has not been innovation in measurement technique, but the beginnings of a series of periodic "social soundings" on national samples, with comparable data gathered on each occasion. If such arrangements can be continued for an extended period of time, the resulting time series data will provide a valuable and hitherto unavailable resource for social analysis. Like other indicators in social science, these social indicators, designed to assess the general welfare of the populace, consist of presumed correlates of dimensions that cannot be directly measured.

Information Access Measurement, whether direct or indirect, requires relevant information about the cases being studied, and such relevant information may be difficult, if not impossible to come by. Gathering particular kinds of data about family life by observation would threaten the sense of privacy that is understandably valued and legally protected. Employees in complex organizations conceal certain of their activities that would reflect unfavorably on them or their organization. Deviance is commonly hidden and attempts to observe deviant behavior directly may so change the circumstances that the behavior itself is changed—or shifted to other times and places.

Public agencies record limited information about their clients because of the expense of recording more complete details, and such records are protected by regulations pertaining to confidentiality. People forget, or remember selectively, the events of their own past and are unable to report accurately many details of events in which they were participants. The victims of crimes commonly fail to report the offense to law enforcement agencies; consequently official crime statistics are far from complete. As these examples attest, social measurement is beset by many problems of access, and the acquisition of complete data on many social phenomena is effectively precluded by reason of social taboos, private interests, norms of personal privacy, and the fallibility of human memory. Social scientists adapt by studying indirectly what cannot be observed as it happens, by devising ways to supplement fragmentary data, and by making do with less than complete information. With the assurance of confidentiality and anonymity, people may be willing to report what they would not permit to be observed. Official crime statistics can be supplemented by surveys inquiring about criminal victimization to obtain a more complete picture of unreported crimes. Individual records, without identifying names or features, may be released for research purposes under certain circumstances. The incomplete report of a single participant may be filled out by the reports of others who were also involved. But the lack of access to appropriate information frequently requires that the social analyst manage with crude approximations and rough classifications in place of more refined measures that can be conceived but not implemented.

Types of Scales Anyone familiar with the varieties of social data readily comes to regard some measures as precise and accurate while other measures seem rough and crude by comparison. It is possible to classify statistical variables not only on the basis of the precision and accuracy of information content they embody, but also on the basis of the formal relations among the elements composing them. Stevens' widely quoted classification of ratio, interval, ordinal, and nominal scales is based on this criterion of possible relations between scale values.

Ratio Scale A ratio scale is a scale as that term is understood by the lay person. The elements of a ratio scale may be expressed as multiples of one another. Age and weight are commonplace examples. If Tom is 10 years old, and Brian is 5, Tom is twice as old as Brian; if Tom is 150 pounds and Brian is 75, Tom is twice as heavy as Brian. Thus in ratio scales, we can express one value as a multiple of another, or determine how many times the one is larger (or smaller) than the other. Moreover, whenever such ratios may be formed, we may also measure the scale distance between them. Tom is 5 years older than Brian; 75 pounds heavier. Additionally, we may array elements in transitive order according to their magnitude. If Tom is taller than Brian and Brian is taller than Karl, then Tom is taller than Karl. In sum: ratio scale permits us (a) to express any two values as a ratio, (b) to compute the scale distance between them, and (c) to put all elements in an array from low to high. The versatility of the ratio scale relative to the others is a consequence of the additivity of scale units and the fixed zero, both of which may be regarded as the essential requirements of a ratio scale.

Interval Scale Now let us suppose that we retain the requirement of additive intervals but lift the restriction of a fixed origin. The result will be an interval scale. If we are free to locate the origin anywhere on the scale, we can no longer express values as ratios, since ratios will change as the origin (zero) changes. Let us consider the following example of elevation above water level: if we find three poles that extend 1, 2, and 4 feet, respectively, above the water level while resting on the bottom at an indeterminate but uniform depth, we may state that the second is 1 foot longer than the first and the third is 2 feet longer than the second. If the water level drops 1 foot, these intervals remain unchanged but the ratios lack such constancy. The heights above the water level are now 2, 3, and 5 feet, respectively, and the ratio of the second to the first is no longer 2 to 1 but 3 to 2. The distances between heights will remain unchanged if we arbitrarily set the origin at 6 inches above the water level or at any other arbitrary point; but with each shift in the origin, the ratios between heights will change. Since the origin from which our measures are taken is arbitrary, ratios have no meaning.

Interval scales are clearly less versatile than ratio scales because they permit us to establish only the distance between values but not the ratios between them. Thus we may regard an interval scale as a defective ratio scale. Although many statistical techniques are applicable to scales having equal intervals, including ratio scales, few if any interval scales (that is, interval scales with an arbitrary origin) have been devised to measure the sociological aspects of events.

Ordinal Scale Let us now lift the requirement that intervals be equal and additive: the distance from 1 to 2 will not necessarily equal the distance from 2 to 3. Under these circumstances, it will be impossible either to express elements as ratios or to measure the distance between them. All we can do is put our elements in rank order. Strictly speaking, we cannot assign numbers to elements at all; rather, we must utilize words—"first," "second," "third," etc.—to describe them. When our scale is not composed of additive intervals, to say nothing of a fixed origin, we refer to the scale on which the elements lie as an ordinal scale. If we regard an ordinal scale as more limited than an interval scale, we may arrange the foregoing scales according to their relative capabilities as follows: ratio scale as most capable, interval scale next most, and ordinal scale as the least capable of all.

Nominal Scale What of nominal scales, the fourth type in Stevens' classification? Nominal scales are scales in name only and correspond to qualitative data. As defined, a variable is any characteristic that may vary on successive trials, but a characteristic, to be a variable, need not vary in amount. It may vary by kind. If our variable varies by quality rather than by quantity, we may refer to the set of attributes which provide the range of variation as a nominal scale. We may even substitute numbers for words, but we do not thereby transmute our nominal scale into a quantitative scale. We may assign "1" to males and "0" to females, but by that operation we have not transformed sex into a quantitative variable.

The foregoing classification has this importance in social statistics: each type of scale is amenable to certain statistical operations but not to others. If we fail to recognize the peculiarities inherent in each type of scale, it is possible that we may employ inappropriate techniques and thus reach false conclusions.

QUESTIONS AND PROBLEMS

1. Define the following concepts:

case	unit of measure
dimension	continuous variable
variable	discrete variable
data	indicator
quantitative variable	social indicator
qualitative variable	ratio scale
variate	interval scale
attribute	ordinal scale
measurement	nominal scale

2. List three dimensions on which colleges and universities vary that you would consider interesting for analysis. For each dimension listed, devise one or more ways of operationalizing the dimension and indicate how you might collect the data to describe (measure or classify) each college or university being studied. Repeat for college and university *students*.

3. In some social surveys, respondents are classified as "favorable" or "unfavorable" with respect to a specific issue on the basis of the respondents' answers to survey questions. Is this a quantitative variable or a qualitative variable? Explain your answer.

4. It has been proposed that people who interact more with each other also like each other more. How might you measure the amount of interaction between pairs of persons? Is the measure you propose a direct measure, with units of measure that constitute "bits" of interaction, or is your measure an indicator? How might you measure the degree to which one person likes another? Is this measure a direct measure with units of measure that constitute "bits" of liking, or is your measure an indicator? Explain your answers.

5. Individual A endorsed 50 out of 100 items reflecting racial prejudice, while B endorsed 25. Is it reasonable to conclude that A is twice as prejudiced as B? Explain your answer.

6. Family income measured in dollars is a ratio variable. If family incomes are classified as "low," "medium," or "high," is this three-valued variable a ratio variable? An interval variable? An ordinal variable? A nominal variable?

7. Check the operations applicable to values in each of the following scales.

Type of Scale	Operation			
	Classify	Order	Subtract	Divide
Ratio				
Interval				
Ordinal				
Nominal				

2 ERRORS OF MEASUREMENT AND CLASSIFICATION

Measurement and classification are human inventions applied by human observers. In social research, they are applied to human institutions and human groups. Social measurement is thus infused with human fallibility, and hence infected with error. In some instances, measurement and classification error are inconsequential, and the conclusions based on social data are little affected. In other instances, errors of measurement and classification are substantial; they probably introduce serious distortions in the findings, and they may be responsible for misleading conclusions. Even though errors of measurement and classification can never be entirely eliminated, the social analyst strives to reduce such errors to a minimum and to limit their impact on substantive conclusions. In order to minimize error, it is necessary to recognize the various sources of misclassification and faulty measurement, and to develop techniques to guard against errors from each such source. Accordingly, we will consider some of the major sources of measurement and classification error.

Errors in Information Processing Information does not transform itself into statistical tables; it is transcribed, coded, scored, punched, counted, and sorted. Such data handling operations introduce error: the coder misreads, the keypunch operator hits the wrong key, an item is omitted in the count, and the inadvertent recording of a "1" instead of a "7" reclassifies the septuagenarian as a preadolescent.

Some protection against errors of processing is afforded by meticulous care in data processing, by clear instructions and careful training of data processors, and by persistently checking and rechecking as the processing goes on. Even so, some errors in processing are likely to remain, and attempts to eliminate them completely in processing masses of data are exorbitantly expensive.

Response Error When responses are unstable or transitory, the data are subject to errors of measurement due to response variation. Social investigators sometimes ask for more exact information than respondents are able to give. They may request opinions on matters about which the respondent is still vacillating. Interviewers may elicit responses that are based more on a transitory mood than on the enduring features of the respondent that the interviewer seeks to measure. For example, if students are asked to report the number of hours per week that they spend studying, most could only guess and such guesses are inaccurate to varying degrees. Respondents may not be sure whether they "strongly agree" or "agree" with an attitude statement, and their successive responses to the same statement may vary as a result.

One of the major ways of guarding against response variation is by the use of multiple responses. Respondents are asked, for example, to respond to a series of attitude statements instead of a single statement. The attitude is then inferred from the total set of responses, and response error on one item is presumably counteracted by response error in the opposite direction on another. Responses may also be made more reliable by asking the respondent to give specific information as opposed to a general summary response. Thus, instead of asking adolescents to indicate how many times they have been delinquent, self-reported delinquency is ordinarily based on the more detailed report of the number of times the respondent has engaged in specific classes of behavior during a specified time period.

Observer Error Instead of depending on the responses of the subjects of study, an investigator or his associates may seek to observe the events of interest. Although more expensive and time consuming than asking subjects to report, such observations are usually more reliable also. Nonetheless, observational data are also subject to measurement and classification error. The observer commonly fails to see all that is going on, and is especially likely to overlook subtle features of interaction. The observer may also misinterpret what he sees; a playful poke may be classified as an act of hostility—or vice versa. An observer trying to see everything at once may omit, record inaccurately, and misclassify.

To minimize observer error, the intensive training of observers and the detailed specification of what is to be observed and how it is to be classified are often helpful. Discrepancies in the reports of multiple observers alert the investigator to problems of observer error, and the discussion of discrepancies clarifies definitions and increases the sensitivity of the observers.

Indicator Error The research instrument or instructions may be responsible for errors of measurement and classification. Attitude items may be so constructed that they reflect experience rather than attitude. Ambiguities in instructions may make responses noncomparable; for example, if officials in complex organizations are asked to indicate the number of people they supervise, are they to count a part-time employee as a whole or as a fraction of a supervisee? If some interpret the instructions one way and some another, the investigator will have error in some reports, whichever construction may have been intended.

Indicators may reflect less than the investigator needs to know, more than was ever intended, or simply be off their mark. Although multi-item indicators provide some protection against the error entailed in a single item of information, the simple accumulation of items may not correct for indicator error. The reduction of indicator error may be accomplished by a careful consideration of, and control for, the other variables that are likely to influence the indicator, and some assurance is gained from knowing that different indicators are in reasonably close agreement; but there is no certain guarantee against errors in indicators and the consequent flaws in conclusions. Ultimately, indicators are judged according to their utility in explaining and understanding the phenomena of interest, but this criterion does not provide clear guidelines for the construction of useful indicators.

Misrepresentation as Error Records may be falsified, behavior may be a façade, and responses to an interviewer's questions may be calculated to mislead. Such misrepresentation may constitute a deliberate falsehood, or consist merely in "stretching" the truth. A public agency may be diligent in searching out evidence suggesting its successes and much more casual in searching out negative instances. Respondents may give the socially desirable response if they can possibly interpret the meaning of the words to permit it—or they may simply give responses that have no basis in fact. Giving the socially desirable (that is, respectable, approved, admired, or expected) response is so widely recognized that instruments have been devised specifically for the purpose of measuring a respondent's inclination to lean in the socially desirable direction. No comparable clues are available to help identify communities that minimize the number of crimes reported or maximize the number cleared by arrest. No device has been developed to help ferret out calculated misrepresentations in the statistics describing nation-states. Nor is there an easily applied procedure for identifying those items of information that have been distorted in an attempt to cast the most favorable light on the actions—and appropriations—of public and private agencies.

Sampling Error as Measurement Error In describing collectivities and aggregates, the measures are commonly summaries of the characteristics of the persons constituting them, and the description may be based on a sample of those persons rather than the totality. Because a sample, no matter how carefully drawn, does not perfectly represent the population of interest, some variation in these measures of collectivities and aggregates is sampling error; in considering the accuracy of such measures, sampling error becomes, or at least may be regarded as, measurement error. Public opinion polls, for example, are based on samples, and the measures they provide (support for the President, sentiment in favor of a public policy, and so forth) are subject to measurement (sampling) error. Similarly, surveys designed to determine the magnitude of criminal victimization are based on samples of respondents; the measures they provide are subject to error due to sampling variation.

The concept of sampling error is discussed later (Chapters 12–15); here we mention only that measurement error due to sampling variation will be minimized if the samples are large and randomly drawn. Indeed, samples may provide aggregate measures that are subject to *less* measurement error than complete enumerations because greater care can be exercised in minimizing *other* sources of error in the sample study, and such care usually more than compensates for the relatively small sampling error, provided the sample is large and randomly drawn.

As the foregoing discussion suggests, measurement and classification error constitute central problems in social research methodology. Techniques for collecting and producing social data are useful only to the degree that they provide the information required with relatively little error. The statistical analyst has a clear interest in securing data with minimum error, and an obligation to consider the implications of measurement and classification error for the conclusions drawn from data.

Random, Constant, and Correlated Measurement Errors Errors of measurement and classification are said to be *random errors* if they vary from case to case in a way that is unrelated to anything else, including the measurement error on other variables. Random errors behave *as if* the amount and direction of error were determined by drawing signed numbers from a hat, with one-half of the numbers in the hat being positive and one-half negative, and the average of the numbers being zero. Random errors of measurement are troublesome in statistical analysis, but they do behave in predictable ways. In some types of analysis, random errors cancel each other and their effect is nullified. For example, if age were subject only to random measurement error, some ages would be recorded too high and others too low, but the average age for a large group would be unaffected (or only trivially affected) because the positive errors would counteract the negative errors in summing. For other types of statistical analysis, random errors do not nullify each other. Random errors of measurement have the expected effect of increasing the standard deviation (Chapter 7), and the expected effect of decreasing the correlation between two variables (Chapter 9).

If measurement error is identical for every measurement, such error is referred to as *constant error*. For example, if everyone over-reported their charitable contributions by $100, the average contribution would also be increased by $100, but the difference in the amount contributed by any two persons would remain the same, as would the difference in average contribution for any two groupings, since each would have been inflated by a constant amount. Exactly identical error for every measurement is rare, but measurement error commonly includes a constant component. It is *as if* the error were determined by drawing numbers from a hat, but the average of the numbers in the hat is not zero; consequently, each measure is inflated (or deflated) by the same amount on the average. Constant errors, or approximately constant errors, evidently entail the possibility of misleading conclusions, but for certain kinds of statistical analysis, constant errors are inconsequential. For example, the product-moment correlation coefficient between two variables (Chapter 9) will not be affected at all by constant errors in either variable.

If measurement error varies from case to case and the magnitude or direction of the error differs depending on other characteristics of the case, such error is referred to as *correlated error*. Correlated error behaves *as if* the error were determined by drawing numbers from hats, but a different hat (containing different numbers) was used for males and females, or for rich and poor, or for other differentiated groupings. Thus correlated error arises if responses are more likely to be miscoded in a given direction for low income respondents than for those more affluent, if ambiguous instructions are interpreted differently by respondents in large as compared to small cities, or if misrepresentation in the socially desirable direction is more marked among older as compared to younger respondents. The effect of correlated errors on statistical conclusions depends on a number of factors, for example, which variables the errors are related to, how strongly the variables and the errors are related, the nature of the relationships between variables if they were measured without error, and how the correlates of error are distributed among the cases being analyzed. For example, if males are more likely than females to guess low in reporting hours of study, males would appear to study less than females even if there were actually no difference between the sexes, while the difference between males and females would be underestimated if males actually studied more, but overestimated if males actually studied less. If the actual difference between males and females were slight and the differential tendency of males to guess low were strong, males might appear in the data to study less even though they actually studied more! The average number of hours of study per week among all students would be much underestimated if most students were males and there was a marked tendency for males to guess low, but it would be less heavily underestimated if most of the students were female. If females predominate among students who live in organized housing units while males predominate among other students, the differential tendency of males and females to guess low would also carry over to the comparison of study time among those who do and do not live in organized housing units, as it would infect any other comparison between groupings having different sex distributions. As this much simplified illustration suggests, the ramifications of correlated measurement error may be difficult to trace, and detailed information on a number of matters is necessary to make even rough estimates of the effect of correlated errors on research conclusions.

The Reliability and Validity of Measures and Classifications Measures are said to be reliable to the degree that they are replicable; estimates of reliability are based on agreement or correlation between replications. A replication may be the application of the same measurement to the same cases at different times (*test-retest reliability*), the application of the same measurement to the same cases by different observers (*inter-observer reliability*), or the application to the same cases of nonidentical measures which purport to measure the same characteristic of those cases (for example, *split-half reliability* or *equivalent-tests reliability*). If the replications yield measures which are highly correlated, the measurement or classification is said to be highly reliable. More complex estimates of reliability are extensions and elaborations of these relatively simple ideas. The extensive technical literature on reliability[3] and procedures for estimating it are beyond the scope of this book and the interested student is referred to the references listed at the end of this chapter for a more extensive but relatively nontechnical discussion of this topic.

The *validity* of a measure refers to its "purity," that is, the degree to which it measures what it is supposed to measure—and only what it is supposed to measure. When measurement proceeds by counting uniform units of measure and those units actually stand for bits of what is being measured (for example, inches as units of length, persons as units of population size), reliability and validity are indistinguishable; the measure is as valid as it is reliable. But validity becomes an issue distinct from reliability when measures are indirect, that is, when indicators are used. An indicator may be highly reliable (replicable) but still measure very "impurely" the dimension of interest. Then, the question "How valid?" is largely a question of how well the stand-in indicator serves as a substitute for direct measurement. Since an indicator may serve well as a stand-in for some purposes and not others, the validity of a measure may be relatively high or low depending on the purpose for which the measure is used.

Determining the validity of measures—and improving measures with low validity—presents a complex challenge in social science research. Technical procedures for estimating validity are typically based on the correlation or agreement between measures that purport to measure the same dimension

[3] The technical definition of reliability is the proportion of measured variation that is true variation. We have used a simpler definition above because the meaning of this technical definition rests on statistical concepts not yet introduced.

in different ways, but may also be based on the consistency with which measures of the same dimension are related to other variables. In either case, they measure only imperfectly the "purity" (validity) of the measures in question, and the validity of such indicators of validity is as problematic as is the validity of other indicators. The basic issue is the degree to which "impure" measures may distort conclusions from data, especially conclusions pertaining to how social events are interconnected with each other, which is the primary purpose of social analysis.

The Implications of Measurement Error in Statistics The results of statistical analysis require interpretation. The thoughtful interpretation of results, in turn, requires a consideration of how measurement and classification error may have affected the data analyzed. No general formula for assessing the effects of measurement and classification error can be stated, but the prudent interpreter of statistical results will attend to the probable sources of measurement error, the probable magnitude of such error, and the relative contributions of random, correlated, and constant components to total measurement error. The following rules of thumb may be useful in assessing the probable impact of measurement and classification errors of each type.

(1) ***Random errors*** of measurement and classification have little effect on central tendency (Chapter 5) but will inflate measured variation (Chapter 7) and also affect other statistics which rest on the concept of variation (for example, correlation, Chapter 9). Errors of information processing commonly behave like random errors, and errors from other sources usually have random components also.

(2) ***Constant errors*** of measurement distort measures of central tendency, but have little if any effect on other statistics. Errors from all sources *may* have a constant component, but processing errors and response errors usually include little if any constant error, while misrepresentation is the source that is most suspect in regard to constant error.

(3) ***Correlated errors*** of measurement and classification *may* affect statistics of all kinds, and the nature and magnitude of the effect depends, not only on the magnitude of error, but also on the correlates of error and their distribution in the cases being analyzed. Correlated errors have especially high potential for affecting statistics which pertain to the relationship between variables and to the differences between subgroups. Since these are central in the statistical analysis of social science data, correlated errors are especially threatening in social statistics, and the careful interpreter of social statistics will attempt to determine the most likely correlates of measurement and classification error and how such errors might affect the findings.

QUESTIONS AND PROBLEMS

1. Define the following concepts:

error of measurement	sampling error
error of classification	random error
response error	correlated error
observer error	constant error
indicator error	

2. The unemployment rate is the percentage of the labor force that is unemployed. To determine the unemployment rate at a given time, it is thus necessary to determine how many are in the labor force (i.e., working or seeking work) and how many are unemployed. Consider ways of determining who is in the labor force and how measurement error might be introduced in this count. Consider ways of determining who is unemployed and how measurement error might be introduced in this count. Focusing on the unemployment rate (rather than on the classification of individuals), which of the errors you anticipate are probably random errors? Correlated errors? Constant errors?[4]

3. Suppose you are an investigator seeking to measure the involvement in delinquency of a set of adolescent boys by asking them to self-report their delinquent acts. Respondents are provided with descriptions of certain classes of delinquent acts (for example, taking things worth more than $50, threatening someone with physical harm unless they surrender their money, and destroying someone else's property for the fun of it) and after each description, each respondent is asked to check one of a set of response alternatives (none, once, twice, three or more times) to indicate how many times he has engaged in an activity like this during the past year. Would you anticipate response error in the replies? Indicator error? Misrepresentation? Would the errors you anticipate probably be random, correlated, or constant? Explain.[5]

4. An investigator wishes to measure the degree of industrialization for a number of nation-states. For this purpose, the investigator might use as indicators: (a) the per capita gross national product (that is, the value of all goods and services produced in that nation-state in a given year divided by the total population); (b) the proportion of the labor force employed in nonagricultural occupations; or (c) the per capita energy consumption (that is, total energy consumption expressed in kilowatt hours divided by the total population). Consider the measurement errors to which each

[4] For a brief description of how such information is collected by the United States Census Bureau, see Conrad Taeuber, "Information for the Nation from a Sample Survey," in Judith Tanur, ed., *Statistics: A Guide to the Unknown*, San Francisco: Holden-Day, 1972.

[5] For a study exploring measurement errors in such devices, see John P. Clark and Larry L. Tift, "Polygraph and Interview Validation of Self-Reported Deviant Behavior," *American Sociological Review* 31 (August, 1966): 516–523.

indicator is probably subject—processing errors, response errors, observer errors, indicator error, misrepresentation, and sampling error. On the basis of your assessment of the likely errors of measurement, which indicator would you consider best? Explain.

3 ROUNDED MEASURES

For reasons previously described, most measurements are destined to remain to a greater or lesser degree imprecise. Available figures on births and deaths, marriages and divorces, public opinion and morale, migration and city size, industrialization and employment, income and prestige can never be more than approximate. In practice, therefore, we must terminate at some convenient stage in the process of measuring the attempt at precision. In colloquial speech, we "round off" when further exactitude is neither possible nor necessary. Besides, rounded figures are more easily manipulated, and therefore represent a justifiable economy of effort, even at the sacrifice of accuracy, provided such accuracy is not essential. Thus, we arrive at two kinds of values: *true* values, which are generally unobtainable, and *rounded* values, which are employed in actual calculation. But rounding should not be capricious; it must be systematically carried out according to established rules.

Rounding Procedures If measures are only approximate, they must obviously be either too high or too low. When values have been rounded to the *next lower*, or last, unit, they are always too low; when rounded to the *next higher* unit, they will always be too high; and when rounded to the *nearest unit*, they will be either too high or too low. We may illustrate these differences in rounding procedures with an example of age. A person may quote his age as 20. If he follows the popular practice of giving his age as of last birthday—a practice the United States Census follows as well—he will convey the information that his true age is somewhere between exactly 20 and not quite 21. He has accordingly rounded his age to the next lower whole number on the assumption that greater accuracy is not called for. This *rounding error*, which may be as much as a whole year, is of course not a mistake but a known consequence of the system of rounding employed.

For actuarial purposes slightly greater precision is required. To an insurance statistician, a quoted age of 20 signifies an age between 19.5 and 20.5 years. This is rounding to the nearest whole year, or nearest birthday. It is a more exact procedure, since the rounding error in this case cannot be more than a half-year from the true age. For that reason, rounding to the nearest unit is the most common statistical procedure, and may be assumed unless it is otherwise noted.

In Oriental countries, age is rounded to the next *higher* unit, but this convention is not current in the West. Nevertheless, it is adopted in certain other familiar instances. For example, postage charges are assessed for "an ounce or fraction thereof," the fraction being rounded up to the full ounce. The reasons are probably: (1) quick calculation according to the first notch that tips the scales; and (2) maximization of charges. Parking garages charge for the full hour; hotels, for the next full day if the guest does not vacate by check-out time. Workers who are paid by the hour, or by the day, are sometimes paid in full units for any fraction of the unit of time.

But whatever the rounding procedure, the rounding unit must always be specified; one, ten, one hundred, a tenth, or a hundredth, or any other unit that satisfies our particular purpose. Infants' ages are often rounded to the week; the ages of adults may be rounded (estimated) in multiples of 5 years; small amounts of money are rounded to the dollar; astronomical national budgets may be rounded to the billion according to our tolerance limits.

All rounding procedures are quite uncomplicated, except when the observed value is at dead center between two adjacent values.[6] In this instance, an amendment to the rule is required. When rounding to the nearest unit, 7.5 could logically be rounded either to 7 or 8, since both are equally proximate; but the convention in such an instance is to round to the nearest even number. Thus, 7.5 would be raised to 8, while 6.5 would be lowered to 6. The justification for this practice is that, theoretically, even and odd numbers occur with equal frequency in the long run. Consequently, in rounding consistently to the nearest even number, half of the values will be raised and the other half lowered, thereby canceling out the errors and leaving the sums free of rounding error (Columns 1 and 2). With uniform raising or lowering, however, the rounding errors would cumulate rather than compensate one another (Columns 3 and 4).

(1) Observed Value	(2) Nearest Even	(3) Next Higher	(4) Next Lower
6.5	6	7	6
4.5	4	5	4
3.5	4	4	3
7.5	8	8	7
22.0	22	24	20

Severity of Rounding According to the United States Census, the 1970 population of the Atlanta Standard Metropolitan Statistical Area (SMSA) was 1,390,164. This figure not only requires an effort to remember and

[6] One solution is never to get stuck on dead center by the simple device of increasing the decimal accuracy of the observed measure. However, this resort will not be available when the data are obtained from secondary sources.

manipulate, but additionally, the last several digits are not even worth such an effort since they are almost certainly unreliable, owing to difficulties in enumerating large human populations. Hence, one would not hesitate to round the figure to 1,390,000 or 1,400,000, which seems sufficiently precise for all practical purposes. With the amount of error inherent in census taking, we could just as well have come out with a count of 1,388,975, which would also have rounded to 1,400,000. In any event, the question of how far to round is almost altogether a substantive issue whose resolution will normally be guided by two criteria: (1) our estimate of the reliability of the last digits, which, if inaccurate, should be suppressed by rounding; and (2) the degree of imprecision we are willing to tolerate. Underlying both these criteria is, of course, the usual desire for economy of effort which rounded values permit.

Significant Digits The digits that have been retained, on the assumption that they are reliable, are called *significant digits*. The digits of doubtful dependability are dropped. In the case of whole numbers, as in the above instance, the discarded digits are replaced by zeros only to preserve the location of the decimal point. Thus, in rounding 1,390,164 to the nearest thousand, or to the nearest one-hundred thousand the vacated places must be held by zeros; otherwise the meaning of the number would be lost. However, other than to indicate the unit of count, such zeros have no function. The significant digits constitute the number. It is the significant digits 1.4 that specify how many millions of persons reside in the Atlanta Metropolitan Area.

Occasionally, even reliable digits are dropped for pure convenience, as in the informal quotation of a bank balance of $500, when the exact amount is $511.74. But the zeros used as substitutes, of course, would not be considered significant. If, on the other hand, there is reason to believe that the last zeros are exact and therefore reliable, as in the salary of $200 per week, such zeros would be regarded as significant. In such a case, the two zeros are part of the dollar count; they signify that 200 is part of the sequence 199, 200, 201, rather than 100, 200, 300.

Confidence and Precision The population of the Atlanta SMSA, quoted as 1,390,164, contains seven significant digits only on the assumption that all digits are reliable and deserve confidence. A student of population, as already intimated, will question that reliability. The foregoing figure appears to be precise, but it is actually inaccurate. The expert therefore will fall back on the less precise but more dependable number of 1,400,000, since it makes no pretense of being accurate in the last five places. Paradoxically, he will have *more* confidence in the *less* precise number. Thus, we may formulate the rule: other things being equal, as precision decreases, confidence increases; as precision increases, confidence decreases. A little reflection and experience will support the simple logic of this generalization.

True Limits of Rounded Numbers It is obvious that in rounding to the nearest whole unit, observed values such as 4.7, 4.9, 5.2, and 5.4 all will be rounded to 5. Hence, when we encounter that rounded value, we must be aware that the observed value, inaccurately represented by 5, may lie anywhere within the interval, 4.5 to 5.5. Such a reconstruction of what are labeled the *true limits* from the rounded number is the reverse of the process of rounding. That is, we infer that the original observed value must have been at a point somewhere along the interval, extending a half-unit on either side of the rounded value, 5. Therefore, the rounded value is the *midpoint* of that interval and lies exactly half-way between the true limits.

However, these true limits will be impossible to establish when the severity of the prior rounding is not indicated, as may be seen in the following illustration:

Rounded Number	Unit of Measure	True Limits
400	Ones	399.5–400.5
400	Tens	395 –405
400	Hundreds	350 –450

From these examples, we may formulate a working rule for establishing the true limits as follows: *add to, and subtract from, the rounded number one-half of the given unit of measure.* If 10 was the rounding unit, we would add and subtract 5, resulting in the true limits of 395–405. If 1.3 is the rounded number given, we would add and subtract .05 to obtain the true limits, 1.25–1.35. Thus, the true limits will always consist of one more significant digit than the rounded value.

The details of the foregoing operations hold only when rounding is to the nearest unit. When other rounding procedures are employed, the same general principles will obtain, although the computational details will vary.

QUESTIONS AND PROBLEMS

1. Define the following terms:

true value	true limits
rounded value	midpoint
significant digit	rounding error

2. Round the following to the nearest whole number:
   4.05
   4.51
   4.50
   5.50
   4.501

3. Round the following (a) to the next higher, and (b) to the next lower whole number:
   5.399
   6.3
   2.25
   0.0001

4. Assume that 10 and 20 have been rounded to the nearest whole number.
   (a) Calculate:
      minimum true product
      maximum true product
      minimum true quotient of 10 ÷ 20
      maximum true quotient of 10 ÷ 20
      minimum true quotient of 20 ÷ 10
      maximum true quotient of 20 ÷ 10
      minimum true sum
      maximum true sum
   (b) What is the statistical significance of the variety in these answers?

5. Write the true limits of the following rounded values:
   30, to nearest whole
   30, to nearest ten
   3.5, to nearest tenth
   400, to nearest whole
   400, to nearest ten
   400, to nearest hundred

6. If a girl's height to the nearest inch is 65, between what true limits would her observed height fall?

7. An array of homicide rates extends from .5 to 17.6, rounded to the nearest tenth. What are the true limits of this interval?

8. Designate the significant digits in each of the following numbers:
   2.3
   0.0203
   .600
   .006
   800
   8,000

9. Distinguish between .65 and .650.

10. (a) Is 51 centimeters necessarily less precise than .51 meter?
    (b) Is 5,280 feet more precise than one mile?
    (c) Is 50 miles less precise than 50.25 miles?

11. Two baseball players have 60 and 59 hits out of 190 times at bat, respectively. How would one determine the number of decimal places in

the batting average? Why does it usually consist of three places? [Note: The "batting average" is defined as the number of hits divided by the number of times at bat.]

12. Explain why rounding to the nearest unit introduces less rounding error than rounding to the last unit.

13. (a) Formulate the rule for calculating the true limits of a number rounded to the last unit.
    (b) What is the maximum error in such rounding; and in what direction does it lie from the rounded value?

SELECTED REFERENCES

Kerlinger, Fred N.
1964 Foundations of Behavioral Research. New York: Holt, Rinehart and Winston. Chapter 23.

Baggaley, Andrew R.
1969 Mathematics for Introductory Statistics. New York: Wiley.

Bonjean, Charles M., Richard J. Hill and S. Dale McLemore
1967 Sociological Measurements: An Inventory of Scales and Indices. San Francisco: Chandler.

Miller, Delbert C.
1970 Handbook of Research Design and Social Measurement. Second Edition. New York: David McKay Company, Inc.

Nunnally, Jum C.
1967 Psychometric Theory. New York: McGraw-Hill. Chapters 6 and 7.

Sheldon, E. B., and W. E. Moore (eds.)
1968 Indicators of Social Change. New York: Russell Sage Foundation.

Statistical Policy Division, U.S. Office of Management and Budget
1973 Social Indicators, 1973. Washington, D.C.: U.S. Government Printing Office.

Walker, Helen M.
1951 Mathematics Essential for Elementary Statistics. Revised edition. New York: Holt.

Grouping of Qualitative Data 3

1 SINGLE CLASSIFICATION

If a given trait, or characteristic, is subject to differentiation but cannot be graded on a scale of more or less, higher or lower, or smaller or larger, we designate it as *qualitative*, or *nominal*, to distinguish it from traits that may be quantitatively measured. The U.S. Bureau of the Census does not differentiate persons according to the magnitude of their sex but according to their answer to the question "Male or Female?" In fact, all qualitative variables answer to the question "Which one?" rather than to "How much?" They answer to questions such as the following: "Married or Divorced?" "Protestant or Catholic?" "French, German, or Italian?" "Republican or Democrat?" Sorting cases into such categories and counting the number of cases in each gives rise to a simple frequency table. An example is provided in Table 3.1.1, where the 1957 United States population is grouped according to reported religion.

Rules of Classification Although the construction of such a tabulation is quite simple, it is still governed by generally accepted rules that guard against ambiguous and misleading compilations. The first rule requires that the classification include every observation in the series; it must exhaust the data so that class frequencies add up to the total frequency. It must be *exhaustive*. If cases are omitted, class proportions, or *relative frequencies*, will be too large: 20 as a proportion of 40 cases (10 omitted) is smaller than 20 as a proportion of 50 (all cases). Hence, if for any reason some items remain

Table 3.1.1 *Adult Population by Religion Reported, United States, 1957*

Religion	Number	Percent
Protestant	78,952,000	66.2
Baptist	*23,525,000*	*19.7*
Lutheran	*8,417,000*	*7.1*
Methodist	*16,676,000*	*14.0*
Presbyterian	*6,656,000*	*5.6*
Other Protestant	*23,678,000*	*19.8*
Roman Catholic	30,669,000	25.7
Jewish	3,868,000	3.2
Other religion	1,545,000	1.3
No religion	3,195,000	2.7
Religion not reported	1,104,000	0.9
Total	119,333,000	100.0%

Source: U.S. Bureau of the Census, *Current Population Reports*, Series P–20, No. 79, Washington, D.C.: U.S. Government Printing Office. February, 1958.

unclassifiable—their identity may be unknown or perhaps, in the interest of brevity, not required—they must still be accounted for in such residual categories as "unknown" or "all other," so that the weight of each category relative to the total number of cases is not overstated. This procedure is illustrated in Table 3.1.1, which makes provision for persons (albeit less than 1 percent) not reporting their religion.

The second rule requires that the classification consist of *mutually exclusive* classes so that the presence of a case in one class precludes its presence in every other class. If it is possible for a case to appear in one and only one category, the total frequency will be equal to the number of cases; otherwise, the total frequency may be greater than the number of cases. It is of course possible to classify cases by two or more criteria, but that procedure requires a two-way classification or higher—a single classification will not do. But even with cross-classifications, the total frequency must be equal to the number of cases in the sample.

Problems in Application Most sociologists would probably accept the above principles of classification as useful guides, even though they sometimes waive them in practice. We cite three common problems.

(1) Mixed Criteria Table 3.1.2 classifies the United States population by race and gives the number of persons in each category in 1970. This classification, regularly used by the Bureau of the Census, is seemingly defective in that the categories do not possess a common denominator. There are at least two criteria competing for the assignment of items: physical race and national origin. It is logically possible for a person to be both white and

Table 3.1.2 *Characteristics of Racial Groups, United States, 1970*

Racial Group	Number (Thousands)	Percent under 16	Percent 65 Years and Over	Percent Urban
White	177,749	29.5	10.3	72.5
Black	22,580	37.8	7.0	81.3
American Indian	793	40.7	5.7	44.6
Asian				
Japanese	591	25.4	8.0	89.1
Chinese	435	28.4	6.2	96.6
Filipino	343	32.7	6.3	85.6

Source: U.S. Bureau of the Census, *Current Population Reports*, Series P–23, No. 49. Table 3.18, p. 93. "General Characteristics of Persons of Selected Ethnic and Racial Groups: United States, 1970." Washington, D.C.: U.S. Government Printing Office. May, 1974.

Asian, or both black and Asian. In spite of this logical problem, the Census Bureau finds this scheme to be practically useful and meaningful to the general public. In this case, little confusion is created by classifying persons as white, black, American Indian, or Asian.

(2) ***Multiple Entries*** When we permit multiple answers to the same question—a not uncommon practice in survey research—our tabulation may have more entries than cases. In that event, there will be a discrepancy between the number of cases upon which our relative frequencies are based and the number of entries we list in the table.

To illustrate: In one study of social class, persons were asked: "In deciding whether a person belongs to your class or not, which of these things is most important to know: who his family is; how much money he has; what sort of education he has; or how he believes and feels about certain things?" The replies, as summarized by the investigator, are shown in Table 3.1.3.

This table gives every appearance of representing the frequency distribution of a single variable. Hence, in a quite natural reaction, the reader will move his eye down the column and spontaneously add up the percentages, which do not sum to 100, but rather to almost 130. Nevertheless, the author treats the tabulation as a demonstration that subjective beliefs, rather than objective circumstances, such as money and education, are the predominant earmark of social class. Multiple entries introduce an element of uncertainty which disallows such a pat conclusion, because plural votes are accorded equal weight, which they may not deserve.

To avoid such a dilemma, the investigator might have tabulated the responses by preferential order, or even could have restricted each subject to the "one best response." If this had been done, it is conceivable that the criterion of "Beliefs and Attitudes," which was mentioned by nearly half the group, and hence collectively takes first place, might rarely have been

Table 3.1.3 *Criteria for Social Class Membership, Sample* (N = *1,097*) *of United States Population*

Criterion	Percent Reporting[a]
Beliefs and attitudes	47.4
Education	29.4
Family	20.1
Money	17.1
Other answers	5.6
Don't know	9.1

[a] Percentages add to more than 100%. People often gave more than one answer.
Source: Richard Centers, *The Psychology of Social Classes*. Princeton: Princeton University Press, 1949, p. 91. Copyright 1949 by Princeton University Press. Reprinted by permission of Princeton University Press.

given first place by the individual respondents. Thus, the tabulation of the "one best response" could have reduced, or even wiped out, the attribute that now stands at the head of the column.

(3) Overlapping Categories In the foregoing examples, the confusion was produced by mixing distinct criteria. In Table 3.1.4, which we have assembled as a case in point, the confusion is created by overlapping categories (that is, most children of laborers will be enrolled in public high school; some children of laborers will be living in nonmetropolitan areas, etc).

Although class frequencies are not cumulative, they may be cumulated anyway. Thus, an unwary reader may erroneously conclude that 76 percent of all college applications come from seniors in public schools and from seniors whose parents are laborers. To reduce the likelihood of such false interpretations, it is advisable to complete the table with a second column

Table 3.1.4 *Percent High School Seniors Planning to Attend College, United States, October 1973*

Category	Percent
Enrolled in public schools	40.4
Living in nonmetropolitan areas	34.9
Children of farmers and farm managers	39.5
Children of laborers	35.7

Source: U.S. Bureau of the Census, *Current Population Reports*, Series P-20, No. 270. Tables 1 and 3. "College Plans of High School Seniors: October 1973." Washington, D.C.: U.S. Government Printing Office. October, 1974.

Table 3.1.5 *Percent High School Seniors Planning to Attend College, United States, October 1973*

Category	Yes	No	Total	Number in Thousands
Enrolled in public schools	40.4	59.6	100	3,139
Living in nonmetropolitan areas	34.9	65.1	100	1,070
Children of farmers and farm managers	39.5	60.5	100	119
Children of laborers	35.7	64.3	100	98

Source: See Table 3.1.4.

giving the percentage not applying to college in each category, in the manner of Table 3.1.5. The "No" column, together with the row totals, will serve as a reminder that percentages are to be read across rows rather than down columns.

Some Rules for the Percentage Table The quality of any set of data is independent of its tabular presentation, but good data may be distorted by a bad table. To forestall that eventuality, social scientists permit themselves in constructing tables to be guided by conventions of proven worth. Some of these rules have been codified by Davis and Jacobs (1968) in *The International Encyclopedia of the Social Sciences*, Volume 15. Since they are particularly pertinent to the tabulation of attributes, they are introduced here:

(1) The total number of cases in the sample must appear in every table; this total frequency will permit the reader to distinguish percentages based on the grand total from percentages based on a subtotal (see Table 3.1.6).

Table 3.1.6 *Respondents by Answer to Question "In general, how do you feel about your job?"*

Answer	Percent
I like it very much	58
I like it somewhat	40
I dislike it	2
Total	100%
Number (base N) answering question	2,834
No answer	8
Not applicable (housewives and unemployed)	314
Total of all cases	3,156

Source: "Tabular Presentation," by James A. Davis and Ann M. Jacobs. Reprinted with permission of the publisher from the *International Encyclopedia of the Social Sciences*, David L. Sills, editor, Volume 15, page 498.

(2) The number of cases (base N) on which percentages are based must appear in every table (see Table 3.1.6). This number may be the grand total or a subtotal.

(3) Any discrepancy between the percentage base and the grand total of all cases must be accounted for by giving the number of "No answer" and "Inapplicable" cases.

(4) Exclude the "No answer" and "Inapplicable" cases from base N. The frequency of "No answer" and "Inapplicable" is a function of the research procedure and is not a characteristic of the population under study.

(5) Omit numerical frequencies corresponding to category percentages; report only the base frequency corresponding to 100 percent.

Although these rules seem uncomplicated, they are often violated, possibly more from indifference to tabular detail than from carelessness.

QUESTIONS AND PROBLEMS

1. Define the following terms:

   qualitative data
   nominal scale
   inclusive (exhaustive) classification
   exclusive classes
   mixed criteria
   multiple entries
   overlapping categories
   base N

2. (a) List the salient features of the tabulation of employed persons by reason for not working (Table 3.1.7).
   (b) Explain the given order of categories.
   (c) How can the percentages be accounted for?

3. (a) List salient features of the tabulation of persons 14 years and over by main reason for not working in 1973 (Table 3.1.8).
   (b) Explain the order of categories.

Table 3.1.7 *Reasons for Not Working, Employed Persons, United States, July 7–13, 1957 (Numbers in Thousands)*

Reason	Number	Percent
Bad weather	17	0.2
Industrial dispute	113	1.6
Vacation	5,577	79.6
Illness	793	11.3
All other	514	7.3
Total with job but not at work	7,014	100.0%

Source: U.S. Bureau of the Census, *Current Population Reports*, Series P–57, No. 181. Washington, D.C.: U.S. Government Printing Office. August, 1957.

Table 3.1.8 Persons 14 Years and Over by Main Reason for Not Working in 1973, by Race (Numbers in Thousands)

Reason	White	Black
Ill or disabled	5,666	1,428
Keeping house	23,526	1,924
Going to school	9,261	2,024
Unable to find work	521	261
Retired	6,910	464
Other	724	93
Total	46,608	6,194

Source: U.S. Bureau of the Census, *Current Population Reports*, Series P–60, No. 98. Table 12, p. 58. "Work Experience by Family Status—Persons 14 Years Old and Over by Low-Income Status in 1973, and Race." Washington, D.C.: U.S. Government Printing Office. January, 1975.

(c) Convert frequencies to percentages of column totals (base *N*s) and compare percentage distributions of whites and blacks.

4. In a national survey, each of 1,513 respondents was asked to assign a prestige rating to the occupation "rehabilitation counselor." Eighty-seven percent assigned a rating while 13 percent made no assignment. Respondents who assigned a prestige rating were asked to describe what a rehabilitation counselor does. Of this group, 87 percent gave a description while 13 percent gave none. Of all descriptions, 59 percent were classified as "wholly inaccurate" and 41 percent as "at least partially accurate." Devise a set of categories for classifying respondents by degree of knowledge about the job "rehabilitation counselor." Count the number of respondents in each category. If the sum of these frequencies exceeds 1,513, your categories cannot be mutually exclusive; if the sum of these frequencies is less than 1,513, your categories cannot be exhaustive.[1]

2 CROSS-CLASSIFICATION

Data analysis is rarely limited to classifying cases by variables one at a time; it usually extends to cross-classifying cases by variables two at a time, or more. Thus, instead of classifying persons by religion or politics separately, we may classify them by religion and politics simultaneously; and from the resulting cross-classification, we may judge whether religion and politics

[1] Adapted from Marie R. Haug and Marvin B. Sussman, "Professionalism and the Public," *Sociological Inquiry* 39 (Winter, 1969): 60.

Table 3.2.1 *Marriages by Religion of Husband and Wife, 437 Married Couples, New Haven, Conn.*

Husband	Wife			Total
	Catholic	Protestant	Jewish	
Catholic	271	20	0	291
Protestant	17	61	0	78
Jewish	1	1	66	68
Total	289	82	66	437

Source: A. B. Hollingshead, "Cultural Factors in the Selection of Marriage Mates," *American Sociological Review* 15, 1950, p. 623. Reprinted by permission.

tend to go together, that is, whether Protestants tend to be Republicans and whether Catholics tend to be Democrats. When variables go together, we say they are *statistically associated,* or that there is a *statistical association* between them. Because the cross-tabulation brings out association, and because association may suggest causation we state that cross-tabulating is a first step in disentangling cause and effect relations. We usually go farther and state that cross-tabulating is essential to analyzing the causal relations among qualitative variables and that without it there could be no such analysis.

Table 3.2.1 gives the cross-classification of 437 marriages by religion of husband and religion of wife. Note that the couple is the unit of analysis rather than the individual. Adding up row and column entries gives marginal totals which correspond to the simple frequencies for each variable separately. Since these frequencies appear in the margin, they are named *marginal frequencies.* Adding up marginal frequencies for all rows (or columns) gives the grand total of all cases.

The pattern of *joint frequencies* (cell entries) permits us to judge the relative frequency of intrafaith marriages in this sample. That judgment may be drawn more easily from Table 3.2.2, where cell frequencies have been converted to percentages. Upon summing the diagonal percentages, left to right, we obtain 92 percent, which compactly summarizes the extent of religious endogamy in this sample.

When our concern is primarily with the causal influence of one variable on the other, we may find it instructive—as a first step—to express cell frequencies as percentages of either the row or the column totals. If the presumed causal variable is displayed in rows and its effect in columns, we base cell percentages on the row marginals; if the causal variable is displayed in columns, cell percentages are based on column marginals. Each category of the causal variable provides a "base N" as in Table 3.2.3. This table suggests that worker morale is affected by type of supervision and that too much supervision tends to have a demoralizing effect on the worker.

Table 3.2.2 *Marriages by Religion of Husband and Wife (Percentage Distribution)*

Husband	Wife			Total
	Catholic	Protestant	Jewish	
Catholic	62	4	0	66
Protestant	4	15	0	19
Jewish	0	0	15	15
Total	66	19	15	100%
				$N = 437$

Source: Table 3.2.1.

Another example is supplied by Table 3.2.4, which subclassifies both blacks and whites as movers or nonmovers. It answers to the speculation that blacks are more likely to move than whites, owing to the greater social push and pull to which they are subject. But the substance of this table is against that speculation. The percentage of movers within each race is virtually identical and we might conclude that blacks are no more prone to move than whites.

Cases may be classified in any number of ways, depending on our objectives. Table 3.2.5 classifies persons 18–24 years old by three variables: marital status, sex, and race. This triple classification permits us to compare the marital status of men and women for blacks and whites and the marital status of blacks and whites for men and women. It shows, among other things, that women are more likely to be married than men for both blacks and whites; that blacks are less likely to be married than whites for both men and women; that the ratio of married women to married men is higher for blacks

Table 3.2.3 *Work Alienation by Supervisor Type*

Supervisor Type	Work Alienation			Total	N
	Low	Medium	High		
Directive	10	33	57	100%	63
Participatory	32	42	26	100%	158
Laissez-faire	35	42	23	100%	176
					397

Source: Adapted from George A. Miller, "Professionals in Bureaucracy: Alienation Among Industrial Scientists and Engineers," *American Sociological Review* 32, 1967, p. 762. Reprinted by permission.

Table 3.2.4 *Mobility Status between March 1970 and March 1973, for the Population 3 Years Old and Over, by Race (Numbers in Thousands)*

Mobility Status	Race		
	White	Black	Both
Same house (nonmovers)	109,833	13,873	123,706
Different house in the U.S. (movers)	54,564	6,782	61,346
Movers from abroad	1,973	147	2,120
No report on mobility status	5,383	805	6,188
Total	171,753	21,607	193,360

Percent of Reported Cases in Three Mobility Statuses

Mobility Status	Race		
	White	Black	Both
Same house (nonmovers)	66.0	66.7	66.1
Different house in the U.S. (movers)	32.8	32.6	32.8
Movers from abroad	1.2	0.7	1.1
Total	100.0% (166,370)	100.0% (20,802)	100.0% (187,172)

Source: U.S. Bureau of the Census, *Current Population Reports*, Series P–20, No. 262. Washington D.C.: U.S. Government Printing Office. March, 1972.

(1.68) than for whites (1.56). Other comparisons may be made according to one's interests.

We note in passing that Table 3.2.5 may be reduced to a double classification by ignoring any one of the three variables. Thus, by ignoring race, one

Table 3.2.5 *Marital Status of Persons 18 to 24 Years Old, by Race and Sex: March 1972*

Marital Status	White		Black	
	Male	Female	Male	Female
Total (in thousands)	10,401	11,050	1,336	1,577
Percent distribution				
Total	100.0	100.1	100.0	100.0
Single	66.5	47.0	75.4	58.2
Married	32.6	50.8	23.9	40.1
Separated	0.8	2.0	2.7	8.4
Other, spouse absent	0.7	1.6	0.6	2.0
Widowed	—	0.2	—	0.3
Divorced	0.9	2.1	0.7	1.4

— Represents zero.
Source: U.S. Bureau of the Census, *Current Population Reports*, Series P-23, No. 44. Table 22. March, 1973.

Table 3.2.6 *Marital Status of Persons 18 to 24 Years Old*

Marital Status	Sex	
	Male	Female
Single	67.5	48.4
Married	31.6	49.4
Widowed	0.0	0.2
Divorced	0.9	2.0
Total	100.0%	100.0%
(Total in thousands)	11,737	12,627

Source: Table 3.2.5.

gets the distribution of persons (Table 3.2.6) by marital status and sex for both racial groups combined. This combined distribution may be compared with the race-specific distributions of Table 3.2.5. The utility of such comparisons is taken up in Chapter 8 under the heading of "Elaboration in the Analysis of Association."

QUESTIONS AND PROBLEMS

1. Define the following terms:

   cross-classification
   cross-tabulation
   statistical association
   marginal frequency
   marginal distribution
   joint frequency
   joint frequency distribution

2. Study Table 3.2.7.
   (a) List the important and conspicuous entries.
   (b) Does occupation appear to be dependent upon sex and race?

3. (a) In Table 3.2.8, which variable—family organization or basic economy—would you consider the causal variable? Explain.
   (b) Express cell frequencies as percentages of row totals. Does this enhance the effectiveness of the table? Explain.

4. A club has 100 members. Among them are 50 lawyers and 50 liars. The number of members who are neither lawyers nor liars is 20. Construct a 2×2 table to show the number of lawyers who are liars, the number of lawyers who are not liars, the number of liars who are not lawyers, and the number who are neither.

5. In a poll of white commissioned and noncommissioned officers during World War II, 17 percent of the commissioned officers ($N = 60$) said that black troops were better than white troops, 69 percent said that they were

Table 3.2.7 *Employment by Occupation, Sex, and Race, 1974*

Occupation	Men		Women	
	Nonwhite	White	Nonwhite	White
Total employed (thousands)	5,179	47,340	4,136	29,280
Percent	100	100	100	100
White-collar workers	24	42	42	64
Professional and technical	9	15	12	15
Managers and administrators, except farm	5	15	2	5
Sales workers	2	6	3	7
Clerical workers	7	6	25	36
Blue-collar workers	57	46	20	15
Craft and kindred workers	16	21	1	2
Operatives, except transport	17	12	17	12
Transport equipment operatives	9	6	0	1
Nonfarm laborers	15	7	1	1
Service workers	15	7	37	19
Farm workers	4	5	1	2

Source: U.S. Bureau of the Census, *Current Population Reports*, Special Studies, Series P–23, No. 54. Table 48 and 49, pp. 73–74. "Occupation of Employed Men: 1964, 1970, and 1974," and "Occupation of Employed Women: 1964, 1970, and 1974." Washington, D.C.: U.S Government Printing Office. 1975.

Table 3.2.8 *Type of Family Organization by Type of Economy, 194 Primitive Societies*

Type of Economy	Type of Family			Total
	Maternal	Paternal	Intermixed	
Hunting	30	18	22	70
Pastoral	1	10	3	14
Agricultural	44	47	19	110
Total	75	75	44	194

Source: L. T. Hobhouse, C. C. Wheeler, and M. Ginsberg, *The Material Culture and Social Institutions of the Simpler Peoples*. London: Chapman and Hall, 1915, p. 153. Reprinted by permission.

the same, 5 percent said they were not as good, and 9 percent gave no answer. Of the 195 noncommissioned officers, the corresponding distribution was 9 percent, 83 percent, 4 percent, and 4 percent.[2] Arrange these results in a 4×2 table and attach an appropriate title. Calculate the percentage marginal distribution for all cases combined and compare with the respective distributions for commissioned and noncommissioned officers. Comment on the probable reason behind this poll.

6. Older people take more interest in political elections than younger persons, as suggested by Table 3.2.9. The difference is especially marked among those having more education, but it is also present among those with less education.
   (a) Collapse Table 3.2.9 to show the percentage distribution of interest in elections by education, disregarding age. (Suggestion: Convert given

[2] Source: Samuel A. Stouffer *et al.*, *The American Soldier*, Vol. 1 of Studies in Social Psychology in World War II. Princeton: Princeton University Press, 1949, p. 589.

Table 3.2.9 *Interest in Political Elections, by Education and Age*

Degree of Interest	Education			
	No High School		Some High School or More	
	Under 45 Yrs.	45 or Older	Under 45 Yrs.	45 or Older
Great	19	25	26	41
Little	81	75	74	59
Total	100%	100%	100%	100%
N	376	869	1,174	439

Source: From P. F. Lazarsfeld et al., *The People's Choice*, 3rd edition. New York: Columbia University Press, 1968, p. 44. By permission of the publisher.

percentages into absolute frequencies, recombine as necessary, and reconvert to percentages.)
(b) What information is lost in collapsing?

7. In Table 3.2.1, what percentage of
   (a) Catholic husbands marry Catholic wives?
   (b) Catholic wives marry Catholic husbands?
   (c) Protestant husbands marry Catholic wives?
   (d) Catholic husbands marry Protestant wives?
   What light do these figures throw on the tendency of men and women to enter into interfaith marriages?

3 GRAPHS OF QUALITATIVE DATA

Graphs are usually more effective than tables in conveying an accurate impression of the pattern of the frequency distribution. With that object in mind, we often convert frequency tables into frequency graphs. In a simple frequency table, frequencies are symbolized by tallies or numbers, whereas in a frequency graph, frequencies are represented by lines or areas; we make lines or areas proportional to frequencies, as in the following examples.

The Simple Bar Chart The bar chart (Figure 3.3.1) is a succession of evenly spaced bars whose lengths are proportional to the frequencies of the categories in the set. In drawing a bar chart, we require a *baseline* and *vertical axis.* One of these—usually the baseline—carries the frequency scale; the other

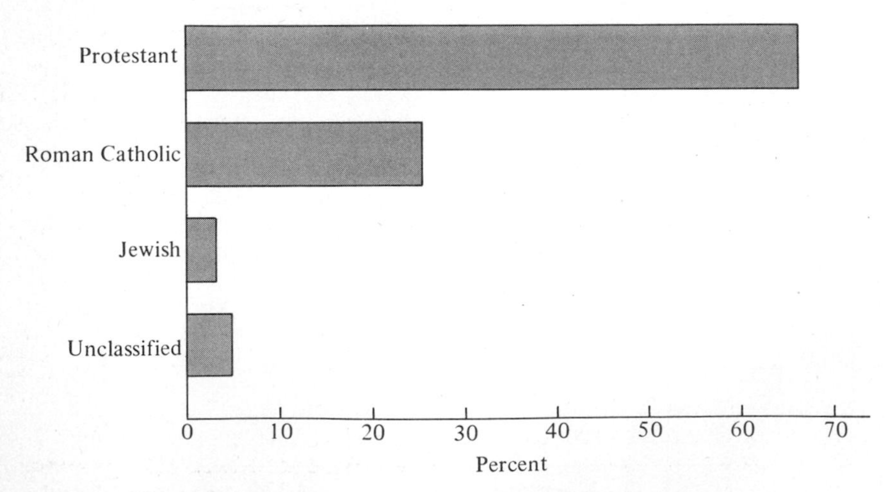

Figure 3.3.1 *Bar Chart, Religion Reported, Sample of United States Population, 1957*

axis carries no scale, since the categories of a qualitative variable form no scale.

The attributes are set down by name alongside the bars, and names are easily read when the bars are drawn horizontally. The bars may be of any convenient width, placed in any plausible order, and left physically unconnected; arranging bars from longest to shortest is tantamount to ranking attributes according to the frequency of their occurrence. Drawing bars to uniform width is justified not only on grounds of style, but also to conform to the principle that equal frequencies be represented by equal areas. The dead space between bars serves to give emphasis to the discrete character of qualitative data.

The Sliding-Bar Chart In presenting our materials, of course we are not limited to the format of the simple bar chart. On the contrary, we may experimentally arrange bars in a variety of ways with the idea of finding the most effective display of our data. We may even design a chart especially for our tabulations when none of the standard models is suitable. To illustrate the flexibility in the manipulation of bars, we briefly describe the sliding-bar chart, so named because the bars slide back and forth across a fixed origin, or cutting point.

The sliding-bar chart is a succession of bars of equal length and width, placed at right angles to a vertical axis which represents the "boundary" between two categories. There will be as many bars as there are populations in the comparison, and each bar will have as many segments as there are categories, with segments proportional to category percentages. Each bar will represent 100 percent. The displacement of each bar relative to the vertical axis will be proportional to class frequencies, and the different displacements will presumably give emphasis to the sociological differences among the populations.

In Figure 3.3.2, Gold has employed the sliding-bar chart to good advantage in displaying differences among boys and girls, by class and color, in respect to the seriousness of their delinquent behavior. It is evident at a glance (which is the test of a good graph) that boys of lower social status, both whites and nonwhites, differ from the other groupings in the seriousness of their delinquent acts. The graph has thrown into relief a finding that might be hardly noticeable in a frequency table.

The Pie Chart As the term suggests, the pie chart (Figure 3.3.3) is a circular figure in which the areas of the respective slices are drawn proportional to the attribute frequencies. In constructing this chart, percentage frequencies are successively measured with a protractor along the circumference of 360°, or 100 percent (360° = 100 percent, 1 percent = 3.6°). From these cutting points, we draw radii to the center of the circle, which lines serve to partition the total area into proportional slices. The effectiveness of the pie chart, and for

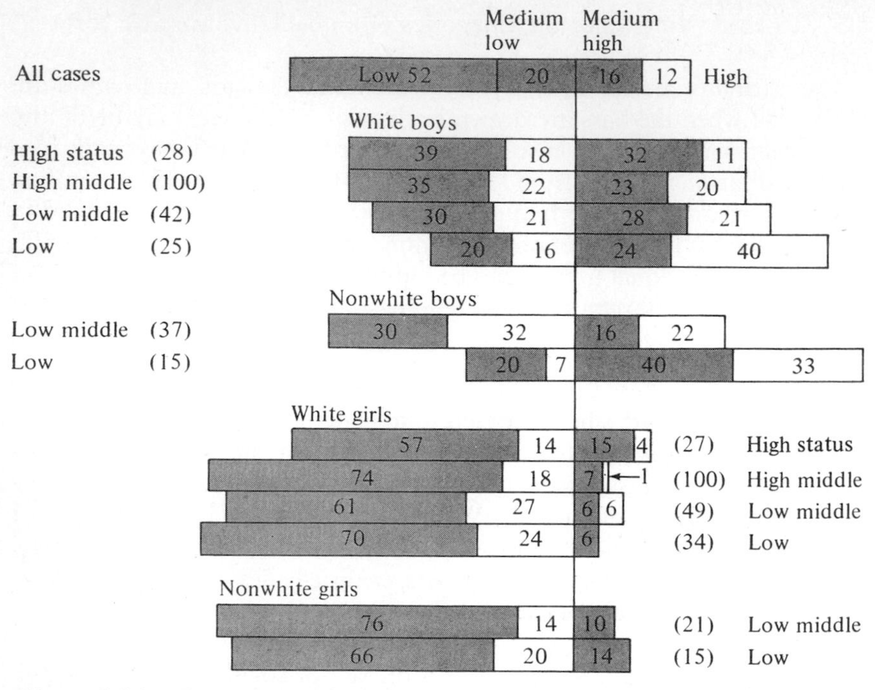

Figure 3.3.2 *Percentage of Youth at Four Levels of Delinquency by Social Status and by Race and Sex* (N = *493*)

Source: Martin Gold, "Undetected Delinquent Behavior," *Journal of Research in Crime and Delinquency*, 3 (January 1966), 43. Reproduced by permission.

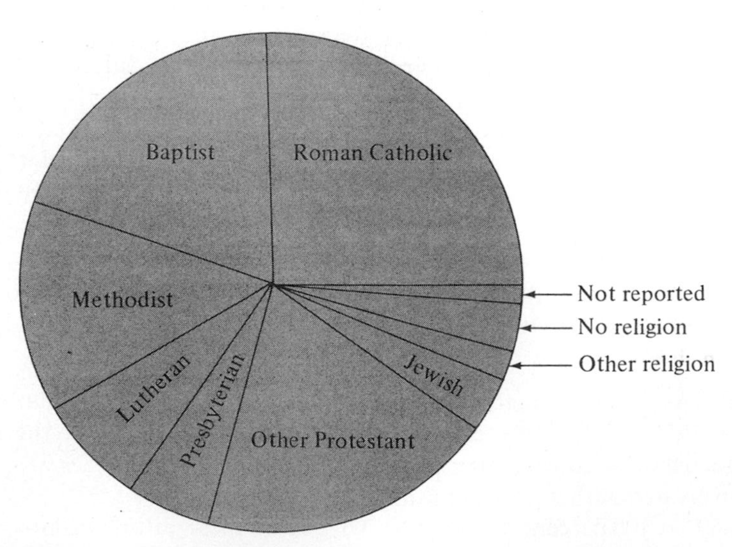

Figure 3.3.3 *Pie Chart, Religion Reported, United States Population, 1957*

that matter all qualitative charts, may often be enhanced by judicious shading and coloring.

Perspective on Qualitative Data In this chapter, we have been mainly concerned with ways of tabulating and presenting qualitative variables with little regard for their importance in sociology. There is no question that such classifications appear regularly in sociological research, but their significance for lasting knowledge has been disputed. In fact, it has been asserted that qualitative variables are an impediment to the development of sociology because they rest on practical social distinctions of little or no scientific value.

The counter to this view is that most technical concepts of general science are an outgrowth of common sense; and that, in any case, sociology has nothing better to offer in their stead at the moment. In adopting this pragmatic attitude, there is some danger that the limitations of qualitative variables will be forgotten and that in consequence little or no effort will be made to repair and refine them.

We should therefore keep in mind that many if not most social classifications (for example, place of birth, cause of death, marital status, and so forth) answer to very practical purposes and were not devised in the first place with the interests of sociologists at heart. For that reason, they differ from place to place and from time to time, according to differing social needs and goals: the classification of persons by nationality has been changed from time to time by the U.S. Bureau of the Census in accordance with the demand for finer distinctions; the classification of workers in a peasant society will differ from that in a modern industrial society since the variety of work in such a simple society is much more restricted. And because social classifications are both local and ephemeral, their long-run value for sociology may at least be called into question. They appear not to have that constancy of meaning that is essential to both scientific replication and generalization.

It should also be borne in mind that social classifications may be and usually are correlated among themselves. The classification of persons by race may be correlated with the classification of the same persons by occupation; the classification of persons by religion may be correlated with a classification of the same persons by social prestige. And, if we neglect these correlations, we may falsely ascribe a causal influence to one qualitative variable that is actually the work of another. For example, we may falsely attribute a difference in average academic achievement between races to race itself, when that difference is actually a product of a difference between racial groups in socioeconomic privilege and advantage.

Although neither of these difficulties—changing definitions or overlapping variables—is peculiar to qualitative data, they are perhaps less easily overcome than in the case of quantitative data. It is therefore understandable that specialists in social statistics (for example, Goodman, 1972a, 1972b) have devoted considerable energy to effective ways of analyzing qualitative variables.

QUESTIONS AND PROBLEMS

1. Define the following terms:
   frequency graph
   bar chart
   sliding bar chart
   pie chart

2. List the distinctive features of graphs of qualitative data.

3. Represent Table 3.3.1 by a bar chart.
   (a) Should the width of the bars be adjusted to the unequal age intervals?
   (b) How should the tabulation be closed?
   (c) Explain the declining percentage of high school graduates with increasing age.
   (d) How should bars be ordered?

Table 3.3.1 *Percentage of Specified Age Groups Completing High School, United States, 1974*

Age	Percent
25–29	41.9
30–34	43.6
35–44	42.3
45–54	40.1
55–64	32.0
65–74	21.2
75 and over	15.3

Source: U.S. Bureau of the Census, *Current Population Reports*, Series P–20, No. 274. Table 1, p. 15. "Years of School Completed by Persons 14 Years Old and Over, By Age, Race, Spanish Origin, and Sex: March 1974 and March 1973." Washington, D.C.: U.S. Government Printing Office. December 1974.

Table 3.3.2 *Percentage of Specified Income Groups Completing High School, Males 25 and Over, United States, 1974*

Income in 1974	Percent
Under $3,000	25.9
$3,000 to $5,999	27.0
$6,000 to $9,999	37.3
$10,000 to $14,999	43.4
$15,000 and over	31.2

Source: U.S. Bureau of the Census, *Current Population Reports*, Series P–20, No. 274. Table 5, p. 55. "Years of School Completed by Employed Males 25 to 64 Years Old, by Income in 1973, Broad Occupation Group, Age, Race, and Spanish Origin: March 1974." Washington, D.C.: U.S. Government Printing Office. December 1974.

4. (a) Plot Table 3.3.2 as a bar chart.
   (b) Income usually increases with age. As of 1974, the higher income groups show higher educational levels, the older age groups show lower educational levels. How, then, do you account for the increasing percentage of high school graduates in the higher income groups?

5. Represent Table 3.3.3 by a bar chart.

Table 3.3.3 *Percent Illiteracy in Population 15 years and Over, circa 1970, by Region*

Region	Percent
Africa	74
Asia[a]	47
Europe and U.S.S.R.	4
Latin America	24
Northern America	2
Oceania	10

[a] Excludes the mainland of China, North Korea, and North Vietnam.
Source: National Center for Educational Statistics. Selected Statistical Notes on American Education. February 1973. Washington, D.C.: U.S. Government Printing Office. 1973.

6. Prepare a bar chart in vertical position from the data of Table 3.3.4. Draw a line parallel to the horizontal axis through the overall average of 1,015.2.

Table 3.3.4 *Rate of Plural Births by Age of Mother, White Population, United States, 1944*

Age of Mother	Rate Per 100,000 Births
10–14	331.1
15–19	549.6
20–24	774.8
25–29	1,019.8
30–34	1,310.7
35–39	1,589.3
40–44	1,292.7
45–49	721.0
50 and over	0.0
All Ages	1,015.2

Source: U.S. National Office of Vital Statistics, *Vital Statistics of the U.S., Special Reports, National Summaries, Plural Birth Statistics, U.S. and Each State, 1944*. Vol. 25, No. 18, Table D. Washington, D.C.: U.S. Government Printing Office. 1947.

7. Construct a bar chart showing percentage voting after 6 P.M. for each region (Table 3.3.5) and construct a sliding-bar chart for all regions with vertical axis located at noon.

Table 3.3.5 *Reported Time of Day Vote Cast for Persons Who Reported Voting for the United States and Regions, November 1972*

Time of Voting	Total	Northeast	North Central	South	West
Before 8 : 00 a.m.	10.6	6.9	13.5	10.6	11.0
8 : 00 a.m. to noon	29.8	28.9	29.5	31.0	29.6
Noon to 4 : 00 p.m.	20.0	19.0	20.6	21.2	18.4
4 : 00 p.m. to 6 : 00 p.m.	19.8	17.6	21.0	21.0	18.8
After 6 : 00 p.m.	13.1	21.2	9.0	9.3	14.9
Absentee ballot	3.7	3.1	3.9	3.1	4.9
Did not know or did not report time voted	3.0	3.3	2.5	3.8	2.3
Total	100.0	100.0	100.0	100.0	100.0

Source: *Current Population Reports.* Series P–20, No. 253. Voting and Registration in the Election of November 1972. Washington, D.C.: U.S. Government Printing Office. October, 1973.

SELECTED REFERENCES

Davis, James A. and Ann M. Jacobs
1968 "Tabular presentation." The International Encyclopedia of the Social Sciences, Vol. 15 : 497–509. New York: Macmillan and Free Press.

Feinberg, Barry M. and Carolyn A. Franklin (eds.)
1976 Social Graphics Bibliography. Washington, D.C.: Council on Social Graphics.

Goodman, Leo
1972a "A general model for the analysis of surveys." The American Journal of Sociology 77, No. 6 (May): 1035–1086.
1972b "A modified multiple regression approach to the analysis of dichotomous variables." American Sociological Review, 37 (February): 28–46.

Schmidt, Calvin F.
1954 Handbook of Graphic Presentation. New York: Ronald Press.

Yule, G. Udny and M. G. Kendall
1950 An Introduction to the Theory of Statistics, Fourteenth edition. London: Charles Griffin.

Zeisel, Hans
1968 Say It With Figures. Fifth edition. New York: Harper and Row.

Grouping of Quantitative Variables 4

1 SINGLE VARIABLE

If a given characteristic can be graded on a scale of more or less, higher or lower, larger or smaller, we designate it as quantitative to distinguish it from qualitative variables, which by definition are not subject to such scaling. Since age, education, the birth rate, and the death rate are things of which there may be more or less, they are quantitative in nature. Cities may vary in amount of suicide, crime, or divorce; individuals may vary in income, weight, or age; marriages may vary in duration and happiness. Such variation usually conforms to some specific pattern which will emerge only after cases have been grouped into intervals suitably spaced along the range. We split up the range into segments and count the number of cases in each.

Compared with the grouping of qualitative data, the grouping of quantitative variables calls for more judgment on the part of the analyst. The analyst must settle on the location of class boundaries; decide whether intervals are to be of uniform width; decide whether the classes at either end are to be closed or left open; establish the true class boundaries and the points midway between them. To illustrate the grouping of quantitative data, we sort the ages of 140 persons (given in Table 4.1.1) into class intervals spaced so as to bring out the main outline of the distribution. The numbers were purposely selected to approximate the 1975 U.S. age distribution. But before sorting them into intervals, we consider ways of making ungrouped data more intelligible, and incidentally, more amenable to grouping.

Table 4.1.1 *Random Listing of Ages of 140 Persons*

7	18	33	58	24
13	22	28	12	42
38	2	39	5	27
41	25	20	54	23
32	17	24	7	6
49	10	24	50	54
22	37	6	4	30
4	76	13	37	70
2	10	64	17	55
1	57	2	14	62
31	43	61	41	61
2	43	10	47	9
26	7	9	16	47
33	17	21	16	39
44	47	48	75	58
8	4	53	69	12
6	19	2	5	33
15	6	31	12	0
13	33	27	67	7
12	23	57	28	14
17	37	65	50	27
29	54	66	69	70
22	15	40	0	55
44	70	15	28	62
18	60	45	14	76
20	49	4	70	14
22	54	57	36	76
49	6	16	39	81

Source: Randomly generated to simulate U.S. age distribution, 1975.

Table 4.1.2 *Ages of 140 Persons in Array*

0	10	22	37	54
0	12	22	37	55
1	12	22	38	55
2	12	22	39	57
2	12	23	39	57
2	13	23	39	57
2	13	24	40	58
2	13	24	41	58
4	14	24	41	60
4	14	25	42	61
4	14	26	43	61
4	14	27	43	62
5	15	27	44	62
5	15	27	44	64
6	15	28	45	65
6	16	28	47	66
6	16	28	47	67
6	16	29	47	69
6	17	30	48	69
7	17	31	49	70
7	17	31	49	70
7	17	32	49	70
7	18	33	50	70
8	18	33	50	75
9	19	33	53	76
9	20	33	54	76
10	20	36	54	76
10	21	37	54	81

Source: Table 4.1.1.

Array The order of variates in Table 4.1.1 is purely arbitrary and carries little or no statistical meaning. From this random listing it is virtually impossible to discern by eye the manner in which these 140 numbers distribute themselves over their range. A first step in the direction of putting these values in orderly form is to arrange them by order of magnitude, from the largest to the smallest, or the reverse. This arrangement is called an *array*; it is comparable to lining up a company of soldiers according to height or apples according to size. Table 4.1.2 orders the ages according to size. In moving from a random listing to an array, we have changed what was no more than a collection of numbers into a statistical order.

This new view of the data reveals the concentration of ages within specific intervals. By counting from the bottom, we find that 14 of the 140, or 10 percent, are 65 years of age or older; by essentially the same method, we find that 7 out of 140, or 5 percent, are 16 or 17 years of age. But even with this simplification, the shape of the distribution is likely to escape all save the most experienced eye. Moreover, if one assumes that many other arrays are still more complex and bulky than this one, it is obvious that something else must be done to bring out the tendency of items to concentrate here, to thin out there. This additional step consists of grouping identical or contiguous values into convenient class intervals. Counting the number of cases in each class interval gives us a *simple frequency distribution.*

Number and Size of Class Intervals Before we can proceed to the actual tabulation of cases, we must first determine the appropriate number and size of the class intervals. In this respect, there are no rigid prescriptions, although obviously the ungrouped array must be compressed sufficiently to achieve the purpose of grouping. Broadly speaking, grouping should be adapted (1) to the nature of the data, (2) to the objectives the classification intends to serve, and (3) to the background and needs of the readers for whom it is intended. Thus, data embracing an immense range, such as the ages of the United States population, would require a different grouping than would the ages of American schoolchildren. The oversimplified statistics in the popular brochures of insurance companies, depicting the relative incidence of heart disease and cancer, may be very instructive for the lay public, but they are inadequate for scientific uses. A grouping of data intended for a professional audience could be more intricate than one planned for popular consumption.

In spite of this flexibility, it is customary to lay down a few technical rules which could normally serve as a guide.

(1) The *number* of class intervals ordinarily should be no more than 15, nor less than 10. (This is not a hard and fast rule and may be abridged under special circumstances.) Fifteen intervals usually are sufficient to reveal the pattern of distribution, and yet not so numerous that the pattern cannot be readily apprehended. On the other hand, if fewer than 10 intervals are employed, the salient features of the distribution may be obscured.

(2) The *size* of the interval should be a whole number, and, whenever practicable, of convenient divisibility such as 2,10, or 25. Additionally, there may be some practical advantage attached to the use of multiples of 10, to be consistent with our decimal system. With units in feet and inches, minutes and hours, the requirement of adhering to the decimal system may be waived.

(3) Whenever possible, intervals should be of *uniform width.* Then they are most readily grasped and greatly facilitate further computation. At times, however, the rule of equal intervals may have to give way in the face of special circumstances. Thus, the school population may call for age intervals of 6–11, 12–14, 15–17, and 18–22; otherwise, information essential to the school administrator would be concealed. Similarly, in the classification of cities by size of population, or families by income brackets, uniformity of class intervals cannot be maintained. At the lower end of the range, variates are numerous and differences are small, while in the upper brackets, cases are

infrequent and gaps are large. Hence, class intervals should increase progressively in size throughout the distribution. However, whenever unequal intervals are used, the larger intervals should be multiples of the smaller ones, especially when graphs will be developed from the statistics.

(4) The ends of a frequency distribution may be either *open* or *closed*. Although the ends are normally closed, it is often convenient to leave one or both extremes open, as in this National Office of Vital Statistics tabulation of births by age of mother:

Age of Mother		
Under 15 years	25–29 years	40–44 years
15–19 years	30–34 years	45 years and over
20–24 years	35–39 years	

Such open ends avoid the need to itemize small, irregular, and essentially trivial frequencies which extend well beyond the limits of the significant range of the data. Nevertheless, for technical reasons it is a good rule to close the ends whenever possible. Otherwise subsequent calculations, such as the arithmetic mean, are impossible, and graphing becomes awkward.

Constructing the Frequency Distribution In grouping the 140 ages of persons given in Table 4.1.2, the first step is to divide the range of 80 years into the optimum number of class intervals. In fixing this number, we must guard against too many or too few classes. A little exploration will reveal that 10 intervals of width 8 would give us too little detail; whereas 40 intervals of width 2 would give us too much. Sixteen intervals of width 5 would seem to be about right for bringing out the main outline of the distribution without too many rough edges. (Exceeding the suggested maximum of 15 intervals is justified by the nature of our data.) Accordingly, we set up 16 intervals of width 5 and proceed with the sorting of the 140 cases.

The resulting frequency distribution (Table 4.1.3) seems quite acceptable: the class frequencies relative to the total ($N=140$) are neither too large nor too small; there are no vacant intervals and the contour of the distribution is easily discernible from the varying class frequencies. Among other things, Table 4.1.3 shows that, while younger people are more numerous than older people, the number of persons of working age (20–65) is about equal to the number of persons of nonworking age—under 20 years and over 65. Insights into social structure are frequently based on such simple social statistics.

In this table, the values have been listed from low to high. This practice, which is that of the United States Census and prevails generally in the tabulation of social statistics, differs from the convention employed by psychologists and educators, who reverse the order of values. The difference, of course, is superficial rather than essential. In the tabulation of test scores and

Table 4.1.3 Frequency Table of Ages

Age	Frequency	% Frequency
0– 4 Years	12	8.6
5– 9 Years	14	10.0
10–14 Years	14	10.0
15–19 Years	13	9.3
20–24 Years	12	8.6
25–29 Years	9	6.4
30–34 Years	8	5.7
35–39 Years	8	5.7
40–44 Years	8	5.7
45–49 Years	8	5.7
50–54 Years	7	5.0
55–59 Years	7	5.0
60–64 Years	6	4.3
65–69 Years	5	3.6
70–74 Years	4	2.8
75 and Over	5	3.6
Total	$N = 140$	100.0

school marks it is perhaps more natural to begin with the highest score, against which all other scores are judged. In general, however, with such variables as age and income, one is likely to begin counting from the zero origin, in conformity with habits acquired in childhood and reinforced through daily use.

Rounded and True Limits The limits of the intervals in Table 4.1.3 have been designated by rounded numbers: 0–4, 5–9, 10–14, etc. These are the rounded limits; but what are the true limits? To answer that question, we must know the prior rounding procedure: whether to the nearest or to the next lower unit.

In either case, a rounded number stands for an interval of unit width. The unit itself is usually a power of $10: 10^0 = 1$, $10^1 = 10$, $10^{-1} = 0.1$, etc. Rounded to the nearest unit, a number stands for an interval extending from one-half unit below itself to one-half unit above; rounded to the next lower unit, a number stands for an interval extending from no units below itself to exactly one unit above.

It follows that with rounding to the next lower (last) unit, as in Table 4.1.3, we get true limits by adding 0.0 to rounded lower limits and 1.0 to rounded upper limits. Applying this rule to the first interval gives true limits of 0.0 and 5.0; applying it to the second interval gives true limits of 5.0 and 10.0, etc.

With rounding to the nearest unit, which is the more conventional operation, the procedure for obtaining true limits from rounded limits differs in detail but not in principle. Table 4.1.4 supplies an example. This table is a frequency tally of 229 suicide rates rounded to the nearest whole

Table 4.1.4 *Tally and Count of 1970 Standard Metropolitan Statistical Area Suicide Rates*

Rounded Class Limits	True Class Limits	Tally	Frequency
3– 4	2.5– 4.5	|||	3
5– 6	4.5– 6.5	𝍸 ||||	9
7– 8	6.5– 8.5	𝍸𝍸𝍸𝍸𝍸𝍸𝍸	35
9–10	8.5–10.5	𝍸𝍸𝍸𝍸𝍸𝍸𝍸𝍸𝍸𝍸 |	51
11–12	10.5–12.5	𝍸𝍸𝍸𝍸𝍸𝍸𝍸𝍸𝍸𝍸 ||	52
13–14	12.5–14.5	𝍸𝍸𝍸𝍸𝍸𝍸	30
15–16	14.5–16.5	𝍸𝍸𝍸𝍸 |	21
17–18	16.5–18.5	𝍸𝍸	10
19–20	18.5–20.5	𝍸 |	6
21–22	20.5–22.5	𝍸 |||	8
23–24	22.5–24.5	|	1
25–26	24.5–26.5	|||	3
			$N = 229$

Source: Appendix, Table IX

number. (The ungrouped suicide rates, together with divorce rates for the same 229 Standard Metropolitan Statistical Areas [SMSAs], are given in the Appendix, Table IX.) Given this information about the rounding procedure used, we construct the true limits by subtracting $(1/2)(1)=0.5$ from rounded lower limits, and adding $(1/2)(1)=0.5$ to rounded upper limits. The true limits are given alongside the rounded limits to bring out the difference between them. Note that the true boundaries of adjacent intervals overlap, and that an interval's width is the difference between its true limits.

Midpoint of Class Interval For purposes of further statistical calculation, it is often necessary to assign a single value to all of the items within a given class interval. The selection of this value must be guided by the principle of representativeness as well as expediency. On the assumption that items are more or less uniformly distributed between the class boundaries, a point halfway between the true boundaries would satisfy our needs. Accordingly, we select the *midpoint* of the interval to represent all the values within the interval.

Assigning the midpoint to each case is equivalent to making the sum of the values in that interval equal to the product of the midpoint and the class frequency. For example, giving the midpoint to 10 cases in an interval ranging from 16.5 to 18.5 is equivalent to making the sum of those 10 values equal to $10 \times 17.5 = 175$. The product obtained by multiplying a midpoint by its corresponding frequency will seldom be exactly equal to the actual sum of the individual variates in that interval. When items are distributed symmetrically over the interval, the discrepancy between approximate and

Table 4.1.5 *Determination of Interval Midpoint, Rounding to Nearest Whole Number and Last Whole Number*

Rounding Procedure	Rounded Limits	True Class Limits	Class Width	One-Half Class Width	Interval Midpoint
Nearest whole number	2–4 5–7	1.5–4.5 4.5–7.5	3 3	1.5 1.5	3 6
Last whole number	2–4 5–7	2.0–5.0 5.0–8.0	3 3	1.5 1.5	3.5 6.5

exact sums will be small; however, when the items are bunched at one end of the interval, the discrepancy will be relatively large. We refer to the discrepancy between sums (or averages) based on ungrouped data and approximate sums (or averages) based on grouped data as *grouping error*. When there is reason to believe that grouping error is large, correction formulas should be applied in order that calculations based on grouped data will more closely approximate those for the ungrouped data.

To find the midpoint of any interval, we add one-half the width of the interval to the true lower limit. This procedure holds for all rounding procedures, and for both continuous and discrete variables. To illustrate, we present in Table 4.1.5 a set of rounded class limits (2–4, 5–7) whose true limits and midpoints differ according to the prior rounding procedure. Since the rounded values are indistinguishable for the two methods of rounding, care must be exercised in reading correctly the rounded value. If, for example, rounded 3 is mistakenly read as 3.0–4.0 instead of 2.5–3.5, the midpoint would be located one-half unit too high.

Discrete variables in the form of integers follow the same rules of grouping that are applied to continuous data. However, in establishing true limits, we make the assumption that the observed values have been rounded to the nearest integer. True limits calculated on this assumption are illustrated in the tabulation of black families by size (Table 4.1.6). Notice that the midpoint will become a decimal (rather than a whole number) upon classifying families into intervals of width 2.

Relative (Percent) Frequencies The tabulation of cases in one population takes on added meaning by comparing parallel tabulations in other populations. But it is difficult to compare two or more different frequency tables that have widely varying totals. To *norm* them, or make them comparable, the absolute distributions must be reduced to relative, or percentage, distributions. The conversion of an absolute to a percentage frequency is accomplished by dividing each class frequency by the total, and multiplying the resulting quotients by 100—a familiar arithmetic procedure. The results of this procedure are illustrated in the last column of Table 4.1.6.

Table 4.1.6 *United States Black Families by Size, March 1974*

Size of Family	True Limits	Numbers in Thousands	Percent
2	1.5–2.5	1,627	29.9
3	2.5–3.5	1,168	21.5
4	3.5–4.5	925	17.0
5	4.5–5.5	655	12.0
6	5.5–6.5	430	7.9
7 or more	6.5–	635	11.7
All families		5,440	100.0

Source: U.S. Bureau of the Census, *Current Population Reports*, Series P–20, No. 276. Table 7. "Household and Family Characteristics: March 1974." Washington, D.C.: U.S. Government Printing Office. February, 1975.

Cumulative Frequency Distribution In some instances, we may require the number or percentage of items above or below a certain cutting point on the scale. For example, a welfare administrator will want to know how many persons are 65 years or older. Such answers may be obtained readily from the *cumulative frequency distribution*, which is derived from the simple frequency distribution by a process of merging successive class frequencies until all have been cumulated. This cumulating procedure reveals the frequency of items below and above each class boundary, and thus permits the cutting point to be moved easily to any convenient position according to the user's interests and needs.

The process of constructing a cumulative tabulation is shown in Table 4.1.7. In cumulating frequencies, it is possible to start with either the highest or the lowest class interval and proceed to the opposite end. When we start with the lowest interval, we obtain a so-called "less than" cumulative distribution (Column 3); when we start with the highest interval, we obtain an "or more" cumulative distribution (Column 4).

The "less than" cumulation gives the percentage of all cases below the true upper limit of each class interval. Reading from Column 3, we find, for example, that 18.6 percent of the sample is less than 10.0 years of age; continuing, we find that 52.9 percent of the cases are less than 30.0 years of age; etc. The "or more" cumulation is read similarly from true lower limits: it gives the percentage of all cases above each lower limit. Reading from Column 4, we find that 6.4 percent of the cases are 70.0 years or older; that 24.3 percent are 50 years or older; etc. From these two tabulations, any number of useful readings may be made: 70.0 percent of the cases are under 45 years of age; 10 percent are over 65 years of age. Similar questions could be answered from comparable tabulations: What proportion of families have yearly incomes over $10,000 or more? What proportion of families have six or more children? What proportion of unemployed families have been without work for six months or more?

Table 4.1.7 Cumulative Age Distribution of 140 Persons

(1)	(2)	(3)	(4)
Age	Percent	"Less Than" Cumulative Percent	"More Than" Cumulative Percent
0–4	8.6	8.6 (Less than 5)	100.0 (0 or more)
5–9	10.0	18.6 (Less than 10)	91.4 (5 or more)
10–14	10.0	28.6	81.4
15–19	9.3	37.9	71.4
20–24	8.6	46.5	62.1
25–29	6.4	52.9	53.5
30–34	5.7	58.6	47.1
35–39	5.7	64.3	41.4
40–44	5.7	70.0	35.7
45–49	5.7	75.7	30.0
50–54	5.0	80.7	24.3
55–59	5.0	85.7	19.3
60–64	4.3	90.0	14.3
65–69	3.6	93.6	10.0
70–74	2.8	96.4	6.4
75 and over	3.6	100.0 (Less than 100)	3.6 (75 or more)

Source: Table 4.1.3.

QUESTIONS AND PROBLEMS

1. Define the following terms:

array	grouping error
frequency distribution	relative (percent) frequency
class interval	cumulative frequency distribution
class width	cumulative frequency
rounded limits	"less than" cumulation
true limits	"or more" cumulation
class midpoint	

2. What considerations govern the choice of the class interval?
3. Why is it good practice to make class intervals equal in size?
4. When are unequal class intervals appropriate? Give several examples.
5. Must the lowest class interval in every tabulation begin with zero?
6. Distinguish between rounded and true class limits.
7. Is the distinction between rounded and true class limits applicable to discrete data? Explain.
8. Give two illustrations of groupings adapted to the nature of the data.
9. How do you determine whether there are too many (too few) intervals?

10. (a) How do you recognize whether class limits are true or rounded?
    (b) From the rounded limits is it possible to determine the manner of rounding? Explain.

11. The delinquency rates listed in Table 4.1.8 are for 140 neighborhoods in Chicago. Each variate is the number of boys per 100 making a court appearance.
    (a) Round numbers to the nearest whole.
    (b) Construct a frequency table: rounded lower limit, 0; class width, 2.
    (c) Write the true class limits.
    (d) Convert class frequencies to percentages; these should sum to 100.

Table 4.1.8 *Delinquency Rates, 140 Local Areas, Chicago*

0.8	1.6	1.8	2.5	18.2	5.2	0.6
0.8	1.9	2.9	3.7	4.9	5.1	0.6
1.5	1.5	4.6	5.5	2.7	1.7	2.1
2.6	0.9	3.5	7.4	2.3	1.5	2.2
1.0	2.4	3.0	11.8	5.8	2.6	3.7
0.7	1.7	1.9	12.3	8.8	4.0	1.0
0.5	2.2	4.4	12.1	18.9	5.8	3.9
1.1	2.2	2.9	3.0	2.3	9.4	0.6
0.8	2.3	3.4	4.3	2.7	2.4	1.9
0.8	4.3	4.8	5.0	3.2	2.2	4.6
1.4	1.6	6.1	5.7	7.0	3.5	1.5
1.9	1.2	7.8	11.9	13.4	2.8	1.3
2.8	2.1	2.7	9.5	17.5	2.2	1.3
1.0	1.9	3.1	14.8	4.5	2.5	6.0
1.1	2.2	5.2	2.5	2.7	3.9	4.2
0.9	2.9	9.0	2.1	3.4	1.5	4.2
0.7	2.5	11.4	5.0	4.5	0.8	2.0
1.6	4.2	9.5	5.1	9.4	0.6	2.8
3.1	3.1	12.1	3.7	14.8	1.2	2.1
2.1	1.9	2.7	5.0	1.6	3.0	2.5

Source: Reprinted from *Juvenile Delinquency and Urban Areas* by Clifford R. Shaw and Henry D. McKay by permission of The University of Chicago Press.

12. Calculate the width and midpoint of each of the following intervals, assuming that data are rounded to:
    (a) the nearest whole number
    (b) the last whole number

20–22	0– 4	90– 99
23–25	5– 9	100–109
26–28	10–14	110–119

13. From the given class limits and midpoints, determine:
    (a) whether class limits are *true* or *rounded*
    (b) the size of the class interval
    (c) the manner of rounding

Limits	Midpoint
10–14	12.5
15–19	17.5
145–195	170.0
195–245	220.0
16–18	17.0
19–21	20.0

14. (a) Rearrange the following class intervals and their corresponding frequencies from high to low.
    (b) Is the meaning of the frequency distribution in any way changed by the operation in (a)? Explain.

Class Limits	Frequency (f)
4– 6	3
7– 9	9
10–12	12
13–15	5

15. Construct an "or more" cumulative frequency distribution from Table 4.1.9.

Table 4.1.9 *Distribution of Families by Size, United States, March 1974*

Size of Family	Percent
2 persons	37.4
3 persons	21.2
4 persons	19.6
5 persons	11.6
6 persons	5.5
7 or more	4.7
Total	100.0

Source: U.S. Bureau of the Census, *Current Population Reports*, Series P–20, No. 276. Table 1, page 17. "Household and Family Characteristics: March 1974." Washington, D.C.: U.S. Government Printing Office. February, 1975.

16. (a) Express the number of American Indians in each age group (Table 4.1.10) as a percentage of the total.
(b) Cumulate percentage frequencies as necessary to compare the age distribution of American Indians with the 1974 United States age distribution of Table 5.3.4.

Table 4.1.10 *Age Distribution, American Indians, United States, 1960*

Age	Number (in Thousands)
0– 4	91
5– 9	76
10–14	63
15–19	50
20–24	40
25–29	33
30–34	30
35–39	28
40–44	23
45–49	22
50–54	21
55–59	32
60–64	12
65–69	10
70–74	7
75 and over	9
Total	547

Source: U.S. Bureau of the Census, *U.S. Census of the Population: 1960, Subject Reports*, Chapter C, *Nonwhite Population by Race*, PC(2)–IC, Table 2. Washington D.C.: U.S. Government Printing Office. 1963.

17. Construct an "or more" cumulative distribution for the percentages given in Table 4.1.11.

Table 4.1.11 *Distribution of Families by Total Money Income, United States, 1974*

Total Money Income	Number of Families	Percent
Under $1,000	701,000	1.3
$1,000 to $1,999	766,000	1.4
$2,000 to $2,999	1,489,000	2.7
$3,000 to $3,999	2,040,000	3.7
$4,000 to $4,999	2,305,000	4.1
$5,000 to $5,999	2,475,000	4.4
$6,000 to $6,999	2,478,000	4.4
$7,000 to $7,999	2,501,000	4.5
$8,000 to $9,999	5,203,000	9.3
$10,000 to $11,999	5,702,000	10.2
$12,000 to $14,999	7,879,000	14.1
$15,000 to $19,999	10,032,000	18.0
$20,000 to $24,999	5,755,000	10.3
$25,000 and over	6,384,000	11.5
Total	55,712,000	99.9

Source: U.S. Bureau of the Census, *Current Population Reports*, Series P–60, No. 99. Table A. "Money Income and Poverty Status of Families and Persons in the United States: 1974, (Advance Report)." Washington, D.C.: U.S. Government Printing Office. July 1975.

2 CROSS-CLASSIFICATION

As previously noted, data analysis is seldom limited to the frequency distribution of a single variable, but usually extends to the joint distribution of two or more variables. A joint distribution is called for when relations among variables are in question. Thus, a cross-tabulation of two quantitative variables will indicate whether high values in one series tend to go with high values in the other; or whether high values in one series tend to go with low values in the other; or whether no such tendencies are present. From such patterns, it is possible to judge whether a statistical correlation is present; and if so, whether it is direct and positive, or inverse and negative. And from such evidence of statistical correlation we may, if required, go on to speculate about the causal connection between variables.

Joint Frequency Table An example of the cross-tabulation of two quantitative variables is presented in Table 4.2.1, where 1,839 marriages are jointly

Table 4.2.1 *Age at Marriage, Husband and Wife, New Haven, Conn.*

Age of Husband	Age of Wife								Total
	15–19	20–24	25–29	30–34	35–39	40–44	45–49	50 +	
15–19	42	10	3						55
20–24	153	504	51	10	1				719
25–29	52	271	184	22	7	2			538
30–34	5	52	87	69	13	5			231
35–39	1	12	27	29	21	2	3		95
40–44		1	9	18	17	8	2	1	56
45–49	1		3	6	16	16	7	1	50
50 and over			1	4	11	15	21	43	95
Total	254	850	365	158	86	48	33	45	1,839

Source: A. B. Hollingshead, "Cultural Factors in the Selection of Marriage Mates," *American Sociological Review* 15, 1950, p. 622. Reprinted by permission.

Table 4.2.2 *Joint Frequency Table: Suicide Rates by Divorce Rates, 229 SMSAs, United States, 1970*

Suicide Rate (Y)	Divorce Rate (X) 10–19	20–29	30–39	40–49	50–59	60–69	70–79	Frequency (f_y)
24–27	1	1					1	3
20–23	1	3	4	1	1	1		11
16–19		5	16	3				24
12–15	5	21	25	2				53
8–11	16	51	27	3				97
4–7	17	15	5	1				38
0–3	2	1						3
Frequency (f_x)	42	97	77	10	1	1	1	229

distributed by age of husband and age of wife. In keeping with the purposes of cross-classification, we examine the joint frequencies in order to determine whether they form a discernible pattern. There is a fairly obvious trend: husbands tend to be slightly older than their wives. For example, of the 719 husbands ages 20–24, 153 (21.3 percent) are married to younger women, 62 (8.6 percent) to older women; of the 56 husbands ages 40–44, 80 percent are married to younger women. We reach the same conclusion if we base our calculations on the column totals. Of the 850 wives ages 20–24, only 10 (1.2 percent) are married to younger men, but 336 (39.5 percent) to older men. Of the 48 wives ages 40–44, 31 (64.6 percent) are married to older husbands. Other entries may be similarly examined for their possible sociological meaning.

The cross-tabulation of suicide and divorce rates for 229 cities (Table 4.2.2) provides another sociologically meaningful example. It is consistent with the nineteenth-century social philosopher Durkheim's notion of a correlation between anomic social conditions and personal disorganization.

Classifying by Attribute and Magnitude The variables in a cross-classification need not be exclusively quantitative or qualitative; it is quite possible to combine quantitative and qualitative variables in the same table. An example is supplied by the cross-classification of SMSAs by region and suicide rate. This breakdown of the suicide rates (Table 4.2.3) permits us to compare regional frequency distributions, and from such comparisons to judge the importance of region as a factor in the suicide rate. These distributions are shown in percentage form in Table 4.2.4. Comparing them, we see that the Northeast, South, and North Central regions do not differ markedly

Table 4.2.3 *1970 SMSA Suicide Rates by Magnitude and Region*

Suicides per 100,000 population	Region				Total
	North East	South	North Central	West	
3–4	1	2	0	0	3
5–6	3	5	1	0	9
7–8	8	14	13	0	35
9–10	9	20	18	4	51
11–12	5	25	19	3	52
13–14	3	11	8	8	30
15–16	1	7	2	11	21
17–18	4	0	2	4	10
19–20	2	2	2	0	6
21–22	0	1	1	6	8
23–24	0	1	0	0	1
25–26	1	0	1	1	3
Total	37	88	67	37	229

Table 4.2.4 *1970 SMSA Suicide Rates by Magnitude and Region (Percent)*

Suicides per 100,000 population	Region			
	North East	South	North Central	West
3–4	2.7%	2.3%	0.0%	0.0%
5–6	8.1	5.7	1.5	0.0
7–8	21.6	15.9	19.4	0.0
9–10	24.4	22.7	26.9	10.8
11–12	13.5	28.4	28.3	8.1
13–14	8.1	12.5	11.9	21.6
15–16	2.7	8.0	3.0	29.8
17–18	10.8	0.0	3.0	10.8
19–20	5.4	2.3	3.0	0.0
21–22	0.0	1.1	1.5	16.2
23–24	0.0	1.1	0.0	0.0
25–26	2.7	0.0	1.5	2.7
Total	100.0%	100.0%	100.0%	100.0%

from one another. However, the cities of the West exhibit a somewhat deviant pattern: roughly 30 percent show suicide rates higher than 16.5 and almost 60 percent are above 14.5; none are under 8.5. From almost any angle, the suicide rates for the cities of the West are higher than for the cities of the other regions. Evidently, there are social and demographic conditions in the western cities that increase the incidence of suicide, and we might logically turn to the characteristics of these cities in an effort to account for the regional differences in rates. The breakdown of suicide rates by region has served its preliminary purpose of bringing out differences that may be of sociological significance.

QUESTIONS AND PROBLEMS

1. Define the following terms:
   cross-classification
   joint frequency
   joint frequency table

2. (a) For husbands 25–29, calculate the percentage married to younger, to older, and to women of the same age (Table 4.2.1).
   (b) For wives 25–29, calculate the percentage married to younger, to older, and to men of the same age.
   (c) Comment on the social significance of the above figures.

3. One may wonder whether the association between the suicide rate and the divorce rate for the smaller SMSAs (say 200,000 or less) is the same as the

association for all SMSAs combined (as represented by Table 4.2.2). To explore this idea, set up one cross-tabulation for SMSAs under 200,000 (Appendix, Table IX); another for SMSAs over 200,000. Compare. Consider alternative methods for testing the idea that the association between the divorce rate and the suicide rate is affected by size of city.

3 GRAPHIC PRESENTATION

Graphic forms are usually more effective in conveying an impression of the pattern of the total distribution than are the more intricate frequency tables. The graph gives greater visibility to the salient features of the frequency distribution and more readily suggests its meaning and interpretation. The graph's principal function is to convey an accurate conception of the shape of the frequency distribution—a conception which even the skilled expert cannot construct easily from the raw tabular data.

Since graphs are the equivalents of tables, they should carry analogous identifying titles and markers. They are subject to the same criteria of intelligibility, simplicity, and clarity. In fact, a graph cannot be planned or constructed until the corresponding table has been prepared. Here we consider graphic forms corresponding, respectively, to the simple frequency table, the cumulative frequency table, and the joint frequency table.

Histogram The histogram corresponds to the simple frequency table. It consists of a set of contiguous columns whose heights are proportional to class frequencies, and whose widths are proportional to the size of the class intervals of the variable. A histogram based on the frequency distribution of the 229 suicide rates tallied in Table 4.1.4 is shown in Figure 4.3.1. It is not only a graphic record of the absolute class frequencies, it also mirrors the size of each frequency relative to all others.

The histogram is constructed on arithmetic graph paper—paper ruled with equally spaced horizontal and vertical guide lines. The first step in the procedure is to draw at right angles on selected grid lines two axes of approximately equal length, with their intersection near the lower left hand corner of the sheet. Usually, class boundaries are plotted on the horizontal axis, and class frequencies are plotted on the vertical axis, although there are occasional exceptions. For example, in graphing an age distribution, frequencies are usually plotted on the horizontal axis, and class intervals on the vertical axis. Figure 4.3.2 supplies an example. When the male population is shown on one side of the vertical axis and the female population on the other side, we get the familiar "age pyramid." Figure 4.3.2 may be regarded as a folded-up age pyramid.

Before marking off the class intervals on the base line, we must fix the appropriate number of linear units to be assigned to each class interval of

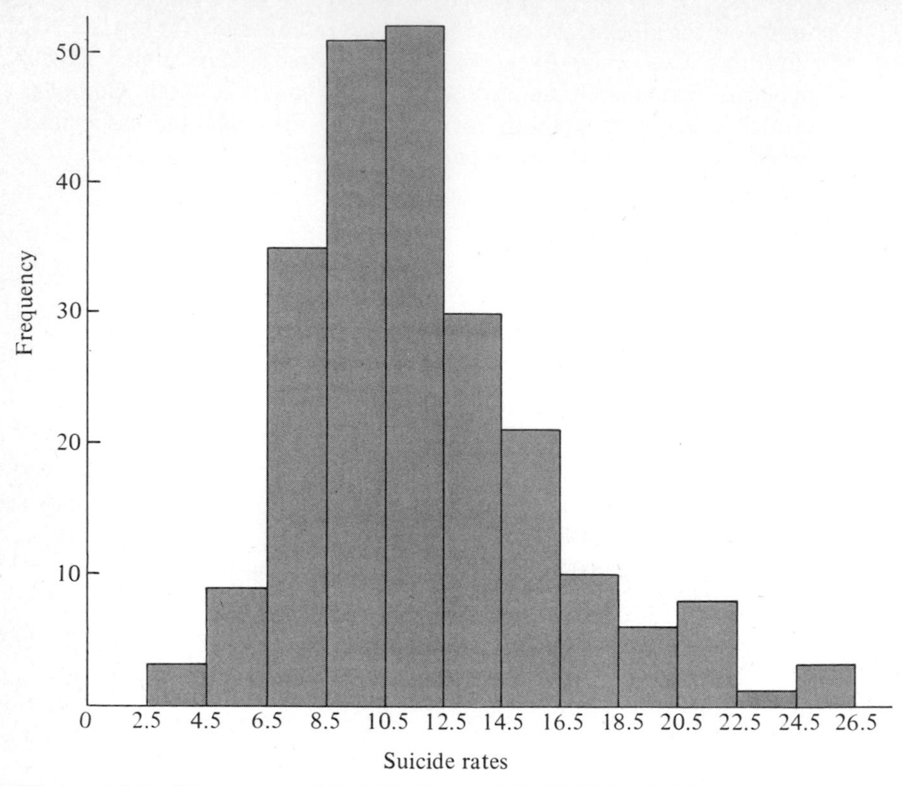

Figure 4.3.1 *Histogram of Suicide Rates, 229 SMSAs, 1970*

the table. It is often good practice to allow for a vacant interval at either end of the horizontal scale, which improves the aesthetic appearance of the graph and promotes readability. Accordingly, we count the total number of units along the length of the axis and divide that total by the number of class intervals (including the two vacant ones) to be plotted. The result will be the number of linear units allotted to each interval. Beginning at the intersection of the two axes, we now can lay down the class intervals and mark their boundaries with tiny upright lines, or *ticks*, appropriately placed inside the axes, since they are vestigial stubs of the original grid lines. Because these ticks represent the contact points of the contiguous intervals, they are designated by markers carrying the true values of the class limits. Such true limits will necessarily be in fractional or decimal form when the values have been rounded to the nearest whole. Thus, in Figure 4.3.1, class limits are designated as 2.5, 4.5, 6.5, and so forth. However, when values have been rounded to the last whole, as in United States Census age data (Figure 4.3.2), the true boundaries may be simply written as integers: 0, 5, 10. For clarity, markers may be spaced uniformly at selected class boundaries or placed at selected midpoints. By these discretionary procedures, the cluttered effect of crowded markers may be avoided.

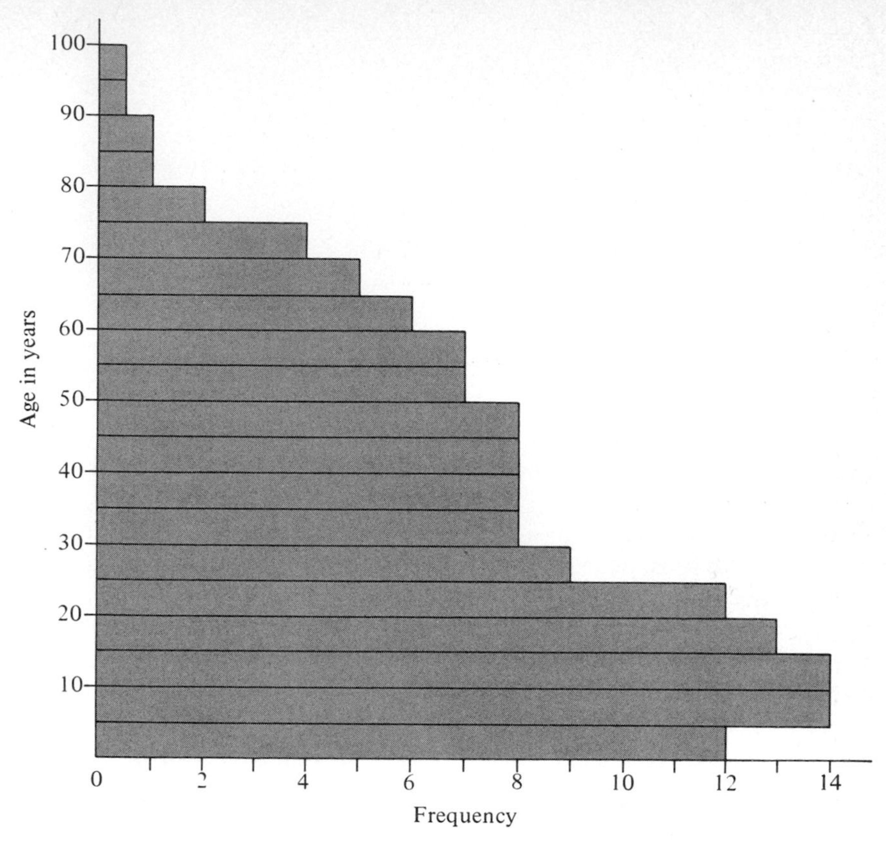

Figure 4.3.2 *Histogram of 140 Ages*

The frequency scale is established on the vertical axis in an analogous manner. We take the largest class frequency to be accommodated in the graph, and divide that frequency by the number of linear units available on the previously drawn axis. This operation gives the number of cases to be assigned to each unit on the upright axis. Thus, in Figure 4.3.1, two cases correspond to one unit on the frequency scale of the graph. Beginning with the *zero origin* at the intersection, markers are now spaced at equal intervals and given in such multiples (5, 10, and so on) as may be clear and intelligible.

Unequal class intervals present a special problem. In Table 4.3.1, made up for present purposes, the last interval is twice the width of the other intervals. Its frequency of 4 is therefore not comparable to the frequencies of the smaller intervals. To establish comparability, and to preserve the proportionality of columns to frequencies, we must divide the last interval into two intervals of equal width, and correspondingly break up its frequency into equal frequencies of two cases each. These adjusted subfrequencies are then plotted (Figure 4.3.3). Had the unadjusted frequency of 4 been plotted (Figure 4.3.4), the frequency column on the extreme right would have enclosed twice as

Table 4.3.1 *Frequency Distribution, Unequal Class Intervals*

Class Interval	Frequency
0–4	3
5–9	6
10–14	14
15–19	5
20–29	4
Total	32

large an area as was its due, and thereby would have created an impression contrary to fact.

In general, before graphing a table in which class intervals are unequal, all larger intervals must be expressed as multiples of the smallest interval; and these obtained multiples must be divided into the corresponding frequencies. The latter step will yield the adjusted frequencies, which are then plotted. This procedure is in accord with the principle that each item in the frequency distribution be represented by an equal area on the graph, so that relative frequencies be proportionately presented in areal form.

Now we can appreciate the practicability of the rule set forth in Section 1 of this chapter, that class intervals be of equal width, or, failing this, that unequal

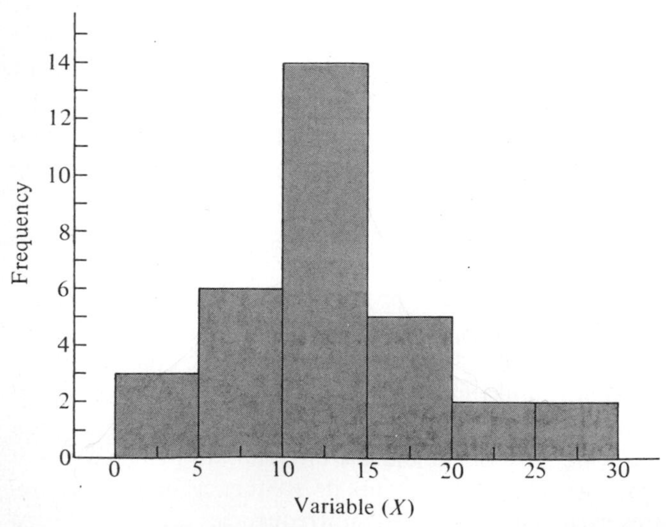

Figure 4.3.3 *Histogram (Correct), Unequal Class Intervals*

Source: Table 4.3.1.

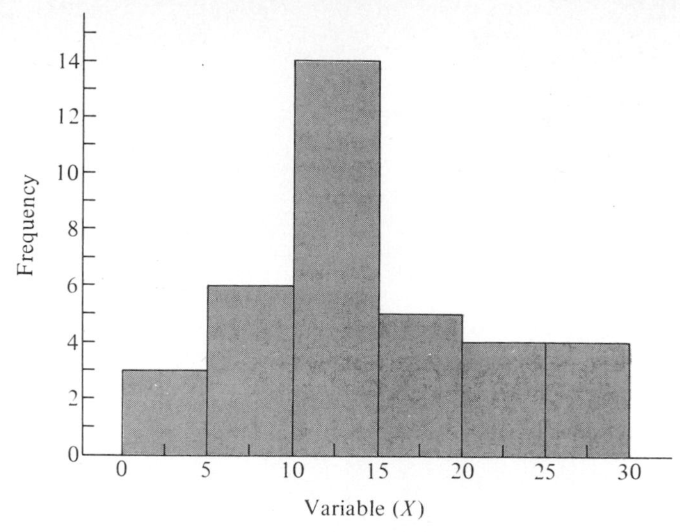

Figure 4.3.4 Histogram (Incorrect), Unequal Class Intervals

Source: Table 4.3.1.

class intervals be convenient multiples of smaller ones. Similarly, the caution against open-ended tables now has acquired new practical meaning. There is no way to adjust the frequency of a class interval of unknown size, since an indefinite interval cannot be expressed as a multiple of a closed interval. Therefore it is impossible to represent an open-ended table by the histogram unless one arbitrarily closes the interval—a measure to which we sometimes may resort.

The distinguishing feature of the histogram is its schematic simplicity. Columns are more easily apprehended than numbers. It clearly reveals the relative concentration of items in each interval and shows the contour of the distribution.

Frequency Polygon When items are compressed into a relatively small number of broad intervals, successive class frequencies tend to jump one after another in an abrupt manner. It is reasonable to suppose, however, that the progression of class frequencies would be considerably smoother if we employed many relatively small class intervals. The function of the *frequency polygon* is to provide an approximation of the smooth curve that would presumably emerge if class intervals were made as small as possible and the number of observations were unlimited.

The frequency polygon consists of lines connecting points plotted over class midpoints at heights proportional to class frequencies. It may be drawn on a histogram by connecting, seriatim, midpoints of adjacent columns at their heights. Such a conversion is illustrated in Figure 4.3.5, where a

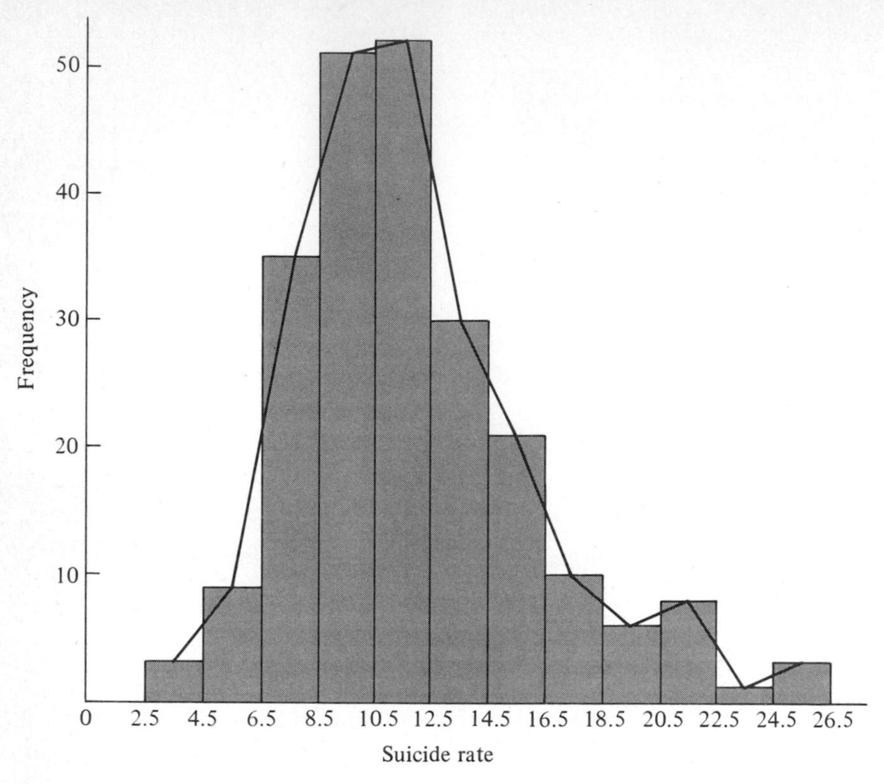

Figure 4.3.5 *Frequency Polygon of Suicide Rates, 229 SMSAs, 1970*

frequency polygon is *superimposed* on the histogram of suicide rates. In practice, only one graph or the other would be presented, depending upon whether the emphasis is to be placed on the tabulated class frequencies or on the hypothetical point frequencies which the frequency polygon provides.

It should be evident that the procedure in constructing the frequency polygon will parallel that of the histogram in all but the final stages. First, two axes are drawn on the graph paper. Then class intervals are marked off on the base line and the frequency scale along the vertical axis. But at this stage, the procedure diverges, in that points and connecting lines are plotted instead of columns. As previously noted, the points are plotted over class *midpoints* at heights called for by the respective class frequencies and then joined by straight lines to form the polygon.

The frequency polygon need not be extended to the base line beyond the range of actual observation. In fact, where the concentration of cases increases at either end, it would be illogical to do so. Therefore, it would appear sound practice to leave the polygon open at either end and thereby avoid the hazard of misrepresenting the frequencies where none were observed. In any event, the general contour in the bulk of cases is in no way affected by open ends, and it is this shape, after all, which the polygon seeks to portray.

When a closed figure is contemplated, the usual procedure is to extend the graph to the base line at the midpoints of the vacant intervals on either end. The rationale of this practice is to preserve the area enclosed by the histogram—which symbolizes the total frequency—by setting up pairs of congruent and compensating triangles. While this rule is reasonable enough, it should not be overlooked that the polygon inevitably disturbs the proportionality between column areas and class frequencies. It will always underrepresent the proportion of items in the interval with the largest frequency.

Therefore, we should remind ourselves again that no statistical device ever perfectly represents the data on which it operates. The frequency polygon pretends to do no more than provide quickly and economically a rough first approximation of the theoretical frequency curve of infinitely many values grouped into the smallest possible intervals. For comparison, frequency polygons representing different distributions of the same variable may be mutually superimposed without blurring their respective outlines, as would be true of histograms.

Cumulative Frequency Graph (Ogive) The cumulative frequency table (Table 4.1.7) gives the percentage of cases below (or above) each class boundary, but not the cumulated frequencies below the innumerable intermediate values within each interval. Yet information on such intermediate values is frequently required. For example, we may require the percentage of the population under 21 years of age, or the age such that 50 percent of the population is younger and older. Although this kind of information could be obtained from the cumulative table by arithmetic interpolation between the class limits, it can be obtained with less effort and sufficient accuracy from the *cumulative frequency graph*, or *ogive*. "Ogive" is an architectural term for the diagonally curved rib of the Gothic vault, which the cumulative frequency curve often resembles; hence, picturesquely, it was so named in 1875 by the English statistician, Francis Galton. Today, the cumulative frequency graph is generally referred to as an *ogive*, regardless of its shape.

Construction of the Cumulative Frequency Graph The construction begins with plotting axes as for the simple frequency polygon: class intervals are laid off on the base line, and frequencies along the vertical axis. Since frequencies are now cumulative rather than simple, the range of the vertical scale will be equal to the total frequency. For quicker comprehension and ready comparability, the cumulative frequencies usually are expressed as percentages, so that the frequency scale extends from 0 to 100.

Corresponding to the two types of cumulative frequency tabulations—the "less than" and the "or more"—there are two types of cumulative graphs. Because the two graphs provide identical information, for all practical

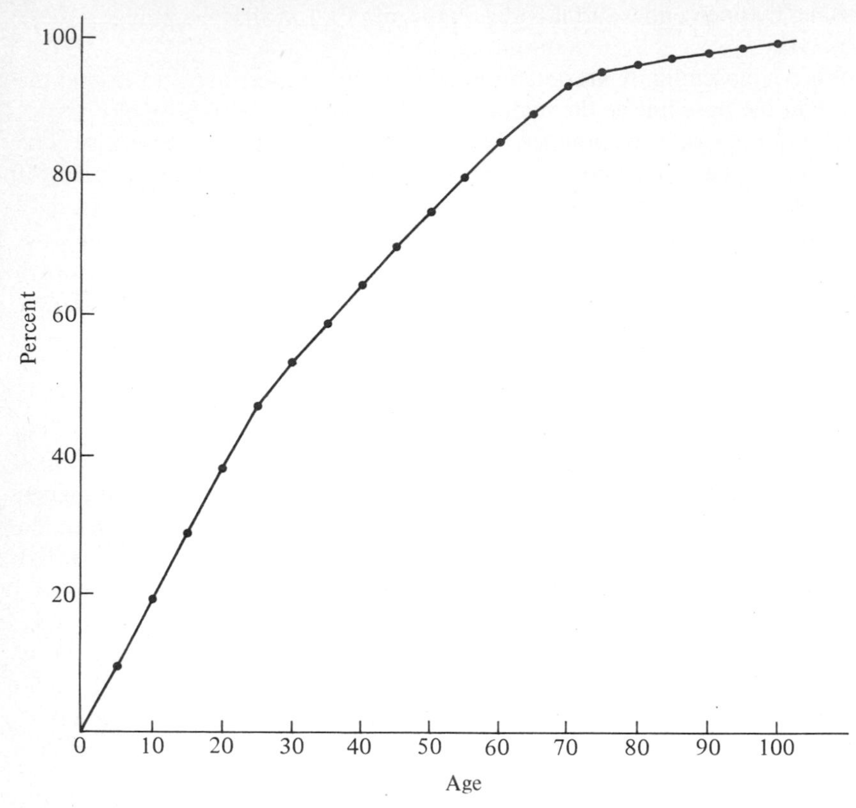

Figure 4.3.6 *"Less Than" Cumulative Graph, 140 Ages*

Source: Table 4.1.7.

purposes only one would ever be constructed. In constructing a "less than" cumulative graph (Figure 4.3.6), the cumulated frequencies are plotted over true upper boundaries of class intervals, whereas in constructing an "or more" graph, the frequencies are plotted over the true lower boundaries (Figure 4.3.7). This procedure differs from that of the simple frequency polygon where class midpoints are plotted. Owing to the manner in which scales are laid off, a "less than" cumulative graph will begin in the lower left-hand corner of the graph and move diagonally across to the upper right-hand corner, while the "or more" cumulative graph will begin in the upper left-hand corner and move diagonally across to the lower right-hand corner. The ogive of the age distribution (Figure 4.3.6) is almost a straight line, whereas the ogive of the distribution of suicide rates (Figure 4.3.8) more nearly resembles the curved rib of a Gothic vault. When the ogive is very nearly a straight line, we may infer that the simple frequency distribution is either rectangular or markedly skewed; when the ogive resembles a stretched out "S," we conclude that the simple frequency distribution is bell-shaped.

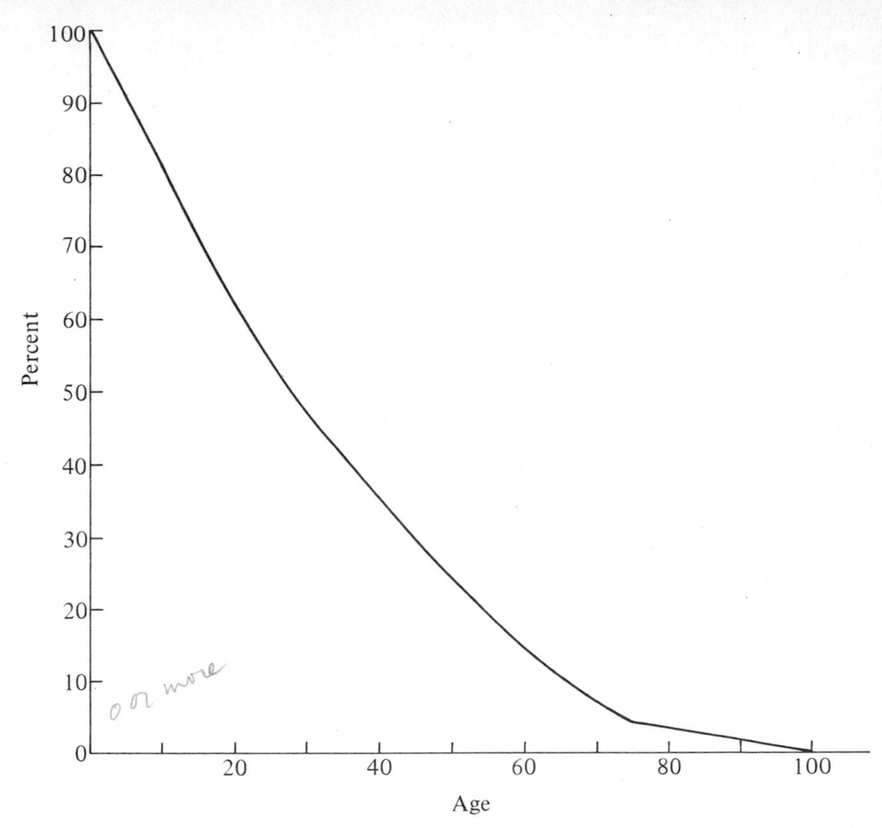

Figure 4.3.7 *"Or More" Graph of Cumulative Age Distribution, 140 Ages*

Source: Table 4.1.7.

When one or both ends of the frequency table are open, the cumulative graph will not extend over the entire range of the frequency scale and thus will appear incomplete. In that event, it may be possible to close the tabular distribution at some convenient point without doing violence to the data, thereby allowing the ogive to be completed. For example, we may close the United States age distribution at 100 years without injustice to the facts, since only a tiny fraction of all individuals exceed the century mark. To close the cumulative graph at 100, we therefore would require only several additional 5-year intervals on the base line. On the other hand, in order to close the tabulation of family incomes (see Table 4.1.11), which reads "$25,000 and over," we would have to add literally hundreds of class intervals since the maximum income is in excess of a million dollars.

Using the Cumulative Graph The cumulative graph permits the frequency distribution to be partitioned at any point whatsoever, according to one's interests and needs. For example, the point that separates the lower 25 percent

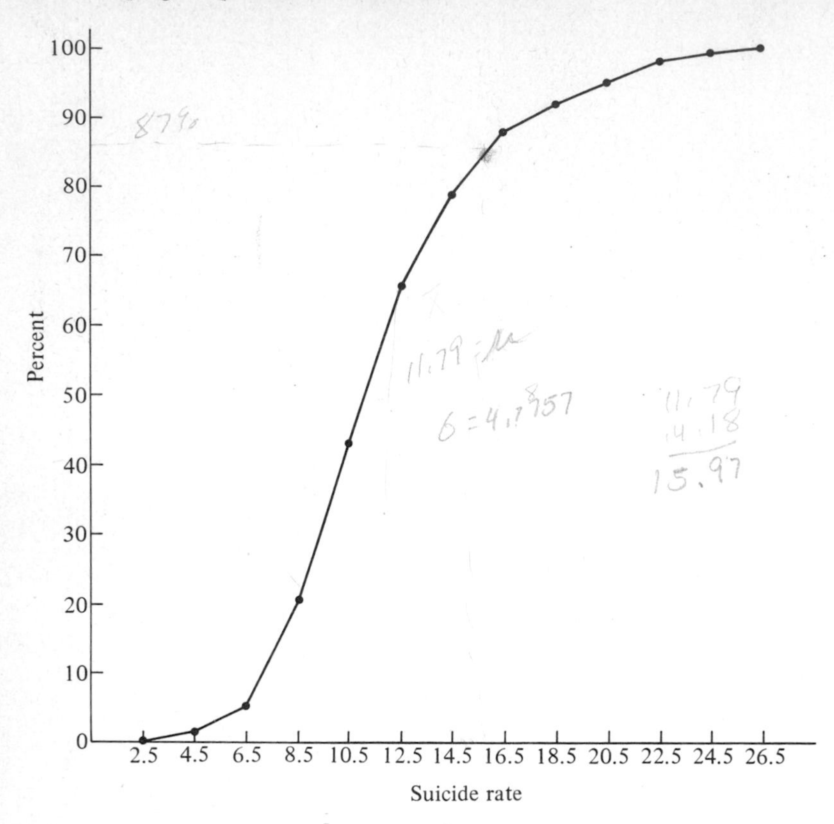

Figure 4.3.8 *"Less Than" Cumulative Graph, Suicide Rates, 229 SMSAs, 1970*

and the upper 75 percent of the items, the so-called *first quartile* (Q_1), may be obtained readily in the following manner (Figure 4.3.9): from the 25 percent marker on the frequency axis, run a line parallel to the base line until it intersects the "less than" ogive; from this intersection drop a perpendicular to the base line, which fixes the first quartile at 13.7 years. This is the age below which 25 percent of the cases lie. The second and third quartiles are found analogously. *Centile* values (points that divide the total frequency into 100 equal parts) may be similarly obtained. Thus, to find the 90th centile value, C_{90}, we drop a perpendicular to the base line from a point on the curve that represents a "less than" cumulated frequency of 90 percent; the required value (64.7 years) is then read off the base line at the junction point.

By reversing the foregoing procedure, the percentage above (or below) any particular value may be quickly established. For instance, we may seek the percentage of the population under 30 years of age. We first locate 30 on the base line, at which point we erect a perpendicular extending to the curve. From that intersection we extend a perpendicular to the frequency axis, where we find that the percentage of persons under 30 years of age is 52.9 percent.

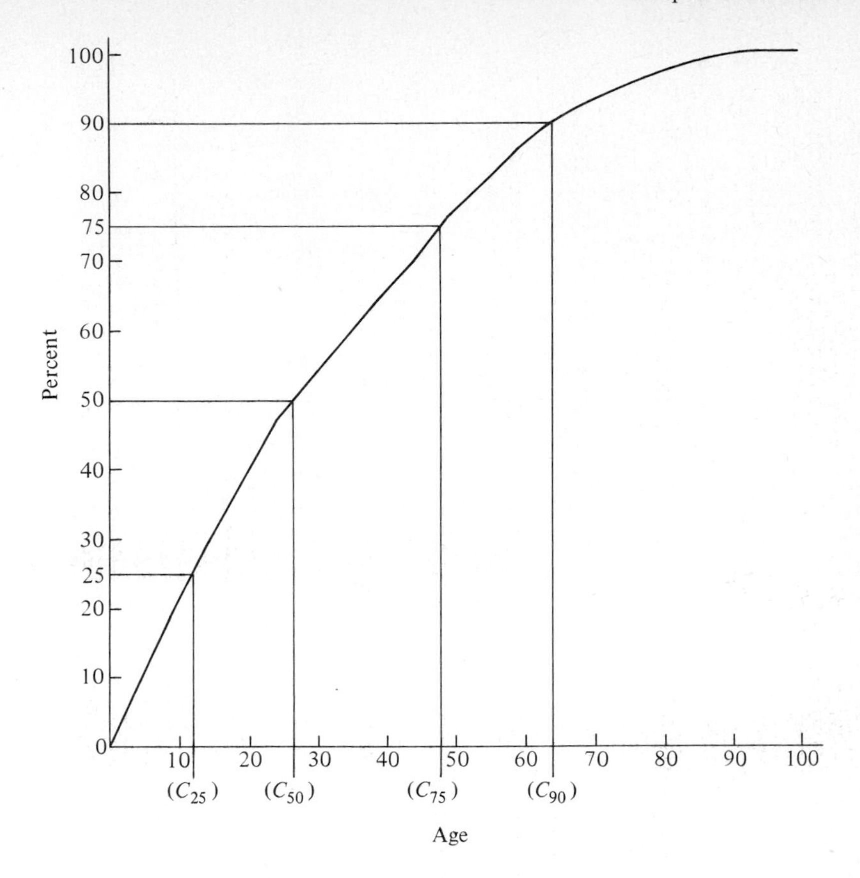

Figure 4.3.9 *Graphic Approximations to Selected Centiles*

The ease with which such graphic solutions may be obtained gives the cumulative graph some advantage over the cumulative frequency table, which necessitates tedious interpolation.

Scatter Diagram In Table 4.2.2 each cell stands for two class intervals, one comprising a group of suicide rates, the other comprising a group of divorce rates; cell entries give the number of tally marks, or cases, in each cell.

Now, if we represent each pair of values, not by a tally mark within a cell, but instead by a dot corresponding to the ungrouped values, we get what is known as a *scatter diagram*. Figure 4.3.10 supplies an example; it is based on the ungrouped data of Table IX, Appendix. The suicide rate fixes the height of a point above the base line, the divorce rate fixes its distance from the vertical axis. Cleveland, for example, is represented by a point at the intersection of guide lines running perpendicularly from a suicide rate of 13.1 and a divorce rate of 30.6; similarly, San Francisco is represented by a point at the intersection of guide lines running perpendicularly from a suicide rate

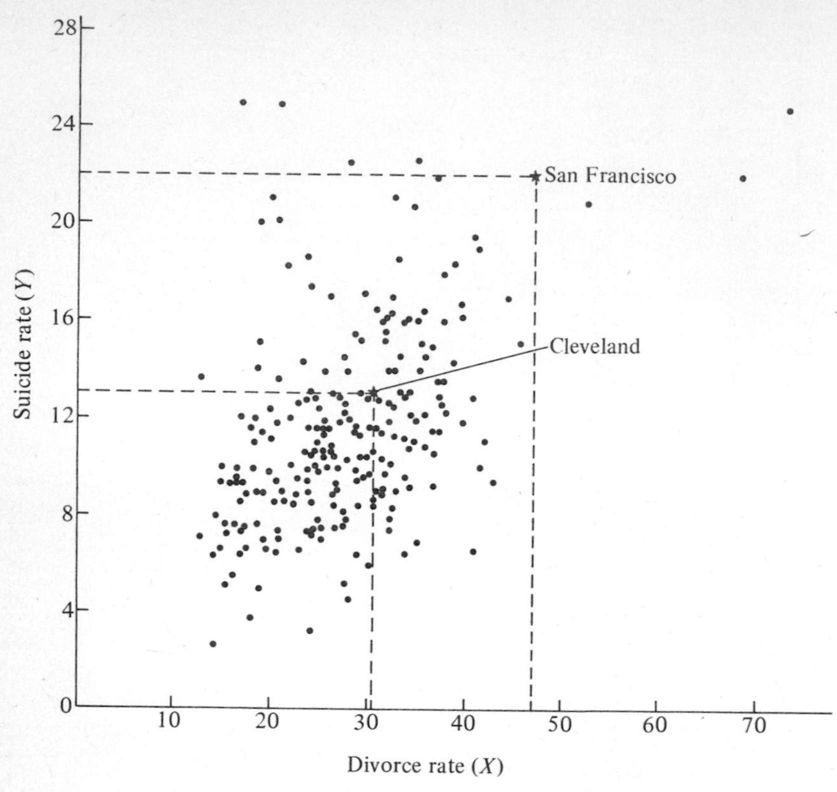

Figure 4.3.10 *Scatterplot of Divorce and Suicide Rates for 229 SMSAs, 1970*

of 22.1 and a divorce rate of 47.2. The swarm of all these points is the scatter diagram.

The scatter diagram is a useful device for judging the kind and degree of relationship between two quantitative variables. It shows (1) whether the plotted points follow a definite trend; (2) whether the trend is consistently upward (or downward); (3) whether the plotted points are widely or narrowly scattered along the trend line; and (4) whether some cases (*outliers*) lie conspicuously outside the swarm of the cases. More detail on the scatter diagram is presented in Chapter 9.

QUESTIONS AND PROBLEMS

1. Define the following terms:

histogram	quartile
frequency polygon	centile
cumulative frequency graph	scatter diagram
ogive	

2. How is the appearance of the histogram changed when the horizontal axis is lengthened in relation to the vertical axis? The vertical in relation to the horizontal?
3. What is the advantage of plotting relative frequencies? Is the appearance of the histogram altered when relative frequencies are plotted instead of absolute frequencies?
4. State the principles governing the closing of the frequency polygon.
5. If you wished to compare two distributions by superimposing the graph of one on the graph of the other, which type would you select: histogram or frequency polygon? Explain.
6. Which figure (histogram or polygon) adheres more closely to the frequency table?
7. Can the polygon be converted as readily into the histogram as the reverse?
8. (a) Represent the frequency distribution of delinquency rates (Table 4.1.8) by a histogram.
   (b) Combine the last four intervals into two intervals of equal width and show by a heavy line how the appearance of the histogram would be altered.
9. (a) Prepare a histogram for the distribution of United States families according to size (Table 4.1.9), placing markers at midpoints instead of true class boundaries.
   (b) What are the true boundaries of the intervals?
   (c) Justify your choice of an upper boundary to close the graph.
10. Plot divorce rates and suicide rates for cities under 200,000 (Appendix, Table IX) as a scatter diagram.

SELECTED REFERENCES

Schmidt, Calvin F.
1954 Handbook of Graphic Presentation. New York: Ronald Press.

Yule, G. Udny and M. G. Kendall
1950 An Introduction to the Theory of Statistics. Fourteenth edition. London: Charles Griffin.

5 Averages

1 MEASURES OF LOCATION AND CENTRAL TENDENCY

The frequency distributions of quantitative data which have been previously analyzed constitute condensations of large masses of observations. Basically, such distributions are nothing more than series of values deployed on a continuous scale. We may therefore declare that the items are *located* on a segment of the scale. Such a description becomes particularly meaningful when we compare two or more distributions which are located in different regions of the same scale. Thus, when husbands and wives are classified by age and located on the same age scale, the husbands, who are usually older, occupy the upper end of the scale, and wives, being younger, occupy the lower end. Similarly, whites and nonwhites in the United States are differently located on the wage continuum. It is clear that the statistical concept, *location*, is concerned only with quantitative variables, and is not applicable to qualitative data, which have no scale to be located on.

But when making these comparisons, it is not always practical to quote or depict the full distribution, however compactly it may be presented in tabular or graphic form. For many purposes, the complete table is unnecessarily detailed, too cumbersome to manipulate, and not always readily comparable to other tables of analogous data. In accordance with the general function of statistics to simplify large masses of data, we need a more condensed statement that will (1) provide information on the locational value in which we are interested, (2) eliminate those values which are at the moment irrelevant, and (3) still faithfully represent the totality with reasonable efficiency.

The limit of such efficient condensation would obviously be the reduction of the multitude of items to one single value which would, in some way, represent the entire aggregate. But it is clear that no single value is sufficiently versatile to reflect every characteristic of location of a distribution; instead, it can reflect only one feature of it. The representative value is not, therefore, a replica or miniature of the total; it is rather a selected value of limited utility which will "do the work" for the totality. When this task is to fix the location of the distribution along the scale, then the value is called a *measure of location*.

Any value in the distribution could serve to represent the totality, if we knew the position of that item in the full array. Hence, we must analyze all of the items in the aggregate in order to assess the representativeness of any one. But, in practice, not all values will be equally serviceable, even though their representativeness has been determined; rather two types of values have been found most convenient and useful to abstract from a tabulation: (1) the extreme values, or *maximum* and *minimum*, and (2) the central or typical values, known as *averages*.

The Maximum and Minimum as Measures of Location If the average depth of a river is one meter (3.28 feet), one should not thereby be encouraged to wade across; the maximum depth may be six meters (19.68 feet). In this and certain similar circumstances, the most useful information about location is one of the extremes of the distribution—the *maximum* or the *minimum* value—rather than the *central tendency* of the distribution. In deciding whether schools should provide some students with individual tutors, the lowest scores on an examination, rather than the average score, may be the deciding factor. In planning an investment strategy, the planner may be more interested in the maximum possible loss and the minimum possible return than in the expected or average return. A desert expedition may carry only the minimum water supply rather than enough to accommodate each member's average consumption. A prospective college student with short resources may find the minimum expenditure by students more useful information than the average of the students' annual expenses. In the fluctuations of traffic volume and weight, it is the peak load which must be safely accommodated on the highway or bridge; the structure designed to accommodate only the average load would collapse under the maximum. In such instances as these, the decisions to be made require maximum or minimum values, and the central tendency provides inappropriate information.

Maximum or minimum values are sometimes set by law or other regulations, and the effect is to locate the corresponding distributions above or below a specified value. Minimum wage laws, for example, have the effect of restricting the lower bound of the hourly wage distribution, although such a lower bound is not very informative in regard to differences in actual wage rates from region to region or from one occupation to another. Elevators are built to accommodate a maximum number of persons, and that number is ordinarily posted in the vehicle. With an occasional rare exception, this maximum sets the upper bound of the number of persons carried, but it does not serve to describe well the actual numbers of persons transported each day. Thus, although maximum and minimum values may provide the crucial information for many decisions, they frequently fail to serve the purposes for which the locations of distributions are summarized. But the student should be cautious about assuming that the central tendency or average value in a distribution is necessarily the most relevant measure of location. The key is the purpose for which a measure of location is to be used.

The Concept of Average Any value in a distribution could be selected as a locational measure. If information is rare or exceedingly expensive, the value for a single case may be used to represent an entire unknown distribution, although it may provide a poor summary. Inferences are made about the height and weight of prehistoric humans from the fragmentary remains of a single person, even though the remains that happen to be available to us may not be typical of the population they are presumed to represent. Hence we would not be justified in stating that the fragmentary remains available tell us about the *average* height or weight of prehistoric humans; they provide only a general and imprecise notion about the location of these distributions—that the smallest was no larger and the largest was no smaller than the single case in hand. If that is the only information available, we "make do" with such imprecision, but we would naturally prefer more complete information. With information about more cases, we would seek a value that summarizes the entire distribution of values by describing the point around which the distribution seems to center or cluster. The point around which a statistical distribution centers or clusters is often termed the *central tendency* of the distribution or the *average value* of the distribution.

In common usage as well as in statistical language, the concept *average* connotes the typical, the ordinary and expected. Like most other statistical measures, the average has its roots in common experience and is indispensable to daily discourse. People summarize and simplify the whole range of experiences by speaking quite casually about the average voter, the average family, the average student, a batting average. We may refer to almost any experience as "just average." The wide variety of traits which are reducible to an average is shown in such popular descriptions as that of the "average" American male who "stands 5 feet 9 inches, weighs 158, prefers brunettes, baseball, beefsteak, and French fries."

Of course, no one is average in every respect: the person of average height will not necessarily be a person of average intelligence or average handsomeness. Therefore, the popular claim that the "average person" does not exist is completely justified. Such unrealistic expressions as "average school" and "average household" do not conform to the stricter statistical concept which applies only to a series of measures on a single variable.

Nevertheless, in whatever manner the concept is employed, the layperson unconsciously implies what every statistician explicitly recognizes: that the average is a kind of norm around which the values tend to vary. But the difference between the layperson and the professional is that the latter requires more precision than is provided by informal folk usage. The statistician therefore devises various averages, each answering to the requirements of a given problem, and utilizes specified procedures to measure each of these diverse averages.

For certain distributions, the point around which the values center or cluster is relatively unambiguous, but for other distributions the central tendency is not so readily apparent. The *bell-shaped curve* sketched in Figure 5.1.1 has an evident central tendency—a point around which the values center and cluster—and this will also be the case for any frequency distribution that is symmetrical around a single peak frequency. For the *U-curve*, however, the values cluster around the two extremes, and neither of these represents a central value for the distribution as a whole. The *bimodal frequency distribution* carries the suggestion of two separate distributions melted into one; whenever this is the case, a single average will not do justice to either one. In the *skewed distribution*, the point around which the values seem to cluster will not give full representation to the values at the opposite

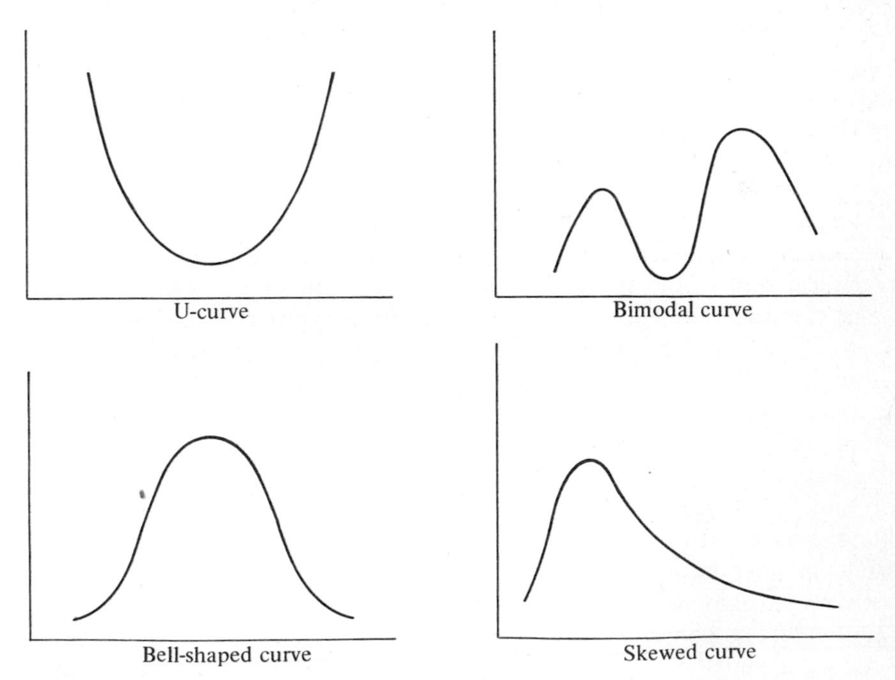

Figure 5.1.1 *Relation of Central Tendency to Frequency Curve*

extreme. The common occurrence of frequency distributions that are not symmetrical around a single peak frequency complicates the task of adequately representing an entire distribution by a single average or measure of central tendency. Such measures cannot be expected to summarize all aspects of a distribution and the purpose of the inquiry will commonly dictate which of the several averages summarizes best the location of the distribution.

Uses of Averages The average of a distribution may be used simply to represent the central tendency or the location of a single distribution, but averages are used even more commonly in social statistics to compare the locations of two or more distributions. An average value for a given population may have limited meaning, or no clear meaning at all except in comparison to an average of the same variable for some other population. The average size of firms in the electronics industry may be a descriptively accurate statistic which, alone, conveys little useful information. It takes on added meaning in relation to the average size of firms in another industry—for instance, the steel industry. The finding that firms in steel are considerably larger than firms in electronics may stimulate us to investigate the characteristics of industries that require relatively large (or small) firms. In a similar vein, the average age of persons arrested for burglary takes on added meaning when it is compared to the average age of the total population. The average age of those arrested for burglary is lower than for the population as a whole, and we may inquire into the reasons for the relative youth of those who are apprehended for this type of crime. Lacking a relevant comparative context, the average productivity of American workers is difficult to interpret. When placed in the context of the average productivity for workers in other nations we have some basis for regarding the American average productivity as relatively high or relatively low and hence can proceed to make an appropriate interpretation. The average score on a statistics examination of students in a given class does not by itself tell us very much about the statistical skill of the students who took the examination. The numerical value of the average will depend on the weighting given each question and the level of difficulty of the examination. But when two classes have very different averages on the same examination, the reason for the difference piques our curiosity—is it because students in the class with the higher average were more skillfully taught, because they were more highly motivated, because they had a more extensive background of relevant prior training, or because of some other reason? According to Figure 5.1.2, the incomes of black families have a strong center a little beyond $3,000, whereas the incomes of white families have a weak center around $11,000. The sharp contrast between central tendencies calls for sociological explanation: Why do black families fare so poorly relative to white families?

In social statistics, the average for a single distribution is generally calculated for the purpose of making such comparisons, because such comparisons

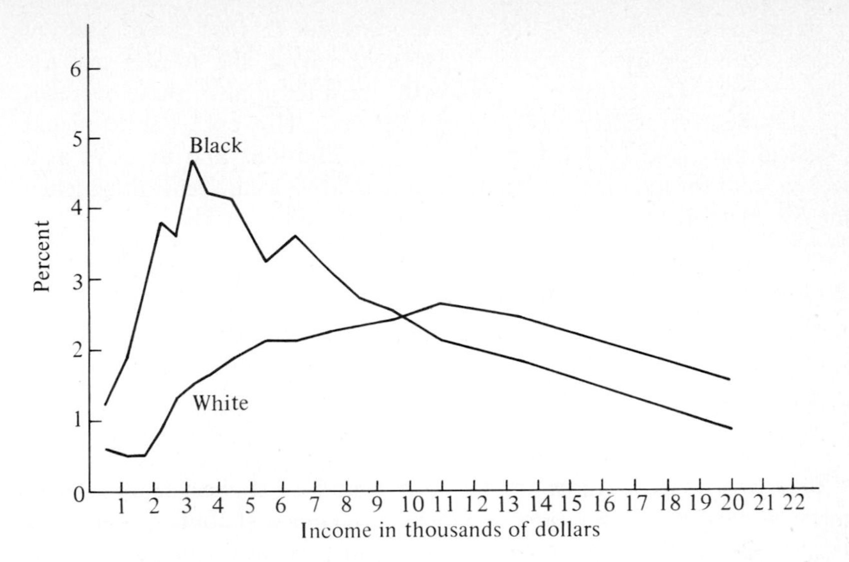

Figure 5.1.2 *Total Money Income, Percentage Distribution, White and Black Families, United States, 1974*

Source: U.S. Bureau of the Census, *Current Population Reports*, Series P–60, No. 99. "Money Income and Poverty Status of Families and Persons in the United States: 1974, (Advance Report)." Table 5. Washington, D.C.: U.S. Government Printing Office. 1975.

give rise to interesting questions or, if the differences between averages observed are in accord with theoretical expectations, because the observation of such a difference provides supporting evidence for the theoretical reasoning that lies behind the expectations. When averages are compared for such purposes, it should be obvious that the same type of average should be used to summarize the location of each of the distributions compared. If those distributions compared have quite different shapes, a fair comparison may require the computation and comparison of more than one type of average.

The average of a single distribution may itself serve as a useful reference point for the individual values in that distribution. In the United States, a family income of $15,000 may seem low relative to the ceiling but high relative to the average. An examination score of 45 may be relatively high, relatively low, or "about average" depending on the value of the average. A score of 45 elicits pride or depression depending on its distance from the average—and in which direction. If the average is the value that is typical or ordinary or expected, we may inquire into the reasons for individual deviations from that average. Do persons with above average incomes, for example, also have above average levels of education, while those with below average incomes have less than the typical number of years of schooling? Averages serve not only to summarize the location of the distribution of values, but they also serve as a useful reference point in describing and exploring the reasons for individual deviations from the typical location.

In subsequent sections of this chapter, we will discuss three types of averages commonly used in social statistics: the *mode*, the *median*, and the *arithmetic mean.* The arithmetic mean is the most familiar of these averages and is also the most widely used in statistical work. However, the mode and the median may also be used to compare distributions and to serve as a reference point for locating individual values relative to the central tendency of the complete distribution.

QUESTIONS AND PROBLEMS

1. Define the following terms:
   measure of location
   central tendency
   average
2. If the lowest score on an examination is 20 and the highest score is 80, is the average necessarily about 50? In what kind of a distribution would the average fall at about 50? In what kind of a distribution would the central tendency be closer to 80 than to 20?
3. In 1970, the average age of the residents in an American city of 25,000 people was 30 years. Against what averages might this be placed to determine whether the residents of this community are relatively young, relatively old, or "about average"?

2 THE MODE

The mode is the most frequently recurring value in the distribution; it is that point at which the values are most densely concentrated. If the distribution has been represented in a frequency polygon, the mode is the value at which the frequency polygon is at its peak.

Etymologically, the mode is related to the notion of the prevailing fashion of dress or etiquette to which a majority of a given social class would be expected to conform. Hence, the mode may also be defined as "the most probable value" and therefore distinctively labeled the *probability average.*

An examination of popular expressions suggests that the mode is, in fact, often implied by the concept *average.* Such usage stems, in part, from the fact that attributes as well as variates may take on predominant frequencies in a series of observations. When politicians refer to "average voters" as wanting their interests protected, they usually mean that most voters are motivated by self-interest; when waitresses remark that the "average customer" does not drink his or her coffee black, they are likely to mean that most restaurant patrons do not drink black coffee. Restaurant patrons and voters, in their turn, speak of the "average waitress" and the "average politician."

Statistically speaking, the mode is the value that is most likely to appear in a random draw from among all the values in the distribution. Inspecting Figure 5.1.2, we find that the most probable (modal) income for black families falls somewhere between $3,000 and $3,500; the most probable (modal) income for whites falls between $10,000 and $12,000. We may take the midpoints of these intervals—$3,250 and $11,000—as approximations of the modal values. Although these values are most probable in the sense that they are most likely to be drawn, this does not mean that they are highly probable. In fact, they are not: only around 4.7 percent of all cases are in the modal interval for blacks; only around 2.6 percent of the cases are in the modal interval for whites. If cases were more heavily concentrated in their respective modal intervals (say 60 percent and 40 percent), the mode would be more likely, but the modal value would remain the same. The mode is the most likely value, whatever the likelihood of that value may be. It is clearly incorrect to state that a majority of the values fall in the modal interval, unless there are only two categories; the largest number of cases, but not necessarily a majority, will fall in the modal interval.

Finding the Mode Since the mode is the most frequently occurring value, it is obviously necessary to count the number of occurrences of each value. In the case of continuous data, it is conceivable that empirical measurements would be so minute and discriminating that no two cases in the set would be found to be identical in value. Hence, the first step in finding the mode is to construct a frequency distribution with intervals appropriately selected for the range of values and the number of cases.

If the intervals for the frequency distribution are *uniform in width* (as, for example, in Table 4.1.5), the modal interval is simply the interval with the highest frequency. If the frequencies of every interval were identical, there would be no mode; this is a possible but uncommon occurrence in social data. If the frequencies of every interval were almost but not quite identical, one might refrain from calculating a mode on the grounds that the mode is so unstable as to be worthless as a measure of central tendency. If a change in the widths of the class intervals or the addition of a few cases in a given interval brings about a drastic shift in the mode, the mode should not be used as a summary of location for that distribution. A frequency distribution based on a reasonably large number of cases however, usually reveals a clear plurality in the frequency for a single interval. The mode is then placed at the midpoint of that interval.

If the intervals for the frequency distribution are *not uniform in width*, appropriate adjustments must be made before locating the modal interval. This circumstance is illustrated in the frequency distribution displayed in Table 5.2.1, where some intervals are one year wide while other intervals are wider than one year. Taking the frequencies as presented, the largest is the count for the interval 25–29 years, and one might proceed to the erroneous conclusion that the mode is the midpoint of that interval, or 27.5 years. But

Table 5.2.1 *Male Population of Prisons and Reformatories in the United States by Age, 1970*

Age	Frequency (f)
Under 5 years	48
5–9	67
10–14	395
15	410
16	923
17	2,150
18	4,673
19	7,040
20	9,098
21	9,720
22	10,971
23	11,174
24	9,227
25–29	38,838
30–34	26,695
35–39	20,293
40–44	15,926
45–49	10,252
50–54	6,278
55–59	3,842
60–64	2,147
65–74	1,705
75–84	177
85 years and over	69
Total	192,118

Source: U.S. Bureau of the Census. Census of Population: 1970. *Subject Reports.* Final Report PC(2)–4E. "Persons in Institutions & Other Group Quarters." Table 3. Washington, D.C.: U.S. Government Printing Office. 1973.

the frequencies for the various intervals must be made comparable before they are compared, and this requires that the intervals be of uniform width. Thus, the 38,838 cases in the interval 25–29 must be distributed over the one-year intervals from 25 to 29 to make the frequencies comparable to those for the one-year intervals preceding 25–29. Lacking information about the distribution of these 38,838 cases over these single years, we simply distribute them evenly over the constituent one-year intervals. Each of the five, one-year intervals constituting this interval is therefore assigned one-fifth of the total cases for the interval: 7,768. A similar distribution for each of the other five-year intervals would clearly yield an even smaller number for each of their constituent one-year intervals. If we examine the frequencies for one-year intervals, it is evident that the modal interval is 23 years, and the mode is the midpoint of that interval: 23.5 years. (We assume rounding to the last whole year in this instance because that is the customary rounding procedure for age.)

The age distribution in Table 5.2.1 may be manipulated to illustrate how the mode fluctuates with regrouping. If this frequency distribution had been originally constructed with all intervals 5 years wide, the modal interval would have been 20–24 years, with a frequency of 50,190 (the sum of the frequencies for the one-year intervals constituting the 20–24 interval). The mode would then have been the midpoint of the 20–24 interval, or 22.5 years. We noted above that for the distribution as tabulated in Table 5.2.1, the mode is 23.5. Had the original investigator elected to use two-year intervals to show in detail the age distribution between 15 and 24 years, the modal interval would have been 21–22 years, and the mode, at the midpoint of that interval, would have been 22.0 years. Which of these is the "true" mode—23.5, 22.5 or 22.0 years? Each is a proper description of the mode; they differ from one another because of differences in the placement of class boundaries.

Because the mode will shift with shifting class boundaries, and because the location of boundaries is arbitrary, the mode is not without an arbitrary element. And because it is generally more sensitive to changing class lines than either the mean or the median (to be considered next), it is sometimes said to be less stable than these averages. Yet such a verbal evaluation, however good its intentions, is subject to abuse: it may be cited for not calculating the mode when it is the mode that is actually required. It is perhaps enough to bear in mind that the mode may change appreciably with a slight shift in the class boundaries.

The mode has been dubbed "unstable" for still another reason—because the loss of a very few cases may change the modal category, and thereby the value or location of the mode. Again, Table 5.2.1 is a case in point. For this table, the mode is not very "modish"—several intervals other than the modal interval are close contenders for the most frequently occurring interval. If a mere handful of cases have been omitted for lack of information about age and if these happen to fall in the same interval, the mode for all cases (including those omitted here) might shift by one year or more. And even though the general pattern in the distribution of male inmates in correctional institutions may remain very similar from one year to the next, the modal interval may shift with relatively minor redistributions of the inmates' ages. Indeed, there are so many close contenders for the modal interval in this distribution that one might reasonably decline to summarize the location of the distribution by the mode, and prefer to use some other measure of central tendency instead.

Bimodality Some distributions display two concentrations or humps and, therefore, are called *bimodal* to distinguish them from *unimodal* distributions. This bimodality of a given distribution may be the result of amalgamating

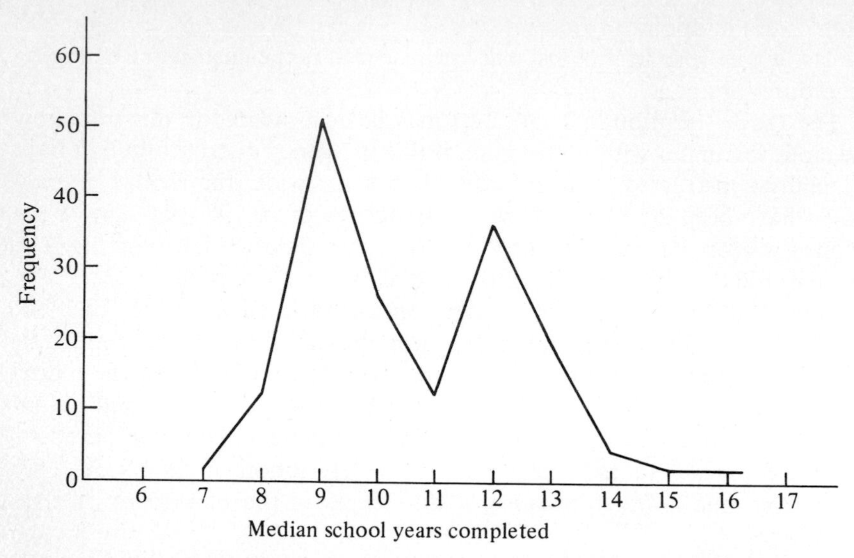

Figure 5.2.1 *Bimodal Frequency Distribution, Median School Years Completed, Persons 25 Years Old and Over, 163 Census Tracts, New Orleans, 1970*

Source: U.S. Bureau of the Census. Census of Population and Housing: 1970. *Census Tracts.* Final Report PHC (1)–144 New Orleans, La. SMSA Table P–2. Washington, D.C.: U.S. Government Printing Office. 1972.

two or more populations which have markedly different locations on the scale. Thus, the bimodality in a frequency polygon of adult heights would be due to the consolidation of groups of males and females, who are characterized by two different sets of heights. A graphic example of bimodality is furnished by the frequency distribution of median school years attended in 163 New Orleans (La.) census tracts (Figure 5.2.1). These two modes probably reflect prevalent terminal points in school attendance: a large number leave school immediately after attaining the legal minimum age; another large concentration take their leave after they have completed high school. When confronted with such bimodality, the worker is often called upon to disentangle the populations which caused it; or, failing that, to accept the dual modality as a valid description of a single population.

Evaluation of the Mode While the mode would appear to recommend itself for measuring representativeness—what is more representative than the most frequent value?—some of its characteristics disqualify it for any but the simplest purposes. (1) It cannot be subsequently manipulated by algebraic rules because of its own derivation—a point which will become clearer when we discuss the mean; (2) it is influenced by the width of the class interval employed and therefore to that extent lacks stability; and (3) it cannot show its degree of modality and is in that sense nonspecific. Legitimately we may raise a question about the serviceability of a measure which indicates the most frequent value, but not its relative weight in the distribution, that is, the *degree of modality*. The mode gives us the most probable value but offers no clue to how probable is that value. If this probability is required, the whole distribution must be examined again. Finally, there is no rule but good practical sense to follow when determining how large the predominant frequency should be in order to dignify it as a modal frequency.

Any criticism that might be leveled against the mode is not to deny that this measure is frequently quite useful. For measures of common behavior, the mode is ordinarily the norm, as the derivation of the term suggests. Thus the modal amount of family income spent on recreation will not be affected by rare extremes but will indicate what is most typical. The modal number of persons per household usually serves better than other averages to indicate the housing needs in a given community. The modal response to a public opinion question reflects the most prevalent opinion even though it may not prevail in public policy. Although the mode does not lend itself to further statistical manipulation, for purposes of straightforward description, readily comprehended by a wide audience, the mode may be usefully employed.

QUESTIONS AND PROBLEMS

1. Define the following terms:
   probability average
   modal frequency
   modal interval
   mode
   degree of modality
   bimodality
2. For a given N, what would be the effect on degree of modality of the following circumstances?
   (a) a wider range of data with more concentration in the tails
   (b) a narrower range, with less concentration in the tails
3. List two instances of bimodal frequency distributions of social data.
4. For what kind of data is the mode the only possible average?
5. Calculate the mode for the following frequency tables:
   (a) Table 4.1.8
   (b) Table 4.1.10
   (c) Table 4.1.11
   (d) Table 4.1.4
6. How would you test the claim that the modal frequency at 8 years of schooling (Figure 5.2.1) is attributable to the legal minimum age and that modal frequency at 12 years of schooling reflects importance attached to high school education?

3 THE MEDIAN OR POSITION AVERAGE

The point which cuts the array into two equal divisions, so that exactly one-half ($N/2$) of the items are below, and one-half are above that point, is called the *median*. For example, the "average" student usually considers himself or herself near the middle of the array with about 50 percent of the students below and above. According to this method of reckoning, although the average student might also display the most frequent score (that is, the mode), the focus here is on position in the rank order rather than on membership in the modal group. Therefore, the student who stands exactly in the middle of the lot, with a grade of 80, for instance, is thought of as the median student, and that student's grade or score is the median value, or simply the median. In 1974, the median age of the total United States population was 29.0 years, signifying that one-half of the population was older, and one-half younger than that age. Since the median clearly denotes position in a sequence of values, it is often referred to as a *position average*. It is also referred to as a *partition value*, since it may be regarded as a partition between the higher and lower values in the distribution.

Calculation of the Median Since the median divides the range into two parts, with each division containing exactly 50 percent of the items, it is nothing more than the common true boundary between those two segments. Therefore, there are three essential steps in determining the median: (1) arranging the values in rank order; (2) finding the $N/2$th rank; and (3) determining the value of the item in that rank.

With ungrouped data involving only a few cases, the median might be found by inspecting the array. Let us suppose that five persons have respectively 5, 2, 7, 4, and 10 dollars each. To find the median, we arrange the values in order of magnitude, and designate the position, or rank, of each, as shown in Table 5.3.1. According to the basic definition, the number of items above and below the median is $N/2$, in this case, $2\frac{1}{2}$ items. Hence, starting at either end, we count through the $N/2$th, or $2\frac{1}{2}$th rank, which places the halfway

Table 5.3.1 *Computation of Median, Odd Number of Items*

X	Rank	
\$2	1	$2\frac{1}{2}$ ranks below
4	2	
--5	-----3	------------ Median Point
7	4	$2\frac{1}{2}$ ranks above
10	5	

Table 5.3.2 Computation of Median, Even Number of Items

X	Rank	
$2	1	
4	2	3 ranks below
5	3	
		Median Point
7	4	
10	5	3 ranks above
13	6	

point exactly in the middle of the third rank—it bisects the third rank.[1] The corresponding value, or the median, would be the midpoint of 5, or exactly 5.0.

When the array contains an *even* number of items, a short additional step is required. For example, if an item of $13 were added to the array of Table 5.3.1, we would have the situation shown in Table 5.3.2. The number of items below and above the median would now be three. Counting off three items from either end, we would meet halfway between ranks 3 and 4, or at 3.5, which is the true upper limit of the third and the true lower limit of the fourth rank. The corresponding point in the array of values is a point midway between 5 and 7. Since this interval extends from 4.5 to 7.5, the median point is exactly 6.0. Actually, any value between the midpoints 5 and 7 would satisfy the definition of the median, since the median is a point such that half the ranked items are above and half below it. However, it is conventional to locate the median exactly halfway between the values, a placement consistent with the even division of ranks.

Procedure in Grouped Data Most statistical problems are not as simple as those cited above; in fact, such simple problems are quite unrealistic because statistical investigations generally involve large masses of grouped data. Although in such cases the procedure for computing the median is fundamentally the same as that for ungrouped data, it must necessarily accommodate itself to the fact that the median almost always will lie somewhere within a class interval. This median interval itself can be readily identified from the cumulative frequency table, but it is still necessary to establish the median point within that interval.

The cumulative frequency distribution of Table 5.3.3 will demonstrate the method of determining the median age of the population of the United States in 1974. (Observe that the frequencies have been rounded to the nearest million, a type of simplification that will not significantly influence the median and will save much tedious work.)

[1] We remind the reader that it is quite conventional to treat discrete data, such as ranks, as if they were continuous.

Table 5.3.3 *Population by Age, Cumulative Distribution, United States, July 1, 1974*

Age	f (in Millions)	cf
Under 5	16	16
5–14	38	54
15–24	40	94
25–34	30	124
35–44	23	147
45–54	24	171
55–64	19	190
65 and over	22	212
Total	212	—

Source: U.S. Bureau of the Census, *Current Population Reports*, Series P–25, No. 529. "Population Estimates and Projections." Table 1. Washington, D.C.: U.S. Government Printing Office. September, 1974.

Conforming to the above directive, we first divide the total frequency into halves: $212/2 = 106$. We then locate the class interval of the $N/2$th, or 106th case, by counting from either end. Cumulating from the lower end, the 106th case is obviously not among the first 16 cases, nor is it among the first 94 cases, which represent the cumulation of the first three frequencies. The fourth cumulation of 124 overshoots the mark by a substantial margin. The 106th case is therefore somewhere within the class frequency of 30, which is conceptualized as being spread uniformly throughout the interval 25–34. Having accounted for 94 cases, we must pass through 12 additional items (106 minus 94) of the 30 in order to include the desired item, and to establish the position of the median. Starting at its true lower limit, therefore, we must penetrate the fourth class interval 12/30th, or 40 percent of its width. Since the class interval is 10, the distance to go to reach the median is $.40 \times 10$, which is 4.00 years. Adding this amount to the true lower limit of the median class interval (25.0), the median is found to be 29.0 years.

The entire procedure for computing the median may be written as follows:

$$Md = L + \left(\frac{\frac{N}{2} - cf}{f} \times w \right), \tag{5.3.1}$$

where Md = median,

L = the true lower limit of the class interval in which the median is located,

$\frac{N}{2}$ = one half of the total frequency,

cf = the cumulated frequency up to but not including the median class interval,

f = the frequency of the median class interval,
and
w = class width.

Substituting in this formula, we arrive at exactly the same result as before:

$$Md = 25.0 + \left(\frac{106 - 94}{30} \times 10\right)$$
$$= 25.0 + 4.0$$
$$= 29.0.$$

Nor would the result differ if we operated on percentage frequencies, instead of absolute frequencies. Since $N/2$ is necessarily 50 when the calculation is performed on percentages, the formula becomes:

$$Md = L + \left(\frac{50 - cf}{f} \times w\right), \tag{5.3.2}$$

it being understood that f and cf stand for the appropriate percentage frequencies (see Table 5.3.4). We note finally that, as in the above example, the frequency table need not be closed in order to obtain the median.

Table 5.3.4 *Computation of Median Age, United States Population, July 1, 1974*

Age	Percent	Cumulated Percentages
Under 5	7.5	7.5
5–14	17.9	25.4
15–24	18.9	44.3
25–34	14.2	58.5
35–44	10.8	69.3
45–54	11.3	80.6
55–64	9.0	89.6
65 and over	10.4	100.0
	100.0	

$$Md = 25.0 + \left(\frac{50 - 44.3}{14.2} \times 10\right)$$
$$= 25.0 + \left(\frac{5.7}{14.2} \times 10\right)$$
$$= 25.0 + 4.0$$
$$= 29.0 \text{ years}$$

Source: Table 5.3.3.

Table 5.3.5 *Computation, Median Family Size, Iowa, 1970*

Size of family	f	Percent	Cumulated Percentages
2	275,604	38.4	38.4
3	137,023	19.1	57.5
4	126,741	17.7	75.2
5	85,745	11.9	87.1
6	49,397	6.9	94.0
7 or more	43,266	6.0	100.0
	717,776	100.0	

$$Md = 2.5 + \left(\frac{50 - 38.4}{19.1} \times 1\right)$$
$$= 2.5 + .61$$
$$= 3.11 \text{ persons}$$

Source: U.S. Bureau of the Census, Census of Population: 1970. *Detailed Characteristics.* Final Report PC(1)–D17 Iowa. Table 157. Washington, D.C.: U.S. Government Printing Office. 1972.

Median of Discrete Data Some writers would limit the use of the median to continuous data, for the reason that by definition discrete data cannot be fractionated as would be required for the median. But such a prohibition does not seem appropriate. In the discrete tabulation of Iowa families by size (Table 5.3.5), there is no observed family size such that exactly 50 percent of the families are larger, and 50 percent smaller. We find that 38.4 percent of all families consist of 2 persons or less, and 57.5 percent consist of 3 persons or less. Hence, some of the 3-person families lie below the 50 percent point, and the remainder lie above it—a circumstance which seems to exclude the possibility of an integral median family size. In this situation, must we then abandon the median, or may we pragmatically treat the data as continuous and proceed to accept a fractional value as the median? The simple answer is that we follow the latter alternative, as we do with every other average. We do this to make finer distinctions than would be possible with whole numbers.

If, for example, the 50 states were to be ranked for 1970 by median size of family, it is obvious that fractional values would have to be employed. Our only option is to treat the data as continuous in order to satisfy the requirements of the problem posed. Operating on this practical principle, we find the median of the above tabulation to be 3.11, a result which would enable us to draw comparisons among the states in respect to median family size.

The median is a simple concept: 50 percent of the items are smaller in value, and 50 percent are larger. Furthermore, by at least one criterion, it is the most representative of the averages: the aggregate distance between the median and each of the values is less than from any other point. Therefore, it is nearer to its companion values than any other average. It is in this sense that the median occupies the most central position in a distribution.

Other Position Measures Instead of reporting a student in the upper half of the class, which requires the median, we may wish for greater precision to locate the student in a smaller interval, such as the highest quarter, tenth, or even hundredth. We may state that Mr. Jones is in the upper half of the income distribution, but to place him in the highest 1 percent is certainly more informative, although both statements may be true. For such added discrimination, smaller subdivisions are required. The computation of *quartiles, deciles,* and *centiles,* which divide the array into fourths, tenths, and hundredths, respectively, is carried out according to the same principle as in the case of the median, except that the appropriate frequency, or percentage, of items below the point in question is substituted for $N/2$ in the formula above.

For example, to find the point below which the lowest quarter of the cases fall, we replace $N/2$ by $N/4$. Thus, the first quartile (or Q_1) in the age distribution shown in Table 5.3.3 is:

$$Q_1 = L + \left(\frac{\frac{N}{4} - cf}{f} \times w\right) \tag{5.3.3}$$

$$= 5.0 + \left(\frac{53 - 16}{38} \times 10\right)$$

$$= 5.0 + 9.7$$

$$= 14.7 \text{ years.}$$

If we wish to locate the point below which 75 percent of the items fall (or Q_3), we make the following substitution in the formula:

$$Q_3 = L + \left(\frac{\frac{3N}{4} - cf}{f} \times w\right). \tag{5.3.4}$$

The 90th centile (or C_{90}) would be found by using the formula:

$$C_{90} = L + \left(\frac{\frac{90N}{100} - cf}{f} \times w\right). \tag{5.3.5}$$

Quantiles as Standardized Measures The median, quartiles, quintiles, deciles, and centiles, which by their definitions indicate the proportion of items located below or above a given value, are collectively referred to as *quantiles*. As has been demonstrated in this section, quantiles may be used to fix the relative position of any given value in its array. A weight of 180 pounds may be located at the 90th centile ($C_{90} = 180$ lbs.) which places that weight in a position such that 10 percent of the population are above that weight, and 90 percent, below it. Similarly, an age of 62, an IQ of 120, a height of 5 ft. 9 in., an income of $10,000, a suicide rate of 22 per 100,000—all conceivably may be located at exactly the same abstract centile point.

It is now evident that quantiles must be recognized as standardized measures of position, independent of the metric system used or of the substantive type of the data. Thus, a person who is at the 90th centile in intelligence may also be around the 90th centile in income, suggesting an affinity between these two social phenomena. By such a transformation, noncomparable absolute measures may be usefully juxtaposed.

As will be more clearly seen in a later section, these measures are also independent of the pattern of distribution—whether normal, skewed, or rectangular. This circumstance enhances the versatility of quantile measures and enables them to represent any variate as a rank position in a set of data, by means of the statistical procedures here set forth.

QUESTIONS AND PROBLEMS

1. Define the following terms:

partition value	median point
median	median class interval
position average	quantile
rank order	decile
median rank	centile

2. (a) Compute the median of all family incomes (Table 5.3.6); of all white and black incomes, respectively, by Formula 5.3.1.
   (b) Could you have obtained the median of all incomes if only the distribution of incomes below $8,000 were given?
   (c) Compute quartiles for both white and black distributions and compare.
   (d) Would any of the above measures be affected if absolute frequencies were used instead of percentages?
3. Is it possible to compute the median of a combined set of values when only the medians and total frequencies of each group are known? When total frequencies are identical?
4. Explain why the median can be calculated even when the frequency table is open-ended.
5. In computing the median of grouped data, what assumption is made concerning the distribution of items within the median interval?

Table 5.3.6 *Total Money Income, White and Black Families, United States, 1974*

Total Money Income	All Races	White	Black
Under $1,000	1.3	1.1	2.3
$1,000 to $1,499	0.6	0.5	1.9
$1,500 to $1,999	0.7	0.5	2.8
$2,000 to $2,499	1.2	0.9	3.8
$2,500 to $2,999	1.5	1.3	3.6
$3,000 to $3,499	1.8	1.5	4.7
$3,500 to $3,999	1.8	1.6	4.2
$4,000 to $4,999	4.1	3.7	8.2
$5,000 to $5,999	4.4	4.2	6.3
$6,000 to $6,999	4.4	4.2	7.2
$7,000 to $7,999	4.5	4.3	6.1
$8,000 to $8,999	4.6	4.5	5.3
$9,000 to $9,999	4.7	4.7	5.0
$10,000 to $11,999	10.2	10.5	8.4
$12,000 to $14,999	14.1	14.6	10.7
$15,000 to $24,999	28.3	29.7	16.2
$25,000 to $49,999	10.4	11.2	3.0
$50,000 and over	1.1	1.2	0.2
Percent	100.0	100.0	100.0
Number (in thousands)	55,712	49,451	5,498

Source: U.S. Bureau of the Census, *Current Population Reports,* Series P–60, No. 99. "Money Income and Poverty Status of Families and Persons in the United States: 1974, (Advance Report)." Table 5. Washington, D.C.: U.S. Government Printing Office. 1975.

6. Is it possible to find the median for grouped data without interpolation when one of the cumulated frequencies is equal to $N/2$? Verify your answer with a simple illustrative calculation.
7. Compute the median suicide rate (Table 4.1.4), by Formula 5.3.1. Compare with the mode of the same distribution and interpret the difference.
8. Compute the median delinquency rate (Table 4.1.8). Compare with the mode.
9. For the following set of values, calculate the midpoint of the range and the median, respectively: 2, 5, 12, 16, 40. Distinguish between these two concepts.
10. In computing the median of discrete data, why is it necessary to treat them as continuous?

4 THE MEAN OR ARITHMETIC AVERAGE

Everyone has had frequent occasion to add a series of figures and divide the sum by the number of items. This is an operational definition of the mean—often used synonymously for "average"—which does not differ from the basic statistical procedure, as evidenced in the formula:

$$\bar{X} = \frac{\Sigma X}{N}, \tag{5.4.1}$$

where $\bar{X}$ = the mean (read: "X-bar"),
Σ = the sum of the variates (read: "summation X"),
X = a variate, or value, and
N = total frequency, or number of items.

This is a very ancient conception of the average, and is consistent with the etymological derivation of the term. The Latin *havaria* once referred to the insured cargo loss which was equally divided among the participating shippers. This actuarial principle of spreading the risks survives, of course, as the basic procedure in mutual insurance practice. Although this arithmetic conception of the mean is quite adequate for most workaday purposes, it does not exhaust all its statistical implications.

In statistical comparisons, the arithmetic mean serves as a measure that is normed for group size. The relative productivity of work teams of varying sizes is not readily revealed by examining the total production of each team but by comparing their production per member, or their mean production. The total taxes collected in a large city and a small city do not accurately reflect the relative tax liability of the citizens who reside in each; a comparison of taxes per taxpayer, or the mean taxes paid, provides a basis for comparison that is not confounded by the varying sizes of the populations. The severity of legal sanctions for burglary and auto theft cannot be inferred from the aggregate years of penal confinement accumulated by persons convicted of each crime, but rather by the years of legal confinement per person convicted of each crime, or the mean sentence served. Thus, the arithmetic mean facilitates the informative comparison of assorted measures for social aggregates of widely differing sizes. All constituent elements in each aggregate contribute to the sum for that aggregate, and hence to the mean, but the mean norms the sum by translating it into an amount per element. The mean therefore yields a measure that is comparable for large and small aggregates.

Another approach to the definition of the mean, necessarily congruent with the foregoing, views the values in the distribution as deviations from a central norm which is considered the true value. The random scatter of shots around the bull's-eye of a target and the bell-shaped distributions of replicated laboratory measurements illustrate this conception. The deviations on both sides of the central value constitute departures, or errors, from that norm. From this point of view, the mean is considered the true value, free of errors.

In line with this trend of thought, deviations often tend to be discounted as being random, temporary, exceptional, or less valid than the central value. The despondent person who has encountered a "streak of bad luck" confidently expects "things to average out." He or she feels that something akin to a law of nature provides that in the long run bad luck must be balanced by good luck. Even someone who has enjoyed a run of good luck does not dare "push his luck too far." The average monthly income of a sales representative, the average of irregularly spaced rainfall, the batting average which is flanked by slumps and sprees, the "picnic finance" where expenses are equally divided—all these are firmly rooted in daily experience and in popular parlance. Statistically speaking, the true value emerges when the deviations cancel out, that is, when their net value is zero.

Therefore, the question is: What value would we observe if there were no errors? The answer is: that central value around which the errors exactly balance one another, so that the positive and negative errors, when added together algebraically, equal zero. But if there were no errors, all values would be equal. Therefore, the mean may be defined as *that value which would occur if all the values in a given aggregate were equal.* Hence, the mean taken N times would equal the sum of the observed values: $N\bar{X} = \Sigma X$.

The basic principle of the balanced deviations is illustrated computationally in Table 5.4.1. The same principle may be illustrated graphically by a sketch of fulcrum, lever, and weights (Figure 5.4.1). In conformity to childhood experience on the seesaw, the aggregate force of the children on either side of the fulcrum is determined not only by their number, but also by their respective distances from the fulcrum. The sum of the two weights (-5 and -3) below the fulcrum balances the three weights ($+1, +3$, and $+4$) above the mean, as any child on a seesaw would appreciate intuitively.

Table 5.4.1 *Computation of the Mean and Deviations from the Mean for Ungrouped Data*

X	$X - \bar{X} = x$
0	$0 - 5 = -5$
6	$6 - 5 = 1$
2	$2 - 5 = -3$
9	$9 - 5 = 4$
8	$8 - 5 = 3$
Sum = 25	0
Mean = 25 ÷ 5 = 5	

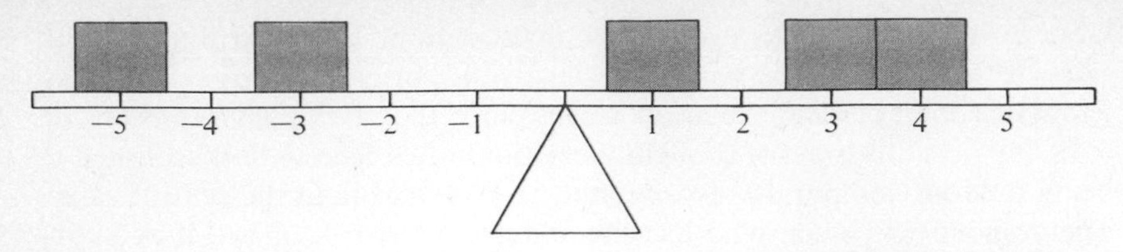

Figure 5.4.1 Diagram of the Mean as the Center of Gravity, Ungrouped Data from Table 5.4.1

Computing the Mean from an Arbitrary Origin The subtraction of a *constant* or *arbitrary origin* (A) from each value in a series of values will leave the intervals between values unchanged. If the arbitrary origin subtracted happens to be the mean, its subtraction will yield the deviations from the mean, and the algebraic sum of these deviations is zero. If the arbitrary origin selected is some value other than the mean, the algebraic sum of deviations from this origin will not be zero, but the mean of these deviations will give the deviation of the arbitrary origin from the mean of the given values. To illustrate, consider three numbers: 102, 104, 109. Their sum is 315 and their mean is 105. If we subtract 100 (an arbitrary origin) from each number we have left: 2, 4, 9. The sum of these deviations is 15 and their mean is 5. The mean of these deviations added to the arbitrary origin gives the mean of the given values: $100 + 5 = 105$. Alternatively, we might select 120 as the arbitrary origin, in which case the respective deviations are -18, -16, and -11. The sum of these deviations is -45 and their mean is -15. Adding this mean of -15 to the arbitrary origin of 120 gives the mean of the series: $120 - 15 = 105$. This principle is sometimes useful in simplifying the computation of the mean of a set of values grouped around an easily subtracted constant (for example, 10, 50, 100); the same principle is utilized in computing the mean of grouped data by the "coding" procedure to be discussed later.

The foregoing procedure for arriving at the arithmetic mean of a series of numbers by way of an arbitrary origin may be put in the form of a statistical argument. The following condensed version may be helpful.

(1) Write any value (X_i) in the series as the sum of itself plus and minus the arbitrary origin (A):

$$X_i = X_i + A - A$$

(2) Rearrange terms on the right-hand side and express $X_i - A$ as a single quantity:

$$X_i = A + (X_i - A)$$

(3) Sum over all N terms:

$$\Sigma X_i = \Sigma[A_i + (X_i - A)]$$

(4) Rewrite the sum (on the right-hand side) as the sum of the sums:

$$\Sigma X_i = \Sigma A + \Sigma(X_i - A)$$

(5) Rewrite the sum of the constant as N times the constant:

$$\Sigma X_i = NA + \Sigma(X_i - A)$$

(6) Divide both sides by N:

$$\frac{\Sigma X_i}{N} = A + \frac{\Sigma(X_i - A)}{N} \tag{5.4.2}$$

On the left-hand side we have the mean of the given values; on the right-hand side we have the mean of the deviations around an arbitrary origin added to that origin. Note that when $A = \overline{X}$, the mean of the deviations around A is equal to zero; when $A = 0$, the mean of the deviations around A is equal to the mean of the given values.

The Weighted Mean The *weight* of a value, statistically speaking, is simply its frequency. Therefore, any collective measure may be considered as having a weight equal to the number of observations it represents. The mean is such a collective measure: it is not itself an observed value, it is a derived figure. Consequently, like ratios and percentages, which are also derived values, the mean should never be manipulated without taking its weight into possible consideration.

Two or more means may be averaged; that is, we may calculate the mean of a series of means. If the means derive from equal Ns, that is, are of equal weight, there is no special difficulty in this averaging operation. Consider the following problem:

42 students from one housing unit have a grade point average of 1.4; 42 students from another housing unit have a grade point average of 1.8. What is the average of all 84 students?

Since the grade point means have equal weight, one need only add the two means and divide by 2:

$$\overline{X} = \frac{1.4 + 1.8}{2}$$

$$= 1.6 \text{ grade points per student.}$$

However, it is obvious that in many problems unequal weights will be encountered. In such a case, the means of the subgroups, prior to their being combined, would have to be weighted by their respective Ns.

Thus, if the two housing units, with grade point averages of 1.4 and 1.8, had resident populations of 26 and 42, respectively, each mean would have to be accorded its proper weight in order to obtain the mean of the combined

Table 5.4.2 *Computation of Weighted Mean*

$\bar{X}$	f	$f\bar{X}$
1.4	26	36.4
1.8	42	75.6
	68	112.0

Weighted $\bar{X} = \frac{112}{68}$

$= 1.65$ grade points per student

group. This is done by multiplication, which constitutes the procedure of weighting. Expressed in symbols:

$$\text{Weighted } \bar{X} = \frac{\Sigma f_i \bar{X}_i}{N} = \frac{\Sigma N_i \bar{X}_i}{N}$$

where $N = \Sigma f_i = \Sigma N_i$.

The computation of the weighted mean is illustrated in Table 5.4.2.

The failure to weight often leads to absurd results. We cite an illustration from baseball:

> A batter, in 75 trips to the plate, has amassed 25 hits for a batting average (mean) of .333; on this day he gets 5 hits in 5 trips, for a day's average of 1.000. What is the combined average?

A naive "grandstand statistician" may add the two averages, and then divide by 2, for a combined average of .667—a patently unreasonable result. The proper calculation is shown in Table 5.4.3.

Unweighted Means The logical requirement of weighting is not always as clear-cut as it is in the illustration of the batting averages. There are occasional instances in which discretion may be employed concerning whether to weight

Table 5.4.3 *Mean of Combined Sets, Unequal* Ns

$\bar{X}$	f	$f\bar{X}$
.333	75	25
1.000	5	5
	80	30

Weighted $\bar{X} = \frac{30}{80}$

$= .375$ (batting average)

in the conventional manner as described above, or whether to disregard the *N*s from which component means have been derived.

Let us suppose that we wish to calculate the grade average of a large number of housing units. Weighting by the number of students in each housing unit will yield the number of grade points *per student*. However, if we desire the grade average *per housing unit*, we would ignore the size of the population of each housing unit and compute the average of the house averages. Such a mean is usually labeled an *unweighted average*.

In a certain sense, of course, there is no such thing as an unweighted average; it is only a question of which weight we select, and this is determined by the nature of the desired information. To put it concisely, it is a question of which average we demand. In the following problem, the choice of the unweighted average again will seem most reasonable.

The ten largest SMSAs in the United States have suicide rates (per 100,000 population) as follows:

City	Rate/100,000
New York City	6.3
Los Angeles	20.9
Chicago	8.4
Philadelphia	9.8
Detroit	11.3
San Francisco	22.1
Washington, D.C.	9.8
Boston	9.4
Pittsburgh	11.9
St. Louis	9.8

f/n

What is the mean suicide rate?

Presumably, we could weight the rate of each city according to its population size and compute the combined rate, as we did for the batting average. However, we are not likely to be interested in the suicide rate of such a population aggregate, since this aggregate of persons in ten individual cities does not constitute a meaningful social unit. Rather, we are more likely to be interested in the rate *per city* in this category; hence, each city would be treated as a unit and not weighted for population size. This average is an unweighted mean of the rates for the individual cities:

$$\overline{X} = \frac{119.7}{10}$$

$$= 12.0, \text{ the rate per city.}$$

Of course, there can be no routine, ironclad rule on the issue of weighting. It is a question of purposes and consequences. For example, consider the consequences of weighting the above list of cities by population. With populations as weights, the overall rate may fall nowhere near the center of the array but rather at one end or the other, depending on the rates of the largest and most influential cities. For this reason, as well as for the even more pertinent one that the unit of interest is the individual city, not the person, the unweighted average is called for.

Calculation for Grouped Data As we have seen, grouped data do not differ from simple ungrouped data except in the fact that similar values are grouped together and the frequency, or *weight*, of the grouped items is indicated for each group in the f column. Therefore, to calculate the mean of grouped items, we obtain the sum of the weighted values, and divide by N.

For example, to determine the number of persons per family within a given population unit, we first determine the total number of persons (ΣfX), and then divide by the number (N) of families. If there are 20 families of 1 person each, as in Table 5.4.4, the weighted total in this class will be 20; 24 families of 2 persons each give a total of 48 persons. Continuing in this way, we find a grand total of 335 persons in 100 families, or 3.35 persons per family.

Although this average, based as it is on discrete data, cannot correspond to any observed size, it is nevertheless a useful measure of magnitude, which, as with the median, no one finds disturbing.

Table 5.4.4 *Computation of Mean Family Size, One Hundred Families*

X	f	fX
1	20	20
2	24	48
3	17	51
4	15	60
5	9	45
6	5	30
7	4	28
8	3	24
9	1	9
10	2	20
	$N = 100$	335

$$\bar{X} = \frac{\Sigma fX}{N}$$

$$= \frac{335}{100}$$

$$= 3.35 \text{ persons per family}$$

Table 5.4.5 *Computation of Mean Family Size (Table 5.4.4 Regrouped)*

Family Size	Midpoint (X)	f	fX
1– 2	1.5	44	66
3– 4	3.5	32	112
5– 6	5.5	14	77
7– 8	7.5	7	52.5
9–10	9.5	3	28.5
		$N = 100$	336.0
$\bar{X} = \frac{\Sigma fX}{N} = \frac{336}{100} = 3.36$			

Source: Table 5.4.4.

In Table 5.4.4, each variate (size of family) is considered to be at the midpoint of an interval of one (for example, .5 – 1.5, 1.5 – 2.5, and so on). These variates are readily weighted by the frequencies since they are already tabulated as midpoints instead of class intervals. However, when the class interval is larger than one, and the class limits are set down, the midpoint must be calculated first. Then each midpoint is weighted as previously, and the mean computed (Table 5.4.5).

A Useful Computing Formula In practice, frequencies as well as magnitudes may be large, and the foregoing method of calculation may become quite cumbersome. A more convenient computing formula can be utilized to avoid tedious calculations when the computing is done by hand. This computing formula entails subtracting the midpoint of one of the intervals—an arbitrary origin (A)—from all of the midpoints and dividing these differences by the interval width. After computing the mean for this much simplified distribution, we multiply the result by the interval width to nullify the earlier division and add the midpoint originally subtracted to obtain the mean of the original distribution. If all intervals are of uniform width, the procedure is simpler in application than this description suggests, as illustrated in Table 5.4.6.

With *intervals of uniform width*, the result of subtracting the midpoint of a *reference interval* (A) and dividing by interval width yields a perfectly predictable outcome that may be referred to as the *codes* for the intervals. The code of the reference interval is always zero, and the codes for other intervals take on increasing integer values as one moves away from the reference interval in either direction, with the codes for intervals smaller than the reference interval having negative signs and the codes for intervals larger than the reference interval having positive signs. Any interval may serve as the reference interval and the result will be of the form described no matter which interval is selected, but computational labor is ordinarily minimized

Table 5.4.6 *Computation of Mean by Coding, Suicide Rates, 1970*

(1)	(2)	(3)	(4)	(5)	(6)
Class Limits	Frequency (f)	Midpoint (X)	$X - A$	$x' = \frac{X - A}{w}$	fx'
3–4	3	3.5	−8	−4	−12
5–6	9	5.5	−6	−3	−27
7–8	35	7.5	−4	−2	−70
9–10	51	9.5	−2	−1	−51
11–12	52	11.5	0	0	0
13–14	30	13.5	2	1	30
15–16	21	15.5	4	2	42
17–18	10	17.5	6	3	30
19–20	6	19.5	8	4	24
21–22	8	21.5	10	5	40
23–24	1	23.5	12	6	6
25–26	3	25.5	14	7	21
	229				33

$$\overline{X} = A + \left(\frac{\Sigma fx'}{N} \times w\right)$$
$$= 11.5 + \left(\frac{33}{229} \times 2\right)$$
$$= 11.5 + .29$$
$$= 11.79$$

Source: Table 4.1.4.

by selecting as the reference interval an interval with a relatively high frequency near the center of the distribution. Since the codes always display the same pattern (provided intervals are of uniform width), their calculation is not necessary and they can simply be written down. Stated otherwise, Columns 3 and 4 in Table 5.4.6 are not necessary, and the Column 5 entries can always be *written without calculation.*

The computing formula for the mean using codes consists of three steps: (1) compute the mean of the codes, $\Sigma fx'/N$; (2) multiply this result by the interval width, w; and (3) add the midpoint of the reference interval, A. Expressed in symbols:

$$\overline{X} = A + \left(\frac{\Sigma fx'}{N} \times w\right). \tag{5.4.3}$$

With *intervals not uniformly wide*, the same computing formula may be used, but the codes will not display the pattern of uniform integer increases with each interval deviation from A. Hence, the codes should be computed as shown in Column 5 of Table 5.4.6. Fractional codes may occur with nonuniform interval widths.

Proportions and Rates as Means Proportions and rates share with the arithmetic mean the property of norming for group size, as well as other of its features. They may be computed by dividing the sum of a set of values—artificial or natural—by the number of values in that set. To be more concrete: Consider a small group of three men and two women. The proportion male is .6 and the proportion female is .4. To show that these proportions are arithmetic means, we construct a score to represent the distinction between male and female, subject to the restriction that the score for one category is 1 and the score for the other category is 0. We may construct either a score of maleness (male = 1; female = 0) or a score of femaleness (female = 1; male = 0). The resulting listing of scores and the computation of means is shown in Table 5.4.7. The mean of the maleness scores is .6, the proportion male; and the mean of the femaleness scores is .4, the proportion female. The calculation of the mean in each instance is identical to the usual calculation of the proportion. If scores of 0 and 100 were assigned to the two categories, the resulting mean would be the percentage male or the percentage female.

Demographic rates are also arithmetic means. The crude birth rate, for example, is the number of births in a given year per 1,000 persons in the general population. This rate may be conceived as the arithmetic mean of as many variates as there are persons in the population, the variate for each person being the number of infants born to that person weighted by 1,000. The sum of these weighted variates is the total number of births times 1,000. The mean of these weighted variates is the crude birth rate per 1,000 population. Demographic rates are conventionally expressed in relation to a numerical base (for example, per 1,000, per 100,000, and so on) to clear the decimals, but whatever the numerical base selected, it does not change the fundamental character of these rates as arithmetic means.

Table 5.4.7 *Proportion Male and Proportion Female as Arithmetic Means*

Person	Sex	Maleness Score (X)	Femaleness Score (Y)
A	Male	1	0
B	Male	1	0
C	Female	0	1
D	Male	1	0
E	Female	0	1
		$\Sigma X = 3$	$\Sigma Y = 2$
		$\bar{X} = \frac{\Sigma X}{N}$	$\bar{Y} = \frac{\Sigma Y}{N}$
		$= \frac{3}{5}$	$= \frac{2}{5}$
		$= .6$	$= .4$
		$= \frac{\text{Number of Males}}{N}$	$= \frac{\text{Number of Females}}{N}$

QUESTIONS AND PROBLEMS

1. Define the following terms:

arithmetic mean	weighting
arbitrary mean (origin)	weighted mean
deviation from mean	unweighted mean
coding	proportion (mean)

2. Discuss the validity of the statement that the extreme values in a skewed distribution make a disproportionate contribution to the arithmetic mean.

3. Outline the procedure for finding the mean of two or more groups when only the total frequency and the arithmetic mean of each group is given. Express this procedure symbolically.

4. A group of 25 families has a mean weekly income of $96.00. A second group of 15 families has a mean weekly income of $132.00. What is the total income of the combined group?

5. If the mean of 100 items is 12, and the mean of 50 items is 21, what is the mean of the combined group? What would the mean be if each group had 50 items? 100 items?

6. A student received letter and percentage grades in his courses as shown in Table 5.4.8. What is his percentage grade per course? per hour? Which average would be used? Would you weight one or both averages? Explain.

Table 5.4.8 *Student's Grade Record*

Subject	Credit Hours	Grade	
		Percentage	Letter
History	5	94	A
English Literature	3	83	B
Psychology Lab	2	77	C
French	5	91	A

7. In Table 5.4.8 if the letter grades are scored as follows: A (3), B (2), C (1), what is the student's grade point average per course? per hour?

8. For the distribution in Table 5.4.6, select an alternative reference interval, assign codes, and verify that the computation of the mean with these alternative codes yields the same result.

9. Compute the mean of the frequency distribution of delinquency rates with and without coding (Table 4.1.11). Which procedure entails less computational labor?

Table 5.4.9 *Age of Male Addict Patients at Time of First Admission to the Lexington Hospital, by Race, Resident Population, 1962*

Age	White	Nonwhite
17–19	29	14
20–29	177	241
30–39	58	137
40–49	28	12
50–59	23	4
60–64	2	0
Total	317	408

Source: John C. Ball and William M. Bates, "Migration and Residential Mobility of Narcotic Drug Addicts," *Social Problems*, Vol. 14 (Summer 1966). p. 60, Table 1 (abbreviated). © 1966. Used by permission of the Society for the Study of Social Problems and the authors.

10. (a) Compute the mean age for white and for nonwhite addicts admitted to the United States Public Health Service Hospital in Lexington, Kentucky (Table 5.4.9). Note that the interval widths are not equal. The mean age of white male admissions based on ungrouped data was 30.0 and the mean age of nonwhite male admissions based on ungrouped data was 28.9. How much grouping error is introduced into the computation of these means by utilizing grouped data?
(b) Compute the median age for each distribution. Note that the medians of the two distributions are almost equal, with the median age for whites being slightly less than the median age for nonwhites even though the mean age of whites is greater than the mean age of nonwhites. Describe the features of the two distributions that lead to a higher mean age for whites even though the medians are almost equal.

5 CRITERIA FOR CHOICE OF AVERAGE

Because we are frequently faced with alternative choices among the various averages, it is necessary to lay down a few principles which will serve as guides in the selection of a central value. To a certain extent, this problem has been clarified already. It has been stated that there is no all-purpose average which can be employed universally. It always should be kept in mind that an average is a single representative value which possesses the convenience of compactness, but also the inconvenience of brevity. Even at best, it will conceal and exclude as much or more information than it reveals of the distribution from which it is extracted. For this reason we should always consider the possibility of using all three averages, on the premise that the three together will tell us more about the underlying distribution than any one of them alone.

Nevertheless, whenever we elect to rely on a single figure, we will want to be assured that we have made the best choice. Three criteria, in order of priority, are pertinent to that choice: (1) the purpose to be served, or the question the average is designed to answer; (2) the pattern of the distribution of the data; and (3) various technical considerations, primarily of an arithmetic nature, which limit the choice of average.

Purpose to Be Served Statistical distributions may be summarized by an average for a variety of purposes, and the average selected should be in accord with the purpose of the inquiry. In a study of marriages that end in divorce, the investigator may summarize the years of marriage prior to divorce by the mode, if the purpose is to locate the most hazardous stage in the marital relationship. But if the purpose of the study is to assess the effect of involvement with the extended family on the duration of marriages that end in divorce, the median or mean will provide a more appropriate comparison since it will be sensitive to the distribution outside the modal category, as the purpose requires. Church attendance among persons in selected congregations might be summarized by the mode, median, or mean of the number of services attended over the last year. If the purpose is to determine the social norm concerning church attendance in each congregation, the mode would serve well as the average, but if the investigator's purpose is to determine which church is most successful in generating high attendance records, the mean will serve better. A comparison of the mean segregation index for United States school districts in 1960 and 1970 will provide a better summary of the trend in segregation over that decade than would a comparison of medians or modes for the distributions in those years, because the mean will give due weight to those districts that experienced a marked change in segregation as well as to those districts that did not change at all—a desirable characteristic in an aggregate measure of change. But the mean segregation index in 1970 probably serves as a less adequate summary description of the general level of segregation in United States school districts than the median or the mode, because the mean may be unduly affected by values at the extremes.

There are no fixed rules for determining which average is best suited for the purpose of a given inquiry, and the investigator is obliged to make an informed judgment as to which average is best for the purpose at hand. The difficulty of making such a judgment is complicated by the fact that the social analyst must frequently design a study to serve multiple purposes, each of which entails its own complex of varied and perhaps contradictory implications for the choice of an average.

Pattern of Distribution The shape of frequency distributions varies from those which are almost flat (that is, where the frequencies are highly similar from one interval to another over most intervals) to those that are sharply peaked. Distributions may also be highly symmetrical or extremely skewed. Symmetry means that the values are distributed identically on either side of the mean, while skew is merely the absence of symmetry. The degree of skewness affects the typicality and representativeness of the average values and hence must be taken into account in judging their relevance.

The mode is neither informative nor stable in a distribution that is almost flat, while it has a unique applicability to distributions that are sharply peaked. In a distribution that is symmetrical around a single peak, the mean, median, and mode will be the same, although each may continue to carry its distinctive imagery. As the distribution becomes more and more skewed, the values of the averages diverge correspondingly, and the choice of the average becomes more critical. In such cases, the mean loses its typicality, in that it is less likely to be empirically encountered. In the U-curve, in fact, the mean may be rarely encountered and by that criterion it may be an unrealistic average. Since many sociologically relevant distributions—wages, sizes of families, and the like—are often severely skewed, it is essential to consider the pattern of the frequency curve in the selection of the most suitable average.

Technical Considerations There are certain purely technical features of a tabulation which may compel the use of one or another average. Thus, since the mean cannot be calculated for a distribution with open-ended intervals (unless those intervals are closed arbitrarily), the median may have to be utilized. But this is often quite satisfactory, provided the distribution is not unduly skewed, in which event mean and median do not differ much in any case. On the other hand, the mean is the only average computable when only the total of the values and N are known, even though another average might have been preferred.

The technical requirements of possible subsequent calculations always should be kept in mind. There is no method for combining or weighting medians and modes; and for that reason they cannot be standardized (processes for which will be discussed in Chapter 6), although subclass medians and modes can be compared directly. In general, when additional computations based on the average are contemplated, the mean (including special cases of the mean such as proportions and rates) should be selected.

Characteristics of Averages Selection of the proper average requires a thorough knowledge of the descriptive characteristics of each average. The characteristics of the different averages are frequently discussed in terms of their "advantages and disadvantages." But these terms are evaluative, rather than descriptive, and therefore have no fixed meaning. An advantage in one context may be a disadvantage in another. Hence, we prefer to set forth the characteristics which are intrinsic in the averages, quite independent of the setting in which they may be used.

Summary of Characteristics of Averages

The Mode

1. It is the most frequent value in the distribution; it is the point of greatest density.
2. The value of the mode is established by the predominant frequency, not by the values in the distribution.
3. It is the most probable value, hence the most typical.
4. A given distribution may have two or more modes. On the other hand, there is no mode in a rectangular distribution.
5. The mode does not reflect the degree of modality.
6. It cannot be manipulated algebraically: modes of subgroups cannot be combined.
7. It is unstable in that it is influenced by grouping procedures.
8. Values must be ordered and grouped for its computation.
9. It can be calculated when table ends are open.

The Median

1. It is the value of the middle point of the array (not midpoint of range), such that half the items are above and half below it.
2. The value of the median is fixed by its position in the array and does not reflect the individual values.
3. The aggregate distance between the median point and all the values in the array is less than from any other point.
4. Each array has one and only one median.
5. It cannot be manipulated algebraically: medians of subgroups cannot be weighted and combined.
6. It is stable in that grouping procedures do not affect it appreciably.
7. Values must be ordered, and may be grouped, for computation.
8. It can be computed when ends are open.
9. It is not applicable to qualitative data.

The Mean

1. It is the value in a given aggregate which would obtain if all the values were equal.
2. The sums of the deviations on either side of the mean are equal; hence, the algebraic sum of the deviations is equal to zero.
3. It reflects the magnitude of every value.
4. An array has one and only one mean.
5. Means may be manipulated algebraically: means of subgroups may be combined when properly weighted.

6. It may be calculated even when individual values are unknown, provided the sum of the values and N are known.
7. Values need not be ordered or grouped for its calculation.
8. It cannot be calculated from a frequency table when ends are open.
9. It is stable in that grouping procedures do not seriously affect it.

Comparability of Averages Like every other statistical quantity, averages are used for comparative purposes. Specifically, averages are used to compare the locations of distinct groups or distributions on the same scale. However, it should be obvious that the different types of averages are not comparable. That is to say, the mean of one distribution cannot be legitimately compared to the mode of another in determining their relative locations on a given continuum. The reason for this prohibition is simply that the various types of averages are not coordinate measures; they reflect different aspects of a distribution: the mode reflects the highest *frequency*, the median reflects the middle *position*, and the mean reflects the centrality of *values*.

These differentiating characteristics become all the more apparent when the effect of *skewness* on the averages is taken into account. Being influenced by each individual value, the mean of a skewed distribution will be pulled in the direction of the extreme values. The mode, of course, is not influenced by the flanks of the distribution, and the median is drawn toward the tail solely by the relative frequency of items in that tail. However, this latter attraction is not very great because the concentration of items in the tail is necessarily sparse. Accordingly, when the skew of a unimodal distribution is to the right (Figure 5.5.1), the order of the averages on the base line will be mode, median, and mean, and the gap getween them will vary according to the severity of the skew. A skew to the left, of course, will reverse the order.

But even two averages of the same type cannot be compared safely unless the patterns of distribution are similar. It is not too reckless to proffer the maxim that two or more averages never should be compared unless "all other things are equal" or at least similar. The comparison of the mean height of males ($\bar{X}$) and that of females ($\bar{Y}$) (Figure 5.5.2) is quite appropriate. Men

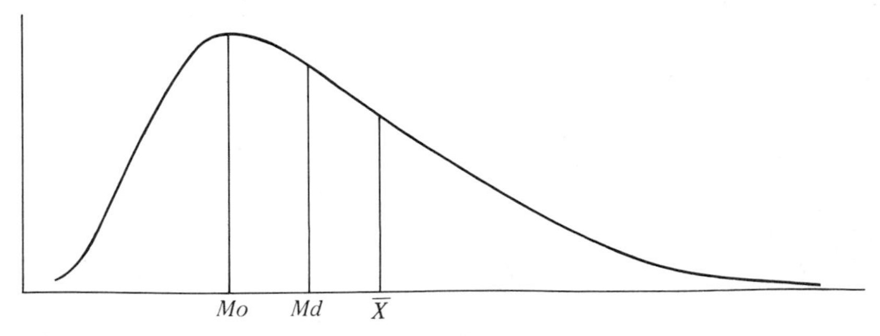

Figure 5.5.1 *Mean, Median, and Mode, Right Skew*

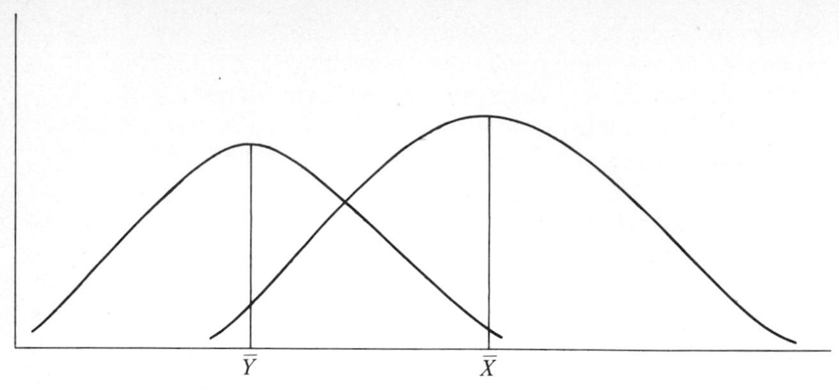

Figure 5.5.2 *Mean Height for Males* ($\bar{X}$) *and Females* ($\bar{Y}$)

are taller than women, and the degree of difference is properly measured by the differences in their means. Males and females are located merely in different segments of the same scale of heights; their patterns of variation are identical.

However, when the pattern of distribution differs markedly, such comparisons may, in fact, be very misleading, especially in the case of means. For example, when two means are quoted as identical, the reader may be misled into reconstructing the curves imaginatively also as identical patterns. In truth, however, the two distributions may present quite different contours, as in Figure 5.5.3.

To be sure, in a gross sense, the locations of the two sets of data are identical, for they cover exactly the same range. But the locations of the predominant segments of the distributions are widely separated and would be more accurately specified by the modes. In this instance, the identical means are a consequence of the overlapping tails, rather than a reflection of the overlapping humps.

In Figure 5.5.4, the locations of the predominant frequencies are within approximately the same range; the means are separated, however, because of the skew in one of the distributions. Here, too, the modes, which are almost identical, would do greater justice to the location of the data. Of

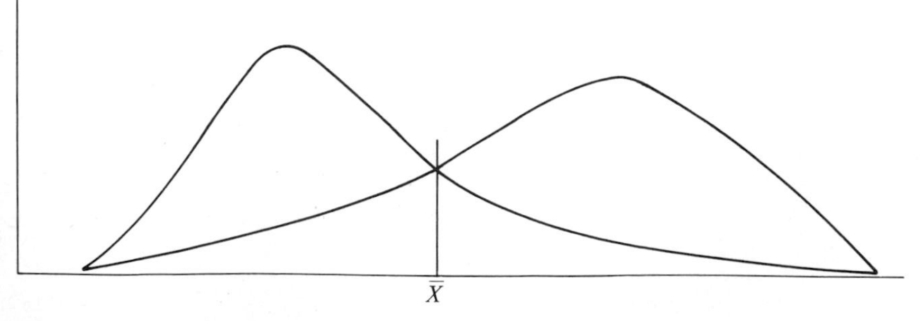

Figure 5.5.3 *Opposite Skews, Identical Means*

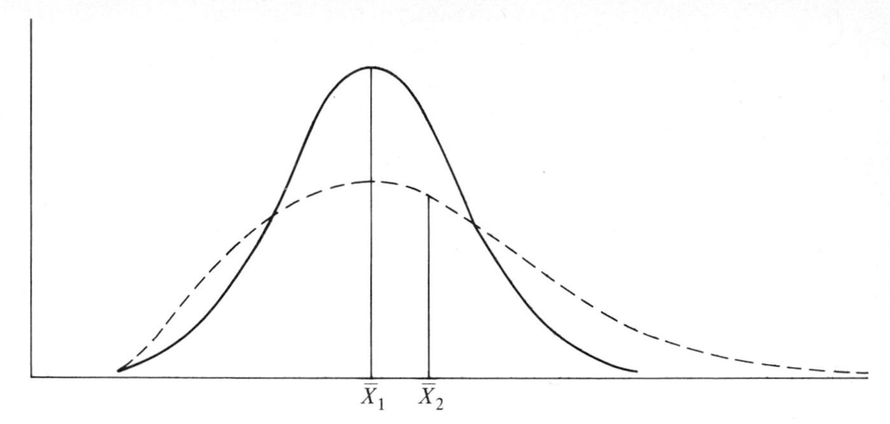

Figure 5.5.4 *Effect of Skew on the Mean*

course, when dominance of location is less pertinent than centrality of values, our comparison would necessarily be based on means.

Conclusion The foregoing discussion inevitably leads to the truism which could have been anticipated: Averages, like every other statistic, are incomplete descriptions of a set of data. They should not be conceived of as autonomous, but rather as being bound to the data from which they are derived. A responsible use of the average—or, for that matter, any other measure of location—must take into consideration the user's tendency to reconstruct mentally the distribution from which only a single value has been extracted. To supplement this incomplete description afforded by the average, it is essential to examine the scatter or dispersion of the distribution relative to the average. This will be the subject of Chapter 7.

QUESTIONS AND PROBLEMS

1. Define the following terms:
   symmetry
   skew
   right (positive) skew
   left (negative) skew
   unimodal
   rectangular distribution
2. State in general terms how right skew affects the three common averages. Left skew?
3. In a frequency distribution of family incomes in the United States (Table 5.3.6), how would the median, mode, and mean rank by order of magnitude?

4. As a result of the mental tests administered to American soldiers during World War I, it was reported that "most Americans were below the average score." Is that possible? Probable? Which average could have been meant?

5. How would the median and mean grades in a certain class be affected if:
   (a) the poorest students in the class withdrew?
   (b) the poorest students were successfully tutored?
   (c) the middle students were successfully tutored?
   (d) an easier examination were given?

6. In what type of distribution would the median student not be included among the modal students? Show graphically.

7. List the absolute minimum information necessary for the calculation of the mean, mode, and median from grouped data.

8. In Community A, the modal length of life is 55, the median is 60, and the mean is 65; in Community B, the modal length of life is 70, the median is 65, and the mean is 60. From this information, reconstruct the frequency curves. Which community is the healthier?

SELECTED REFERENCES

Moroney, M. J.
1954 Facts from Figures. Baltimore: Penguin Books. Chapter 4.

Yule, G. Udny, and M. G. Kendall
1950 An Introduction to the Theory of Statistics. Fourteenth edition. London: Charles Griffin. Chapter 4.

Norming Operations 6

1 ELEMENTARY NORMING

Rates and indexes, as they are used in sociology, may be regarded as *normed measures*; and the processes by which these normed measures are constructed may be referred to as *norming operations*. Although these operations are hardly folk knowledge, they are nevertheless rooted in the requirements of everyday living. People seldom respond to a measure as an isolated number, but usually in terms of some norm based on either informal experience or formal training, or both. They thereby supply the mental backdrop against which the measure is interpreted. For example, an annual income of $12,000 is not perceived as a detached figure, but rather in relation to a standard based on the experience of the observer. One hundred births in a community in a given year have meaning only in terms of a comparison with such relevant data as the size of the population, the number of births the preceding year, or the number of births in comparable communities.

One important function of the statistical method is to furnish the techniques by which single measures may be properly interpreted, or two or more meaningfully compared. There are many such devices, several of which are discussed in this chapter. Collectively, they belong to a family of procedures which we may call *norming operations*, because they set up an appropriate statistical norm in terms of which the raw data are expressed and thereby rendered comparable. They represent the well-known principle that comparisons are ambiguous to a greater or lesser degree unless "all other things are equal." All involve forming the *ratio* of one quantity to another; in particular, the ratio of the quantity to be normed to the norming quantity.

Ratio A ratio is by definition the quotient of one quantity divided by another; hence, any measure arrived at by that operation may be regarded as a ratio. Percentages and rates are no more than ratios; the percent unemployed and the death rate are cases in point. Still, the term *ratio* is used only now and then in social statistics to label the magnitude of one quantity relative to that of another, possibly because it is so general, possibly because other terminology is more suitable. Nevertheless, there are exceptions.

The number of males in the population divided by the number of females (conventionally multiplied by 100 to clear the decimal point) is generally known as the *sex ratio*. As a formula:

$$\text{Sex Ratio} = \frac{\text{Number of Males}}{\text{Number of Females}} \times 100.$$

Without the multiplier, this ratio gives the number of males per one female; with the multiplier, it gives the number of males per 100 females. For example, on July 1, 1974, there were approximately 103,454,000 males and 108,455,000 females in the United States population. The relation between these unwieldy numbers is retained much more easily when given as .95 males per 1 female, or 95 males per 100 females, or simply 95, as is customary.

The *child-woman ratio* supplies another exception. This is the number of children under five years of age divided by the number of women between the ages of 15 and 50 inclusively (multiplied by 1,000 to clear the decimal). As a formula:

$$\text{Child-Woman Ratio} = \frac{\text{Children under 5 Years}}{\text{Women between Ages 15 and 50}} \times 1{,}000.$$

Notwithstanding such exceptions, the term *ratio* has generally been set aside in favor of more restricted terms: proportion, percentage, rate, index. For example, the *cephalic index* is the ratio of cranial width to cranial length; the *intelligence quotient* is the ratio of mental age to chronological age. Both are essentially ratios in spite of the fact that they do not go under that name.

Proportion Although the term *proportion* has several related meanings, we attempt to use it consistently to denote the frequency of cases in one category divided by the frequency of cases in all categories combined. It is a part–whole ratio. In this sense, the number of high school seniors divided by the number of all students, including seniors, is a proportion, but the number of seniors divided by the number of all other students (excluding seniors) is not. The latter would be a part–part ratio.

Because the frequency in a single category cannot be larger than the frequency for all categories combined, proportions cannot exceed 1.00. Moreover, since categories must either have some cases or none, proportions cannot fall below zero. Proportions therefore range from 0 to 1. Furthermore, since the sum of the proportions is equal to the sum of the category frequencies divided by the total frequency, the sum of the proportions is necessarily equal to 1.00.

We have already encountered instances of proportions (as defined above): class frequencies divided by the total frequency yield a set of proportions; cell frequencies within a row divided by the marginal frequency yield a set of proportions (p. 42). The arithmetic mean of a 0,1 variable is a proportion (p. 113); in getting a weighted mean, we attached proportions (N_i/N) to subclass means as weights.

We will presently learn that all proportions may be converted to percentages, although not all percentages are convertible to proportions. Later we shall learn that every probability may be construed as a proportion, and every proportion as a probability. The usual answer to the question "What is the probability of living past 60?" is the proportion of persons living beyond that age; similarly, the answer to the question "What is the probability of a male birth?" is the proportion of male births. We shall also learn later that the strength of the association between two variables is usually measured by the proportion of variation in one that is attributable to the other. There is hardly any topic in social statistics that is free of proportions in the sense that some units may be divided by all.

Percent Since most persons gain familiarity with percentages during the process of growing up, it would hardly seem necessary to discuss them here. But this very familiarity may obscure the system on which they rest. The reduction of one quantity to parts per hundred of another is dictated by our practice of counting in multiples or fractions of ten. Thus, the statement that 50 percent of the population (50 per 100 population) is male is readily understood; whereas an equivalent statement in the binary system would strike practically everyone as esoteric.

Since a percentage is a normed quantity, it is especially convenient in comparative analysis. The relative number of women in college may be declining over time, while their absolute number increases. To portray change in the relative number of women in college, we would want percentages for a succession of years. Similarly, if our concern was with the trend in the relative number of married men, we would want percentages rather than absolute frequencies. That trend could be judged from absolute numbers, but such judgments would be troublesome to form. Percentages are convenient, if not essential, when our concern is with relative magnitudes.

Percentages are invaluable in analyzing population growth and social change. By way of example, consider the growth of the U.S. population 65 years of age and over between 1910 and 1970:

	Population over 65 (thousands)	Total Population (thousands)	Percentage over 65
1910	3,986	92,407	4.3
1970	20,177	204,879	9.9

In analyzing these figures, leaving aside the change in absolute numbers, we may calculate: (1) the percentage increase in the number of persons 65 years and over; (2) the increase in the percentage of persons 65 and over; (3) the percentage increase in the percentage of persons 65 years and over.

To find the percentage change, we divide the numerical increase of 20,177 − 3,986 = 16,191 by the number of persons over 65 in 1910, and multiply that ratio by 100:

$$\text{Percentage Increase} = \frac{16{,}191}{3{,}986} \times 100$$

$$= 406\%.$$

Between 1910 and 1970 the population 65 years and over increased by 406 percent.

To find the change in the percentage of persons 65 and over, we simply subtract the percentage in 1910 from the percentage in 1970: 9.9 − 4.3 = 5.6. Note that this difference is itself not a percentage; rather it is a difference between percentages. Because such differences are expressed in terms of percentage points, they are referred to as *percentage point differences.*

The percentage increase in the percent 65 years and over is simply the percentage point difference (5.6) divided by the base of 4.3:

$$\text{Percentage Change} = \frac{5.6}{4.3} \times 100$$

$$= 130\%.$$

The 130 percent reminds us again that percentages may exceed 100; if the percent over 65 years had been smaller in 1970, the percentage change would have carried the minus sign.

We have already remarked the convenience of converting class frequencies to percentages of the total frequency. But that procedure carries some hazard. When the total frequency (N) is not given (or is perhaps even suppressed), we have no means of knowing whether the several percentages rest on 25, 50, or 500 cases, and to that extent, therefore, we cannot judge their dependability. Percentages based on small absolute numbers are of doubtful reliability. For this reason it is good practice to include the base frequency with every percentage distribution, so that actual frequencies can be reconstructed. In addition, actual frequencies may be recombined, whereas percentages on different bases may not be so manipulated.

Index In statistics, the term *index* usually pertains to relatively complicated ratios or sets of ratios, although it is used colloquially as well as technically for other types of measures. As a derived measure, it is designed to express simply the variation in a given set of values which in the raw form would be quite unintelligible. In its more formalized version, it usually refers to a ratio between two quantities, one of which is taken as the norm, or as an expectation, against which the other is measured.

Thus, the cost-of-living index compares the average of prices in a particular year with the average for the "normal" year. An index of 150 in 1970, on a base year of 1960, indicates that the cost of living increased 50 percent over the base year taken as 100. Although such a deceptively simple index may be glibly quoted by every columnist, nevertheless the many bits and pieces of information of which it is composed and the method by which these bits and pieces are weighted—as well as the choice of the base period—all testify to its statistical complexity.

Table 6.1.1 illustrates the construction and use of an index to measure group differences in degree of suburbanization. It addresses the question of whether American males (25 years of age or older) with more education show a greater propensity toward suburbanization than American males

Table 6.1.1 *Indexes of Suburbanization for the Detroit Metropolitan Area, 1950 and 1960*

SMSA Males, 25 Years of Age or Older, by Years of School Completed	Percent in Suburb		Index of Suburbanization		Differences in Index
	1950	1960	1950	1960	1950–60
All School Levels	36.6	53.1	100.0	100.0	
None	26.7	30.7	73.0	57.8	−15.2
Grade: 1–4	28.7	35.4	78.4	66.7	−11.7
5–6	30.7	40.7	83.9	76.6	− 7.3
7–8	37.2	48.6	101.6	91.5	−10.1
High: 1–3	37.8	54.1	103.3	101.9	− 1.4
4	38.8	59.5	106.0	112.1	6.1
College: 1–3	36.7	59.9	100.3	112.8	12.5
4+	40.8	65.1	111.5	122.6	11.1

Source: Adapted with permission of Macmillan Publishing Co. Inc. from *The Urban Scene* by Leo F. Schnore. Copyright © 1965 by The Free Press, A Division of The Macmillan Company.

with less education. If there were no differences in degree of suburbanization, the same percentage of each group would live in the suburbs (out of the core city but in the metropolitan area). This single percentage of suburban dwellers across all educational groupings would be a weighted mean, and could materialize only if the degree of suburbanization of American males was equal for all groups regardless of schooling. The weighted mean would be the percentage expected on the assumption that educational groupings do not differ in their tendency to move out of the city.

For example, to gauge the tendency of males 25 years of age or older with no schooling to reside in the suburbs rather than in the city in 1960, we calculate:

$$\text{Index} = 30.7/53.1 \times 100$$
$$= 57.8.$$

This result may be interpreted to mean that there are roughly 42 percent (or 100 − 57.8) fewer persons in this category than we would expect on the assumption that educational classes are identical in their inclination toward suburban life. From all such results we deduce that there is a tendency for men of superior schooling to flee the city leaving a disproportionately large residue of relatively uneducated men in the city. By comparing 1950 and 1960, we may judge the degree to which differences in suburbanization among educational groupings are becoming larger or smaller.

The *seasonal index*, which is not elaborated here, is similarly calculated as a ratio between a given monthly measure and the average measure for the twelve-month series and is used to measure the fluctuation in births, deaths, production, and certain other economic phenomena. The so-called *aggregative index*, which is the ratio of one weighted mean to another, is taken up at a later point in this chapter.

Demographic Rates Measures such as the death rate and the birth rate are arrived at by dividing the number of events in a given period by the population at risk during that period; hence, they might just as well be called *ratios*. However, because they give the number of events per person, or the average, and because the term *rate* connotes the process of averaging, their labeling as rates is not without some linguistic justification.

We have regular recourse to rates in social analysis because they in effect make all populations equal in size. One population may experience many more deaths than another, not because it is less sanitary and hygienic, but simply because it is much larger. If it were no larger than the other, it might experience no more or even fewer deaths. By standardizing all populations to the same size, we eliminate one factor that may confound our comparisons.

The base of a demographic rate may be the total population or some subpopulation. For example, the total population is the base of the *crude birth rate*:

$$\text{CBR} = \frac{E}{N} \times 1{,}000$$

where E is the number of births in a given year and N is the population at the middle of the year. This is roughly the average population during the year.

By way of contrast, the *general fertility rate* is based on the number of women between the ages of 15 and 44 years:

$$\text{GFR} = \frac{E}{N_{15-44}} \times 1{,}000$$

where N_{15-44} is the midyear female population 15–44 years of age.

The base population of a rate should logically consist of those individuals who are liable to the event in question—collectively known as the "exposed group." Although anyone may die, only women may have children and only married persons may get divorced. Hence, a rate based on a judiciously selected exposed population is less subject to distortion by extraneous factors. Birth rates, marriage rates, and divorce rates calculated on a total, unselected population are usually referred to as a "crude" rate; others are labeled "specific" or "refined." The advantage of the crude rate lies in its ready comprehensibility, the economy of tradition, and its utility for rough unspecialized purposes. The advantage of the specific rate lies in its precision and its serviceability in technical and professional research.

We note in passing that the ratio of events to population is often a relatively small decimal. To clear the decimal point with room to spare, we usually multiply such ratios by some suitable power of 10. In some cases (for example, the sex ratio) 100 will do; in other cases (for example, the birth rate) 1,000 is required; with rare events (such as suicides) we may have to use 100,000 to clear the decimal point. In this latter case, we give the number of suicides per 100,000 population, instead of the number per capita.

Specific Rates We have already noted that rates may be affected by the composition of the population on which they are based. The crude birth rate, other things being equal, will be raised by the disproportionate presence of young women; the crude death rate will be lowered by the disproportionate presence of young men and women. Whenever rates are subject to compositional effects, differences between them are ambiguous in their meaning, at best. Thus, a difference between crude death rates may reflect a constant difference in the rate of dying at every age, or a difference between age structures, or both. The student therefore may rightly wonder whether, in drawing such comparisons, there is some method for controlling for differences in population composition, that is, for holding them constant.

We have already hinted at one solution to this problem, namely: limiting comparisons to relatively homogeneous subpopulations. For example, instead of comparing crude birth rates, we might compare birth rates for women of specific ages.

In setting up specific rates, we do not partition the population in every conceivable way; rather we select those subpopulations that will mitigate the shortcomings of rates based on total populations. For example, since vital processes—living and dying—are affected by sex and age, it is common practice in social demography to calculate birth and death rates for populations homogeneous in respect to age and sex. In other inquiries, it may be necessary to set up subpopulations homogeneous in respect to race, education, and socioeconomic background.

The calculation of a specific rate is in accordance with the general formula: number of events divided by number of persons in subpopulation, multiplied by 1,000 (or a larger or smaller multiple of 10):

$$R = \frac{E}{N} \times 1{,}000$$

where R = rate, E = number of events, N = population size.

Specific rates often show their utility by throwing light on puzzling differences between overall rates. By way of example: Consider that in 1970 (Table 6.1.2) California had 35.8 births per 1,000 women of all ages, whereas Kansas had only 33.4 (approximate figures). This difference is at odds with the popular notion that rural populations have higher birth rates than urban populations. It is a sociological oddity.

Perhaps the overall rate in Kansas is depressed by the disproportionate presence of women either too young or too old to have children. To explore this possibility, we turn to age-specific birth rates (Table 6.1.3). We see that

Table 6.1.2 *Births per 1,000 Female Population: United States, California, Kansas, 1970*

	Total Female Population	Number of Births	Births per 1,000
United States	104,299,734	3,731,386	35.76
California	10,136,449	362,756	35.79
Kansas	1,145,005	38,204	33.37

Source: U.S. Bureau of the Census, *Census of Population: 1970*. Vol. I—Characteristics of the Population. Part 1, U.S. Summary—Section 1, Table 49, Part 6, California and Part 18, Kansas, Table 20. Washington, D.C.: U.S. Government Printing Office. 1973.
U.S. Department of Health, Education and Welfare, *Vital Statistics of the United States: 1970*. Volume I—Natality. Tables 1–52 and 1–54. Washington, D.C.: U.S. Government Printing Office. 1975.

Table 6.1.3 Age-Specific Birth Rates: California and Kansas, 1970

Age	California			Kansas		
	Female Population	Number of Births	Births per 1,000	Female Population	Number of Births	Births per 1,000
15–19	886,495	60,864	68.7	107,492	6,932	64.5
20–24	868,710	137,304	158.1	91,159	15,316	168.0
25–29	730,640	99,128	135.7	68,951	9,984	144.8
30–34	608,157	43,142	70.9	59,789	3,834	64.1
35–39	574,773	16,988	29.6	59,931	1,564	26.1
40–44	616,220	4,364	7.1	63,922	480	7.5
45–49	638,743	232	.4	65,029	26	.4
Total	4,923,738	362,022	73.5	516,273	38,136	73.9

Source: See sources of Table 6.1.2.

the number of births per 1,000 women ages 20 to 29 is higher in Kansas; the number of births per 1,000 women ages 30–39 is higher in California; rates for women between the ages of 40 and 49 are about the same. All in all, these age-specific rates suggest that the women of Kansas are at least as productive as the women of California, and their crude rate of 33.4 suffers (relative to California) from a disadvantageous age distribution. Indeed, when women under 15 years and over 49 years are excluded from the base of our calculations, Kansas has a slight edge—73.9 to 73.5. Our refined rates have thus helped to throw light on what was at a first glance a somewhat puzzling figure.

QUESTIONS AND PROBLEMS

1. Define the following terms:
   norming operation
   ratio
   proportion
   percentage
   rate
   index
   crude rate
   specific rate
2. In 1970, the state of Washington had a population of about 3,414,000, and the number of births in that year was 61,000. What was the birth rate per 1,000?
3. In the eighteenth century, European churches maintained birth and death records, which were later used by population scholars to estimate

the sizes of cities. Thus, in Berlin around 1700, there were 178 deaths. The estimated ratio of deaths to total population was 1:35. Calculate the death rate per 1,000. Estimate the size of the city of Berlin at that period.

4. Between 1940 and 1972 the birth rate per 1,000 declined from 19.4 to 15.6. Calculate the permillage-point decrease (the decrease per 1,000); the percentage decrease.
5. In a city of 300,000, the sex ratio is 98. Calculate the number of males.
6. If a college has an enrollment of 1,236, of which 692 are males, what is the sex ratio for the school? Explain briefly what this ratio means.
7. In 1940, coffee cost 21.2 cents per pound. In 1950, coffee was 79.4 cents per pound; in 1955 its price had risen to 93.0 cents per pound; in 1965, the price per pound was 83.3 cents; and in 1973, its price was up to 100.2 cents per pound. If 1940 is considered the base year, what would be the price index for 1950, 1955, 1965, and 1973? Interpret your answer.
8. The sex ratio in City A is 100, in City B, 50. The total population in the two cities is identical. Does City A have twice as many men as City B? Explain.
9. Verify that California's overall birth rate of 73.5 (Table 6.1.3) is the weighted mean of its age-specific birth rates. Carry out the same calculations for Kansas.
10. Calculate births per 1,000 for women between the ages of 20 and 29 in Kansas and California, respectively. Compare and interpret (Table 6.1.3).
11. Convert differences (last column, Table 6.1.1) to percentages, using 1950 as the base. Which set of numbers—differences or percentages—lends itself more readily to interpretation?

2 STANDARDIZED RATES

Although specific rates permit stricter comparisons than overall rates, they usually involve many more comparisons. With two populations, there will be as many comparisons as there are specific rates. And in the end, we may wonder whether there is some method for reducing the differences between specific rates to an average of all such differences. The answer to that question brings us to the subject of *standardization*.

In standardization, we proceed as if two or more populations were identically composed on a characteristic (for example, age) thought to affect the frequency of events (for example, deaths) in question. The object of standardization is to eliminate from the differences between crude rates

Table 6.2.1 *Age Distributions and Age-Specific Death Rates, United States and Japan* [a]

Age	United States		Japan	
	Population 1970	Death Rate 1971	Population 1970	Death Rate 1971
Under 1	3,485	19.6	1,877	13.2
1– 4	13,669	.8	6,928	1.0
5– 9	19,956	.4	8,159	.4
10–14	20,790	.4	7,858	.3
15–19	19,070	1.1	9,064	.7
20–24	16,371	1.6	10,660	1.0
25–29	13,477	1.2	9,089	1.0
30–34	11,430	1.8	8,372	1.3
35–39	11,107	2.4	8,207	1.9
40–44	11,981	3.6	7,340	2.7
45–49	12,116	5.7	5,878	3.9
50–54	11,104	8.7	4,805	5.9
55–59	9,973	13.0	4,425	9.4
60–64	8,617	20.1	3,726	16.0
65–69	6,992	29.6	2,984	26.5
70–74	5,444	43.1	2,134	45.5
75–79	3,835	67.5	1,268	77.9
80–84	2,284	104.1	650	126.1
85 and over	1,511	170.9	296	215.5
Total	203,212	9.5	103,720	6.9

[a] Populations are in thousands and death rates are per 1,000 population.
Source: United Nations, *Demographic Yearbook, 1972*. United Nations Publishing Service, New York, 1973. Tables 6 and 24.

the effect of differences in population composition on one or more characteristics. In order to carry out a standardization, we require a schedule of specific rates (the r_i) on the population whose crude rate is to be adjusted; and the number of persons (the N_i) for each subclass of the standardizing population—the so-called *weight schedule*. For example, to standardize Japan's crude death rate on the age distribution of the United States, we require age-specific death rates for Japan and the distribution of the United States population by age. These series are given in Table 6.2.1.

The general method consists of calculating the expected number of deaths in a hypothetical population having the age-specific death rates of Japan and the age-specific population frequencies of the United States. Dividing this expected number of deaths by the total U.S. population gives the adjusted, or standardized, rate. As a general formula:

$$AR = \sum_{1}^{k} N_i r_i / N, \tag{6.2.1}$$

Table 6.2.2 *Expected Number of Deaths (in Hundreds) in Japan and United States*

Age	Japan[a]	U.S.[b]
Under 1	460	368
1– 4	137	55
5– 9	80	33
10–14	62	31
15–19	133	100
20–24	164	171
25–29	135	109
30–34	149	151
35–39	211	197
40–44	323	264
45–49	473	335
50–54	655	418
55–59	937	575
60–64	1,379	749
65–69	1,853	883
70–74	2,477	920
75–79	2,987	856
80–84	2,880	677
85 and over	3,256	506
Expected Deaths	18,751	7,398
Age-Adjusted Death Rate	9.2	7.1

[a] U.S. age-specific population proportions multiplied by Japan's age-specific death rates.
[b] Japan's age-specific population proportions multiplied by U.S. age-specific death rates.
Source: Table 6.2.1.

where AR = adjusted rate; k = number of subclasses; r_i = rate for ith subclass of standardized population; N_i = number of members in ith subclass of standardizing population; $N = \Sigma N_i$. From this formula, it is clear that a standardized rate is nothing more than a weighted mean[1]: it is the sum of the weighted r_i divided by the sum of weights $\Sigma N_i = N$.

Upon multiplying Japan's age-specific death rates (from Table 6.2.1) by the U.S. age-specific frequencies (from Table 6.2.1) and summing products, in Table 6.2.2 we obtain 18,751 deaths (in hundreds). Dividing this expected

[1] The crude rate is also a weighted mean; upon replacing the standardizing weights by a population's own weights, we recover the crude rate. Japan's crude rate, expressed as a weighted mean is:

$$CR = \frac{\Sigma n_i r_i}{n}$$

where CR = crude rate; n_i is the size of the ith subpopulation; r_i is its rate; $\Sigma n_i = n$.

number of deaths (in thousands) by the total frequency (in thousands) gives Japan's adjusted death rate per 1,000 population:

$$\frac{1{,}875}{203{,}212} \times 1{,}000 = 9.2.$$

Assessing the Effect of Standardization Upon completing a standardization, it is natural to wonder whether it made much of a difference. In assessing the effect of standardization, it is necessary to examine first the difference between the crude rate (CR) and the adjusted rate (AR); and second, the difference between the adjusted rate (AR) and the standardizing rate (SR).

The first of these differences, $CR - AR$, reflects the difference in population weight schedules. It will be zero in the event that no differences between category-specific weights are present. To demonstrate this claim, we set up the difference between the crude rate and the adjusted rate as the difference between the weighted means:

$$CR - AR = \frac{1}{n}\Sigma n_i r_i - \frac{1}{N}\Sigma N_i r_i, \tag{6.2.2}$$

which we then rewrite as the sum of the weighted differences

$$CR - AR = \Sigma r_i(p_i - P_i) \tag{6.2.3}$$

where $P_i = N_i/N$ and $p_i = n_i/n$. This last expression makes clear that the adjusted rate (AR) cannot differ from the crude rate (CR) unless some of the age-specific population proportions differ; any difference between the adjusted and the crude rate must be attributed to a difference between weight schedules (since the rate schedule is constant).

Consider by way of example the difference between Japan's crude rate and adjusted rate:

$$\begin{aligned} CR - AR &= 6.9 - 9.2 \\ &= -2.3. \end{aligned}$$

Since Japan's crude and adjusted rates of 6.9 and 9.2, respectively, are based on the same age-specific death rates, the difference between them (-2.3) must be ascribed to the differences between the proportions of persons in the various age categories—the p_i and the P_i. In this case we might say that Japan would experience 2.3 more deaths per 1,000 population under the disadvantage of the U.S. age distribution.

The second aforementioned difference, namely, the difference between the adjusted rate (AR) and the standardizing rate (SR), reflects the differences between category-specific rates in the population whose rate is to be standardized, and category-specific rates in the population serving as the standard.

To demonstrate this point, we first set up the difference between the adjusted rate and the standardizing rate as the difference between weighted means:

$$AR - SR = \frac{1}{N} \Sigma N_i r_i - \frac{1}{N} \Sigma N_i R_i \tag{6.2.4}$$

which we then rewrite as the sum of the weighted differences:

$$AR - SR = \Sigma P_i (r_i - R_i). \tag{6.2.5}$$

This last formulation makes clear that the adjusted rate will be equal to the standardizing rate unless the populations differ in their specific rates; in other words, an inequality between AR and SR must be due to differences between the specific rates (the r_i and the R_i).

Consider the difference between Japan's adjusted death rate (AR) and the standardizing rate (SR), which in this problem is the U.S. crude rate (Table 6.2.1):

$$\begin{aligned} AR - SR &= 9.2 - 9.5 \\ &= -.3. \end{aligned}$$

Since the age-adjusted death rate of 9.2 and the standardizing rate of 9.5 are alike in their weights (the P_i) but not in their age-specific rates, the difference between them ($-.3$) must be attributed to differences between the age-specific death rates—the r_i and the R_i. In this case, we might say that Japan, owing to its relatively favorable rate schedule, has 30 fewer deaths per 1,000,000 population than the United States.

This difference of $9.2 - 9.5 = -0.3$ supplies an answer to the question with which we began this discussion: "Is it possible to summarize the k (k = number of subclasses) differences between specific rates by means of a single, summary measure?" The answer is that the quantity, $AR - SR$, constitutes one such index. It is just the weighted mean of the differences between the specific rates, with the N_i of the standardizing population serving as weights (Equation 6.2.5).

Although the difference between AR and SR will be zero when specific rates are equal, a difference of zero is not proof that specific rates are equal. Because the differences between the r_i and R_i may be positive or negative, the sum of the weighted algebraic (signs regarded) differences may be small, even though the sum of the weighted numerical (signs disregarded) differences is large. To investigate whether the $r_i - R_i$ are large numerically, even though their weighted sum is small, we may either examine them directly, or analyze statistically the *interaction* between category-specific rates and population proportions. This latter method is beyond the scope of this book; interested students should consult Kitagawa (1955).

Component Analysis The student may have noticed that the difference between the crude rate and the standardizing rate is equal to the sum of the difference between the crude rate and the adjusted rate and the difference between the adjusted rate and the standardizing rate. This is necessarily the case, since a quantity is unchanged if we put back into it what we previously took out of it. In our case:

$$CR - SR = (CR - AR) + (AR - SR).$$

The right-hand side differs from the left-hand side only in the addition (putting in) and subtraction (taking out) of the adjusted rate. By this operation, the overall difference has in effect been broken up into two component differences, or simply components; hence, we refer to this procedure as *component analysis.* The first component reflects the differences between population weights $(p_i - P_i)$; the second component reflects the differences between population rates $(r_i - R_i)$. The sum of these two components represents the net effect of both tendencies combined.

By way of example, consider first the overall difference between Japan's crude death rate and the rate for the standardizing population (the U.S. population):

$$CR - SR = 6.9 - 9.5 = -2.6.$$

Upon expressing this overall difference of −2.6 as the sum of its components, we get:

$$\begin{aligned} CR - SR &= (6.9 - 9.2) + (9.2 - 9.5) \\ &= (-2.3) + (-0.3) \\ &= -2.6. \end{aligned}$$

This decomposition of the difference between the death rates of the United States and Japan shows that Japan has an advantage both in its age distribution and in its age-specific death rates. Japan's death rate is lower by 2.3 deaths per 1,000 because of its more favorable age distribution; it is lower by an additional .3 death per 1,000 because of its more favorable age-specific death rates. The effect of both factors combined is to give Japan a lower crude death rate by 2.6 deaths per 1,000.

Component analysis thus answers to the question of whether an overall difference between the crude rate and the standardizing rate is due principally to a difference in category-specific weights or to a difference in category-specific rates. Because the difference between the crude rate and the adjusted rate $(CR - AR)$ reflects differences in population composition, it is sometimes referred to as a *compositional effect*; the difference between the adjusted rate and the standardizing rate $(AR - SR)$, which reflects what is left over after composition has been taken into account, is then referred to as a *residual effect*. Any difference between a crude rate and a standardizing rate may be partitioned into its components—compositional and residual—provided that the adjusted rate is given. That partitioning consists simply of finding the respective differences between the crude rate and the adjusted rate and the adjusted rate and the rate of the population serving as a standard.

The Standard Million Quite frequently, the crude rates to be converted into standardized rates are scattered over a large territory and cover an extended duration of time. Interstate vital statistics and international statistics involve cumbersome comparisons unless there is a generally recognized and available standard. To satisfy that need on an international basis, the *Standard Million* of England, 1901, was often employed by common consent at the beginning of this century. Since the age distribution of a population is one of the most distorting factors in the interpretation of vital statistics, this Standard Million was composed of the age distribution of the British population given in terms of a million population.

The United States National Office of Vital Statistics at present employs, for the computation of its vital statistics, a Standard Million based on the population census of April 1, 1940 (Table 6.2.3). Such a procedure standardizes the vital rates for age, thereby rendering them comparable on fertility

Table 6.2.3 *Standard Million as Determined from the Population of the United States, Enumerated as of April 1, 1940*

Age	Million
Under 1 year	15,343
1– 4 years	64,718
5–14 years	170,355
15–24 years	181,677
25–34 years	162,066
35–44 years	139,237
45–54 years	117,811
55–64 years	80,294
65–74 years	48,426
75–84 years	17,303
85 years and over	2,770
All Ages	1,000,000

Source: U.S. Department of Health, Education, and Welfare, National Office of Vital Statistics, *Vital Statistics: Special Reports*, Vol. 49, No. 34. Washington, D.C.: U.S. Government Printing Office. May, 1949.

and mortality. Also, we avoid the confusion that may result from em
one rather than another of the actual distributions as a standard.

Aggregative Index An adjusted rate standing alone is likely to be o
interest to the social analyst. It takes on significance only in relation to o
more rates based on the same set of weights. The significance of Japa
adjusted death rate of 9.2, as we have seen, lies in its relation to the standard
izing rate of 9.5 deaths per 1,000. The comparison between these rates is not confounded by a compositional effect. We carried the analysis one step farther and attempted to show that the difference between Japan's crude rate of 6.9 and the standardizing rate of 9.5 is the algebraic sum of two terms: one based on differences between age-specific death rates, the other based on differences between age-specific population proportions.

In all of these comparisons, the difference was the absolute number of deaths per 1,000 population. But for some purposes, we will want not absolute differences, but rather the magnitude of one rate relative to another. It may be pertinent to determine not whether one rate is 1 death per 1,000 higher than another, but rather whether it is 10 or 15 percent higher. In that case, we form ratios between rates based on the same set of weights.

In social science writing the ratio of one weighted mean to another has come to be known as an *aggregative index*: "index" because one rate is normed on another, "aggregative" because the rates themselves are weighted means. Corresponding to the difference between a standardized rate and the rate for the standard population, we have the index

$$\frac{AR}{SR} = \frac{\Sigma N_i r_i}{\Sigma N_i R_i}. \tag{6.2.6}$$

The sum in the denominator (right-hand side) is the number of events in the standard population; it serves as the norm. The quantity in the numerator is the number of events in a population having its own rates but the weights of the standard population; it is the quantity to be normed.

Norming Japan's adjusted rate on the standardizing rate (in this case the U.S. crude rate) gives:

$$\frac{AR}{SR} = \frac{9.2}{9.5}$$

$$= .968.$$

Converting this index to a difference score in percentage form gives:

$$\frac{9.2 - 9.5}{9.5} \times 100 = -3.2\%.$$

Table 6.2.4 *Age-Specific Arrest Rates, White and Black Populations*

Age	White			Black		
	Population	Arrests	Rate per Thousand	Population	Arrests	Rate per Thousand
15–24	27,000	378	14.0	5,000	65	13.0
25–44	36,000	324	9.0	4,000	32	8.0
45–74	27,000	108	4.0	1,000	3	3.0
Total	90,000	810	9.0	10,000	100	10.0

This latter result may be interpreted to mean that with age controlled Japan's death rate per 1,000 would be lower than the U.S. rate (the norm) by approximately 3 percent.

Whether we draw comparisons in relative terms by means of index numbers or in absolute terms by means of differences will depend on our objectives. An absolute difference of two deaths per 1,000 may correspond to a relative difference of 5, 10, or 20 percent; on the other hand, a relative difference of 10 percent may correspond to absolute differences of 5, 10, or 20 deaths per 1,000. Since ratios and differences provide different perspectives on the same set of data, both might be routinely obtained in a given investigation.

Versatility of Standardization Standardization is not limited to birth and death rates; in fact, it may be applied to any weighted mean, whether that mean is a proportion or an interval measure. In support of this point and also for greater familiarity with standardization, we offer the following examples:

(1) ***Arrest Rates*** Table 6.2.4 shows blacks as having a higher overall arrest rate than whites; the age-specific entries suggest, however, that the higher rate does not reflect a greater propensity to crime, but rather a larger proportion of younger ages. To test this possibility, we have recourse to standardization and component analysis.

Adjusting the crude rate for blacks by the age-specific weights of whites gives (Table 6.2.5):

$$\text{Adjusted Rate} = 720/90{,}000 \times 1{,}000$$
$$= 8.0.$$

To get the compositional effect, we subtract the adjusted rate from the crude rate:

$$\text{Compositional Effect} = 10.0 - 8.0$$
$$= 2.0.$$

Table 6.2.5 Standardized Black Crime Rate

Age	White Population	Black Arrest Rate	Arrests Expected
15–24	27,000	13.0	351
25–44	36,000	8.0	288
45–74	27,000	3.0	81
Total	90,000		720

$$\text{Standardized Rate} = \frac{720}{90{,}000} \times 1{,}000$$
$$= 8.0$$

Our conclusion is that the blacks would experience 2 less arrests per 1,000 population with the benefit of the age distribution of the whites.

To get the residual effect, we subtract the standardizing rate (in this case, the crude rate of the whites) from the adjusted rate for blacks:

$$\text{Residual Effect} = 8.00 - 9.00$$
$$= -1.00.$$

This negative quantity may be interpreted to mean that the blacks would experience 1 more arrest per 1,000 population with the disadvantage of the rate schedule for whites. Putting components together, we get the overall difference as the sum of its component differences:

$$10.0 - 9.0 = (10.0 - 8.0) + (8.0 - 9.0)$$
$$= 2.0 + (-1.0)$$
$$= 1.0.$$

The overall difference is thus the resultant of two opposing tendencies: the rate of the blacks is pushed up by 2 per 1,000 population by their unfavorable age distribution (relative to that of whites); it is pulled down by 1 per 1,000 by their relatively favorable rate schedule. Together these two components account for the overall difference of 1.00.

(Note that the same conclusion would have been reached if we had reversed our procedure: the adjusted rate for whites would have advanced from 9.0 to 11.0 for a compositional effect of −2.00 and a residual effect of +1.00).

(2) ***Percent Males Married*** In the preceding example, the standing of the two populations was reversed when age was standardized; the blacks,

Table 6.2.6 *Percentage of Males Married in Specified Age Groups, United States, 1890*

Age	Males		Males Married[a]	
	Number	Percent	Number Married	Percent
15–19	3,248,711	15.7	16,746	.5
20–24	3,104,893	15.0	585,748	18.9
25–29	2,698,311	13.1	1,421,407	52.7
30–34	2,425,664	11.7	1,728,930	71.3
35–44	3,705,648	17.9	2,997,030	80.9
45–54	2,627,024	12.7	2,213,901	84.3
55–64	1,630,373	7.9	1,342,414	82.3
65 and over	1,233,719	6.0	869,925	70.5
Total	20,674,343	100.0	11,176,101	54.1

[a] Excludes widowed and divorced.
Source: U.S. Bureau of the Census, *Census of the Population: 1950*. Vol. II, Characteristics of the Population. Part I, United States Summary, Table 102. Washington, D.C.: U.S. Government Printing Office. 1953.

initially showing a higher rate, emerged with a lower rate after standardization. But such an extreme reversal is perhaps exceptional. In the following example, which compares the proportion of American males 15 years of age and over, who were classified as married in 1890 and 1970, the original difference is reduced when age is controlled, but it is not eliminated or reversed.

In 1890, only 54.1 percent of American men were classified as married (Table 6.2.6), whereas in 1970 the percentage had risen to 67.8 (Table 6.2.7),

Table 6.2.7 *Percentage of Males Married in Specified Age Groups, United States, 1970*

Age	Males		Males Married	
	Number	Percent	Number	Percent
15–19	9,718,189	14.0	381,500	3.9
20–24	7,761,209	11.2	3,329,772	42.9
25–29	6,569,934	9.5	5,066,314	77.1
30–34	5,607,593	8.1	4,803,203	85.7
35–44	11,261,731	16.2	9,895,931	87.9
45–54	11,138,181	16.1	9,813,513	88.1
55–64	8,858,893	12.8	7,587,085	85.6
65 and over	8,433,330	12.2	6,103,326	72.4
Total	69,349,060	100.1	46,980,644	67.8

Source: U.S. Bureau of the Census. *Census of Population: 1970*. Vol. I, Characteristics of the Population, Part 1, U.S. Summary—Section 2. Table 203. Washington, D.C.: U.S. Government Printing Office. 1973.

an increase of 13.7 percentage points. We ask: Does this appreciable increase in the proportion of married men demonstrate an increasing propensity among American men to marry? Or, could it merely be due to a higher average age more favorable to marriage? Such a shift to a higher average age could induce a rise in the percentage of married men, with the basic propensity to marry remaining unchanged.

To answer this question we turn to standardization and component analysis. Since our question is whether the change in the age distribution accounts for the change in percent married, we adjust the marriage rate for 1970 on the age distribution for 1890. (Our conclusions would be the same if we reversed this procedure.)

The adjusted rate (see Table 6.2.8 for details) is

$$AR = 65.4/100.0 \times 100$$
$$= 65.4\%.$$

Subtracting this adjusted rate from the actual rate for 1970 gives us the compositional effect:

$$\text{Compositional Effect} = 67.8 - 65.4$$
$$= 2.4.$$

Table 6.2.8 *Percentage of Males Married in 1970 Standardized on 1890 Age Distribution*

(1)	(2)	(3)	(4)
Age	Percentage Males, 1890	Percentage Married, 1970	Expected Percentage Married, 1970[a]
15–19	15.7	3.9	.6
20–24	15.0	42.9	6.4
25–29	13.1	77.1	10.1
30–34	11.7	85.7	10.0
35–44	17.9	87.9	15.7
45–54	12.7	88.1	11.2
55–64	7.9	85.6	6.8
65 and over	6.0	72.4	4.3
Total	100.0		65.4

$$\text{1970 Standardized Rate} = \frac{65.4}{100.0} \times 100 = 65.4\%$$

[a] Expected percent married 1970 is the product of percent males 1890 and actual percent married 1970. That is, Column (4) = Column (2) × Column (3).

Source: Tables 6.2.6 and 6.2.7.

Subtracting the standardizing rate (in this case the actual rate for 1890) gives the residual effect:

$$\text{Residual Effect} = 65.4 - 54.1 = 11.3.$$

Together these two components equal the overall difference:

$$67.8 - 54.1 = 2.4 + 11.3 = 13.7.$$

Since both effects carry the same sign, they supplement rather than offset one another. The higher percent of males married in 1970 is the result of both a more favorable age distribution and an apparently greater tendency for males to marry.

(3) ***Mean of Interval Measures*** As noted above, the process of standardization may be applied to any weighted mean, including those based on interval measures, as in the following case. Let us suppose that the "campus activists" at some college have a grade point average (GPA) of 1.98, while the "nonactivist" students have a mean of 1.88 grade points (Table 6.2.9). From these averages alone, it would appear that the activists are scholastically superior. However, the nonactivists might object to such an interpretation on the ground that activists contain a larger proportion of upperclassmen, who normally receive higher grades than freshmen and sophomores. The superior grades of the activist group may be attributed in part to the concealed factor of class composition rather than mere academic ability. If this hypothesis is plausible, why not standardize grades of nonactivist students on the class composition of activists as a test?

Pursuing this suggestion, we reweight the class-specific averages of nonactivists by the class percentages of the activists (Table 6.2.10). In this

Table 6.2.9 *Grade Point Averages by Class, Campus Activists and Nonactivists (Hypothetical)*

Class	Activists		Nonactivists	
	Percent	$\bar{X}$	Percent	$\bar{X}$
Freshman	20	1.8[a]	40	1.8
Sophomore	35	2.0	30	1.8
Junior	25	2.0	20	2.0
Senior	20	2.1	10	2.2
Total	100%	1.98	100%	1.88

[a] Based on a grading system in which grade points per credit hour are as follows: A = 3, B = 2, C = 1, D = 0, F = −1.

Table 6.2.10 Computation of Standardized Mean: Nonactivists' Average Standardized on Activist Class Distribution

Class	Activists	Nonactivists	
	f	$\bar{X}$	$f\bar{X}$
Freshman	20%	1.8	36
Sophomore	35	1.8	63
Junior	25	2.0	50
Senior	20	2.2	44
Total	100%	1.93	193

$$\text{Standardized Grade Point Average of Nonactivists} = \frac{193}{100} = 1.93$$

reweighting, the grade average of nonactivist freshmen is multiplied by the percentage of activist students classified as freshmen ($1.8 \times 20 = 36$). In identical manner, we calculate the remaining products. The sum of such products, divided by 100, yields the sought-after standardized mean: the grade average of nonactivist students standardized on the class composition of activist students.

The mean of the nonactivist group, when standardized (1.93), is still somewhat below that of the activist group (1.98). Thus, standardization has narrowed but has not wiped out the original difference between the two groups. Even when the nonactivists are accorded the benefit of activist class composition, they still do not quite attain the level of activist students. Evidently other factors besides class composition have affected the grade averages, such as courses of study, motivation, and the like. These traits, as well as others, could be employed for an indefinite refinement of the data, but at the same time they suggest the impracticality of carrying standardization very far. The infinite potentialities of standardization remind us once again how remote "ultimate truth" is from the data which lie before us and on which belief and action must be based.

Conclusion While standardization gives every appearance of being a neat and objective routine, there is no formula which prescribes how fine the classification should be, or for that matter what classification to employ. Therefore, standardization does not release the social analyst from critical subject matter decisions.

Furthermore, it should not be forgotten that an adjusted rate is hypothetical rather than descriptive; it is a single composite index that is weighted by a set of subclass frequencies used as a standard. It hypothesizes what the measure would be if the selected subclass frequencies did in fact obtain. Thus, we calculated the percentage of men married in 1970 as if they had had the same age distribution as males in 1890.

Since the standardized (adjusted) rate is a hypothetical rather than an observed measure, a certain amount of caution is always necessary in its interpretation. It must be remembered that standardization is basically a process of reweighting, the logic of which always must be appraised in relation to the standardized results. In fact, when category-specific rates are not independent (in a statistical sense) of their corresponding relative frequencies, standardization probably should not be undertaken at all. But all derived measures have their problems. To interpret them will always require a reserve of statistical and subject matter sophistication, the more so as the chain of derivations increases in length and complexity.

QUESTIONS AND PROBLEMS

1. Define the following terms:

rate schedule	adjusted rate
weight schedule	component analysis
standardization	compositional effect
standardizing population	residual effect

Table 6.2.11 *Age-Specific Death Rates, Connecticut and Arizona, 1970*

Age	Connecticut		Arizona	
	Population	Age-Specific Death Rates	Population	Age-Specific Death Rates
0– 4	253,262	3.9	158,675	5.3
5–14	600,177	.4	378,856	.5
15–24	493,847	.9	317,923	1.6
25–34	374,765	1.2	216,843	1.9
35–44	355,947	2.4	195,323	3.6
45–54	381,691	5·8	189,873	7.2
55–64	283,112	14.7	151,933	16.3
65–74	174,810	34.1	107,740	33.6
75–84	91,676	76.7	44,233	75.4
85 and over	22,422	160.0	9,501	147.2
All Ages	3,031,709	8.56	1,770,900	8.39

Source: U.S. Bureau of the Census, *Census of Population: 1970*, Vol. I, Characteristics of the Population, Part 4, Arizona and Part 8, Connecticut. Table 20. Washington, D.C.: U.S. Government Printing Office. 1973.

U.S. Department of Health, Education and Welfare, *Vital Statistics of the United States: 1970*, Vol. II, Mortality, Part B. Table 7–3. Washington, D.C.: U.S. Government Printing Office. 1974.

Table 6.2.12 *Fertility of Women 40 and Over by Farm Background and Income: Detroit, 1952–1958*

Income of Head of Family	Two-Generation Urbanites		Farm Migrants	
	Average No. of Children	No. of Cases	Average No. of Children	No. of Cases
Under $3,000	2.36	134	3.21	99
$3,000–$4,999	2.04	276	2.85	158
$5,000–$6,999	2.15	292	2.61	100
$7,000 +	2.19	329	2.24	72

Source: David Goldberg, "The Fertility of Two-Generation Urbanites," *Population Studies*, Vol. 12 (1958–59), p. 217. Reprinted by permission.

Table 6.2.13 *Religion by Self-Esteem, Controlling on Father's Education*

Self-Esteem	I				II				III			
	8th Grade or Less				Some High School				High School Graduate			
	Catholic		Jewish		Catholic		Jewish		Catholic		Jewish	
	No.	%	No.	%	No.	%	No.	%	No.	%	No.	%
High	245	68.1	28	71.8	330	68.5	89	70.6	388	71.7	102	74.5
Medium	93	25.8	10	25.6	126	26.1	33	26.2	127	23.5	27	19.7
Low	22	6.1	1	2.6	26	5.4	4	3.2	26	4.8	8	5.8
Total	360	100.0	39	100.0	482	100.0	126	100.0	541	100.0	137	100.0

Self-Esteem	IV				V				VI			
	Some College				College Graduate				Post-Graduate			
	Catholic		Jewish		Catholic		Jewish		Catholic		Jewish	
	No.	%	No.	%	No.	%	No.	%	No.	%	No.	%
High	100	70.9	67	78.8	77	67.5	87	87.9	51	72.9	62	82.7
Medium	34	24.1	15	17.6	34	29.8	11	11.1	18	25.7	11	14.7
Low	7	5.0	3	3.5	3	2.6	1	1.0	1	1.4	2	2.7
Total	141	100.0	85	100.0	114	100.0	99	100.0	70	100.0	75	100.0

Source: Reprinted from *Social Forces*, 41 (October 1962). "Test factor standardization as a method of interpretation," by Morris Rosenberg. Copyright © The University of North Carolina Press.

2. Do standardized rates necessarily differ from crude rates? Explain.

3. Explain why the effect of standardization varies according to the choice of the standard population.

4. Standardize the Connecticut death rate on the Arizona age distribution (Table 6.2.11). Run a component analysis. Interpret.

5. From Table 6.2.12, compute weighted means for both urban and migrant women. Standardize the mean for migrants on the income distribution for urbanites. Do a component analysis. Interpret.

6. Compute the proportion of all Catholic students with high self-esteem, and the proportion of all Jewish students with high self-esteem. (Table 6.2.13).

7. In a study of parents' anticipations for their ninth-grade children, 34 percent of the Anglo respondents expected their child to complete college, whereas 29 percent of the Black and 13 percent of the Mexican-American parents held such expectations. The Anglo, Black, and Mexican-American samples differed, however, in their social class composition and these variations may account for all or part of the racial-ethnic differences in educational expectations. Percentages by social class and by racial-ethnic grouping are shown in Table 6.2.14. How would the overall percentages of

Table 6.2.14 *Parental Anticipations that Their Ninth-Grade Child Will Complete College, for Anglo, Black, and Mexican-American Parents, by Social Class, California, 1963*

	Percent Indicating that They Expect Their Child Will Complete College	*N*
Anglo	34	1937
Middle class	54	336
Working class	32	1145
Lower class	23	456
Black	29	191
Middle class	78	9
Working class	26	72
Lower class	27	110
Mexican-American	13	350
Middle class	42	19
Working class	19	134
Lower class	7	197

Source: Table I (adapted) from "Class and race differentials in parental aspirations and expectations," by Aubrey Wendling and Delbert S. Elliott is reprinted from *Pacific Sociological Review* Vol. 11, No. 2 (Fall, 1968) by permission of the Publisher, Sage Publications, Inc.

parents anticipating that their child would complete college compare if the Black and Mexican-American samples had the same social class composition as the sample of Anglo parents (for instance, after standardization)? Interpret your results.

SELECTED REFERENCES

Althauser, Robert P. and Michael Wigler
1972 "Standardization and component analysis." Sociological Methods & Research 1, No. 1 (August): 97–135.

Barclay, G. W.
1958 Techniques of Population Analysis. New York: John Wiley.

Duncan, Otis Dudley
1961 "A socioeconomic index for all occupations." Pp. 109–138 in Albert J. Reiss, Occupations and Social Status. Glencoe, Illinois: Free Press.

Duncan, Otis Dudley and Beverly Duncan
1955 "A methodological analysis of segregation indexes." American Sociological Review 20: 210–217.

Kitagawa, Evelyn M.
1955 "Components of a difference between two rates." Journal of the American Statistical Association 50 (December): 1168–1174.
1964 "Standardized comparisons in population research." Demography, Vol. 1, No. 1: 296–315.

Shyrock, H. S. et al.
1973 The Methods and Materials of Demography. Vol. 2, pp. 418–422. Washington, D.C.: U.S. Government Printing Office.

7 Variation

1 MEASURES OF SPREAD

Variation[1] is the foundation for all statistics. If all the values in a given set were identical, it would be superfluous to calculate an average—or for that matter, any other statistical measure—because any single measure already would represent all. Of course, the very purpose of averaging is to provide a single value to represent a group of unlike values. In fact, averages were invented to suppress the differences among values whenever such differences are not pertinent to our problem.

However, under certain circumstances, these differences may be of as much interest, or even more, than the average itself. In writing a letter of recommendation, a teacher will take into account not only the average of the student's test scores, but the spread of those scores as well. A student with marks of 100, 90, and 50 will be evaluated differently from one with scores of 80, 83, and 77, although both present the same mean score of 80. In choosing between two players with the same scoring average, a basketball coach may be inclined to use the consistent player who is seldom off that average, rather than the erratic performer who is generally low but now and again sensationally high. Two occupational groups having approximately

[1] Many textbooks do not distinguish between the terms *variability* and *variation*. We restrict ourselves to the concept of variation, defining variability as the capacity to vary, and variation as the manifestation of the capacity which we endeavor to describe and measure. Without variation, there would be no statistics in the first place. For that reason, statistics is referred to now and then as the "science of variation."

the same average annual income—school teachers and insurance agents, for example—may nevertheless present very different patterns of variation around that average. In the teaching profession, salaries are more standardized around the average with few in the lowest and highest brackets; whereas, in the insurance business, incomes are more widely dispersed, with relatively more reaching the extreme levels.

In general, therefore, the degree of variation among the observed values—what statisticians call *scatter* or *dispersion*—is as relevant as the location of the distribution. Various devices are available for the measurement of variation and are usually included in every well-provided kit of statistical tools. Although these devices differ in detail, all fall into one of three broad categories corresponding to the procedure on which they rest: (1) measurements of range that include all or a specific percentage of the items; (2) averages based on deviations of variates from a selected central value; and (3) averages based on the number of qualitative differences among the cases in a given set.

Measurements of Range The crudest and simplest measure of variation is the *total range,* or simply the *range*. Since the range is by definition the smallest interval encompassing all the values, it is calculated by taking the difference between the true lower and upper limits of the array. Thus, to find the range of the 229 suicide rates (Appendix, Table IX), we identify the true extremes and subtract the smaller from the larger. The smallest rounded suicide rate is 3 and the largest is 26; therefore, the range of the suicide rate is $26.5 - 2.5 = 24.0$.

For discrete data, the procedure is the same as that for continuous data, except that the true limits are now a necessary fiction. Thus, an array of family sizes of 2 through 12 has a range of 11, which is the difference between 12.5 and 1.5. This means that the variable can take 11 and only 11 consecutive values: 2, 3, 4, 5, 6, 7, 8, 9, 10, 11, or 12. Some texts label this the *inclusive range* and define it as $(H - L) + 1$, or, in this case, $(12 - 2) + 1 = 11$.

Simple in conception and calculation, the range, like every other statistic, provides only limited information. Because its significance is relative to its location on the scale, it will generally be more serviceable when quoted in conjunction with its boundary points. Everyday usage recognizes the soundness of this principle in such expressions as "Prices on new cars will range from $4,000 to $5,000," or "Tomorrow's temperatures will range from a low of 42° to a high of 78°." A salary schedule that extends from $5,000 to $10,000 has the same absolute range as one that extends from $20,000 to $25,000, but these two ranges have very different connotations. In choosing a vacation site, it is not enough to know that the range in temperature is 30°; it is equally necessary to know the scale location of the extreme temperatures.

The range has the further characteristic that it disregards the pattern of variation between the extremes, and yet at times this pattern may be of great import. The range of annual family incomes in the United States, which is greatly in excess of one million dollars, gives no clues to whether the incomes are compactly bunched in the middle, concentrated at one end, or uniformly spread over the entire scale.

Furthermore, in most observed distributions, the extreme values are infrequent, erratic, and unstable; hence, the overall range, which rests exclusively on these extremes, may leave the impression of a greater volume of variation than actually exists. By basing the age range of college students on the singular 14-year-old prodigy and the 64-year-old mother who wishes to attend school with her grandchildren, we obtain a range of 51 years. But this result obscures the fact that most students differ from one another by only a few years, and the total range is therefore misleading as an index of variation.

Intermediate Ranges This dependency upon the almost unique extreme observations may be overcome by computing an intermediate range which excludes a minor fraction of the cases at either end, but which still includes a significant proportion of cases. By basing it on the less exceptional items, the range acquires greater stability and dependability. One common practice is to take the difference between the 90th and 10th centiles, and thereby establish a range that includes the middle 80 percent of the cases. Applying this procedure to the distribution of United States family incomes (Table 7.1.1) we obtain the 10–90 range:

$$C_{90} - C_{10} = \$24{,}771 - \$3{,}972 = \$20{,}799.$$

A still more restricted range is the span between the first and third quartiles, or the interval that subtends the middle 50 percent of the items. It is naturally called the *interquartile range*. The interquartile range of family incomes is:

$$Q_3 - Q_1 = \$19{,}046 - \$7{,}194 = \$11{,}852.$$

This calculation answers the question of whether the middle cases are narrowly bunched or widely spread.

A little reflection will indicate that there is no limit to the construction of intermediate ranges, such as the 10–90 or the interquartile range. Any intermediate range will bring into clearer focus the relative degree of concentration or scatter among the items, particularly when viewed against the total range. The fact that the range of United States family incomes is considerably greater than one million dollars, but that approximately one-half of all families in the United States receive less than $12,000 income annually is suggestive of the essentially high degree of homogeneity among family incomes in the United States. Frequently, the use of strategically placed *quantiles* (the general term for partition values between 0 and 100) will be an altogether satisfactory approach to the analysis of variation, and more complicated methods need not be pursued.

Table 7.1.1 Distribution of United States Family Incomes, 1973 (N = 55,053,000)

Family Income	Percent	Cumulated Percent
Under $2,000	2.9	2.9
2,000–3,999	7.2	10.1
4,000–5,999	9.1	19.2
6,000–7,999	9.7	28.9
8,000–9,999	10.0	38.9
10,000–11,999	10.7	49.6
12,000–14,999	14.8	64.4
15,000–24,999	26.2	90.6
25,000–49,999	8.3	98.9
50,000 and over	1.1	100.0
	100.0	

$$Md = \$12{,}000 + \left(\frac{50.0 - 49.6}{14.8} \times 3{,}000\right) = \$12{,}081$$

$$Q_1 = \$6{,}000 + \left(\frac{25.0 - 19.2}{9.7} \times 2{,}000\right) = \$7{,}194$$

$$Q_3 = \$15{,}000 + \left(\frac{75.0 - 64.4}{26.2} \times 10{,}000\right) = \$19{,}046$$

$$C_{10} = \$2{,}000 + \left(\frac{10.0 - 2.9}{7.2} \times 2{,}000\right) = \$3{,}972$$

$$C_{90} = \$15{,}000 + \left(\frac{90.0 - 64.4}{26.2} \times 10{,}000\right) = \$24{,}771$$

Source: U.S. Bureau of the Census, *Current Population Reports*, "Money Income in 1973 of Families and Persons in the United States," Series P–60, No. 97. Washington D.C.: U.S. Government Printing Office. January, 1975.

QUESTIONS AND PROBLEMS

1. Define the following terms:

variation	intermediate range
variability	interquartile range
dispersion	centile
scatter	quantile
total range	

2. How would you calculate the range when the ends of the frequency table are open?
3. Explain why the quartiles and the median practically never divide the

total range into four equal intervals. In what type of curve will they partition the range equally?

4. The youngest member of a family is 13 years old; the oldest member is 50 years old. What is the age range?

5. The scores on a true–false sociology test of 100 questions ranged from 50 through 90. How would the range be affected if:
   (a) a bonus of 10 had been given to each student?
   (b) the poorest students had been successfully tutored?
   (c) an easier examination had been used? a more difficult exam?
   (d) 200 items had been used instead of 100?

6. Calculate the nine decile points of the percentage distribution of family incomes (Table 7.1.1).

7. From the list of 229 suicide rates (Appendix, Table IX), we draw a sample of 20. Will the sample range be larger or smaller than the overall range of 24.0? Explain your answer.

2 VARIATION AS MEASURED BY ABSOLUTE DEVIATIONS

It has been shown that the range, or any segment of it, will supply an impression of the span of a distribution. Although such a measure has some utility, especially when associated with its location, it represents merely the limits of variation, rather than the aggregate variation within those limits. It does not reflect the variation of the individual items, only the variation of the observed extremes, ignoring the many intermediate values. Therefore, it may be reiterated that the range measures the boundaries of scatter, but not the total amount of variation between those boundaries. Hence, some measure must be devised which will reflect the extent of diversity among all the items.

Deviations from a Norm Now, variation must be variation *from* something. It must be variation from some value. For example, we could subtract every value in the array from every other value—as we subtract the low from the high extreme to obtain the range—and then condense these differences into some type of index. However, a person contemplating the variation in a series of values intuitively considers them in relation to some fixed standard that has been built up in his or her cumulated social experience. A "high" salary, an extremely "high" birth rate, or a "low" intelligence is an assessment of the relative magnitude of a variate as measured from a fixed base. We recognize a high birth rate as deviating toward the positive side, similarly, a low birth rate, toward the negative side of a norm. Thereby this norm becomes a central value around which we perceive and measure the values spontaneously.

It is also quite conceivable to think of death rates, student grades, or teacher salaries in relation to a maximum or minimum—especially when they are in close proximity to these observed extremes. A teacher will perceive his or her salary as near the ceiling of this professional category or far from it. Nevertheless, it is more usual to employ a central value as a point of reference, because most values actually are in proximity to that central point. We tend to arrange our observations around an average as a norm, on the assumption that this average is a representative value and therefore is worthy of being used as a base of comparison. Statistically speaking, we shift the origin of our scale from zero to some suitable central value. Once having settled upon a central value, it is necessary only to calculate the deviations from that value and to reduce them to an index of variation.

Choice of Origin Obviously, with symmetrical distributions that are unimodal, it makes no difference which average is used as origin, since mode, median, and mean are equal in numerical value. But concrete data often fall short of even approximate symmetry around a single peak. Hence, the problem is one of selecting a representative value in those distributions where the mode, median, and mean diverge one from another.

Unquestionably, many an observer informally employs the mode—the most frequent of observations—as a base of comparison. But on theoretical grounds both the mean and the median can present stronger claims to the title of representativeness, and for that reason they are employed extensively in statistical calculations of this type.

As we have learned, the mean is the value from which the deviational variation on the two sides are in balance (Table 7.2.1); hence, the mean would appear to be the most reasonable point of origin of the deviations. However, the median's claim is at least equally impressive. Because the median splits the cases evenly, the sum of the absolute deviations around it is less than from

Table 7.2.1 *Algebraic and Absolute Deviations from Mean and Median,* $\overline{X} = 12$ *and* Md = *10*

Value	Algebraic Deviation		Absolute Deviation	
X	$X - \overline{X}$	$X - Md$	$\lvert X - \overline{X} \rvert$	$\lvert X - Md \rvert$
6	−6	−4	6	4
8	−4	−2	4	2
10	−2	0	2	0
15	3	5	3	5
21	9	11	9	11
60	0	10	24	22

any other point of origin (for a proof, see Yule and Kendall, 1950, p. 138). Expressed in another way, the median is that point around which the arithmetic "errors" are least. Hence, the median would appear to be the more logical origin whenever the measure of variation is based on the absolute, or arithmetic, deviations.

Average Deviation (AD) The sum of the arithmetic deviations is useless as an index of variation, since it will vary with the total number of cases (N) in the distribution. One thousand cases will have a larger sum than one hundred cases, other things being equal. To eliminate such adventitious differences, we divide the sum of the numerical deviations by N and thereby obtain the deviation per case. This result is termed the *mean deviation* or *average deviation* (AD). In symbols,

$$AD_{Md} = \frac{\Sigma |d|}{N}, \tag{7.2.1}$$

where $|d|$ = deviation from median, sign ignored. Applying this formula to Table 7.2.1, we get:

$$AD_{Md} = \frac{22}{5}$$
$$= 4.4.$$

The average deviation may also be based on the mean, in which case the above formula would read:

$$AD_{\bar{X}} = \frac{\Sigma |x|}{N}, \tag{7.2.2}$$

where $|x|$ = a deviation from the mean, sign ignored. Thus,

$$AD_{\bar{X}} = \frac{24}{5}$$
$$= 4.8,$$

which is larger than that based on the median. Our conclusion from this example would be that, whenever skew is present, the median always serves better as the origin around which to express the amount of variation in the distribution, since it lies closer (on the average) to all the values.

Coefficient of Relative Variation (CRV)[2] It is now clear that any deviation takes on significance only when compared to its origin or norm. Although the deviation of \$8 from an average of \$10 and the deviation of \$98 from an

[2] Sometimes written CV. Broadly defined, a *coefficient* is a measure of relationship between two variables expressed as a ratio.

average of $100 are numerically equal, they convey very different meanings. The deviation of one inch from average arm length is no more than would be expected but a deviation of an inch from average nose length would probably call forth astonishment. This principle is recognized and implemented in the *coefficient of relative variation*, which expresses the measure of variation as a fraction or multiple of its origin, be it mean or median. It thereby eliminates the extraneous factor of scale location. In the case of an *AD* based on the median, the formula would read:

$$CRV = \frac{AD_{Md}}{Md}. \tag{7.2.3}$$

If the *AD* is measured from the mean, we replace the median by the mean in Formula 7.2.3.

Let us use this procedure to compare the relative variation between homicide rates in the New England and South Atlantic states (Table 7.2.2). After comparing the respective *AD*s for the New England states (.78) and the South Atlantic states (3.60), we might conclude that the states of the Northeast are much more homogeneous, since their average divergence from the norm is so much narrower. But such a conclusion would be premature, for it is possible that, relative to their origin, the deviations within the South Atlantic group are no larger than those within the New England group. To determine whether this is the case, we express each *AD* as a proportion of its base median:

$$CRV = \frac{.78}{2.75} = .284.$$

$$CRV = \frac{3.60}{12.15} = .296.$$

These results demonstrate that the two groups of states do not differ markedly in their relative variation: Whereas the South Atlantic group has an average deviation almost five times larger than the New England group, its coefficient of relative variation is not even one-tenth again as large.

The coefficient of relative variation is simple to compute and is particularly useful in comparative work, since it has the effect of norming for differences in absolute magnitudes and in substantive units of measure. It makes comparable sets of small and large values of the same kind, as well as of values that are qualitatively different. It is not applicable, however, unless (1) the observed measures have an absolute zero, and (2) scale intervals are equal—that is to say, unless we have ratio scales. Accordingly, the *CRV* is not to be used to gauge the relative variation, for example, in measures of social distance, intelligence, and attitudes. The reasoning is as follows. The *CRV* is designed to standardize for differences in location of absolute measures, or more specifically, for differences in absolute magnitudes of central values such as the mean or median. But when an absolute zero is

Table 7.2.2 *Computation of Average Deviation, Homicide Rates per 100,000 Population, New England and South Atlantic States, 1973*

New England			South Atlantic		
State	Rate	$\lvert d \rvert$	State	Rate	$\lvert d \rvert$
Connecticut	3.3	.55	Delaware	5.9	6.25
Maine	2.1	.65	Florida	15.4	3.25
Massachusetts	4.4	1.65	Georgia	17.4	5.25
New Hampshire	2.1	.65	Maryland	11.3	.85
Rhode Island	3.4	.65	North Carolina	13.0	.85
Vermont	2.2	.55	South Carolina	14.4	2.25
		Sum = 4.70	Virginia	8.5	3.65
			West Virginia	5.7	6.45
					Sum = 28.80
Median = 2.75			Median = 12.15		
AD = .78			*AD* = 3.60		

Source: U.S. Department of Justice, *Uniform Crime Reports, 1973*, Table 3, p. 60. Washington, D.C.: U.S. Government Printing Office. 1974.

nonexistent, we necessarily must assign arbitrary values to a series of observations. Thus, a social distance measure of 30 may represent the same objective fact as a measure of 120, according to the scale origin arbitrarily set up by the investigator. Such scales are not anchored to an empirical zero which represents the absence of a specific phenomenon. If, then, locational measures are arbitrary, any standardization would also be arbitrary and devoid of meaning. (For a discussion of the uses of the *CRV* in social research, see Martin and Gray, 1971.)

QUESTIONS AND PROBLEMS

1. Define the following terms:

absolute deviation	absolute zero
algebraic deviation	arbitrary zero
average deviation (*AD*)	interval scale
coefficient of relative variation (*CRV*)	ratio scale

2. When is it more logical to base the *AD* on the median rather than on the mean?
3. Eleven houses are located on a street that runs due east and west. Which house is connected by the shortest aggregate distance to all others? Demonstrate graphically. How would the question be reworded if there were 10 houses?

4. Compute the AD from the median for the following set ($N = 20$):

7	4	6	5	10
8	14	7	5	8
4	21	11	0	10
17	7	3	13	5

   (a) How would the AD be affected if each value were increased by 10?
   (b) How would the CRV be affected?

5. Subtract each smaller value from every larger value in the following array: 4, 5, 6, 7, 10. Would the sum of these differences be a meaningful index of variation? Explain.

6. It has been said that equal CRVs are consistent with the principle that big things vary more absolutely than little things, but not relatively. Explain and illustrate.

3 VARIATION AS MEASURED BY SQUARED DEVIATIONS

The measurement of variation by simple arithmetic deviations from a central value is a straightforward procedure. If a simple statement of dispersion were all that was wanted, the easily comprehended AD would serve quite well. However, variation may be, and commonly is, measured by squared deviations, which are universally taken from the mean of the series. The fundamental logic of this procedure is identical with that which argues for the simple deviations from the median; the principle of best fit. Just as the sum of deviations from the median is minimal, so the sum of the squared deviations from the mean is also minimal. This is an exemplification of the *principle of least squares*, which is one of the most vital principles in all statistics, known and practiced for 150 years.

At first glance, the technique of squaring deviations may seem unnecessarily circuitous and superfluous. If variation can be measured satisfactorily by simple deviations, what additional information and insight can be gained by squaring them? A completely adequate answer to this question will be possible only at a later stage of the student's statistical studies. Here it must suffice to state that the practical utility of such a measure is incomparably greater than that of the AD, which is less frequently called into use. The squared deviations may be expressed in several ways, each of which serves its own purpose: *sum of squared deviations*, *variance*, and *standard deviation*.

Sum of Squared Deviations Just as we computed the sum of the arithmetic deviations from the median and the mean in Table 7.2.1, now we may compute the sum of the *squared* deviations from both averages (Table 7.3.1). Although the sum of simple arithmetic deviations is minimal from the

Table 7.3.1 *Sum of Squares of Deviations from Mean and Median, where* $\bar{X} = 12$, Md = *10*

X	$\lvert X - \bar{X} \rvert$	$\lvert X - Md \rvert$	$(X - \bar{X})^2$	$(X - Md)^2$
6	6	4	36	16
8	4	2	16	4
10	2	0	4	0
15	3	5	9	25
21	9	11	81	121
60	24	22	$SS = 146$	$SS = 166$

Source: Table 7.2.1.

median, we see in Table 7.3.1 that the sum of the *squared* deviations is minimal when these deviations are measured from the mean.

This result may seem paradoxical. Mathematical proof for this apparent inconsistency cannot be offered here, but a partial explanation lies in the fact that squaring is a geometric computation which has the effect of weighting the deviations disproportionately as they increase in magnitude. Since the mean equalizes the negative and positive deviations, it operates to reduce to the limit the relative frequency of the large deviations, and thereby minimizes the sum of the squared deviations.

Variance Although the sum of the squared deviations is employed in various statistical procedures, it is not a meaningful index of variation, for the reason that its magnitude depends, not only on the dispersion of the values around the mean, but also on the number of cases in the distribution. To eliminate the accidental factor of frequency, we get the squared deviation per case; or the sum of the squared deviations divided by the number of cases. This mean is the *variance* by definition. Since its square root is represented by the lower-case Greek letter "σ" (sigma), the variance itself is represented by the symbol "σ^2" (read "sigma-square"). With these definitions in mind, we may write:

$$\text{Variance} = \sigma^2 = \frac{1}{N}\Sigma(X - \bar{X})^2 \qquad (7.3.1)$$

where N = the total number of cases.

The operational definition of the variance—the sum of the squared deviations from the mean divided by their number—contains no hint of its utility in substantive investigations. About all we can say here—and we risk saying too little—is that the variance of a composite variable (one consisting of two or more components) may be split up so as to reflect the relative strength of each component. For example, if environment and heredity are components in IQ, the variance of IQ may be partitioned so as to reflect the relative strength of each of these factors (provided they have been accurately measured). Similarly, if we regard socioeconomic status (SES) as having its source in education and occupation, then the variance of SES may be decomposed, so to speak, so as to reflect the relative contribution of each of these factors. In general, we may split up the variance of a set of measures so as to reflect the strength of one or more factors that we have explicitly taken into account, relative to those factors that we have not taken into account. Some of the more elementary techniques for doing such an analysis are taken up in Chapter 9 on regression and correlation; the subject is broached again in Chapters 14 and 15 on hypothesis-testing.

The Standard Deviation (*SD*) Since the variance is derived from the squared deviations, it does not provide a description of the given measures. If such a description is required, we have only to unsquare the variance to restore the original scale. In this form, it is known as the *standard deviation*, symbolized by the lower-case Greek σ (sigma).

$$\sigma = \sqrt{\frac{\Sigma(X - \overline{X})^2}{N}}. \tag{7.3.2}$$

As such, it is used as a measure of variation quite analogous to the *AD* from which it differs primarily in the fact that the deviations are squared, and their mean unsquared. However, the unsquaring does not cancel the total effect of previous squaring; in part, the weighting effect remains. If it had been completely nullified, the result would have been the simple *AD* from the mean.

The Coefficient of Relative Variation (*CRV*) Like the *AD*, the standard deviation may be converted into a measure of relative variation by *norming* it on its origin. In symbols,

$$CRV = \frac{\sigma}{\overline{X}}. \tag{7.3.3}$$

Applying Formula 7.3.3 illustratively to the 229 suicide rates (see Table 7.3.4), where $\overline{X} = 11.644$ and $\sigma = 4.176$, we obtain:

$$CRV = \frac{4.176}{11.644}$$

$$= 0.359.$$

This result carries the interpretation that the *SD* is approximately one-third as large as the mean of the distribution.

In some problems, instead of the *CRV*, we may be required to calculate the *rel-variance*. The rel-variance is by definition the variance of a set of measures divided by the squared mean of those measures. In symbols:

$$\text{Rel-Variance} = RV = \frac{\sigma^2}{\bar{X}^2}. \tag{7.3.4}$$

For the above example, the rel-variance is:

$$RV = \frac{(4.176)^2}{(11.644)^2}$$

$$= \frac{17.439}{135.583}$$

$$= 0.129.$$

signifying that the variance of the suicide rates is approximately one-eighth as large as the mean squared.

As previously stated, a measure of relative variation is not likely to be used in isolation from other comparable measures, except occasionally to indicate the extent of scatter in relation to the magnitude of the mean. Thus, a small standard deviation relative to a large mean represents great homogeneity of the data, and consequent typicality of the mean, which may be an important piece of information under certain circumstances. A mean of $125 with a standard deviation of $5, giving a *CRV* of 4 percent, is more representative of the array than a mean of $125 with a standard deviation of $25 and a *CRV* of 20 percent. A *CRV* of zero indicates no variation at all, and that the mean is utterly typical.

However, the *CRV* is likely to be used in outright comparisons between series of related data. We may wish to investigate the relative variation in wages between occupations; similarly, we may wish to compare crime rates in respect to their relative variation. The *CRV* is designed to answer questions such as these. It bears repeating that the *CRV* is inapplicable when the observed measures have no absolute zero.

Computation of Variance: Ungrouped Variates As with the mean, it is possible to compute the variance (and the standard deviation) from an arbitrary origin (A). That procedure will be especially handy when no computer is available, or when the computer is practically inaccessible to the student. In getting the variance from an arbitrary origin, our first step is to get the mean of the squared deviations from that arbitrary origin; second we get the squared deviation of the arbitrary origin from the mean of the series. Third, subtracting the latter term from the former gives the variance. In symbols:

$$\sigma^2 = \frac{\Sigma(X - A)^2}{N} - (A - \bar{X})^2. \tag{7.3.5}$$

Table 7.3.2 *Formulas for Sum of Squared Deviations, Variance, and Standard Deviation*

Name	Notation	Basic Formula	Computing Formula
Sum of Squared Deviations	$N\sigma^2$	$\Sigma(X - \bar{X})^2$	$\Sigma X^2 - \frac{(\Sigma X)^2}{N}$
Variance	σ^2	$\frac{\Sigma(X - \bar{X})^2}{N}$	$\frac{\Sigma X^2}{N} - \left(\frac{\Sigma X}{N}\right)^2$
Standard Deviation	σ	$\sqrt{\frac{\Sigma(X - \bar{X})^2}{N}}$	$\sqrt{\frac{\Sigma X^2}{N} - \left(\frac{\Sigma X}{N}\right)^2}$

Calculating the variance from an arbitrary origin is thus closely similar to calculating the mean from an arbitrary origin: First carry out a calculation on the basis of what may be regarded as a guess (A) and then correct that calculation for the error in that guess ($A - \bar{X}$).

In the special case where our arbitrary origin is zero, the formula may be simplified:

$$\sigma^2 = \frac{\Sigma X^2}{N} - \bar{X}^2. \tag{7.3.6}$$

Since this formula sidesteps the calculation of the deviations, it is very convenient to use; the student will encounter various exercises in which it may be used to good advantage.

We multiply the variance by N to get the sum of the squared deviations from the mean; we take its square root for the standard deviation. These conversions are shown in Table 7.3.2, where computing formulas are given alongside basic formulas in order to emphasize the correspondence between them. To show that they give the same answer, we apply both to the Northeast homicide rates (Table 7.3.3). While the calculation of the variance and the standard deviation will normally be carried out on the basis of the computing formulas, their interpretation will always be in terms of the basic definitional formulas.

Computation of Variance: Grouped Variates The convenience of both an arbitrary origin and coding is evident in the following procedure for calculating the variance (or standard deviation) of grouped measures. We first express deviations of class midpoints from the arbitrary origin—in this case one of the class midpoints—as multiples of class width:

$$x_i' = \frac{X_i - A}{w}.$$

***Table 7.3.3** Computation of Variance and Standard Deviation, Northeast Homicide Rates (X)*

State	X	X^2	$X - \bar{X}$	$(X - \bar{X})^2$
Connecticut	3.3	10.89	.383	.147
Maine	2.1	4.41	−.817	.667
Massachusetts	4.4	19.36	1.483	2.199
New Hampshire	2.1	4.41	−.817	.667
Rhode Island	3.4	11.56	.483	.233
Vermont	2.2	4.84	−.717	.514
Total	17.50	55.47	−.002	4.427
Mean	2.917	9.245	—	.738

$$\sigma^2 = \frac{\Sigma(X - \bar{X})^2}{N} = \frac{4.427}{6} = .738 \qquad \sigma = .859$$

$$\sigma^2 = \frac{\Sigma X^2}{N} - \bar{X}^2 = \frac{55.47}{6} - (2.917)^2 = 9.245 - 8.507 = .738 \qquad \sigma = .859$$

Source: Table 7.2.2.

X_i = midpoint of ith class, A = arbitrary origin, and w = class width. (In practice, we simply set up a column of x_i', as in Table 7.3.4.)

Second, we get the variance of these twice-coded measures:

$$\sigma_{x'}^2 = \frac{\Sigma f_i x_i'^2}{N} - \left(\frac{\Sigma f x_i'}{N}\right)^2 \tag{7.3.7}$$

Note that the second term on the right is the square of the mean of the x_i', whereas the first term is the mean of the squared x_i'. Note too that class frequencies (the f_i) are attached to the coded values as weights, as in the calculation of the mean of grouped data (p. 112).

At this juncture, we have the variance of the coded values but not the variance of the given measures. Since w is the common denominator of the x_i', and since $\sigma_{x'}^2$ is derived from the squared x_i', to get from $\sigma_{x'}^2$ to σ_x^2, we simply multiply the former by w^2. In symbols,

$$\sigma_x^2 = w^2 \sigma_{x'}^2 \tag{7.3.8}$$

where σ_x^2 is the variance of the original variates. To get the standard deviation, we merely take the square root of the variance. In symbols:

$$\sigma_x = \sqrt{\sigma_x^2}\,. \tag{7.3.9}$$

***Table 7.3.4** Computing Variance and Standard Deviation of Grouped Suicide Rates*

Limits	f_i	x_i'	$f_i x_i'$	$f_i x_i'^2$
3– 4	3	−4	−12	48
5– 6	9	−3	−27	81
7– 8	35	−2	−70	140
9–10	51	−1	−51	51
11–12	52	0	0	
13–14	30	1	30	30
15–16	21	2	42	84
17–18	10	3	30	90
19–20	6	4	24	96
21–22	8	5	40	200
23–24	1	6	6	36
25–26	3	7	21	147
	229		33	1003

$$\sigma_{x'}^2 = \frac{1003}{229} - \left(\frac{33}{229}\right)^2$$

$$= 4.3799 - .0208$$

$$= 4.3591$$

$$\sigma_x^2 = (2^2)(4.3591)$$

$$= 17.4364$$

$$\sigma_x = 4.1757$$

Source: Table 4.1.4.

The entire procedure is illustrated in Table 7.3.4; computing formulas for grouped data are displayed as a convenience in Table 7.3.5.

Averaging Variances In averaging the variances of two or more groups, our procedure will depend on the question we have in mind. If our question is "What is the mean variance, without regard to differences in the N_i?" we would simply add up the several variances and divide by their number. In symbols:

$$\bar{\sigma}^2 = \frac{1}{k}\Sigma\sigma^2 \qquad (7.3.10)$$

where $\bar{\sigma}^2$ = the unweighted mean variance, and k = the number of groups.

On the other hand, if our question is "What is the variance of all k groups combined?" then we must adopt a somewhat complicated weighting procedure. As with the weighted mean, we employ the N_i as weights; but before attaching them to the variances, we add to each variance the squared

Table 7.3.5 *Formula for Computing Sum of Squared Deviations, Variance, and Standard Deviation for Grouped Variates*

Name	Notation	Basic Formula	Computing Formula
Sum of Squared Deviations	$N\sigma^2$	$\Sigma f_i(X_i - \overline{X})^2$	$w^2 \left\{ \Sigma f_i x_i^{\cdot 2} - \frac{(\Sigma f_i x_i^{\cdot})^2}{N} \right\}$
Variance	σ^2	$\frac{\Sigma f_i(X_i - \overline{X})^2}{N}$	$w^2 \left\{ \frac{\Sigma f_i x_i^{\cdot 2}}{N} - \left(\frac{\Sigma f_i x_i^{\cdot}}{N} \right)^2 \right\}$
Standard Deviation	σ	$\sqrt{\frac{\Sigma f_i(X_i - \overline{X})^2}{N}}$	$w \sqrt{\left\{ \frac{\Sigma f_i x_i^{\cdot 2}}{N} - \left(\frac{\Sigma f_i x_i^{\cdot}}{N} \right)^2 \right\}}$

X_i = Midpoint of ith class interval

f_i = Frequency of ith class interval

A = Selected midpoint, or arbitrary origin

w = Uniform class width

$x_i^{\cdot} = \frac{X_i - A}{w}$

deviation between the mean on which that variance is based and the overall mean. Weighting these composite terms, and summing and dividing by N gives the variance for all k groups combined:

$$\sigma_T^2 = \frac{1}{N} \Sigma N_i(\sigma_i^2 + d_i^2) \tag{7.3.11}$$

where σ_T^2 = the total variance, $d_i = \overline{X}_i - \overline{X}$. Breaking up the sum on the right-hand side into its constituent sums gives:

$$\sigma_T^2 = \frac{1}{N} \Sigma N_i \sigma_i^2 + \frac{1}{N} \Sigma N_i d_i^2. \tag{7.3.12}$$

In some problems, instead of combining group variances to get the total variance, we may be required to split up the total variance into its component parts. In that case, we calculate $(1/N)\Sigma N_i d_i^2$ and subtract from σ_T^2 for $(1/N)\Sigma N_i \sigma_i^2$; or vice versa. Such partitioning is common practice in sociological research; it permits the analyst to determine the extent to which the overall variation is inflated by differences between group means, or what the variation would be if all groups had the same mean. These matters come within the jurisdiction of the analysis of variance, which we take up in Chapter 15 on hypothesis-testing.

Variance of 0,1 Variates In many statistical problems, we are required to get the variance of a 0,1 variable (p. 113). We could get that variance by

Table 7.3.6 *Mean of 0,1 Variable*

Value (X)	Frequency (f)	fX	fX^2
1	N_1	N_1	N_1
0	N_0	0	0
Sum	N	N_1	N_1
Mean		N_1/N	N_1/N

subtracting the mean from each variate, squaring deviations and summing, and dividing that sum by number of cases (N). However, it is much more convenient to make use of the computing formula with zero as an arbitrary origin.

In this analysis, we suppose that N_1 cases have a value of 1, and that N_0 cases have a value of 0. Since the sum of the squared values and the sum of the values both equal N_1, the mean of the squared values and the mean of the values both equal N_1/N. This latter ratio is usually symbolized by the letter P, since it is the proportion of all cases with a value of 1. These points are summarized in Table 7.3.6.

Substituting the mean of the squared values P and the mean of the values squared P^2 into Formula 7.3.6 gives the variance of a set of 0,1 values. In symbols:

$$\begin{aligned} \sigma^2 &= P - P^2 \\ &= P(1 - P) \\ &= PQ \end{aligned} \qquad (7.3.13)$$

where $Q = 1 - P$. Our conclusion is that when the variance of a 0,1 (binary) variable is required, we need only multiply the proportion of cases with value 1 by the proportion of cases with value 0. This formula may be used to good advantage in a number of problems encountered in this course and elsewhere.

QUESTIONS AND PROBLEMS

1. Define the following terms:
   principle of least squares
   variance
   standard deviation
   weighted variance
   variance of 0,1 variable

2. Explain why the standard deviation, instead of the variance, is ordinarily used as a measure of dispersion.
3. For the same series, why is the *SD* generally larger than the *AD*?
4. Using small sets of illustrative data, verify that the sum of the squares is least from the mean.
5. Why is it impossible for a distribution to have more than one *SD*?
6. Is it generally possible to calculate the *SD* of a qualitative variable? Explain.
7. Is it possible to calculate the *SD* of a 0,1 dummy variable? What is the formula?
8. Compute the *SD* of the series given in Problem 7.2.4. Add a constant (for example, 20) to each value and compute the *SD* of this transformed series. How is the *SD* affected?
9. If college seniors have a mean age of 22 years and a standard deviation of 2 years, what will their mean age and standard deviation be 25 years later? The *CRV*?
10. Compute the *SD* of the frequency distribution of delinquency rates (Table 4.1.8) by the computing formula for grouped data.
11. The variance is said to be an *areal measure*, the standard deviation a *linear measure*. Explain.
12. When will the rel-variance be smaller than the *CRV*? Larger? Illustrate.
13. Calculate the variance of the following series by Formula 7.3.5, using 4 as an arbitrary origin: 11, 4, 6, 5, 9.
14. Compute the variance and rel-variance of the South Atlantic homicide rates (Table 7.2.2).
15. Calculate the variance of the suicide rates (Table 7.3.4) using the midpoint of the smallest interval as arbitrary origin (A). Does this arbitrary origin have any particular advantage? Discuss.
16. Calculate the means and variances of the following groups:

B_1	B_2	B_3
1	4	7
3	2	7
5	8	1
3	7	5
5	9	
7		

Substitute these answers in Formula 7.3.11 to get the variance of all three groups combined. Check your answer by applying Formula 7.3.5 to the unsorted measures.

17. Calculate the variances of the following 0,1 variables by Formula 7.3.13:

Value	(1)	(2)	(3)	(4)	(5)
0	93	65	19	74	53
1	11	31	18	16	46
N	104	96	37	90	99

4 THE NORMAL DISTRIBUTION AS A PATTERN OF VARIATION

The utility and importance of the standard deviation depends in large part on its relation to the *normal curve.* In fact, it may be claimed that, as a measure of dispersion, the standard deviation has meaning only insofar as the pattern of variation is normal. Hence, an adequate discussion of the standard deviation must specify its fundamental relation to the normal curve, the most illustrious of all patterns of statistical variation.

The history of the normal curve dates back to 1733, when Abraham de Moivre first established it in the course of his investigation into games of chance. Later, in his more pious moments, he contended that it was a manifestation of divine order in the universe.

Since that time it has served various purposes. To astronomers, it described the distribution of measurement errors around the "true" value; hence, it was often characterized as "the curve of error." During the 1830s, the Belgian statistician Quételet was the first to apply this curve to social, psychological, and anthropometric data. His own concept, *l'homme moyen*—the average man—which for him described the norm, flanked by "nature's errors," is the logical basis for the now prevalent terminology of "the normal distribution." The curve also was employed to represent a sampling distribution of all possible sample values, which is today one of its most significant uses—a topic which will be treated in a subsequent chapter.

Quételet felt that social and moral data tend to array themselves on the normal curve, which is thereby given the sanction of nature. However, the normal curve no longer carries this eulogistic connotation. Today it would be a mistake to consider nonnormal distributions as unusual or unnatural. The distribution of raw, empirical data may conform naturally to any one of a number of curves. Nevertheless, as a statistical model in fitting nature's variation, the normal curve is still unrivaled in the scope of its application, notwithstanding the fact that it does not possess the universality for social data which Quételet attributed to it.

Characteristics of the Normal Curve By definition, the ideal normal distribution contains an infinite number of cases, is unimodal and symmetrical, and is unbounded at either end. Consequently, mean, median, and mode are identical in value and divide the array into two equal parts. The graphic version of this distribution is a smooth, bell-shaped curve (Figure 7.4.1), with a characteristic slope that never touches the base line. Starting at its peak, the curve falls more and more rapidly to the *point of inflection* and then gradually levels off, extending indefinitely in either direction. This point of inflection is exactly one standard deviation distant from the zero (mean) origin. Graphically, the standard deviation is the linear distance along the base line from the mean to the ordinate defining the point of inflection.

From Figure 7.4.1, it is evident that the distance from the mean in units of the standard deviation may be converted into a percentage of the total area (frequency). Hence, such distances serve as convenient measures of relative position. If we travel one *SD* from the mean, we leave behind approximately 34 percent of the cases; if we continue on to a point 2 *SD*s beyond the mean, we would leave behind, not 68 percent of the cases, but rather 48 percent, owing to the steady decline in frequency as distance from the mean increases (Figure 7.4.1). Because of its symmetry, approximately two-thirds (68.26 percent) of the cases in a normal distribution lie in an interval extending from 1 *SD* below, to 1 *SD* above the mean; similarly, approximately 95 percent of the cases fall within 2 *SD*s on either side of the mean.

Because such calculations have so many applications in statistical work, reference tables have been prepared for the convenience of statistical

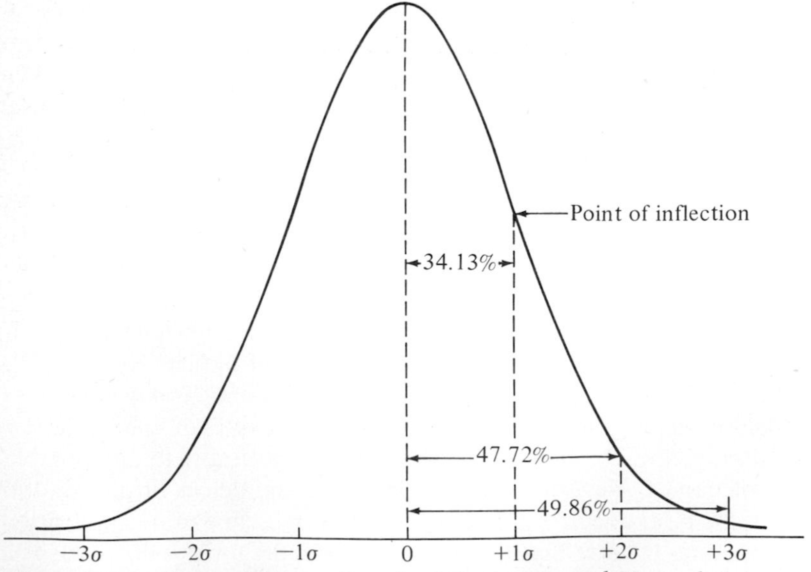

Figure 7.4.1 *Normal Curve, Standard Deviation, and Normal Areas*

Table 7.4.1 *Table of Normal Areas*

x/σ	Area	x/σ	Area	x/σ	Area
.1	.0398	1.1	.3643	2.1	.4821
.2	.0793	1.2	.3849	2.2	.4861
.3	.1179	1.3	.4032	2.3	.4893
.4	.1554	1.4	.4192	2.4	.4918
.5	.1915	1.5	.4332	2.5	.4938
.6	.2257	1.6	.4452	2.6	.4953
.7	.2580	1.7	.4554	2.7	.4965
.8	.2881	1.8	.4641	2.8	.4974
.9	.3159	1.9	.4713	2.9	.4981
1.0	.3413	2.0	.4772	3.0	.4986

Source: Appendix, see Table I.

workers. Table 7.4.1 supplies an example; it gives the proportion of area (cases) between the mean and selected multiples of the standard deviation. Like its parent table (Appendix, Table I), it gives only the distribution of cases above the mean—which suffices because the normal distribution is symmetrical. Examination of Table 7.4.1 indicates that .4953 cases lie between the mean and a point 2.6 *SD*s above the mean, so that .9906 cases fall on an interval extending from a point 2.6 *SD*s below, to a point 2.6 *SD*s above the mean.

Normal Deviates The foregoing description of the normal curve is an abstract one, given in terms of the standard deviation measured from the mean as an origin. As such, the unit of measure is independent not only of diverse measurement systems, but also of the concrete values themselves. It makes no difference whether we are dealing with incomes of a hundred or a million dollars, with durations of 10 seconds or 10 years, or with varying intensity of attitudes.

But sets of data always come to us as raw measures and give every appearance of being noncomparable. For instance, we cannot readily compare teachers' salaries and years of service, even though both variables may be normal in their distributions. The solution to this problem of comparability lies in converting the raw measures into sigma units of measure, which *are* comparable. We express raw deviations from the respective means as multiples of their standard deviations. Hence, such measures are called *standard measures.* When standard measures are distributed normally, they are referred to as *normal deviates* and are conventionally symbolized z.

$$z = \frac{X - \bar{X}}{\sigma} \tag{7.4.1}$$

$$= \frac{x}{\sigma}.$$

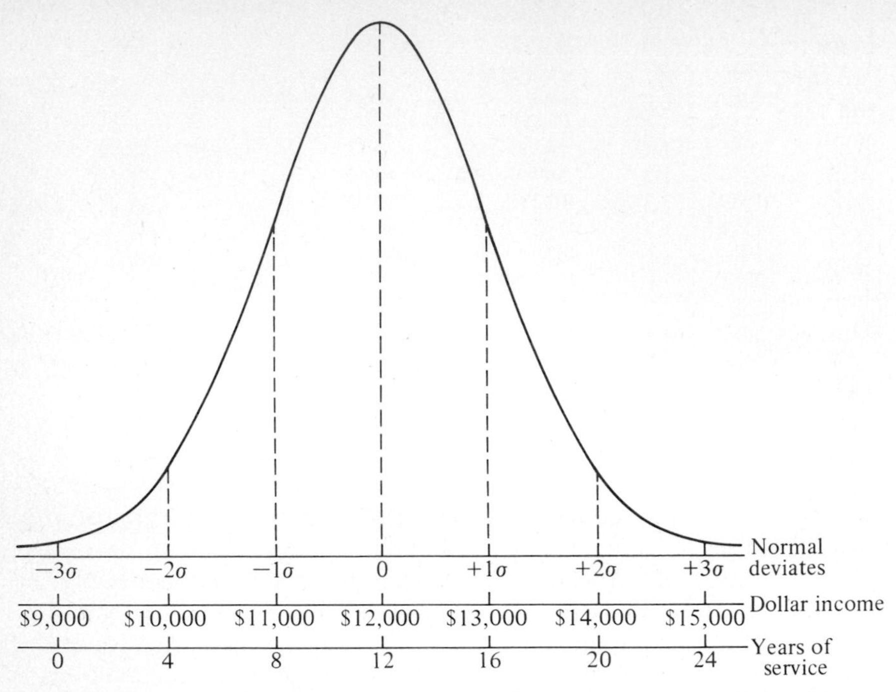

Figure 7.4.2 *Teacher Income and Years of Service as Sigma Units (Hypothetical Data)*

By plotting absolute and sigma scales on the base line of a normal frequency graph, it is possible to display visually the equivalence of given measures and standard measures. In Figure 7.4.2, for example, it becomes plain that a teacher's salary of $13,000 coincides with a standard measure of + 1.00; both values represent the same objective fact. Moreover, by this device of multiple scales, it is possible to exhibit conveniently the essential identity between normally distributed variables which are seemingly very different. Thus, the illustrative graph makes clear that a teacher's salary of $13,000 is statistically identical with 16 years of teaching service, both lying 1 *SD* above the mean.

This transformation to standard form, which may seem awkward at first, will gradually become second nature to every student of statistics, since it finds such a wide variety of applications. Whether familiar with statistical processes or not, every student well knows that in a given population raw total scores of, say, 300 and 500 in English and Mathematics, may or may not be equivalent in relative grade value. But upon discovering both to be 2 *SD*s above the mean, the statistically-trained student will correctly judge them to be identical, because in both instances 98 percent of the grades are presumed to be lower. Thus, the z-measure serves to establish the relative position of an item in a normal array and thereby renders corresponding items in two or more normal arrays comparable.

Converting Given Measures to Standard Measures To convert any series of variates into standard measures, it is first necessary to compute the mean and the standard deviation of the series. These computations are a requirement of Formula 7.4.1. Here, they are carried out illustratively on a set of 10 measures (Table 7.4.2) drawn randomly from a normally distributed variable. The first variate, 8.72, deviates by -2.88 from the mean value of 11.60. Since the standard deviation is 2.1871, this variate deviates from the mean by -1.32 (to nearest hundredth) standard deviations. In other words, the value of 8.72 translated into standard form is equal to -1.32. The remaining standard measures are identically calculated.

These standard measures—in this case, normal deviates—enable us to determine the proportion of cases above (or below) each item by referring to a table of normal areas (frequencies). From Table 7.4.1, we read that the interval between the mean and a point 1.32 *SD*s beyond the mean, corresponds to approximately 40 percent of the total frequency, which leaves about 10 percent of the cases beyond that point. (Although this latter figure cannot be read directly from the table, it may be obtained by subtracting the table entry from 50 percent, as in the above example.) Our conclusion is that $8.72 = 11.60 + (2.19)(-1.32)$ is located close to the 10th centile point. It should be emphasized that the conversion of standard measures to centiles by means of a table of normal areas is valid only if the standard measures themselves have a normal distribution. And we conclude this section by reiterating that the utility of the standard deviation depends in large measure on its relation to the normal distribution.

***Table 7.4.2** Transforming Given Measures to Standard Measures*

	X	$X - \bar{X}$	$(X - \bar{X})^2$	$z = \frac{X - \bar{X}}{\sigma}$
	8.72	−2.88	8.2944	−1.32
	13.19	1.59	2.5281	.73
	12.54	.94	.8836	.43
	11.15	− .45	.2025	− .21
	15.80	4.20	17.6400	1.92
	8.29	−3.31	10.9561	−1.51
	12.43	.83	.6889	.38
	13.23	1.63	2.6569	.75
	9.72	−1.88	3.5344	− .86
	10.93	− .67	.4489	− .31
Sum	116.00	0.00	47.8338	0.00
Mean	11.60		4.7834	
SD			2.1871	

Source: Randomly generated.

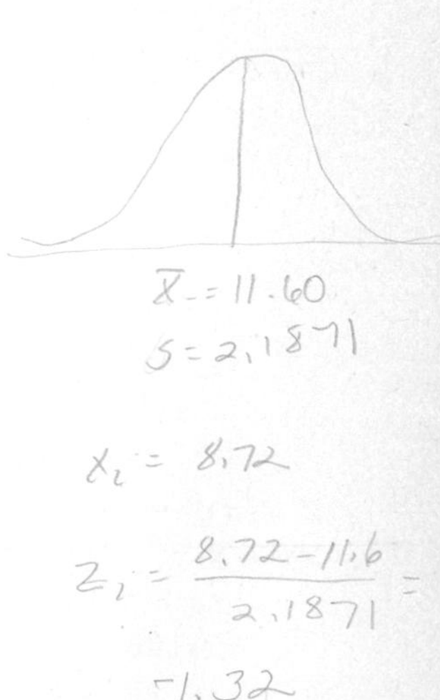

QUESTIONS AND PROBLEMS

1. Define the following terms:
   normal distribution
   normal curve
   standard measure
   normal deviate

2. Explain in what sense the normal curve is "normal."

3. On the base line of a normal curve, sigma units are equal. Would this be the case for a skewed curve? Do quartile points divide the range of a normal distribution into equal intervals?

4. Plot a normal curve as follows:
   (a) From the zero origin placed at the midpoint of the horizontal axis, mark off 3 sigma units in both directions.
   (b) Above each marker, including zero, plot the ordinate (Appendix, Table II).
   (c) Draw a smooth curve through these plotted ordinates.
   (d) From the resultant figure, summarize what seem to be the main features of the normal curve.
   (e) Is it possible to determine visually from a graph whether or not a distribution is normal?

5. Use the Table of Normal Areas (Appendix, Table I) to determine the proportion of cases (area) between the mean and the following normal deviates:

$\pm$ .67	$\pm$1.96	$\pm$2.33	$\pm$3.00
$\pm$1.00	$\pm$2.00	$\pm$2.58	

6. Find the proportion of cases between each pair of normal deviates:

0.3 to 1.6	1.1 to 1.2	-2.58 to $+2.58$
0.3 to -1.6	0.0 to 1.0	1.5 to 3.0
.1 to .2	1.0 to 2.0	-2.3 to $+2.3$

7. Between which normal deviates do the middle 50 percent of the cases lie? The middle 30 percent? The middle 20 percent?

8. The mean of a normal distribution is 75 and the standard deviation is 3.
   (a) What proportion of values lies betwen 72 and 78?
   (b) What values are in the upper tenth of the distribution?
   (c) Approximately what proportion lies between 69 and 81?

9. Of the incomes in a normal distribution, 20 percent are below $50 and 30 percent are above $60. Determine the standard deviation and the mean of the distribution. (Hint: First express $50 and $60 as standard deviates.)

10. By plotting on a cumulative frequency graph of suicide rates (Figure 4.3.8), find:
    (a) the proportion of rates between the mean and a point 1 *SD* above the mean;
    (b) the proportion of rates between the mean and a point 1 *SD* below the mean;
    (c) the proportion of rates between points 1 *SD* below and above the mean.
    From this information, may we conclude that the distribution of suicide rates is normal?

5 VARIATION OF QUALITATIVE VARIABLES

The variation of qualitative variables cannot be measured in the same manner as that of quantitative data. Qualitative variables do not exist in magnitudes; they do not range from low to high on a continuum; there is no central value, nor total nor intermediate ranges. Since there are no measurable intervals between cases, it is impossible to calculate the average interval from any origin.

But that is not to say that a group of qualitative items are necessarily identical, that there is only homogeneity and no heterogeneity. Two items differ when they do not possess the same attribute. Even though these differences are qualitative, it is still possible to devise some measure which effectively summarizes them. But instead of measuring deviations, we count differences.

It is a truism that the greater the number of differences among a set of items, the more heterogeneous is the aggregate, and consequently the more variation there is within it. Similarly, the smaller the number of differences, the greater the homogeneity within it, and the less the variation. For example, there can be no sex differences in a girls' school but in a school for both boys and girls there will always be a smaller or larger number of sex differences among the students, depending upon the sex ratio of that group. Therefore, it appears reasonable to base an index of qualitative variation on the total number of differences among the items in the given set. It is only a question of (1) how to compute the total number of qualitative differences, and (2) how to convert this total into a meaningful index.

Counting Differences To find the total number of differences, we count the differences between each item and every other item and sum these observed differences. For example, in a set of 6 boys and 6 girls, each of the 6 boys will differ in attribute from each of the 6 girls, thereby making a total of 36 differences. If there were 9 boys and 3 girls, each of the 9 boys would differ from each of the 3 girls, producing 27 differences. In a group of 12 boys, the obvious result of no differences would be obtained by multiplying 12 by zero.

Evidently the procedure for determining the total number of actual differences reduces to the following simple rule: Multiply every category frequency by every other category frequency and sum these products. For k categories, there will be

$$\frac{k}{2}(k - 1)$$

such products. In symbols:

$$S_o = \sum_{i=1}^{k} \sum_{j=1}^{k} N_i N_j, \qquad i \neq j \tag{7.5.1}$$

where N_i = number of cases in ith category, k = number of categories, S_o = sum of observed differences. For example, in a set of 4 Catholics, 5 Protestants, and 6 Jews, there would be: $(4 \times 5) + (4 \times 6) + (5 \times 6) =$ 74 differences.

Index of Qualitative Variation (*IQV*) It is necessary to point out that observed differences take on meaning only in relation to some well-defined norm. The number of observed differences may be normed in different ways for different purposes. Norming on the maximum number of possible differences, given some fixed number (k) of categories, has the effect of controlling on k categories. The maximum number of differences occurs when all the frequencies are equal, or when there are no differences between the k frequencies. To get this hypothetical maximum (1) find the mean frequency, (2) square this result, and (3) take this square as many times as there are possible pairs of attributes. In symbols:

$$S_m = \frac{k}{2}(k - 1)\,\overline{N}^2 \tag{7.5.2}$$

where $\overline{N} = \dfrac{N}{k}$.

(By replacing the N_i and N_j in Formula 7.5.1 by $\overline{N}$, we get Formula 7.5.2.) In the aforementioned example involving 9 boys and 3 girls, the maximum number of possible sex differences in a group of 12 would be 6 (boys) $\times$ 6 (girls) $= 36$, or, in this special case, the mean frequency multiplied by itself.

Now the relative amount of variation may be measured by the ratio between the total number of observed differences and the hypothetical maximum:

$$IQV = \frac{S_o}{S_m}. \tag{7.5.3}$$

For 9 boys and 3 girls:

$$IQV = \frac{27}{36} = .75.$$

Among the 15 members of the three religious groups mentioned above, the mean number of members is 5. Multiplying each frequency of 5 by every other frequency of 5 and summing these three products, we find the maximum number of differences to be 75. The observed differences, as already calculated, equal 74. Hence,

$$IQV = \frac{74}{75} = .99.$$

This index will always vary between zero and unity. If the numerator is zero, the index will likewise be zero and will reflect the complete absence of variation. In the event that the observed frequencies are actually equal, the numerator and denominator will be identical, and the index will be 1.00, reflecting maximum heterogeneity, or maximum variation. Index values intermediate between 0 and 1 represent some variation, but less than the maximum.

In some problems, it may be reasonable to norm the number of pairs that are different (S_o) on all possible pairs instead of the maximum number of different pairs (S_m). Since it is possible to form $[N(N - 1)]/2$ pairs from N cases, the total number of possible pairs is $N(N - 1)/2$. Norming on all possible pairs yields an index of heterogeneity that does *not* control for the number of categories as does the *IQV*. However, it may be neither desirable nor necessary to control for the number of categories in some instances, as when different populations are being compared with respect to qualitative variation and the number of categories differs between populations. For example, in comparing the religious heterogeneity of two communities, we may wish to take into account the larger number of denominations in one community than in the other, together with the distribution of the two populations over their respective denominations. This objective may be accomplished by norming the number of observed differences on the total number of possible pairs instead of norming on the maximum number of possible differences for the given number of denominations in each community. Norming on the total number of possible pairs yields an index that, in general, does not vary from zero to one, but varies from zero (all cases in a single category) to an upper limit that depends upon the number of categories with nonzero frequencies; the greater the number of such categories, the higher the upper limit of such an index until that upper limit reaches 1.0, when each case occupies a separate category.

Use of the ***IQV*** Since the *IQV* is a descriptive measure of qualitative variation, it may be used whenever such a measure is called for. For example, it may be used to compare the relative amount of racial homogeneity in two or more communities. In 1970 in Indianapolis there were approximately 969,710 whites and 137,364 blacks. Therefore:

$$IQV = \frac{(967{,}710)(137{,}364)}{(553{,}537)(553{,}537)}$$

$$= \frac{13{,}292{,}851}{30{,}640{,}321}$$

$$= .434\ .$$

In Louisville, Kentucky, in the same year, there were 724,120 whites and 100,683 blacks, so that:

$$IQV = \frac{(724{,}120)(100{,}683)}{(412{,}402)(412{,}401)}$$

$$= \frac{7{,}290{,}657}{17{,}007{,}499}$$

$$= .429\ .$$

Thus, as gauged by the *IQV*, there is no difference in racial heterogeneity between Indianapolis, which has a northerly location, and the border city of Louisville.

An examination of the arithmetic of the formula will disclose that when observed frequencies are expressed as percentages, which are sometimes more convenient, exactly the same result will be obtained. Thus, for Indianapolis, the equation would read:

$$IQV = \frac{(.876)(.124)}{(.500)(.500)}$$

$$= \frac{.108624}{.250000}$$

$$= .434\ .$$

For another example: Table 7.5.1 lists the mother tongues of Switzerland and the percentage of persons in each category of speakers in 1960. To gauge the linguistic diversity of Switzerland, we compute:

$$IQV = \frac{(.70)(.19) + (.70)(.10) + (.70)(.01) + (.19)(.10) + (.19)(.01) + (.10)(.01)}{\frac{4 \times 3}{2} \times .25^2}$$

$$= \frac{.232}{.375}$$

$$= .62\ .$$

Table 7.5.1 *Mother Tongues of Switzerland, 1960*

Mother Tongue	Number	Proportion
German	3,765,203	.70
French	1,025,450	.19
Italian	514,306	.10
Romanche	49,823	.01
Total	5,354,782	1.00

Source: Stanley Lieberson, "National and Regional Language Diversity." *Actes du Xe Congrès International des Linguistes*, Bucharest: Éditions de l'Académie de la Republique Socialiste de Roumanie, 1969, 769–773.

We interpret this outcome to mean that language diversity in Switzerland is .62 of the maximum possible, given $k = 4$ mother tongues.

As defined here, qualitative variation is strictly a statistical characteristic and should not be confused with the sociopsychological state that characterizes social disorganization, anomie, or social conflict. Of course, the degree of social disorganization may be related to the degree of statistical heterogeneity in regard to race, religion, ethnic background, or nativity, because this heterogeneity may be one of the conditioning factors in the attitudes of the population. Thus, it has been hypothesized that social tension increases as conflict groups approach equality in power, of which numerical parity is one element. The *IQV* is one tool for its measurement, and it enables us to study such hypotheses more systematically.

Relation of *IQV* to Variance The *IQV* is obviously a function of the distribution of N cases in the categories of the k-fold classification. If the cases are evenly distributed between categories, *IQV* is equal to 1.00; when all cases fall in a single category, then *IQV* is equal to 0. Since it is the differences among the frequencies that are indexed by the *IQV*, one may wonder whether the variance of those frequencies might serve equally well as an index of qualitative variation.

Let us consider this possibility: To calculate the variance of the k category frequencies, we subtract the mean frequency ($\overline{N}$) from each, square these deviations, sum and divide by their number. In symbols,

$$\sigma_o^2 = \frac{\Sigma(N_i - \overline{N})^2}{k} \qquad (7.5.4)$$

where σ_o^2 = variance of observed frequencies. This variance will be zero when the N_i are equal; it will attain its maximum value when $k - 1$ of the N_i are equal to zero. Therefore, to get an index ranging from 0 to 1, we need only to divide the observed variance by its maximum possible value:

$$\sigma_o^2/\sigma_m^2 \tag{7.5.5}$$

where σ_m^2 = the maximum variance, and σ_o^2 is the observed variance.

In any problem, it would be possible to calculate the maximum variance by setting $k - 1$ frequencies equal to zero, and one frequency equal to N. Subtracting $\bar{N}$ from these frequencies gives the deviations; summing the squared deviations and dividing by k gives the maximum variance. However, that laborious procedure is obviated by a general formula in terms of k and N, which the student will have to take on trust. This is:

$$\sigma_m^2 = (k - 1)\bar{N}^2. \tag{7.5.6}$$

Accordingly, to calculate the ratio of the observed variance to the maximum variance, we need only to divide the former by $(k - 1)\bar{N}^2$.

Like the *IQV*, this ratio runs from 0 to 1, and will be larger or smaller according to the degree of qualitative variation as defined. Since we mentioned earlier that this ratio may do as well if not better than *IQV* as an index of qualitative variation, it will be instructive to consider their relationship.

A heavy mathematical analysis would be out of place here, and we will arrive at our conclusion from numerical examples. Consider first the case of $k = 3$, $N = 90$, and different sets of N_i:

N_1	N_2	N_3	σ_o^2/σ_m^2	*IQV*	Sum
30	30	30	.000	1.000	1.000
40	25	25	.028	.972	1.000
50	20	20	.111	.889	1.000
60	15	15	.250	.750	1.000
70	10	10	.444	.556	1.000
80	5	5	.694	.306	1.000
90	0	0	1.000	0.000	1.000

Second consider the case of $k = 4$, $N = 100$:

N_1	N_2	N_3	N_4	σ_o^2/σ_m^2	*IQV*	Sum
25	25	25	25	.000	1.000	1.000
50	50	0	0	.333	.667	1.000
75	25	0	0	.500	.500	1.000
100	0	0	0	1.000	0.000	1.000

From these examples, we surmise that $\sigma_o^2/\sigma_m^2 + IQV = 1.000$. If this is generally the case, as it is, then we may always obtain *IQV* by subtracting σ_o^2/σ_m^2 from 1.000, or vice versa. Also, we deduce that for each unit increase in *IQV*, there is a unit decrease in σ_o^2/σ_m^2. (That is, there is a perfect linear relation between *IQV* and σ_o^2/σ_m^2.)

Because of this convertibility, the student may wonder which operation is preferred in practice, or whether it makes no difference which one is employed. From a purely arithmetical standpoint, it makes no difference (since they are perfectly related); therefore, the choice will have to be made on extrastatistical considerations. The *IQV*, defined as actual differences to maximum possible differences, has a definite intuitive appeal as an index of qualitative variation. On the other hand, the variance is a familiar concept in statistics. Moreover, the ratio σ_o^2/σ_m^2, since it is the complement of *IQV*, may be construed as the proportion of possible differences that failed to materialize, and might therefore be regarded as an index of qualitative uniformity. It reaches 1.00 when none of the possible differences materialize. Evidently, from a purely statistical point of view, it makes no difference which index is used, and the choice will depend on such practical considerations as convenience and familiarity.

QUESTIONS AND PROBLEMS

1. Define the following terms:
   index of qualitative variation
   variance of frequencies (σ_o^2)
   maximum variance of frequencies (σ_m^2)

2. A given population is 50 percent male and 50 percent female, and 70 percent white and 30 percent black. Is it possible to represent both variables by a single *IQV*? Comment.

3. Is it possible for a group of persons to have more than one *IQV*? Explain.

4. At one college, 80 percent of the students are men; at another, 67 percent. Compute and compare *IQV*s.

5. A population is distributed into four ethnic groups as follows:

German	60%
French	20
Swedish	15
Irish	5
	100%

   Calculate the *IQV*.

SELECTED REFERENCES

Lieberson, S.
1969 "Measuring population diversity." American Sociological Review 34, 850–862.

Martin, J. David and Louis N. Gray
1971 "Measurement of relative variation: sociological examples." American Sociological Review 36 (June): 496–502.

Moroney, M. J.
1954 Facts from Figures. Baltimore: Penguin Books. Chapter 5.

Ray, J. L. and J. D. Singer
1973 "Measuring the concentration of power in the international system." Sociological Methods and Research 1, 403–437.

Theil, H.
1969 "The desired political entropy." American Political Science Review 63, 521–525.

Yule, G. Udny and M. G. Kendall
1950 An Introduction to the Theory of Statistics. Fourteenth edition. London: Charles Griffin. Chapter 6.

Measurement of Association: Qualitative and Ordinal Variables 8

1 CONCEPT OF STATISTICAL ASSOCIATION

Principle of Contingency It is an axiom of science as well as common sense, that no event in nature "just happens" but that events always occur under very specific circumstances, either known or unknown. Therefore an event never should be viewed in isolation. It must be considered as a product of the joint operation of many forces, each of which contributes a variable element to the observed outcome. Thus, family size may depend upon such factors as age at marriage, level of income, extent of the mother's employment outside the home, and the religious ideology entertained by the parents. Parole success of released prisoners may be related to the type of crime committed, age, and history of recidivism. Some of these factors tend to promote and accelerate the occurrence of the event, others tend to retard it in varying degrees or to inhibit it, or at least tend to modify its character or magnitude. The outcome is the *net result* of opposing variables. In any case, we cannot predict the degree of parole success or explain the size of the family unconditionally; we can make our predictions only on the basis of specifically designated factors, or variables, on which the outcome is contingent. The human observer does not possess absolute knowledge; we must use one event as a cue to anticipate another. Therefore, our understanding is grounded in the *principle of contingency*.

Perceptions of the association between events are continuously revised in the trial-and-error of daily experience. In the process, the observer mentally quantifies and summarizes observations: (1) by noting the factors which seem to "cause" or to be linked with the event, and (2) by noting the frequency with which one can successfully anticipate or forecast it. In this casual manner, people begin to cultivate the habits of association and become informal statisticians. We practice intuitively the principle of correlation. In fact, much sociology effectively employs such intuitive statistics, skillfully put together by alert, widely-traveled scholars, and made without benefit of pencil-and-paper calculation or technical procedures. Indeed, many who are critical of the utility of statistical procedures unwittingly employ them in this unofficial manner.

For some purposes, a rough subjective approximation of a correlation is fairly satisfactory, but for scientific purposes more accurate measurements are desired. Such precision is not a simple matter to achieve. The difficulty lies in the complexity of the patterns of relationship in terms of which we view and organize the world. Some of the salient and necessarily interrelated features of this complexity in patterns may be formulated as follows: (1) every event is the outcome of multiple factors; (2) the force of these respective factors varies in intensity; (3) the effect of an increase or change in any given factor may be either to promote or to inhibit the event in question; (4) the separate factors may reinforce, counteract, or cancel one another; and (5) the effect of one factor may be to serve as a catalyst or as an inhibitor of the effects of another. Furthermore, (6) the event may feed back on the factors that produced it; or (7) one may assume that the event is affected by a given factor when, in fact, it is the other way around. We illustrate with a popular example. The football halfback, whose yardage record tends to be associated positively with running ability, is either obstructed or aided in actual performance by the condition of the field, fatigue, and the type of plays called, to say nothing of the varied activities of the other twenty-one players on the field. Therefore, this association between gaining power and running ability may not be visible to the fans in the stadium. So is the correlation between income and size of family beclouded by such factors as religion, age at marriage, occupation of breadwinner, and education. At first it may seem a hopeless task to disentangle these networks of relationships and to subject them to statistical analysis. Nevertheless, from a large number of observations, the essential relation can be inferred with the aid of appropriate techniques of analysis.

This problem is probably less difficult in the physical than in the social sciences. The physical scientist, by means of available laboratory controls, is able to segregate and manipulate elements to a certain extent and to replicate careful and undisturbed observations, whereas the social scientist often is obliged to accept data which are like unrefined ores from "nature in the raw" and to assemble materials from widely dispersed sources and a variety of settings. Therefore, the social scientist is compelled to employ statistical controls, since laboratory controls are frequently difficult or even impossible to impose.

What Are the Evidences of Relationship? How can we be sure that factors are interconnected? And, having discovered a relationship, how may we determine the degree of intensity of that relation? Broadly speaking, there are two earmarks of such linkage: (1) joint occurrence of attributes; and (2) parallel changes in two or more series of quantitative observations.

The relative frequency with which certain attributes happen together is probably the most elementary basis of lay judgment of association. This is the *principle of joint occurrence*. Statistical variables, like human beings, usually are judged "by the company they keep." For example, if delinquency is more often found in boys than in girls, we conclude that delinquency is associated with "boyness." The strength of this association will vary according to other factors such as the boy's age, the type of delinquency, and many other elements in the pattern, all of which will complicate the statistical application or apparently simple principle. Hence, it need hardly be reiterated here that some system of tabulation and classification is necessary as an aid not only in establishing an association, but also in determining it's strength.

Second, in two series of quantitative data, if a unit change in one variable is paralleled with some degree of regularity by a comparable change in the other series—that is, if they move together—we conclude that they are somehow tied together, and that there is an association between the two sets of data. For example, as income declines, the family size tends to increase; and, if the observations endure through a rather extensive range—that is, for the entire range of families of all sizes and of incomes of varying amounts—the evidence of a relation is strengthened. This is called the *principle of covariation*.

Devices for the Measurement of Association Techniques for the measurement of association must be adapted to the nature of the data and to the substantive problem being investigated. Hence, a variety of different measures have been devised, each appropriate under certain circumstances. The measures of association presented herein have been selected for their applicability to data regularly encountered by sociologists, and for the simplicity and clarity of their meaning. These measures may be grouped conveniently on the basis of the presumed level of measurement. First we consider those measures which presume only the classification of cases by attributes, or sets of unordered classes. Such measures of association are based on the aforementioned principle of joint occurrence. Next we consider the measurement of association between ordered classifications where the distance between points on each variable is unknown. The gauging of association between such ordered classifications is based not on the principle of joint occurrence but rather on a primitive form of the principle of covariation. Finally, we shall consider measures of association for quantitative variables subject to interval or ratio measurement. The correlation between such variables is measured by statistical operations which are closely similar to those we employ in measuring central tendency or dispersion for one of the variables.

2 MEASURING ASSOCIATION: SOME GENERAL CONSIDERATIONS

Inadequacy of the 2 × 1 Table for Estimate of Association Many persons untrained in quantitative analysis naïvely succumb to the temptation of drawing conclusions on the degree of association from a 2 × 1 table, instead of from a 2 × 2 table, which is the minimum for that purpose. The deceptive ease with which such erroneous deductions can be made is illustrated in the four 2 × 1 tables depicted in Table 8.2.1. Since only 10 boys out of 100 are delinquent, one may be misled into the conclusion that there is an association between boyness and nondelinquency. Similarly, since most women are classified as having no accidents, one might be tempted to conclude that women are not accident prone. And since Republicans divide 50–50 on isolationism, there would seem to be no striking tendency for political affiliation to be associated with one or the other stand with regard to foreign policy.

However, the fact that most boys are nondelinquent does not imply an association between boyness and nondelinquency. Delinquencies are distributed between boys and girls, and we must inquire in which proportion the limited supply of violations are divided between the sexes. To which sex do delinquencies or accidents differentially attach themselves? To estimate the degree of affinity for *either* sex, we must know the delinquency rate or the accident rate for both sexes; a delinquency rate for boys can be said to be high or low in comparison to some standard such as the delinquency rate for girls. A similar comparison is necessary to determine the degree of association between political party and isolationism, sex and accident rate, and, for that matter, between any two variables. Therefore, let us provide comparative rates for each of the 2 × 1 tables and consider the results.

Table 8.2.1 *2 × 1 Tables*

Boys	Percent
Delinquent	10%
Nondelinquent	90
Total	100%

Republicans	Percent
Isolationist	50%
Internationalist	50
Total	100%

Freshmen	Percent
A and B Grades	60%
Other	40
Total	100%

Female Drivers	Percent
Accident	20%
No Accident	80
Total	100%

Table 8.2.2 *2 × 2 Tables*

	Boys	Girls
Delinquent Nondelinquent	10% 90	0% 100
Total	100%	100%

	Rep.	Dem.
Isolationist Internationalist	50% 50	30% 70
Total	100%	100%

	Freshmen	Others
A and B Grades Other	60% 40	70% 30
Total	100%	100%

	Men	Women
Accident No Accident	10% 90	20% 80
Total	100%	100%

As Table 8.2.2 shows the delinquency rate of boys now turns out to be rather high, since even a 10 percent delinquency rate is higher than no delinquency at all. Similarly, the grades of college freshmen are below the norm; the Republicans display a marked propensity toward isolationism; and women drivers show a marked susceptibility to mishaps.

The fact is that we now have introduced a standard of judgment against which the 2 × 1 table may be compared. Inevitably and unwittingly, some standard is introduced by every observer, and the function of statistical procedures is to make that standard explicit.

To speak of statistical association is to speak of an association between variables, either qualitative or quantitative. In a 2 × 1 table there is only one variable; in place of the second variable, there is but a single category in which there is no variation. The existence of statistical association between a variable and a constant cannot be determined; in fact, it is a contradiction in terms. Statistical association is manifest only if cases divide themselves into at least two classes for each variable. Hence, a 2 × 2 table is the minimum for a dependable conclusion on association.

Statistical Association and the Cross-Classification Table A 2 × 2 table, or any table with *at least* two subclasses of each variable, is a two-way rather than a one-way classification. From our previous discussion of cross-classification (Chapter 3) the student already has acquired the sense of inferring association, and even cause-and-effect relationships, between variables. But at that point we did not seek to compute a single overall measure that would reflect the strength of the association between variables. *Measures of association* are summarizing measures that represent the degree of association in a single value. They are therefore analogous to the mean, median, or mode—single values which represent the central tendency of a distribution. Like measures of central tendency, measures of association cannot reflect in a single value all details and aspects of the distribution they summarize. Measures of association describe that particular aspect of a joint distribution produced by the tendency of two variables to occur together or vary together. More specifically, measures of association gauge the degree to which the variation in one variable can be accurately inferred from the variation in the other. This aspect of bivariate data, summarized in a measure of association, may be called the *degree of predictability* in the joint distribution. The relative accuracy in predicting the one variable from the other variable is the *degree of association.*

An intuitive appreciation of varying degrees of association may be achieved by considering Tables 8.2.3 and 8.2.4. If the association between religious affiliation and political party affiliation were perfect (Table 8.2.3a), knowledge of religious affiliation would permit one to infer without error the political party affiliation of any particular case by following the rule: "If Catholic, then Democrat; if Protestant, then Republican." Similarly, if the

Table 8.2.3 *Varying Degrees of Association: Religion and Political Affiliation*

a. Perfect Association

	Catholic	Protestant	Total
Democrat	50	0	50
Republican	0	50	50
Total	50	50	100

b. Moderate Association

	Catholic	Protestant	Total
Democrat	40	10	50
Republican	10	40	50
Total	50	50	100

c. Weak Association

	Catholic	Protestant	Total
Democrat	27	23	50
Republican	23	27	50
Total	50	50	100

association between income and education, both trichotomized as in Table 8.2.4a, were perfect, knowledge of the educational level would allow one to infer without error the income level of any particular case by following the rule: "If low education, then low income; if medium education, then medium income; if high education, then high income." Perfect associations occur rarely in empirical data because: (1) factors other than the one independent variable in the table will necessarily affect the dependent variable (for example, factors other than religion affect political affiliation), and (2)

Table 8.2.4 *Varying Degrees of Association: Income and Education*

a. Perfect Association

Income	Education			
	Low	Medium	High	Total
Low	33	0	0	33
Medium	0	34	0	34
High	0	0	33	33
Total	33	34	33	100

b. Moderate Association

Income	Education			
	Low	Medium	High	Total
Low	27	4	2	33
Medium	3	25	6	34
High	3	5	25	33
Total	33	34	33	100

c. Weak Association

Income	Education			
	Low	Medium	High	Total
Low	14	11	8	33
Medium	9	13	12	34
High	10	10	13	33
Total	33	34	33	100

there are always observed errors of classification in data (for example, some persons who represent especially ambiguous cases may be erroneously classified). Since perfect association is seldom if ever observed, measures of association are useful in describing the degree to which association is present.

If the association between religious affiliation and political party affiliation were not perfect, but only moderately high (Table 8.2.3b), knowledge of religious affiliation would enable one to infer political party from religion, but with some error, by following the same rule as above: "If Catholic, then Democrat; if Protestant, then Republican." If we draw this inference in each case, we would be correct 80 percent of the time and in error 20 percent of the time. However, the value .80 is not a satisfactory index of association, because it does not take into account the degree of accuracy we could achieve in inferring political party without utilizing any knowledge of religious affiliation. In short, it is an unnormed measure. Since the cases in this table are split 50–50 between Democrats and Republicans, we could correctly classify one-half the cases on the basis of this information alone. If we classify all cases as Democrats, we would be wrong for the half that are actually Republicans. If we classify all as Republicans, we would be wrong for the half that are Democrats. Applying either rule, we would make 50 errors in the 100 tries. But we would make only 20 errors in 100 tries if we predicted the political party from religion according to the aforementioned rule. In predicting political party, we can do better knowing religion than not knowing it. More specifically, we make 20 errors in 100 tries instead of 50, which represents a 60 percent reduction in error,

$$\frac{50 - 20}{50} \times 100 = 60\%.$$

The percentage reduction in error attributable to information on a second variable is generally a more useful measure of association than the percentage of correct predictions from the second variable, for reasons we will examine later.

Now consider the table showing a moderate association between income and education (Table 8.2.4b). If we apply the rule, "If low education, then low income, . . ." etc., we would accurately predict the income level for each of the cases in the cells on the diagonal running from upper left to lower right ($27 + 25 + 25 = 77$ cases). Again, however, our interest lies in the percentage reduction in error rather than simply in the proportion of correct predictions. Since the cases represented in Table 8.2.4b are divided into three categories—low, medium, and high—with marginal frequencies of 33, 34, and 33, respectively, with this knowledge of the marginal distribution alone, we cannot anticipate the correct income classification for even one-half of the cases. The best we can do is to predict "medium" income for all cases. We would then be correct in 34 of the 100 cases but incorrect in 66 cases. Therefore, without knowledge of education, we would make 66 errors, whereas with knowledge of education we would make only 100 − 77 = 23 errors. Since we have eliminated 66 − 23 = 43 of the original errors, we have achieved a 65 percent reduction in error;

$$\frac{66 - 23}{66} \times 100 = 65\%.$$

It is noteworthy that the percentage reduction in error in Table 8.2.3b is less than in Table 8.2.4b—60 percent as compared to 65 percent, although we could predict with higher accuracy in Table 8.2.3b than in Table 8.2.4b (80 percent correct as compared to 77 percent). If this should seem strange, recall that we desire a measure of association that will reflect not only the accuracy in inferring one varible from another, but also the degree of increase in the accuracy of such predictions over the accuracy possible on the basis of the marginal distribution of the predicted variable. A measure of association based on the concept of relative reduction in error reflects the accuracy attributable to the relationship between variables, disregarding the accuracy that could be achieved by utilizing the marginal distributions alone.

Now consider Table 8.2.3c and 8.2.4c, in which there is little or no association. If we were predicting political affiliation from religion by the same rule as above, we would predict political affiliation accurately in 27 + 27 = 54 cases out of the 100 in Table 8.2.3c, leaving 46 cases in error. This represents a low degree of association, not because 54 percent accuracy is inherently low but because it represents only a very slight improvement in accuracy over that obtained on the basis of the marginal distribution alone. On the basis of the marginals, we would make 50 errors; in this table, knowledge of religion would reduce the number of errors by only 4 to yield an 8 percent reduction in error:

$$\frac{50 - 46}{50} \times 100 = 8\%.$$

Similarly, in Table 8.2.4c, we could give the correct income classification from knowledge of education for 40 cases, leaving 60 cases in error. This represents a very low proportional reduction in error because, from knowledge of the marginal distribution alone, we would make only a few more errors, 66 as compared to 60. For practice, the student may wish to compute the relative reduction in error for this table. (Answer: .09 or 9 percent.)

General Features of Relative Reduction in Error Measures of Association

Many common measures of association describe the relative reduction in prediction error resulting from the strength of the relationship between variables. Such measures differ from one another in their prediction rules and in their definitions of prediction error; hence they find their usefulness in different situations. Although they differ in detail, their essential elements are the same, namely, (1) a marginal prediction rule, (2) a prediction rule based on the association between variables, (3) a definition of error, and (4) a formula for computing the relative reduction in error. Our subsequent discussion of specific measures of association will be facilitated by an explicit description of each of these elements.

(1) A Rule for Predicting a Dependent or Predicted Variable from Its Own Distribution In analyzing Tables 8.2.3 and 8.2.4, we employ such a rule, designating the category with the largest marginal frequency as the prediction for all cases. In predicting income without reference to education, we predict "medium income" for all cases, since that category had the largest marginal frequency. In subsequent sections, different rules for predicting the dependent variable from its own distribution will be considered.

(2) A Rule for Predicting the Dependent Variable from an Independent Variable In each of the examples above we formulated a rule for prediction based on the independent or prediction variable—"If Catholic, then Democrat; if Protestant, then Republican," "If low education, then low income; if medium education, then medium income; if high education, then high income." In subsequent sections we will consider different rules for predicting the dependent variable from the independent variable. Although a grasp of these rules is useful for an understanding of the association measures to which they apply, it is not ordinarily necessary to carry out the procedures implied in these rules in order to compute the measure of association. Computational formulas generally enable us to compute the measures of association without computing the errors directly.

(3) A Definition of What Constitutes Error and How Error Shall Be Measured In our discussion above, we adopted a very simple and intuitively comprehensible definition of error: Any case that was misclassified constituted an error. But with interval measures our predictions will not be categorically right or wrong; they will miss the true value by a larger or smaller amount. Hence, with interval measures our errors will be magnitudes rather than attributes. We mention this to anticipate that the definition of error will change according to the nature of our data as interval, ordinal, or nominal.

(4) A Definition of Relative Reduction in Error Measure of Association

$$\text{Proportional reduction in error measure of association} = \frac{\text{Error by Rule (1)} - \text{Error by Rule (2)}}{\text{Error by Rule (1)}}$$

$$= \frac{E_1 - E_2}{E_1},$$

where E_1 denotes error by Rule (1) and E_2 denotes error by Rule (2).

All relative reduction in error measures of association fit this general form.

However, the content of Rule (1) and (2) will differ from one particular measure of association to another.

While the procedure applied to Tables 8.2.3 and 8.2.4 is simple and logical, it is insensitive to some patterns of association and hence it is used in social research less frequently than some other measures of association to be discussed below. In the next section of this chapter, we will describe a measure of association for nominal variables that has the same general format,

$$\frac{E_1 - E_2}{E_1},$$

but is more sensitive to various patterns of association and hence more serviceable as a measure of association in social research. We defer that discussion briefly to note that the measures of association computed for Tables 8.2.3 and 8.2.4 are known as lambda (λ) measures. In Table 8.2.3, the measure we computed is usually symbolized λ_r (lambda sub-*r*), which serves as a shorthand expression for indicating that we have computed lambda for the row variable (political party affiliation) as the predicted variable. Had the predicted or dependent variable been displayed as a column variable in Table 8.2.3, we would have computed λ_c (lambda sub-*c*), indicating that the predicted variable is the column variable. The symbol λ is a shorthand expression that indicates the general nature of the prediction rules utilized (that is, predict the modal category) and the definition of error (misclassification = error).

To illustrate the lack of sensitivity of the lambda measures to specific patterns of association, we consider the data of Table 8.2.5, in which birth order is cross-classified with high and low agreement with the mother's orientation toward the feminine role, for a sample of college women. We might anticipate that, since first-born children are typically subjected to a more intensive socialization process by their parents, first-born children would be more likely to adopt their parents' perspectives. Examining the table, we find that the data are consistent, to some degree, with this anticipation: 74 percent (101 of 137) of the first-born daughters showed high agreement with the mother as compared to 63 percent (45 of 71) of the later-born daughters. Although no one would claim that the data in this table show a high degree of association, the differential distribution for first-born and later-born daughters suggests that there is *some* degree of association in the anticipated direction.

The predicted or dependent variable—"Agreement" in this instance—is displayed as the column variable in Table 8.2.5, and we therefore proceed to compute λ_c. From knowledge of the marginal distribution of the dependent variable, we would predict "High Agreement" (the modal category for all cases combined) and thereby make 62 errors. Thus, $E_1 = 62$. Predicting agreement from knowledge of birth order, for first-born daughters we would predict "High Agreement" (the modal category for first-born daughters)

Table 8.2.5 *Birth Order and Agreement with Mother's Orientation toward the Feminine Role*

Birth Order	High Agreement	Low Agreement	Total
First-born	101	36	137
Later-born	45	26	71
Total	146	62	208

Source: Adapted from Kenneth Kammeyer, "Birth Order and the Feminine Sex Role Among College Women," *American Sociological Review*, 31 (August 1966), p. 513. Reprinted by permission.

and make 36 errors, and for later-born daughters we would predict "High Agreement" (the modal category for later-born daughters) with 26 errors. Thus, $E_2 = 36 + 26 = 62$. Hence:

$$\lambda_c = \frac{E_1 - E_2}{E_1} = \frac{62 - 62}{62} = 0 .$$

The lack of sensitivity of the lambda measure to the differential distribution of first-born and later-born daughters in this instance is a consequence of the lack of sensitivity of the mode to other features of the distributions. Substantively, it appears that the general tendency toward "High Agreement" with the mother is so strong that, even though the later-born daughters exhibit this tendency to a lesser degree, "High Agreement" has remained the mode for both subclasses of daughters. In the next section of this chapter, we consider a measure of association for nominal variables that is based on a more sensitive prediction rule and which will therefore measure the limited degree of association that is present in this table.

QUESTIONS AND PROBLEMS

1. Define the following concepts:

statistical association	prediction rule
2 × 1 table	prediction error
2 × 2 table	proportional reduction in error
marginal distribution	

2. In a study of mental patients "94 percent showed evidence of status conflict before onset of mental illness."[1]

   (a) Does this observation demonstrate an association between status conflict and mental illness? Explain your answer.

   (b) What would be your conclusion if 94 percent of the normal population also had status conflict?

[1] Source: Kingsley Davis, "Mental hygiene and class structure," Psychiatry, I, 1938, 55–65.

(c) If 50 percent of the normal population had status conflict, what would you conclude?
(d) Is a 2 × 1 table sufficient to prove a relationship?

3. (a) Form a complete 2 × 2 frequency table based on the following data: A total of 650 young people; 80 percent are boys, and 10 percent are delinquents; 10 percent of the boys are delinquent.
(b) What percentage of the girls are delinquent? How does this compare to the percentage of boys who are delinquent? Would you conclude from this table that there is some association between sex and delinquency?

4. (a) Construct a 2 × 2 table for the following data: 960 of 1,500 community leaders replied "Yes" to the question "If a person wanted to make a speech in your community against churches and religions, should he be allowed to speak?" Of 897 nonleaders from the same communities, 350 answered "Yes."[2]
(b) What conclusion on the relation between leadership and tolerance is suggested by this table?

5. For Table 8.2.6

Table 8.2.6 *Nonliterate Societies Cross-Classified by Presence or Absence of Slavery and Degree of Power Vested in the Chief*

Power Vested in the Chief	Slavery		
	Present	Absent	Total
High	23	19	42
Low	0	16	16
Total	23	35	58

Source: Leo W. Simmons, "Statistical Correlations in the Science of Society," in G. P. Murdock (ed.), *Studies in the Science of Society Presented to Albert G. Keller*. New Haven: Yale University Press, 1937. Copyright © 1937 by Yale University Press.

(a) Compute λ_c. What does the computed value of λ_c mean?
(b) Compute λ_r. What does the computed value of λ_r mean?
(c) Under what political circumstances is it possible to predict without error the presence or absence of slavery in that society? Under what conditions of servitude is it possible to predict without error whether the power of the chief is high or low? Since perfect prediction is possible under specific circumstances, why are the values of λ_c and λ_r so low?

[2] Source: S. A. Stouffer, *Communism, Conformity, and Civil Liberties*. Garden City: Doubleday, 1955, p. 33.

6. Respondents in a community survey were presented with a list of "reasons different people have given for wanting to have their children finish a certain amount of education" and asked to select the one they considered most important. Selections were grouped into two categories: "instrumental perception of education" (i.e., as a means to other ends) and "noninstrumental perception of education" (i.e., as an end in itself). The results, cross-classified by five social class levels, are shown in Table 8.2.7.

Table 8.2.7 *Instrumental and Noninstrumental Perception of Education by Social Class*

Perception of Education	Social Class					
	(High) I	II	III	IV	(Low) V	Total
Noninstrumental	10	13	26	31	27	107
Instrumental	1	5	21	37	34	98
Total	11	18	47	68	61	205

Source: Adapted with permission of Macmillan Publishing Co., Inc. from *Success and Opportunity: A Study of Anomie*, by Ephraim Mizruchi. Copyright © 1964 by The Free Press, a Division of The Macmillan Company.

(a) Compute λ_r for Table 8.2.7.
(b) Transform the table into a 2 × 2 table by combining Classes I, II, and III into "high" and Classes IV and V into "low." Compute λ_r for the resulting table.
(c) Now transform the original table into a 2 × 2 table by combining Classes I and II into "high" and Classes III, IV, and V into "low." Compute λ_r for the resulting table.
(d) Why does the "cutting point" between Classes III and IV leave λ_r unchanged, while the "cutting point" between Classes II and III reduces λ_r when the original table is transformed into a 2 × 2 table?

3 GOODMAN AND KRUSKAL'S TAU

The concept of prediction does not require that we make the same prediction for every case; indeed, if the outcomes are varied, our predictions must also be varied in order to be highly accurate. In predicting political party affiliation from religion in the previous section (Table 8.2.3), we predicted the same party affiliation for all persons in a given religious category, but different party affiliations for persons in different religious categories. Such differential prediction may be carried one step further to permit two or more predictions within the same predictor category. Thus, instead of predicting that all Catholics would vote Democratic in Table 8.2.3b, we might attempt to predict the specific forty Catholics who vote Democratic and the specific ten Catholics who vote Republican. And instead of predicting that all Protestants vote Republican in that same table, we might attempt to predict the specific forty Protestants who vote Republican and the specific ten Protestants who vote Democratic. More succinctly stated, we might distribute our predictions so as to reconstruct the actual distributions and then proceed to determine the relative reduction in prediction error we would expect to achieve by such a strategy. This strategy is the basis for *Goodman and Kruskal's tau*[3] which is sensitive to even minor differences in distributions, which, as we have seen, is not a general characteristic of the simpler and cruder lambda measures.

Since Goodman and Kruskal's tau is a measure of the relative reduction in prediction error, its calculation must rest on (1) a rule for predicting the dependent variable from knowledge of its own distribution; (2) a rule for predicting the dependent variable from knowledge of the independent variable; (3) a definition of error; and (4) the computation of the relative reduction of error in the form:

$$\frac{E_1 - E_2}{E_1}.$$

Error is defined for Goodman and Kruskal's tau simply as the misclassification of a case, and the computation of the relative reduction in error will be unproblematic if we can determine the errors of prediction in applying each of the prediction rules. We consider each of these rules as they apply to the hypothetical Table 8.3.1, showing an oversimplified cross-classification of offenses and court dispositions.

Table 8.3.1 *Hypothetical Cross-Classification of Offense and Court Disposition for a Sample of Convicted Offenders*

	Disposition			
Offense	Fined	Probation	Imprisoned	Total
Auto theft	5	30	5	40
Burglary	0	30	20	50
Forgery	5	0	5	10
Total	10	60	30	100

[3] It is necessary to refer to this measure as Goodman and Kruskal's tau to distinguish it from a measure of association devised by Kendall, which is also called tau. Kendall's τ (tau) is not treated in this book (see Maurice Kendall, *Rank Correlation Methods*, Charles Griffin: London, 1948). Kendall's τ (tau) is more closely related to the measure gamma (Section 4) than to Goodman and Kruskal's τ (tau).

In predicting court disposition from knowledge of its own distribution, we know that 10 of the 100 cases were fined, but we have no way of knowing *which* ten. Similarly, we know that 60 were placed on probation, but we have no basis for judging which 60 were given that disposition. Finally, those who were imprisoned are known to number 30, but we don't know which thirty they are. It is *as if* we were given a large stack of envelopes, each containing a disposition we cannot see. We are told the stack contains 10 percent fined, 60 percent probation, and 30 percent imprisoned as the disposition, but we have no further information except the requirement to put ten of the envelopes in the fined category, 60 in the probation category, and 30 in the imprisoned category. Lacking any information on which to base our decision as to which envelope to place in which category, we "shuffle" the envelopes and "deal" them at random, placing the required number in each disposition category.

How many errors will we expect to make by this procedure in the long run? In the fined category, to which we will "deal" ten envelopes, the probability of being wrong is the probability that an envelope drawn at random is a probation or imprisoned disposition instead of fined, which is required for a correct placement. This probability is $60/100 + 30/100 = 90/100$. If we "deal" 10 envelopes into the fined category with a probability of error of 90/100, we will expect, on the average, to make $90/100 \times 10 = 9$ errors. We proceed to make a similar computation for each of the remaining dispositions. We deal 60 envelopes to the probation category and the probability of being wrong is 40/100. Hence in that category we expect to make $40/100 \times 60 = 24$ errors. Finally we deal 30 envelopes to the imprisoned category and the probability of being wrong is 70/100; the expected number of errors is $70/100 \times 30 = 21$. For the three disposition categories combined, we expect to make $9 + 24 + 21 = 54$ errors; this sum gives E_1. Expressed symbolically:

$$E_1 = \sum_{1}^{c} \frac{(N - N_{.j})N_{.j}}{N}, \tag{8.3.1}$$

where E_1 = the expected number of errors in predicting column category from observed distribution of cases in columns,

N = total number of cases in the table,

$N_{.j}$ = marginal total for the jth column, and

c = number of columns.

$\sum_{1}^{c}$ is the instruction to sum the quantities in parentheses over all columns, 1 through c.

In predicting court disposition from knowledge of offense, we are given information about the offense for each case, and the distribution for each offense category, but we have no other information about specific cases on which to base a prediction of court disposition. For example, for the 40 offenders who were convicted of auto theft, we know that the cases should be distributed so that 5 fall into the fined category, 30 into the probation category, and 5 into the imprisoned category. But we have no further information to help us place the right case in the right disposition category. Hence, we may proceed as before, but distributing the cases in each offense category to reconstruct the distribution of court dispositions for that offense.

Continuing with the imagery suggested above, it is as if we were given a large stack of envelopes, each containing a disposition we cannot see, but each having "auto theft" visible on the outside of the envelope. We are told that the stack of envelopes contains $5/40 = 12.5$ percent fined, $30/40 = 75$ percent probation, and $5/40 = 12.5$ percent imprisoned, but we have no further information except the requirement to distribute 40 envelopes so that 5 are fined, 30 are on probation, and 5 are imprisoned. If all persons convicted of auto theft were concentrated in a single disposition, we could distribute them with no errors. Since that is not the case, we expect to make some errors and proceed to compute the expected number of errors. The computation of the expected error for the offenders convicted of auto theft will proceed in a manner analogous to the computation of error from the marginal distribution (column totals) alone. In placing the five envelopes into the fined category, the probability of being wrong is 35/40; hence we expect to make $35/40 \times 5 = 4.375$ errors here; in placing the 30 envelopes into the probation category, the probability of being wrong is 10/40; hence we expect to make $10/40 \times 30 = 7.500$ errors; in placing the 5 envelopes into the imprisoned category, the probability of being wrong is 35/40; hence we expect to make $35/40 \times 5 = 4.375$ errors. Among the forty cases in the first row, we therefore expect to make $4.375 + 7.500 + 4.375 = 16.250$ errors. Making analogous computations for the second and third rows and summing results gives the expected number of errors in predicting court disposition from knowledge of offense, E_2.

It may appear that in assigning cases at random so as to reconstruct the frequency distribution within each offense category we are not fully utilizing the knowledge of offense category. Actually, that is precisely what we are doing—utilizing that knowledge and none other. It would clearly be misleading to use any other information in this prediction process if our purpose is to find the relative reduction in prediction error made possible by the association between offense and disposition, which is precisely our purpose here. The outcome of our rather involved computations will be a measure of the relative reduction in prediction error made possible by the differential distribution within each of the rows. The errors of prediction will be reduced in shifting from the marginal predictions to the row-specific predictions to the degree that the row distributions differ from each other, and this reduction will be subject only to the qualification that the most frequent rows will weigh most heavily in the tabulation of expected errors.

Expressing in symbols our procedure for computing the expected number of errors in predicting the dependent variable from knowledge of the independent variable, we have the following:

$$E_2 = \sum_1^r \sum_1^c \frac{(N_{i\cdot} - N_{ij})N_{ij}}{N_{i\cdot}}, \tag{8.3.2}$$

where E_2 = the expected number of errors in predicting column category from knowledge of the row category;

$N_{i\cdot}$ = the number of cases in the ith row;

N_{ij} = the number of cases in ith row, jth column;

$\sum_1^r \sum_1^c$ is the instruction to get the sum of the r row sums, each composed of c terms;

r = number of rows; and

c = number of columns.

Following this formula, the completed calculation of E_2 for Table 8.3.1 is given below:

$$E_2 = \frac{35}{40} \cdot 5 + \frac{10}{40} \cdot 30 + \frac{35}{40} \cdot 5 + \frac{50}{50} \cdot 0 + \frac{20}{50} \cdot 30 + \frac{30}{50} \cdot 20$$

$$+ \frac{5}{10} \cdot 5 + \frac{10}{10} \cdot 0 + \frac{5}{10} \cdot 5$$

$$= 4.375 + 7.5 + 4.375 + 0 + 12 + 12 + 2.5 + 0 + 2.5$$

$$= 45.25.$$

Having found E_1 and E_2 for Table 8.3.1, we may now proceed to find τ_c (tau sub-c, that is, tau with the column variable predicted) for this table, which will give the relative reduction in prediction error for predicting court disposition from offense.

$$\tau_c = \frac{E_1 - E_2}{E_1}$$

$$= \frac{54 - 45.25}{54}$$

$$= .16\,. \tag{8.3.3}$$

We may interpret this result to mean that the association between offense and court disposition permits us to reduce prediction error by 16 percent in shifting from the prediction of court disposition from its own distribution to the prediction of court disposition from offense specific distributions.

Had our substantive interest in Table 8.3.1 required that we conceive of the row variable as the dependent variable, we would have computed τ_r to find the appropriate measure of the relative reduction in prediction error.

$$\tau_r = \frac{E_1 - E_2}{E_1}, \tag{8.3.4}$$

where E_1 and E_2 are defined and computed in the same manner as above except that rows replace columns in all instructions and vice versa. The student may wish to test his or her understanding of this transformation by verifying that, for Table 8.3.1, $\tau_r = .14$, with $E_1 = 58$ and $E_2 = 50$.

Computing Formula for Tau The computing procedure given above is often tedious and cumbersome, especially if the cross-classification table has many rows and many columns, and if the frequencies are large. For that reason we often have recourse to an algebraically equivalent computing formula which is easier to use. Substituting the definitions for E_1 and E_2 in the relative reduction in error formula for τ_c, simple algebra will show that:

$$\tau_c = \frac{\sum_1^r \sum_1^c \frac{N_{ij}^2}{N_{i\cdot}} - \frac{\sum_1^c N_{\cdot j}^2}{N}}{N - \frac{\sum_1^c N_{\cdot j}^2}{N}}. \tag{8.3.5}$$

This set of instructions indicates that a few basic operations, less complicated than those described above, are sufficient for the computation of τ_c. The first term in the numerator instructs us to square each cell entry, divide by its row total, and sum the resulting quotients over all cells. The second term in the numerator instructs us to square each column total, sum over all columns, and divide this sum by the total number of cases in the table. The difference between these two sums yields the numerator of τ_c. The denominator is obtained by subtracting the second term in the numerator from the total number of cases in the table. For Table 8.3.1:

$$\sum_1^r \sum_1^c \frac{N_{ij}^2}{N_{i\cdot}} = \frac{25}{40} + \frac{900}{40} + \frac{25}{40} + \frac{0}{50} + \frac{900}{50} + \frac{400}{50} + \frac{25}{10} + \frac{0}{10} + \frac{25}{10}$$

$$= .625 + 22.5 + .625 + 0 + 18 + 8 + 2.5 + 0 + 2.5$$

$$= 54.75.$$

$$\frac{\sum_1^c N_{\cdot j}^2}{N} = \frac{100 + 3600 + 900}{100}$$

$$= 46$$

$$\tau_c = \frac{54.75 - 46}{100 - 46}$$

$$= \frac{8.75}{54}$$

$$= .16.$$

This alternative formula, convenient for computation, has been so modified that the essential meaning of τ_c is no longer immediately evident. But τ_c, of course, has the same meaning no matter which of these formulas is used.

In the special case of a 2 × 2 table, $\tau_r = \tau_c$ and an alternative computing formula may be used:

$$\tau_r = \tau_c = \frac{(ad - bc)^2}{(a + b)(c + d)(a + c)(b + d)}$$

where the letters represent cell frequencies in the 2 × 2 table as follows:

a	b
c	d

The square root of this quantity was devised as a measure of association for 2 × 2 tables before the more general tau measures were invented, and the Greek letter ϕ (phi) is sometimes used as the name for this older measure whose square is identical to tau for a 2 × 2 table. In fact the measure ϕ is a special case of r, the product-moment correlation coefficient for bivariate data. The relation of ϕ to r will be noted again in the next chapter.

The Use and Interpretation of Tau The tau measures, τ_r and τ_c, indicate the relative reduction in prediction error made possible by the association between variables in predicting from columns to rows (τ_r) and in predicting from rows to columns (τ_c). Although it might at first seem unnecessary to have two measures of the association between a single pair of variables, the need for two measures arises because the predictability of one variable from another will sometimes differ depending on which is designated the independent (predictor) variable and which the dependent (predicted) variable. Each of the tau measures is sensitive to differences in the percentage distributions within the table, but the student will recall (Chapter 3, Section 2) that the percentages may be computed relative to row totals or relative to column totals. If the rule to percentage in the direction of the presumed cause has been followed, the appropriate tau (whether τ_r or τ_c) will then reflect the relative reduction in prediction error made possible by the association roughly indicated by the differences between the corresponding percentages

Table 8.3.2 *Hypothetical Cross-Classification of Offense and Court Disposition for a Sample of Convicted Offenders: Frequencies as a Percentage of Row Totals*

Offense	Disposition			Total	
	Fined	Probation	Imprisoned	%	*N*
Auto theft	12.5	75	12.5	100	40
Burglary	0	60	40	100	50
Forgery	50	0	50	100	10
$\tau_c = .16$					

thus computed. We illustrate by displaying the percentages of row totals and the percentages of column totals for Table 8.3.1 in Tables 8.3.2, and 8.3.3. The two sets of percentages, both accurate, give rise to superficially contradictory statements; for instance, the majority of those imprisoned were convicted of burglary but the majority of those convicted of burglary were not imprisoned. However, the more general point is that the two sets of percentage distributions are different ways of examining the same set of data. There is a corresponding utility in having two measures for summarizing the association when one or another variable is designated as the dependent or predicted variable. The two tau measures answer to this need.

Tables 8.3.2 and 8.3.3 should also make evident the difficulty of providing a single summary measure of the degree of association on the basis of a visual examination of the several percentages and their differences one from another, even in this relatively simple cross-classification table. It is clear from Table 8.3.2, for example, that persons convicted of auto theft, burglary, and forgery are quite differently distributed over the disposition categories, but is

Table 8.3.3 *Hypothetical Cross-Classification of Offense and Court Disposition for a Sample of Convicted Offenders: Frequencies as a Percentage of Column Totals*

Offense	Disposition		
	Fined	Probation	Imprisoned
Auto theft	50	50	16.67
Burglary	0	50	66.67
Forgery	50	0	16.67
Total %	100	100	100
N	10	60	30
$\tau_r = .10$			

the association between these two variables low, moderate, or high? Summary measures of association such as τ_r and τ_c provide a useful summary of the degree of association exhibited, while the comparison of specific sets of percentages may be useful in highlighting special points of interest in the table.

It has become common practice in social science research, especially when the data are summarized in the form of 2 × 2 tables, to use the percentage difference as a summary measure of association. When one of the variables has been designated as the dependent variable and the percentages have been computed in accord with the rule to percentage in the direction of the presumed cause, a single percentage difference is used to summarize the degree of association, whereas a more complicated series of comparisons of percentages would be necessary in tables larger than 2 × 2. While the familiarity of percentages and the intuitive cogency of the difference between percentages as a measure of association recommend this practice, the usual effect is to give the impression of a higher degree of association than other measures would indicate. This is illustrated in Table 8.3.4, which shows the association between education and voting in the South in 1960. The percentage voting among those who completed high school or more is 78 percent, while voters among those with less than a high school education constituted 56 percent. Hence, the percentage difference is 22 *percentage points*, which suggests a moderate degree of association. Cell frequencies as well as percentages have been included in Table 8.3.4 to facilitate the computation of τ_r, which is low (.05). In general, the difference between percentages in a 2 × 2 table gives an impression of stronger association than would be suggested by the appropriate tau measure.

In a 2 × 2 table, Goodman and Kruskal's tau may be obtained as the product of the percentage difference in columns and the percentage difference in rows. In symbols:

$$\tau = \left[\left(\frac{a}{a+c}\right) - \left(\frac{b}{b+d}\right)\right]\left[\left(\frac{a}{a+b}\right) - \left(\frac{c}{c+d}\right)\right].$$

For Table 8.3.4

$$\begin{aligned}\tau &= |.56 - .78| \cdot |.48 - .72| \\ &= .22 \cdot .24 \\ &= .0528.\end{aligned}$$

Thus τ for a 2 × 2 table is the product of both percentage differences; except in the special case of two diagonally opposite zero cells, each percentage difference is less than 1.0—typically much less. Hence, τ—and the relative reduction in prediction error—is ordinarily much smaller than a single percentage difference in a 2 × 2 table.

Table 8.3.4 *Voting by Education in the South, 1960*

	Education				
	Part High School or Less		High School Graduate or More		Total
	N	%	*N*	%	
Voted	124	56	136	78	260
Did not vote	98	44	38	22	136
Total	222	100%	174	100%	396
$\tau_r = .05$					

Source: Adapted from Anthony M. Orum, "A Reappraisal of the Social and Political Participation of Negroes," *American Journal of Sociology*, 72 (July 1966), p. 44.

It is not only permissible but also frequently useful to report *both* the percentage difference and the summary measure of association. Thus the association in Table 8.3.4 can be well summarized by stating that 78 percent of the high school graduates voted while only 56 percent of those with less than a high school education did so, indicating that this educational classification improves the prediction of voting by 5 percent, as measured by Goodman and Kruskal's tau.

Tau may range from 0 to 1.0, in accord with its status as a proportion. If marginal frequencies are regarded as fixed, the upper limit of tau may be considerably less than 1.0, however. Referring again to Table 8.3.4, it is clear that even if all persons in the higher education category had voted, there would still be 86 voters who must have completed less than high school. Hence, with the marginals fixed, the cell frequencies in this table cannot be set to yield two diagonally opposite zero cells, which is the required condition for τ to reach its maximum value of 1.0 in a 2×2 table. It has sometimes been suggested that when the maximum value of tau is less than 1.0 by reason of disproportionate marginal frequencies, as in this instance, the obtained tau should be expressed relative to the maximum possible for the given marginals. As a general rule, this practice seems unwise and serves only to give an unrealistic impression of the degree of association exhibited in the data. Although one might reason that, with the marginals fixed as they are, perfect association is not possible, one can also reason that if perfect association obtained, the marginals would not be fixed as they are.

QUESTIONS AND PROBLEMS

1. In Table 8.2.5, which is the independent variable and which is the dependent variable? Verify that tau is sensitive to association that lambda is not by computing the appropriate tau (τ_r or τ_c) for this table.

Table 8.3.5 *Soldier Morale, United States (1943) and Europe (1945), Percentage Distribution*

1945	1943		
	Good Spirits	Low Spirits	Total
Good Spirits	27	17	44
Low Spirits	9	47	56
Total	36	64	100

Source: Samuel A. Stouffer *et al., The American Soldier: Adjustment During Army Life*, Vol. I of *Studies in Social Psychology in World War II*. Princeton: Princeton University Press, 1949, p. 163. Copyright 1949 by Princeton University Press. Reprinted by permission of Princeton University Press.

2. While in training in the United States (1943) and later in Europe (1945), the same group of 100 soldiers were asked: "In general, how do you feel most of the time, in good spirits or in low spirits?" Their replies are shown in Table 8.3.5.
   (a) What percentage of the total group responded in the same way on both occasions?
   (b) Compute λ_r.
   (c) Compute τ_r.
   (d) Interpret these results: Compare the information which the respective answers yield.

3. The following question was put to 624 high school boys: "Which boy in the senior class seems to you the most poised in social situations?" The votes, classified by religious background of chooser and chosen, are shown in Table 8.3.6.

Table 8.3.6 *Sociometric Choices by Religious Background*

Chooser	Chosen		
	Jews	Non-Jews	Total
Jews	239	44	283
Non-Jews	77	264	341
Total	316	308	624

Source: Jackson Toby, "Universalistic and Particularistic Factors in Role Assignment," *American Sociological Review*, 18 (1953), p. 134. Reprinted by permission.

(a) Compute λ_c and τ_c.
(b) What proportion of the respective choosers chose their own social group?
(c) What proportion of the total chose their own group?
(d) What do these results indicate about religious ethnocentrism?

4. (a) In Table 8.3.7, one variable is a composite of race, class, and gang status. To determine the relative predictability of response to the interview question from this composite variable, compute τ_r.
(b) Table 8.3.7 may be rearranged in a number of ways to show different aspects of the data. Construct 2 × 2 tables showing the association between race and response, class and response, and gang status and response. Compute τ_r for each of these 2 × 2 tables to determine which of these variables, considered separately, is most closely associated with the dependent (row) variable.
(c) Consider Table 8.3.7 as two separate tables, one showing the association for blacks only and the other showing the association for whites only between the row variable and the three-category column variable. As measured by τ_r, for which subsample, blacks or whites, is the association greater?

Table 8.3.7 *Perception of College Aspirations by Race, Class, and Gang Status*

"In our area there are a lot of guys who want to go to college"	Black			White			Total
	Lower Class Gang	Lower Class Non-Gang	Non-Gang Middle Class	Lower Class Gang	Lower Class Non-Gang	Non-Gang Middle Class	
True	77	42	22	15	35	52	243
False	129	47	4	75	44	1	300
Total	206	89	26	90	79	53	543

Source: Adapted from James F. Short, Jr., Ramon Rivera, and Ray A. Tennyson, "Perceived Opportunities, Gang Membership and Delinquency," *American Sociological Review* 30 (February, 1965), p. 60, Table 1. Reprinted by permission.

4 THE PREDICTION OF ORDER: GAMMA

Gamma (γ) measures the relative reduction in errors of predicting order on the dependent variable, whereas tau measures the relative reduction in errors of predicting the correct category of the dependent variable. If high school students have been classified by their teachers into high, medium, and low levels of academic promise and classified by their peers into high, medium, and low levels of popularity, it is natural to ask whether there is a tendency for students to be ordered similarly on both variables. Although tau might be computed for such a cross-classification, it will not indicate the predictability of order and consequently may provide a misleading answer to our question; association as measured by tau does not necessarily imply the predictability of order on one variable from knowledge of order on the other. When the categories for both variables in a cross-classification are ordered from low to high, the variables are not fully quantified but neither are they simply nominal variables. Accordingly, we have recourse in this circumstance to a measure of association that does not require full quantification but takes advantage of the limited degree of quantification that ordered categories represent. Gamma is such a measure.

To understand the logic of gamma, it is necessary to think, not in terms of single cases, but in terms of pairs of cases and their order relative to each other on each of the two variables. Because this is not a familiar way of thinking about cases and variables, it will be helpful to begin with a concrete and deliberately simplified example. Suppose we have three students, each with a different level of academic promise as judged by their teachers and each with a different level of popularity as judged by their peers. We have provided names for the three students and shown their position relative to each other on each variable in Table 8.4.1. From the listing of pairs of

Table 8.4.1 *Academic Promise and Popularity of Three Students*

Popularity	Academic Promise		
	High	Medium	Low
High		John	
Medium	Peter		
Low			George

Pair	Order on Academic Promise	Order on Popularity	Summary of the Two Orders for Each Pair
John and Peter	P > J[a]	J > P	Differently-ordered
John and George	J > G	J > G	Same-ordered
Peter and George	P > G	P > G	Same-ordered

[a] Read "Peter greater than John." If the symbol were reversed (P < J), it would be read "Peter less than John."

students, it is evident that two of the three pairs fall in the same order relative to each other on both variables, while one of the three pairs is differently ordered on the two variables. If we are to predict order on popularity from knowledge of order on academic promise, predicting the same order on the dependent variable as on the independent variable will, in this instance, provide the best prediction rule. This prediction rule will yield a predicted order that is correct for two of the three pairs for which the prediction is made.

This simple example illustrates four fundamental features of gamma: (1) a prediction error is made when the predicted order is not the actual order; a prediction of order is either correct or not correct; (2) the prediction rule for predicting order on one variable from knowledge of order on the other is either to predict same order or different order, depending on whether same-ordered pairs or differently-ordered pairs predominate among the pairs of cases for which a prediction is made; (3) the number of prediction errors is simply the number of differently ordered pairs when the prediction rule is to predict same order, or alternatively, the number of same-ordered pairs when the prediction rule is to predict different order; and (4) the number of errors is the same whether the prediction is from knowledge of order on the row variable to a prediction of order on the column variable or the reverse, that is, from column order to row order.

We now ask about the prediction of order for each pair on one variable without knowledge of order for that pair on the other variable. Suppose we are to predict order on popularity for the pair Peter and John. We do not know the order of this pair on academic promise, nor do we have any other relevant information about the pair. Under these circumstances, our prediction may as well be made on the basis of chance. We can imagine repeating the prediction of order many times by drawing one of these two names from a hat at random and predicting that the name drawn is higher than the one not drawn. If we repeat this operation many times, we would expect to draw the name John one-half of the time, and on those occasions we would predict that John is higher than Peter in popularity—the correct prediction. But we would also expect to draw the name Peter one-half of the time, and on those occasions we would predict that Peter is higher than John in popularity—an incorrect prediction. Hence we expect to be wrong one-half of the time in predicting order on popularity without knowledge of order on academic promise, not only for the pair composed of Peter and John but also for all other pairs for which a prediction is made. Our expected number of errors in predicting order on the dependent variable without knowledge of order on the independent variable is therefore one-half of the number of pairs for which a prediction is made; in this instance, our expected number of prediction errors is $.5 \times 3 = 1.5$.

We are now ready to compute the relative reduction in prediction error. The expected number of errors in predicting order on the dependent variable without knowledge of order on the independent variable (E_1) is 1.5. Given knowledge of order on the independent variable, we predict the same order on the dependent variable (since same-ordered pairs outnumber differently-ordered pairs in this instance), and we make one error in so doing; this is E_2. Gamma (γ) is the relative reduction in error:

$$\gamma = \frac{E_1 - E_2}{E_1} = \frac{1.5 - 1.0}{1.5} = \frac{.5}{1.5} = .33 \, .$$

We would expect to make a 33 percent reduction in errors of predicting order as we shift from predicting order at random to predicting order on one variable from knowledge of order on the other. This gain in the predictability of order is made possible by the association of order on the two variables and thus it serves as a measure of that association. The relative reduction in prediction error will be the same whichever of the two variables is designated the independent or predictor variable. Although the distinction between independent and dependent variables is still appropriate in considering the predictability of order, the relative reduction in prediction error as measured by gamma is not affected by the designation of one or the other variable as dependent.

We now elaborate this oversimplified example slightly by including one additional student in the tabulation. The relative positions of each of the four students are displayed in Table 8.4.2. Since we distinguish only three levels of

Table 8.4.2 *Academic Promise and Popularity of Four Students*

Popularity	Academic Promise		
	High	Medium	Low
High		John	
Medium	Peter	Bob	
Low			George

Pair	Order on Academic Promise	Order on Popularity	Summary of the Two Orders for Each Pair
John and Peter	P > J	J > P	Differently-ordered
John and Bob	Unknown (tied)	J > B	Unknown
John and George	J > G	J > G	Same-ordered
Peter and Bob	P > B	Unknown (tied)	Unknown
Peter and George	P > G	P > G	Same-ordered
Bob and George	B > G	B > G	Same-ordered

each variable, adding a fourth case implies that this added case must be tied with at least one other case on each variable; when the number of cases exceeds the number of distinct levels (the common circumstance), such ties cannot be avoided. A pair of cases tied in the sense of falling in the same category does not necessarily imply that the two cases in that pair are identical on the variable in question; more commonly it means that the ordered categories are not sufficiently precise to distinguish between them. For this reason, in Table 8.4.2, we have listed the order of tied pairs as "unknown"—with a more precise set of ordered categories, the two cases might fall in different levels but we do not know the order in which they would then fall.

When the order of a pair of cases is unknown on one or both variables, we cannot determine whether a prediction of order for that pair is correct or incorrect, and such a pair should not be counted in tabulating errors of prediction. Hence all pairs that are tied on either variable are excluded in computing prediction errors. In Table 8.4.2, six pairs can be formed, but two must be excluded because of ties, and the computation proceeds on the basis of the four remaining pairs. In general, the computation of gamma is based on the number of pairs untied on either variable.

In any table, the expected number of errors in predicting order on one variable without knowledge of order on the other is one-half of the pairs for which a prediction is made. For Table 8.4.2, we make predictions only for the four pairs untied on either variable, that is, on the sum of the same-ordered pairs and the differently-ordered pairs. Hence, $E_1 = .5 \times 4 = 2$. In predicting order on one variable with knowledge of order on the other, the best prediction rule is to predict same order, since same-ordered pairs predominate in Table 8.4.2. Such a prediction will be correct for three pairs and incorrect for one pair. Hence, $E_2 = 1$. The relative reduction in prediction error (γ) is thus .5.

$$\frac{E_1 - E_2}{E_1} = \frac{2 - 1}{2} = \frac{1}{2} = .5.$$

From the foregoing it should be evident that we may compute gamma from knowledge of the number of same-ordered pairs and the number of differently-ordered pairs in a given table. One-half of the sum of same-ordered pairs and differently-ordered pairs will be E_1; the smaller of those two quantities will be E_2. Gamma can then be readily computed in the usual way for a measure of the relative reduction in prediction error. To illustrate the computation of the number of same-ordered pairs and the number of differently-ordered pairs without listing each pair by name—an exceedingly tedious operation with even one-half dozen cases—we expand our previous illustration to include a larger number of cases and drop the practice of providing a name for each. The resulting cross-classification is displayed in Table 8.4.3.

Table 8.4.3 Academic Promise and Popularity of 107 Students

Popularity	Academic Promise			Total
	High	Medium	Low	
High	6	2	0	8
Medium	15	22	6	43
Low	3	34	19	56
Total	24	58	25	107

$\gamma = .69.$

With 107 cases, there are

$$\frac{107 \times 106}{2} = 5{,}671$$

pairs of cases and it would be unreasonably tedious to list each such pair to determine whether the pair is tied on one or both of the variables, and, if not, whether the ordering of the pair is the same or different on the two variables. Such a laborious operation is not necessary. It will suffice for the computation of gamma if we can compute the number of same-ordered pairs and differently-ordered pairs without listing them.

To find the number of same-ordered pairs, we consider first the 6 persons who are high on both variables. If we paired any of these 6 with any of the 22 persons who are medium on both variables, we would have a same-ordered pair, that is, a high-high with a medium-medium. The same holds for pairings of any of the 6 high-highs with any of the 6 medium-lows, any of the 34 low-mediums, or any of the 19 low-lows; any of these pairs will yield a same-ordered pair. Thus, the 6 high-highs may be paired with any of the $22 + 6 + 34 + 19 = 81$ below and to the right of them in Table 8.4.3 to yield a same-ordered pair. On the other hand, if we paired any of the 6 high-highs with any of the other 2 cases in the same row, or with any of the other 18 cases in the same column, or with any of the other 5 in the same cell, we would obtain a pair tied on one or the other variable (or both). Hence, the pairs formed by combining any one of the 6 high-highs with any one of the 81 cases below and to the right constitute the total number of same-ordered pairs in which the 6 high-highs will appear. This yields $6 \times 81 = 486$ same-ordered pairs. We next consider the 2 cases high on popularity and medium on academic promise. They may be paired with any of the $6 + 19 = 25$ cases below and to the right to yield same-ordered pairs; these 2 cases are therefore included in $2 \times 25 = 50$ same-ordered pairs. When this process of multiplying each cell frequency by the sum of the cell frequencies below and to the right has been completed for the entire table, we will have computed the total number of same-ordered pairs. In an analogous fashion, we may compute the number of differently-ordered pairs by multiplying each cell frequency by the sum of the frequencies below and to the left. For example, the 2 cases high on popularity and medium on academic promise may be paired with any of the 18 cases below and to the left to yield 36 differently-ordered pairs. These computational instructions may be expressed in symbols as follows:

$$N_s = \sum_{i}\sum_{j} N_{ij} \sum_{k}\sum_{l} N_{kl} \qquad \text{for } k > i \text{ and } l > j \tag{8.4.1}$$

where N_s = the number of same-ordered pairs;
N_{ij} = the frequency of the cell at the intersection of the ith row and the jth column;
$\sum\sum_{ij}$ means sum over all rows and columns, that is, over all cells in the table;
N_{kl} = the frequency in cells for which $k > i$ and $l > j$; and
$\sum\sum_{kl}$ means sum over all cells for which the row is higher than row i and the column is higher than column j.

$$N_d = \sum_{i}\sum_{j} N_{ij} \sum_{k}\sum_{m} N_{km} \qquad \text{for } k > i \text{ and } m < j \tag{8.4.2}$$

where N_d = the number of differently-ordered pairs;
N_{ij} = the frequency of the cell at the intersection of the ith row and the jth column;
$\sum\sum_{ij}$ means sum over all rows and columns, that is, over all cells in the table;
N_{km} = the frequency in cells for which $k > i$ and $m < j$; and
$\sum\sum_{km}$ means sum over all cells for which the row is higher than row i and the column is lower than column j.

Although this operation may seem formidable as it is expressed symbolically, it can be expressed rather simply for Table 8.4.3: For N_s multiply each cell frequency by the sum of the cell frequencies below and to the right, and then sum all of these products. For N_d, multiply each cell frequency by the sum of the cell frequencies below and to the left, and then sum all of these products. For Table 8.4.3, the complete calculations are as follows:

$$\begin{aligned} N_s &= 6(22 + 6 + 34 + 19) + 2(6 + 19) + 15(34 + 19) + 22(19) \\ &= 486 + 50 + 795 + 418 \\ &= 1{,}749; \end{aligned}$$

$$\begin{aligned} N_d &= 2(15 + 3) + 6(34 + 3) + 22(3) \\ &= 36 + 222 + 66 \\ &= 324. \end{aligned}$$

The student should note that if the table had been arranged so that the cell in the upper left corner were a combination of high-low, instead of a combination of high-high or low-low, the verbal instructions would need to be reversed, that is, for N_s, multiply each cell frequency by the sum of the cell frequencies below and to the left and sum all of these products. And for this alternative arrangement, to compute N_d multiply each cell frequency by the sum of the cell frequencies below and to the right and sum these products.

Having obtained N_s and N_d we may proceed to compute gamma. As indicated above, the number of errors in predicting pair order on one variable from knowledge of the order for that pair on the other variable will be the smaller of the two values N_s or N_d. We may represent this quantity by the symbol

$$min\ (N_s, N_d)\ .$$

The general formula for a relative reduction in error measure may be specialized to a formula for gamma:

$$\gamma = \frac{E_1 - E_2}{E_1}$$

$$= \frac{.5(N_s + N_r) - min\ (N_s, N_r)}{.5(N_s + N_r)}\ . \qquad (8.4.3)$$

For Table 8.4.3, we therefore compute gamma as follows:

$$\gamma = \frac{.5(1749 + 324) - 324}{.5(1749 + 324)}$$

$$= \frac{1036.5 - 324}{1036.5} = \frac{712.5}{1036.5} = .69\ .$$

This result indicates that errors in the prediction of order in pairs can be reduced by 69 percent because of the association between academic promise and popularity.

A somewhat simpler computing formula may be obtained from Equation 8.4.3 by simple algebra. Multiplying both the numerator and denominator by 2, we obtain

$$\gamma = \frac{N_s + N_r - 2\ min\ (N_s, N_r)}{N_s + N_r}\ . \qquad (8.4.4)$$

If N_s is larger than N_r (positive association of order), Formula 8.4.4 becomes

$$\gamma = \frac{N_s - N_r}{N_s + N_r}\ . \qquad (8.4.5)$$

If N_s is smaller than N_r (negative association of order) Formula 8.4.4 becomes

$$\gamma = \frac{N_r - N_s}{N_s + N_r} \, . \tag{8.4.6}$$

In general we may compute

$$\gamma = \frac{N_s - N_r}{N_s + N_r} \, , \tag{8.4.7}$$

and the resulting quantity will be positive if the association is positive (if $N_s > N_r$) and negative if the association is negative (if $N_s < N_r$). The numerical value of the resulting quantity, disregarding sign, represents the relative reduction in errors of predicting order for pairs.

Formulas 8.4.7 and 8.4.3 are equivalent, but 8.4.7 more readily lends itself to application.

Using this alternative and simpler formula, we again compute gamma for Table 8.4.3:

$$\gamma = \frac{1749 - 324}{1749 + 324}$$

$$= \frac{1425}{2073} = .69 \, .$$

Note that gamma does not indicate the proportion of pairs for which a correct prediction of order on one variable would be made from knowledge of order on the other variable. In Table 8.4.3, we predict same order on each trial. Since 1,749 of the 2,073 untied pairs are same-ordered, our proportion of correct predictions is 1749/2073 = .84. We could obtain the same result, except for rounding error, by the formula:

$$\text{Proportion of correct predictions of order for untied pairs} = 1/2 + |\gamma|/2 \, .$$

Marginal Distributions and Gamma Although marginal distributions do not affect the magnitude of γ and do not change its meaning as a measure of relative reduction in error, marginal distributions do affect the proportion of pairs involving ties. Marginal distributions with heavy concentrations in a few categories yield many tied pairs. And gamma is based on the predictability of order for untied pairs only. Hence, as the proportion of tied pairs increases,

Table 8.4.4 *The Effect of Marginal Distributions on the Maximum Number of Untied Pairs*

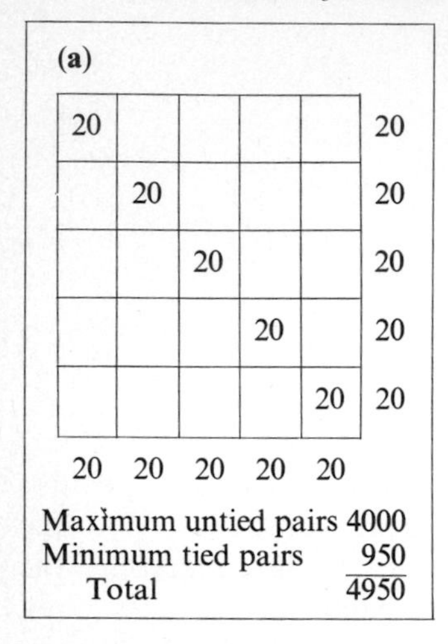

(a)

20					20
	20				20
		20			20
			20		20
				20	20
20	20	20	20	20	

Maximum untied pairs	4000
Minimum tied pairs	950
Total	4950

(b)

15					15
5	10				15
	10	20	10		40
			10	5	15
				15	15
20	20	20	20	20	

Maximum untied pairs	3400
Minimum tied pairs	1550
Total	4950

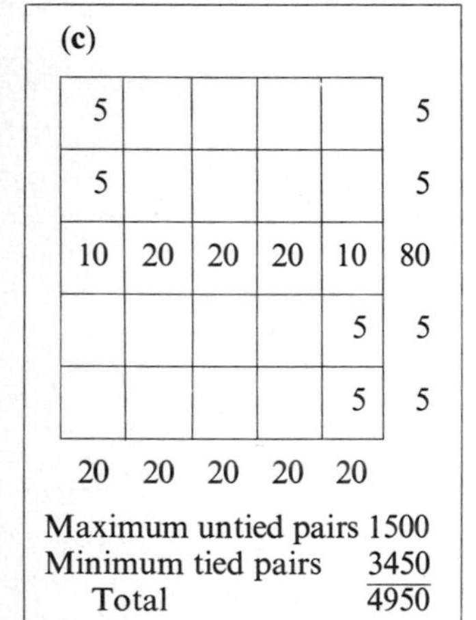

(c)

5					5
5					5
10	20	20	20	10	80
				5	5
				5	5
20	20	20	20	20	

Maximum untied pairs	1500
Minimum tied pairs	3450
Total	4950

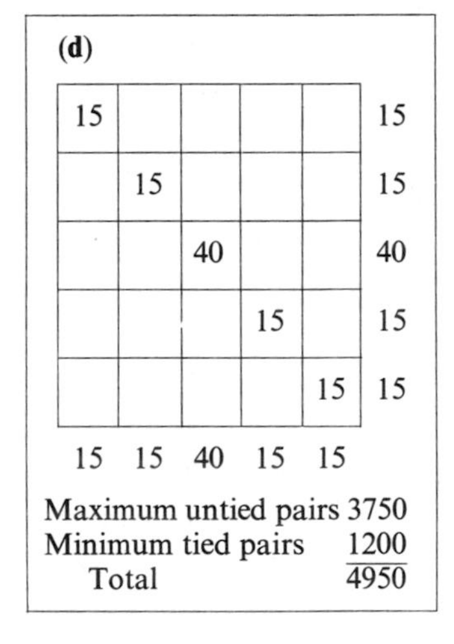

(d)

15					15
	15				15
		40			40
			15		15
				15	15
15	15	40	15	15	

Maximum untied pairs	3750
Minimum tied pairs	1200
Total	4950

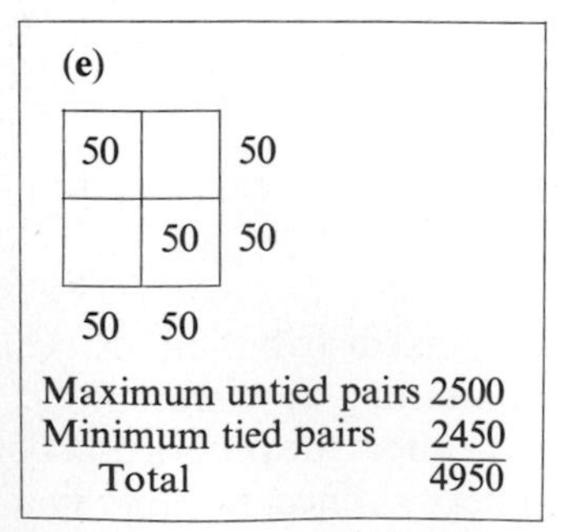

(e)

50		50
	50	50
50	50	

Maximum untied pairs	2500
Minimum tied pairs	2450
Total	4950

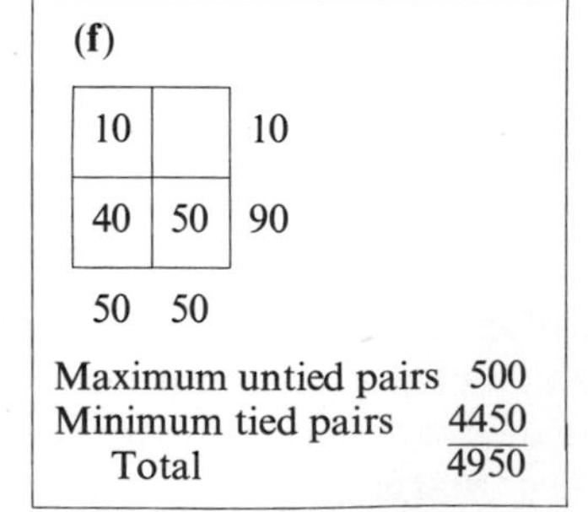

(f)

10		10
40	50	90
50	50	

Maximum untied pairs	500
Minimum tied pairs	4450
Total	4950

the proportion of pairs on which gamma is based decreases. We illustrate this point in the series of marginal frequencies in Table 8.4.4.

Given marginal distributions only, it is possible to arrange the cell frequencies so as to yield the maximum possible number of untied pairs and thereby explore the effect of different marginal distributions on the proportion of all pairs that can enter into the computation of gamma. Untied pairs will be at a maximum if the frequencies in the cells of the positive (or negative) diagonal are the maximum frequencies permitted by the marginals, and if the cases not placeable in the diagonal cells are entered as close to the diagonal cells as the marginals permit, proceeding in a "stairstep" pattern. For selected marginals, the cell frequencies that yield the maximum number of untied pairs have been entered in Table 8.4.4. This table then shows the computed number of untied pairs, along with the minimum number of tied pairs (total minus maximum untied pairs). The major effects of marginal distributions, as illustrated in this table, are:

(1) An increase in the marginal concentration for only one variable results in a decrease in the maximum possible number of untied pairs; compare a, b, and c in Table 8.4.4.

(2) A uniform marginal distribution coupled with a concentrated marginal results in a lower maximum number of untied pairs than concentrated marginals for both variables; compare (b) and (d) in Table 8.4.4.

(3) A decrease in the number of categories for both variables results in a decrease in the maximum number of untied pairs, other things being equal; compare (a) and (e) in Table 8.4.4.

(4) In a 2 × 2 table, one heavily skewed marginal and one uniform marginal reduces the proportion of untied pairs to a small fraction of all pairs; see (f) in Table 8.4.4.

To maximize the potential number of untied pairs and thereby maximize the number of pairs on which gamma is based, it is desirable to have as many levels of both variables as possible, with nearly uniform marginal distributions. However, the sociologist is sometimes constrained by the necessity to utilize data already classified into crude ordinal classes and by the inability of informants and respondents to make anything other than rough ordinal distinctions. In these instances, the use of γ may be inadvisable, since it will necessarily be based on a relatively small proportion of all possible pairs of cases.

The Use and Interpretation of Gamma The numerical value of γ, disregarding sign, gives the percentage of guessing errors eliminated by using knowledge of a second variable to predict order. The sign of gamma indicates which of two possible predictions of order is more accurate: a positive sign indicates that a prediction of *same* order on the predicted variable as on the predictor variable is more accurate, while a negative sign indicates that a prediction of *reverse* order is more accurate. Thus, the numerical value of

Table 8.4.5 *Selected Arrangements of Zero and Nonzero Frequencies that Yield a Gamma of 1.0*

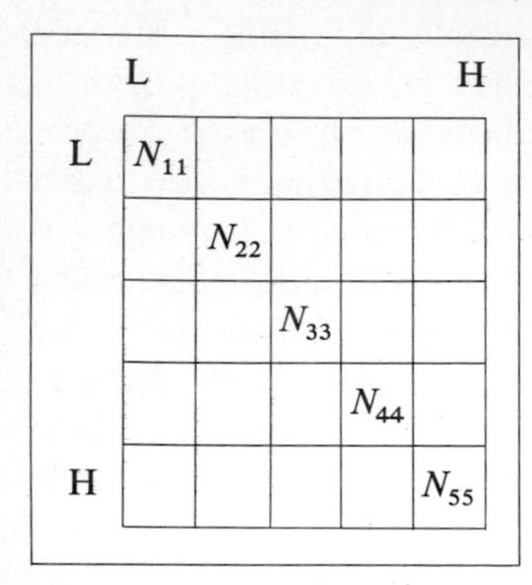

	L				H
L	N_{11}				
		N_{22}			
			N_{33}		
				N_{44}	
H					N_{55}

	L				H
L	N_{11}	N_{12}	N_{13}	N_{14}	N_{15}
					N_{25}
					N_{35}
					N_{45}
H					N_{55}

	L				H
L	N_{11}				
	N_{21}				
	N_{31}				
	N_{41}				
H	N_{51}	N_{52}	N_{53}	N_{54}	N_{55}

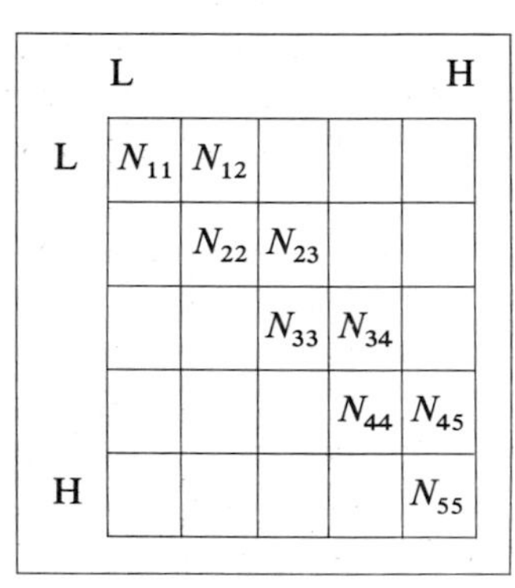

	L				H
L	N_{11}	N_{12}			
		N_{22}	N_{23}		
			N_{33}	N_{34}	
				N_{44}	N_{45}
H					N_{55}

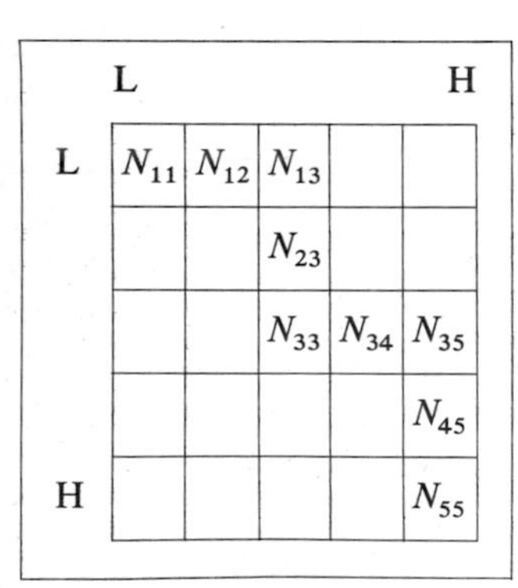

	L				H
L	N_{11}	N_{12}	N_{13}		
			N_{23}		
			N_{33}	N_{34}	N_{35}
					N_{45}
H					N_{55}

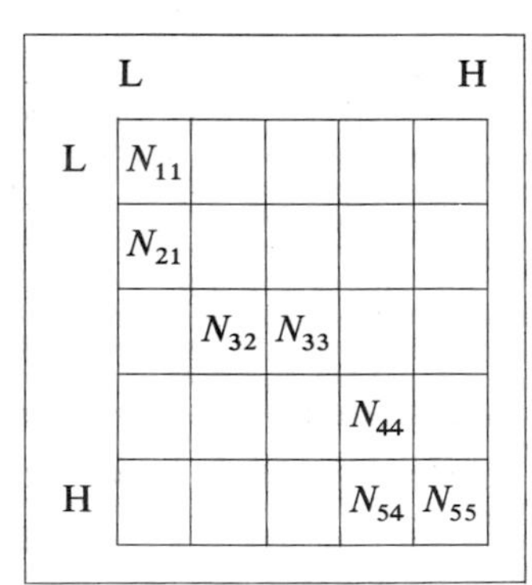

	L				H
L	N_{11}				
	N_{21}				
		N_{32}	N_{33}		
				N_{44}	
H				N_{54}	N_{55}

gamma represents the degree of association, while the sign represents the association as predominantly negative or positive. A positive sign indicates that the variables increase together, whereas a negative sign indicates that, as one variable increases, the other decreases.

Gamma may vary from -1.0 to $+1.0$. A gamma of -1.0 indicates that, for untied pairs, the order on one variable is always the reverse of the order on the other variable. Given a gamma of -1.0, we get perfect predictability of order for all untied pairs by following the rule: predict reverse order. When $\gamma = +1.0$, order is the same on both variables for all untied pairs. Given a gamma of $+1.0$, we get perfect predictability of order for all pairs involving no ties by following the rule: predict same order. A gamma of zero indicates that among all untied pairs there are exactly as many pairs with reversed orders on the two variables as there are pairs with the same order. Values intermediate between 0 and $|1.0|$ indicate the degree to which guessing errors may be reduced by utilizing knowledge of order on a second variable.

More than one arrangement of zero (or near zero) and nonzero frequencies in a joint distribution will yield a γ of 1.0 (or approximately 1.0). This is illustrated in Table 8.4.5.

The student may wish to compute gamma for each cross-tabulation in Table 8.4.5 and verify that γ is 1.0 for all of them. Corresponding arrangements yielding a γ of -1.0 may be obtained by reversing the order of the columns (or rows). These illustrations should make clear that although a given concentration of cases can yield only one value for γ, the same γ may have been produced by many different arrangements. This may be regarded as a limitation.

Yule's Q as a Special Case of Gamma Yule's Q (after Quételet) is a very simple measure of association for a 2×2 table which has enjoyed wide currency in sociological research. To obtain Q, we divide the difference between the diagonal cross-products by their sum. In notation:

$$Q = \frac{ad - bc}{ad + bc},$$

where the letters represent cell frequencies in the 2×2 table as follows:

a	b
c	d

It may be readily demonstrated that Q is a special case of γ. If we compute N_s for a 2×2 table, it contains but a single term, ad. The product ad gives the number of pairs that have the same order on both variables. The diagonal

Table 8.4.6 *Stability of Q, When One or Two Cells Is Empty*

Delinquency Involvement	a			b		
	Academic Achievement			Academic Achievement		
	High	Low	Total	High	Low	Total
High	0	15	15	0	50	50
Low	100	35	135	100	0	100
Total	100	50	150	100	50	150
	$Q = -1.0$			$Q = -1.0$		

cross-product bc yields N_r, the number of pairs having reverse orders on the two variables. Thus, for a 2 × 2 table,

$$\gamma = \frac{N_s - N_r}{N_s + N_r}$$

$$= \frac{ad - bc}{ad + bc}.$$

In tables of all sizes, including a 2 × 2 table, gamma is computed on the basis of all pairs not involving ties on either variable. In a 2 × 2 table, even under the most favorable circumstances—equal frequencies in all marginals—the maximum number of pairs not involving ties on either variable will be approximately one-half of all possible pairs.

Whenever the frequency of a single cell in a 2 × 2 table is zero, Q reaches its maximum value of 1.0. It also reaches the maximum whenever two diagonally opposite cells have zero frequency (Table 8.4.6). Although it might appear intuitively that association is stronger if two of the four cells are empty than if only one cell has zero frequency, this difference is not reflected in the gammas for the two tables. This is an instance of the foregoing conclusion that the same value of γ may be produced by different frequency patterns.

Even if only one cell is empty, as in *a* of Table 8.4.6, all pairs of cases not involving ties are ordered in the reverse direction on the two variables, just as in *b* of Table 8.4.6. Hence for both tables, Q is -1.0, indicating perfect predictability of order for untied pairs. The tables differ in that there are only 1500 such pairs in the table with one empty cell, whereas there are 5000 such pairs in the table with two diagonally opposite zero cells. Prediction for untied pairs is perfect in both tables, but the prediction applies to more pairs and hence is more "complete" in the *b* table than in the *a* table. Gamma (Q) is not designed to be sensitive to differences in the completeness of association in this sense and, consequently, does not distinguish these two types of distribution within a 2 × 2 table.

QUESTIONS AND PROBLEMS

1. Define the following concepts:
   same-ordered pair
   reverse-ordered pair
   tied pair

2. Compute γ for Table 8.4.7. On how many pairs of cases is this gamma based? What proportion of all possible pairs does this constitute? If racial attitudes and reference group support were each represented by five ordered categories instead of two, would you expect this proportion to increase or decrease?

Table 8.4.7 *Reference Group Support and Racial Attitudes*

Reference Group Support	Racial Attitudes		
	Favorable	Unfavorable	Total
Strong	64	30	94
Weak	32	63	95
Total	96	93	189

Source: James M. Fendrich, "Perceived Reference Group Support: Racial Attitudes and Overt Behavior," *American Sociological Review*, 32 (December 1967), pp. 960–970. Reprinted by permission.

3. Compute γ for Table 8.4.8 as given. Combine Columns 1 and 2, and combine Columns 3 and 4. Compute γ for the 3×2 table and compare with γ for the original table.

Table 8.4.8 *Peer Ratings by Social Class, High School Students*

Peer Rating	Social Class				
	I and II	III	IV	V	Total
Elite	27	30	9	0	66
Good Kids	8	114	133	4	259
Grubby Gang	0	2	41	22	65
Total	35	146	183	26	390

Source: Reprinted with permission from A. B. Hollingshead, *Elmtown's Youth*, p. 222. Copyright 1949. John Wiley & Sons, Inc.

Table 8.4.9 *Marital Adjustment of Husbands by Degree of Attachment to Father*

Degree of Attachment	Marital Adjustment			
	Poor	Fair	Good	Total
Little or none	32	28	15	75
Moderate	41	47	69	157
A good deal	26	41	61	128
Very close	28	22	59	109
Total	127	138	204	469

Source: Ernest W. Burgess and Leonard S. Cottrell, Jr., *Predicting Success and Failure in Marriage*. Englewood Cliffs, N.J.: Prentice-Hall, Inc., 1939, p. 377. Reprinted by permission.

4. Compute γ for Table 8.4.9 and interpret.
5. Compute Q, τ_c, and λ_c for the following 2×2 table. Compare results and discuss the differences.

8	2	10
2	8	10
10	10	20

5 ELABORATION IN THE ANALYSIS OF ASSOCIATION

The association between variables in social science research is of interest primarily because such association provides clues about the way social processes operate, how the characteristics and behavior of individuals are affected by their social situations and experiences, and how social forces lead to structural change. In brief, the social scientist probes for clues as to how things "work" and how social events are connected with each other. Measuring the association between variables is a way of linking the search for such clues to relevant data, and a way of checking against relevant data the plausibility of insights based on other kinds of observations. The search for clues about how things "work" does not end with measuring the association between variables, two at a time. The search is often continued in the exploration of association among three or more variables and the interpretation of the resulting patterns in the light of substantive reasoning.

In sociological writing, *elaboration* has come to mean the analysis of the relationship between two categorized variables within subdivisions of a third (and possibly a fourth or fifth) variable. The discussion of elaboration in this sense has been concentrated on the exploration of patterns of relationship among three dichotomous variables, but the basic concept of elaboration is not limited to dichotomous variables, nor to the consideration of only three variables at a time. Nonetheless, the basic features of the process of elaboration can be illustrated most readily with three dichotomous variables, and we therefore focus our discussion on such an illustration.

In Table 8.5.1 the hypothetical data show a low or modest association between income and attitude toward busing to achieve school integration. The relative reduction in error as measured by tau (.05) is low but the relative reduction of error as measured by gamma (−.41) is relatively high. The common practice of judging the degree of association by examining the percentage difference would also lead to the conclusion that the association in this table is moderately high; the percentage difference is 21 percentage points, with 58 percent of low income respondents indicating favorable attitudes toward busing as compared to 37 percent of high income respondents.

In Table 8.5.2, this cross-classification of income and attitude is presented in a modified form. Entries in each cell are no longer a simple count of the number of respondents in that cell; rather each respondent is represented as being either black or white. The original cell frequencies are broken down into subfrequencies. This alternative mode of presentation—introducing the racial category of the respondents as a third variable—permits one to perceive what was concealed in the cross-classification limited to two variables. Examination of Table 8.5.2 reveals the following points: (1) whites are more likely than blacks to be in the high income category (examine column

Table 8.5.1 *Association between Income and Attitude toward Busing to Achieve School Integration (Hypothetical Data), "Original" Table*

Attitude toward Busing	Income		
	High	Low	Total
For	10	14	24
Against	17	10	27
Total	27	24	51

Table 8.5.2 *Table 8.5.1 in Modified Form: Tally Distinguishing Black Respondents and White Respondents*

Attitude toward Busing	Income		
	High	Low	Total
For:	10	14	24
Black	𝍸 I	𝍸 𝍸 II	𝍸 𝍸 𝍸 III
White	IIII	II	𝍸 I
Against:	17	10	27
Black	I	II	III
White	𝍸 𝍸 𝍸 I	𝍸 III	𝍸 𝍸 𝍸 𝍸 IIII
Total	27	24	51
Black	𝍸 II	𝍸 𝍸 IIII	𝍸 𝍸 𝍸 𝍸 I
White	𝍸 𝍸 𝍸 𝍸	𝍸 𝍸	𝍸 𝍸 𝍸 𝍸 𝍸 𝍸

marginals); (2) blacks are more likely than whites to be in the favorable attitude category (examine the row marginals); (3) there is a heavy concentration of high income whites in the unfavorable attitude category and a heavy concentration of low income blacks in the favorable attitude category (examine the internal cells of the table). This examination suggests the possibility that income and attitude are associated because of the association of both variables with race. Since blacks are lower in income than whites and since blacks have more favorable attitudes towards busing than whites, the association between income and attitude may occur largely, or even entirely, because of the effect of race on both. To explore this possibility more explicitly, we turn to Table 8.5.3, in which the original table has been partitioned into two separate tables, one for blacks and one for whites.

Table 8.5.3 *Partial Tables: Income Cross-Classified with Attitude, for Whites and Blacks Separately*

Whites			
Attitude	Income		
	High	Low	Total
For	4	2	6
Against	16	8	24
Total	20	10	30

Blacks			
Attitude	Income		
	High	Low	Total
For	6	12	18
Against	1	2	3
Total	7	14	21

These are called *partial tables*, and the association within each is known as a *partial association* or *partial relation*, because each table shows the association for a part of the total, or, alternatively stated, the association with one variable held constant. The partial tables, of course, are simply the representation in frequency table form of the "White" and "Black" tallies in Table 8.5.2; when both sets of tallies are combined, the original frequency table is the result.

An examination of both partial tables quickly reveals that there is no association between income and attitude in either one; the original association has completely disappeared when race is held constant. The subdivision of cases into categories of a third variable imposes a *control* for that third variable, that is, in each partial table, all respondents are alike with respect to that third variable. Thus we conclude (from the hypothetical data in Table 8.5.3) that the association between income and attitude toward busing disappears when race is controlled.

The control for a third variable does not necessarily, as in our example, result in the disappearance of the association exhibited in the original table; several outcomes are possible. The association being analyzed may remain unchanged in the partials; it may be reduced in the partials but not vanish; it may be reduced in one of the partials and increased in the other; or it may be increased in both partials. The original frequency table alone provides no clues as to what the outcome will be in the partials, but knowledge of the *marginal associations* or *marginal relations* will provide clues even though the outcome in the partials is not completely determined by these marginal relations. The *marginal tables* are so named because the frequencies that constitute them appear as marginal frequencies in the partial tables. We should emphasize, however, that the marginal tables, like the "original" table, include the total set of cases. Hence the marginal relations are "total" relations and the designation "marginal" serves simply to differentiate these "total" relations from the "total" relation being analyzed. Marginal tables (for race and income and for race and attitude) are shown in Table 8.5.4. It will be evident by inspection that the cell frequencies in the cross-classification

Table 8.5.4 *Marginal Tables: Income Cross-Classified with Race, All Respondents, and Attitude Cross-Classified with Race, All Respondents*

Race	Income		
	High	Low	Total
White	20	10	30
Black	7	14	21
Total	27	24	51

Race	Attitude Toward Busing		
	For	Against	Total
White	6	24	30
Black	18	3	21
Total	24	27	51

of race and income appear as column marginals in the partial tables, and that the cell frequencies in the cross-classification of race and attitude appear as row marginals in the partial tables. Knowledge of the marginal associations permits the investigator to anticipate what may and may not happen in the partials. For example, if the association between the control variable and either of the variables in the original table is zero, the original relationship cannot disappear in *both* of the partial tables (although it might disappear in one of the partials). The elaboration of the analysis of association in ways that are substantively useful is, to a large extent, a matter of identifying the control variables that will modify the original relationship, and this task is frequently facilitated by knowledge of the marginal relationships.

The outcome of controlling for a third variable is always empirically problematic, even though the marginal relations are known. Nonetheless, substantive reasoning may suggest an expected result. For example, in examining the association between religion (Protestant and Catholic) and the use of contraceptives, we not only anticipate that contraceptives are more widely used among Protestants than among Catholics but we also anticipate that this association should be stronger for individuals who are actively involved in their respective churches, and weaker among those who are less active. Should the relevant partials fail to show that the association is higher among actives than among inactives, our reasoning is called into question. We might then suspect that the original association between religion and contraceptive use does not result from an effect of religious beliefs but rather that the association arises as a result of some other third factor associated with both religion and contraceptive use, for example, social class. By this process of putting forth a tentative hypothesis, exploring that hypothesis by elaborating an original association, interpreting the outcome of that elaboration, and possibly elaborating again with an alternative variable controlled, the social scientist may discard assumptions about how things "work" that are not substantiated in the data and arrive at other assumptions that are in accord with the results of elaboration. Social analysts sometimes find themselves in a blind alley attempting to understand how a specific association between variables comes about, and it would be unrealistic to expect every hunch to be substantiated by analysis. But the process of elaboration may also provide illumination regarding what would otherwise remain puzzling and obscure.

The statistical outcomes in the partial tables do not, by themselves, indicate the appropriate interpretation; those findings must be supplemented by substantive reasoning to yield conclusions about how things "work." For example, if the association disappears in the partials, the implication of this outcome depends on assumptions about the causal ordering of the three variables. Two possible orderings are displayed in Figure 8.5.1.

In Figure 8.5.1a, the variable controlled is causally antecedent to both variables in the original table. Given this assumption about causal order and the disappearance of the original relationship in the partials the investigator may conclude that the original association is *explained* by the control

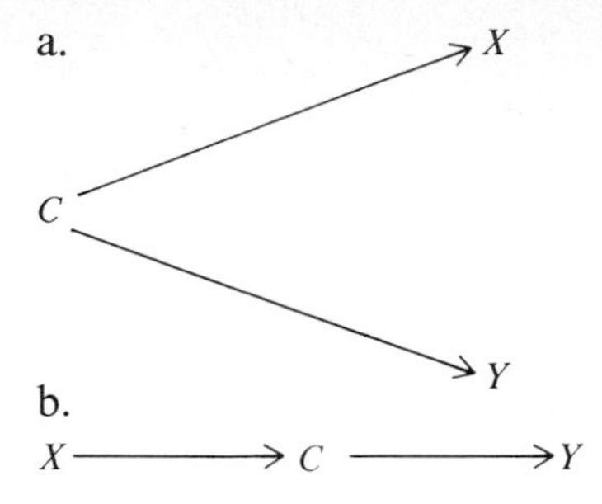

Figure 8.5.1 *Alternative assumptions about the causal ordering of the independent variable* (X), *the dependent variable* (Y), *and the control variable* (C)

a. XY *association* explained *by* C *or spurious because of* C.

b. XY *association* interpreted *by* C *or the effect mediated by* C.

variable or, alternatively, that the original association is *spurious* because of the control variable. In Figure 8.5.1b, the variable controlled is causally antecedent to the dependent variable in the original table but not to the independent variable. Given this assumption about causal order and the disappearance of the original association in the partials, the investigator may conclude that the original association is *interpreted* by the control variable or, alternatively, that the effect of the independent variable on the dependent variable is *mediated* by the control variable. The implications of the two conclusions are quite different. If we conclude that the original association is spurious, we thereby deny the claim of a cause-and-effect relationship between the variables in the original table. Concluding that the control variable mediates the effect entails no such denial. Yet although the substantive meaning is crucially different, the statistical outcomes are identical and it is only the investigator's assumptions about causal order that distinguish one from the other.

When the association in the partials is not uniform—for example, when the association in one (but not the other) is zero, when the association is positive in one but negative in the other, when the association is high in one partial but low in the other, etc.—the original association is said to have been *specified* by the control variable. Otherwise stated, the investigator may use the categories of the control variable to specify the circumstances under which the original association will be high or low, positive or negative. The causal ordering of the control variable with reference to the two variables in the original table is then of no great consequence, and substantive interest in

Table 8.5.5 *Association between Income and Attitude toward Busing to Achieve School Integration (Hypothetical Data)*

Attitude toward Busing	Income		
	High	Low	Total
Favorable	100	140	240
Unfavorable	170	100	270
Total	270	240	510

nonuniform partial associations focuses on the reasons for the difference in association under the two (or more) conditions of the control variable. If control for race revealed that low-income blacks were more likely to be positive toward busing than high-income blacks, while low-income whites were more likely to be negative than high-income whites (as illustrated in Tables 8.5.5 and 8.5.6), the investigator might then search for additional variables that would explain or interpret the association among whites but would be irrelevant among blacks. For example, the investigator might control (among white respondents) for central-city and suburban residence, to see if this variable would *interpret* the relationship for whites. Since the original table and the marginal tables are identical in Table 8.5.5 to those in Tables 8.5.1 and 8.5.4 (except that each frequency has been multiplied by 10) we note in passing that this table also serves to illustrate the possibility of different outcomes in the partials even when the marginals are fixed.

The elaboration of the association between variables offers numerous possibilities for the social analyst to gain insights into the "hows" and "whys" of the interconnections between social events. It is important for the student to recognize, however, that substantive reasoning always enters into the selection of variables to be controlled and into the conclusions that are

Table 8.5.6 *Partial Tables: Income Cross-Classified with Attitude, for Whites and Blacks Separately*

Whites			
Attitude	Income		
	High	Low	Total
Favorable	50	10	60
Unfavorable	150	90	240
Total	200	100	300

Blacks			
Attitude	Income		
	High	Low	Total
Favorable	50	130	180
Unfavorable	20	10	30
Total	70	140	210

drawn from the results of elaboration. Without such substantive reasoning, the results of elaboration would lack meaning and importance in social science research. We should also note that the subject of elaboration has been very briefly treated here and our attention has been limited to the simplest features of elaboration. The student interested in pursuing this topic in greater detail should consult Lazarsfeld (1955) and Rosenberg (1968).

QUESTIONS AND PROBLEMS

1. Define the following concepts:

elaboration	spurious relation
partial table	interpreted relation
partial relation	mediating variable
marginal relation	specified relation
explained relation	

2. If the association between X and Y vanishes when Z is controlled, should one therefore conclude that the association between X and Y is spurious? Why or why not?

3. (a) Using the data of Table 8.3.7, construct a 2 × 2 table showing the association between race and perception of college aspirations for lower-class boys only. What is the difference between the percentages answering "True" for blacks and whites?

   (b) Now construct the partial tables, controlling for gang status. Compute the percentage answering "True" for black lower-class gang boys and for white lower-class gang boys and compare. Compute the percentage answering "True" for black lower-class nongang boys and for white lower-class nongang boys and compare. Compare black-white differences for gang boys and nongang boys. Interpret your results.

4. For a sample of 60 social workers the relationship between worker orientation and worker attitude was as shown in Table 8.5.7. When

Table 8.5.7 *Worker Attitude by Worker Orientation, 60 Social Workers*

Worker Orientation	Worker Attitude	
	+	−
+	16	10
−	13	21

Source: Adapted from Peter Blau, "Structural Effects," *American Sociological Review*, 25 (1960), 178–193. Reprinted by permission.

Table 8.5.8 *Worker Orientation by Worker Attitude by Work Group Orientation, 60 Social Workers*

Worker Orientation	Work Group Orientation			
	+ Worker Attitude		− Worker Attitude	
	+	−	+	−
+	12	4	4	6
−	8	5	5	16

workers were divided into two subgroups according to the orientation of their work as + or −, the relationships were as shown in Table 8.5.8. Compare the partial relations with the original. Interpret your results.

5. (a) In a national sample of American college students, an association was found between academic cheating and perceived peer disapproval of cheating (Table 8.5.9). However, there was also an association between cheating and personal disapproval of cheating as well as an association between personal disapproval of cheating and perceived peer disapproval. Therefore, it is possible that the association between cheating and perceived peer disapproval is "spurious," that association is "explained" by the effect of personal disapproval on the selection of peers with similar attitudes and by the effect of personal disapproval on the tendency to cheat. Fill in the missing portions of the partial tables (Tables 8.5.10 and 8.5.11) to check this possibility. Does the original relationship disappear

Table 8.5.9 *Self-Report of Cheating by Perceived Peer Disapproval of Cheating among College Students at 99 American Colleges and Universities, 1963*

Perceived Peer Disapproval of Cheating	Self-Report of Cheating					
	Yes		No		Total	
	N	%	*N*	%	*N*	%
Strong	618	32	1333	68	1951	100
Moderate or Weak	1961	61	1241	39	3202	100

Source: Adapted from William J. Bowers, *Student Dishonesty and Its Control in College*. New York: Bureau of Applied Social Research, Columbia Univeristy, 1964, p. 149, Table 8.6.

Table 8.5.10 *Cheating by Perceived Peer Disapproval for Respondents with* Strong *Personal Disapproval of Cheating*

Perceived Peer Disapproval of Cheating	Self-Report of Cheating					
	Yes		No		Total	
	N	%	*N*	%	*N*	%
Strong	286				1100	100
Moderate or Weak	407				899	100

Table 8.5.11 *Cheating by Perceived Peer Disapproval for Respondents with* Moderate or Weak *Personal Disapproval of Cheating*

Perceived Peer Disapproval of Cheating	Self-Report of Cheating					
	Yes		No		Total	
	N	%	*N*	%	*N*	%
Strong						100
Moderate or Weak						100

in the partial tables? Is there any indication that "specification" is called for? Discuss the implications of your results.
(b) Using the marginal frequencies you obtain in Tables 8.5.10 and 8.5.11, construct tables showing the association between (1) personal disapproval and perceived peer disapproval, and (2) personal disapproval and self-report of cheating.
(c) By rearrangement of the cells of Tables 8.5.10 and 8.5.11 form tables showing the association between cheating and personal disapproval of cheating, with perceived peer disapproval controlled. Does this relationship disappear with control of peer disapproval? Is there any indication of "specification"? Discuss the implications of your results.

6. From the partial tables, construct the table showing the relationship between Paternal Emotional Support and Self-Reported Delinquency (Table 8.5.12). Does control for the number of delinquent friends reduce the relationship between paternal emotional support and self-reported delinquency? Using the data supplied in Table 8.5.12, construct tables showing the relationship between number of delinquent friends and self-reported delinquency and the relationship between number of delinquent

Table 8.5.12 *Self-Reported Delinquency by Number of Delinquent Friends and Paternal Emotional Support*

Number of Self-Reported Delinquent Offenses	No Delinquent Friends			One or More Delinquent Friends		
	Paternal Emotional Support			Paternal Emotional Support		
	Low	High	Total	Low	High	Total
None	100	101	201	131	251	382
One or More	203	118	321	73	79	152
Total	303	219	522	204	330	534

Source: Gary F. Jensen, "Parents, Peers and Delinquent Action: A Test of the Differential Association Perspective," *American Journal of Sociology* 78 (November 1972): 562–575; adapted from Table 4, page 570.

friends and self-reported delinquency, controlling for paternal emotional support. Does control for paternal emotional support reduce the relationship between number of delinquent friends and self-reported delinquency? Interpret your results.

SELECTED REFERENCES

Costner, Herbert L.
1965 "Criteria for measures of association." American Sociological Review 30 (June): 341–353.

Davis, James A.
1971 Elementary Survey Analysis. Englewood Cliffs, New Jersey: Prentice-Hall.

Goodman, Leo A. and William H. Kruskal
1954 "Measures of association for cross classifications." Journal of the American Statistical Association 49 (December): 732–764.

Lazarsfeld, Paul F.
1955 "Interpretation of statistical relations as a research operation." In Paul F. Lazarsfeld and Morris Rosenberg (eds.), The Language of Social Research. Glencoe, Illinois: The Free Press.

Rosenberg, Morris
1968 The Logic of Survey Analysis. New York: Basic Books.

Bivariate Regression and Correlation 9

1 THE CORRELATION RATIO

Is the sex of an employed person more predictive or less predictive of that person's salary than race? To what degree is special vocational training for criminal offenders related to the subsequent regularity of their employment? How closely related are region and suicide rates for cities in the United States? To what extent, if any, are the governments in former British colonies more stable than governments in the former colonies of other nation-states? Is religious affiliation (Protestant, Catholic, Jew, None) predictive of support for civil liberties? Is the status as a leader or nonleader among high school students a good predictor of grade point averages? Despite the diversity in the substance of these questions, the common element is that all inquire about the degree of association between a qualitative variable and a quantitative variable. In this section, we describe a measure of association for such combinations of variables based on the concept of the relative reduction in prediction error. In the previous chapter, this concept served as the basis for measuring the association in cross-classification tables. In the present chapter, the concept is extended, first, to measure the association between a qualitative variable and a quantitative variable, and in subsequent sections of the chapter, to the correlation between two quantitative variables. These relative reduction in error measures, like those discussed in the previous chapter, may be symbolized in the now familiar way:

$$\textit{Relative reduction in prediction error} = \frac{E_1 - E_2}{E_1}.$$

The specific measures discussed in this chapter will be described by giving specific meaning to the elements of this general formula.

Interval Errors When we predict whether a person voted or did not vote in an election, the prediction is either right or wrong. Similarly, when we predict whether a convicted offender will be fined, imprisoned, or placed on probation, the prediction is either correct or incorrect; we do not claim that some predictions are close to the actual outcome while others are far from it. In contrast, when the predicted variable is a quantitative variable—differing in amount rather than in kind—we will find that prediction errors vary in magnitude; we therefore recognize some prediction errors as being close to the observed value while other prediction errors are far off. If the actual income of a family is $12,000, for example, predicted incomes of $10,000 and $20,000 are both wrong, but the first prediction is much closer than the second. Devices for measuring the association between variables, based on the relative reduction of errors of prediction for a quantitative variable, will therefore be designed to reflect not the reduction in the *number* of predictions that are wrong, but the reduction in the *magnitude* of the aggregate prediction error.

Common sense reasoning suggests that the measure of prediction error might be based on the sum of the absolute distances between the observed and the predicted outcomes in a manner analogous to finding the average deviation (Chapter 7). But the common practice in statistics is to base that measure on the sum of the *squared* discrepancies between actual and predicted outcomes. The student is already familiar with the variance and the standard deviation (Chapter 7), which are based on squared deviations. The mean of the squared deviations as a measure of prediction error entails the same basic concept, and we will therefore refer to the prediction error as *error variance* and will call the sum of the squared discrepancies the *error variation*, coordinate with the variance and variation.

The advantage of the sum of squared deviations is that it is subject to partitioning. In brief, the sum of squared deviations from the overall mean (E_1) may be subdivided into two components: (1) the error variation around the predicted values (E_2), and (2) the *explained variation*, that is, the variation of the predicted values around the overall mean. Thus:

$$\Sigma(X - \bar{X})^2 = \Sigma(X - \hat{X})^2 + \Sigma(\hat{X} - \bar{X})^2 \qquad (9.1.1)$$

where the $\hat{X}$s are predicted values. The addition of these components may also be expressed succinctly in words:

Total variation = *Error variation* + *Explained variation*.

This principle—the additive nature of the components of the sum of squared deviations—will be illustrated with the data of Table 9.1.1, showing hypothetical salaries for men and women librarians.

We first consider the prediction of salary level from our knowledge of the distribution of all salaries combined, without reference to the sex of the

Table 9.1.1 *Hypothetical Salary Levels of Librarians by Sex (in Thousands of Dollars)*

Men	Women	Means	Variation
19			
16			
15	15		
14		$\bar{X}_M = 14$	$\Sigma(X_i - \bar{X}_M)^2 = 64$
	12		
11	11		
		$\bar{X}_T = 10.5$	$\Sigma(X_i - \bar{X}_T)^2 = 257$
	10		
	10		
9	9		
	9	$\bar{X}_W = 9$	$\Sigma(X_i - \bar{X}_W)^2 = 88$
	9		
	8		
	8		
	8		
	6		
	6		
	5		
Total variation $= \Sigma(X_i - \bar{X}_T)^2 = 257$			
Error variation $= \Sigma(X_i - \bar{X}_M)^2 + \Sigma(X_i - \bar{X}_W)^2$ $= 64 + 88 = 152$			
Explained variation $= n_1(\bar{X}_M - \bar{X}_T)^2 + n_2(\bar{X}_W - \bar{X}_T)^2 = 105$			

employee. The overall mean ($\bar{X}_T$) provides the best prediction, since the sum of the squared deviations around the mean will be less than the sum of the squared deviations around any other value. Defining aggregate prediction error (E_1) for such a prediction as the sum of squared deviations, we compute the aggregate prediction error as follows:

$$E_1 = \sum_{i=1}^{N} (X_i - \bar{X}_T)^2 \tag{9.1.2}$$

where $N = n_1 + n_2 =$ the total number of cases in the two groupings. For the illustrative data of Table 9.1.1, we obtain $E_1 = 257$ (a result the student may wish to verify by applying the formula for the sum of squared deviations from Chapter 7).

We next turn to the prediction of salary from knowledge of the sex-specific salary distributions, that is, predicting salary from knowledge of sex. The sex-specific means provide the best predictions and, again defining aggregate prediction error as the sum of the squared deviations from these subgroup means, we compute the error variation as follows:

$$E_2 = \sum_{i=1}^{n_1} (X_i - \bar{X}_M)^2 + \sum_{i=1}^{n_2} (X_i - \bar{X}_W)^2. \tag{9.1.3}$$

For the data of Table 9.1.1, we obtain $E_2 = 64 + 88 = 152$. Again, the student may wish to verify this result by applying the formula for the sum of squared deviations within each sex grouping and summing the two sums of squared deviations to obtain E_2.

We now have the requisite quantities for computing the relative reduction in prediction error for these data. This computation is called the *correlation ratio* and is symbolized by η^2 (read: eta squared).

$$\eta^2 = \frac{E_1 - E_2}{E_1}$$

$$= \frac{\textit{Total variation} - \textit{Error variation}}{\textit{Total variation}}. \qquad (9.1.4)$$

For the data of Table 9.1.1:

$$\eta^2 = \frac{257 - 152}{257}$$

$$= \frac{105}{257}$$

$$= .41.$$

This result may be interpreted as indicating that the sex-specific salary means predict individual salaries with 41 percent less prediction error than the overall mean.

We now proceed to consider the explained variation in Table 9.1.1. We compute the sum of squared deviations of the predicted values (the subgroup means) around the overall mean, weighting each of these squared deviations by the number of cases in the subgroup.[1] Expressed in symbols:

$$\textit{Explained variation} = n_1(\bar{X}_M - \bar{X}_T)^2 + n_2(\bar{X}_W - \bar{X}_T)^2. \qquad (9.1.5)$$

Applying this formula to the illustrative data on the salary levels of men and women librarians, we obtain:

$$\textit{Explained variation} = 6(14 - 10.5)^2 + 14(9 - 10.5)^2$$

$$= 6(12.25) + 14(2.25)$$

$$= 73.5 + 31.5$$

$$= 105.$$

[1] In Chapter 7, we applied this same principle to get the total variance from two or more subgroup variances. Here, we add the weighted mean of the subgroup variances to the weighted mean of the squared deviations between subgroup means and the overall mean. We will apply the same principle again in the analysis of variance (Chapter 15).

We may now note that the total sum of squared deviations, $\Sigma(X - \bar{X}_T)^2$, is precisely equal to the error sum of squared deviations, $\Sigma(X - \hat{X})^2$, plus the explained sum of squared deviations, $\Sigma(\hat{X} - \bar{X}_T)^2$. For the illustrative data:

$$257 = 152 + 105.$$

This result is not peculiar to the illustrative data presented here; it is generally true that the total variation is the sum of the error variation and the explained variation. It is therefore evident that:

$$\textit{Explained variation} = E_1 - E_2$$

and the relative reduction in prediction error can thus be described as the ratio of explained variation to total variation. In symbols:

$$\eta^2 = \frac{E_1 - E_2}{E_1} = \frac{\textit{Explained variation}}{\textit{Total variation}}.$$

Computing Formula for the Correlation Ratio The correlation ratio describes the relative reduction in prediction error in a quantitative dependent variable from knowledge of a categorized predictor with any number of categories; its use is not limited to a dichotomized predictor variable as in the illustration above. It is commonly more convenient to compute the explained variation than to compute the error variation within each subgroup, and for this reason the correlation ratio is ordinarily most readily computed by the following formula:

$$\eta^2 = \frac{\sum_{j=1}^{k} n_j(\bar{X}_j - \bar{X}_T)^2}{\sum_{i=1}^{N} (X_i - \bar{X}_T)^2} = \frac{\sum_{j=1}^{k} n_j(\bar{X}_j - \bar{X}_T)^2}{\sum_{i=1}^{N} X_i^2 - \frac{\left(\sum_{i=1}^{N} X_i\right)^2}{N}} \qquad (9.1.6)$$

where k = the number of categories of the predictor variable
n_j = the number of cases in the jth category
$\bar{X}_j$ = the mean of the jth category
N = the total number of cases in all categories
$\bar{X}_T$ = the mean of all cases in all categories.

Applying this formula to the illustrative data of Table 9.1.1, we obtain:

$$\eta^2 = \frac{105}{257} = .41 .$$

This result is identical, of course, to the result obtained above by computing the relative reduction in prediction error (Formula 9.1.4). Hence, the following two statements are equivalent: (1) The sex-specific salary means predict individual salaries with 41 percent less prediction error than the overall mean; and (2) 41 percent of the variation in individual salaries is explained by variation between the sex-specific means.

Use and Interpretation of the Correlation Ratio The correlation ratio is most commonly used to measure the association (relative reduction in prediction error) between a qualitative predictor and a quantitative dependent variable. This use is illustrated in Table 9.1.2, showing the residential segregation indexes of 60 United States cities classified by region. Even though the indexes for most cities are clustered in the upper end of the possible range of the index (maximum value = 100), the cities within the four major regions are not uniformly segregated. The mean segregation index (not weighted for city population size) in the West is 76.1 as compared to a mean of 91.3 for southern cities. The degree of association between region and segregation cannot be judged by the magnitude of the differences between regional means alone; the degree of association depends also on the variation around those regional means, that is, on the error variance. If the segregation index for each city were located precisely at the mean for its region, the error variation would be zero and the association would be perfect. On the other hand, if the variance within each region were equal to the total variance (that is, if the regional means were all equal) there would be no explained variation and the association would be zero. The actual degree of association depends on the magnitude of the explained variation (the variation among the regional means) in relation to the total variation (the sum of the variation among the regional means and the variation around the regional means).

As indicated in Table 9.1.2, the correlation ratio for region and residential segregation indexes is .56. This may be interpreted to mean that 56 percent of the total variation in residential segregation indexes is constituted by variation among the regional means; in other words, 56 percent of the variation in segregation indexes is "explained" by region. Alternatively stated, the regional means predict the segregation indexes of cities with 56 percent less prediction error than does the overall mean. The remaining 44 percent of the variation (the error variation) must be attributed to intraregional variation in such factors as the recent rate of city growth, black-white differences in income, and the vigor of enforcement of laws and regulations pertaining to civil rights in housing. The error variation is "unexplained" by variation in region, but would presumably be explained by other factors.

Table 9.1.2 Residential Segregation Indexes of 60 United States Cities by Region, 1960

Region	City	Residential Segregation Index	Regional Mean	Overall Mean
Northeast $n_1 = 9$			81.33	
	Boston	84		
	Buffalo	87		
	Jersey City	78		
	Newark	72		
	New York City	79		
	Philadelphia	87		
	Pittsburgh	85		
	Rochester	82		
	Yonkers	78		
North Central $n_2 = 17$			89.06	
	Akron	88		
	Chicago	93		
	Cincinnati	89		
	Cleveland	91		
	Columbus	85		
	Dayton	91		
	Des Moines	88		
	Detroit	85		
	Indianapolis	92		
	Kansas City	91		
	Milwaukee	88		
	Minneapolis	79		
	Omaha	92		
	Saint Louis	91		
	Saint Paul	87		
	Toledo	92		
	Wichita	92		
South $n_3 = 23$			91.35	
	Atlanta	94		
	Austin	93		
	Baltimore	90		
	Birmingham	93		
	Charlotte	94		
	Corpus Christi	89		
	El Paso	81		
	Fort Worth	94		
	Houston	94		
	Jacksonville	97		
	Louisville	89		
	Memphis	92		
	Miami	98		
	Nashville	92		
	New Orleans	86		

(*continued*)

Table 9.1.2 (*continued*)

Region	City	Residential Segregation Index	Regional Mean	Overall Mean
	Norfolk	95		
	Oklahoma City	87		
	Richmond	95		
	Saint Petersburg	97		
	San Antonio	90		
	Tampa	95		
	Tulsa	86		
	Washington, D.C.	80		
West			76.09	
$n_4 = 11$	Denver	86		
	Long Beach	84		
	Los Angeles	82		
	Oakland	73		
	Portland, Ore.	77		
	Sacramento	64		
	San Diego	81		
	San Francisco	69		
	San Jose	60		
	Seattle	80		
	Tucson	81		
				86.40

Total variation $= \Sigma(X - \bar{X}_T)^2 = 3724.4$

$$\begin{aligned}\text{Explained variation} &= n_1(\bar{X}_1 - \bar{X}_T)^2 + n_2(\bar{X}_2 - \bar{X}_T)^2 \\ &\quad + n_3(\bar{X}_3 - \bar{X}_T)^2 + n_4(\bar{X}_4 - \bar{X}_T)^2 \\ &= 9(81.33 - 86.40)^2 + 17(89.06 - 86.40)^2 \\ &\quad + 23(91.35 - 86.40)^2 + 11(76.09 - 86.40)^2 \\ &= 231.34 + 120.29 + 563.56 + 1169.26 \\ &= 2084.45\end{aligned}$$

$$\eta^2 = \frac{2084.45}{3724.4} = .56$$

Source: Reynolds Farley and Alma F. Taeuber, "Racial Segregation in the Public Schools," *American Journal of Sociology* 79 (January 1974): 888–905. Adapted from Table 2, pp. 895–896.

The correlation ratio may also be used to measure the association between two quantitative variables, and this latter application is especially appropriate when the relation of interest is not rectilinear, that is, when it does not conform to a straight line (see Section 2 following). If the relation between two quantitative variables is rectilinear, the square of the product-moment correlation, r^2, would be preferred as the measure of the relative reduction in prediction error (see Section 3 following); but for a curvilinear relationship,

r^2 will understate the degree of association and η^2 would then be more appropriate. To apply η^2 as a measure of association between two quantitative variables, it is necessary to divide the predictor variable into categories, taking care not to create categories with such small frequencies that the subgroup means will be grossly unstable (in other words, the means should not vary appreciably if the boundaries of the categories were slightly narrowed or extended), nor so wide that the essential curvilinear pattern of the relationship would be obscured.

QUESTIONS AND PROBLEMS

1. Would it be possible to calculate η^2 between two qualitative variables? Why or why not?
2. Is the numerical value of η^2 affected by the order of columns (rows)?
3. Compute the correlation ratio between skill level and anomie scores using the hypothetical data of Table 9.1.3. Note that the skill levels are represented by different numbers of respondents. To check the effect of this uneven

Table 9.1.3 *Anomie Scores of Industrial Workers at Three Skill Levels*

Unskilled	Semi-skilled	Skilled
0	0	0
1	0	0
1	1	0
2	1	1
2	2	1
3	2	2
3	2	2
4	2	3
5	2	3
5	2	4
	2	
	3	
	3	
	3	
	3	
	4	
	4	
	4	
	5	
	5	

Source: Hypothetical. Score distributions approximate findings reported for workers in an American industrial plant in William H. Form, "The Social Construction of Anomie: A Four-Nation Study of Industrial Workers," *American Journal of Sociology* 80 (March 1975): 1165–1191. See Table 2, p. 1172.

distribution of cases over the three categories of the independent variable, make the following calculations and compare results.

(a) Assume that each score in the unskilled level represents two cases instead of one (for a total of 50 cases instead of 40). What is the correlation ratio?

(b) Assume that each score in the skilled level represents two cases instead of one (for a total of 50 cases instead of 40). What is the correlation ratio?

In your own words, tell why doubling the number of cases in the skilled level has a greater effect on the correlation ratio than doubling the number of cases in the unskilled category.

2 THE SCATTER DIAGRAM

A relationship between two quantitative variables may appear in a variety of forms, and relationships having the same form may afford varying degrees of accuracy in predicting values of one variable from knowledge of values of the other. Income increases as education increases—a positive and approximately linear relationship. Idealism decreases with advancing age—a negative and approximately linear relationship. The internal differentiation of positions in complex organizations increases as the size of the organization increases, but only up to a point, after which further increases in organizational size imply no greater differentiation—a nonlinear but monotonic relationship. The intensity of feeling about an attitude object is typically highest among those whose attitudes are either very positive or very negative, while intensity of feeling is lower among those whose attitudes are between the extremes—a nonmonotonic relationship, that is, a relationship that changes direction within the range of values explored. Such variations in the form of the relationship between two quantitative variables commonly reflect potentially important social and social-psychological processes, and they may be no less important than the degree of association that the variables exhibit. Before proceeding to the discussion of measures of association for two quantitative variables, we therefore consider in this section a useful device for making visible the form of a relationship. The *scatter diagram* permits the social analyst to examine the form of the covariation between two variables and to make a preliminary and informal assessment of the degree of association between them.

To illustrate the construction and use of the scatter diagram, we take as our point of departure Table 9.2.1, which presents yearly income averages and suicide rates for the Central states, 1929–1950. To reduce the data of this table to a scatter diagram, we first draw horizontal and vertical axes, as in the construction of a histogram. Axes are drawn approximately equal in length, unless there is good reason to deviate from this convention. Also, as a matter

Table 9.2.1 *Suicide Rate by Annual per Capita Income, Central United States, 1929–1950*

Year	Per Capita Income (X)	Suicide Rate (Y)
1929	604	15.4
1930	532	17.9
1931	486	19.1
1932	399	19.3
1933	400	18.1
1934	437	16.9
1935	482	15.5
1936	549	15.2
1937	567	16.3
1938	515	16.7
1939	566	15.3
1940	602	15.2
1941	708	14.0
1942	800	12.8
1943	914	11.0
1944	972	11.0
1945	985	12.3
1946	951	12.3
1947	897	12.1
1948	928	11.6
1949	878	11.9
1950	947	11.9

Source: Donald Faigle, *Suicide in Relation to Income, Urbanization and Race*, unpublished master's thesis, Department of Sociology, Indiana University, 1956, Table 5.

of convention, the independent X-variable is plotted along the base line, and the dependent Y-variable along the vertical axis. The establishment of one variable as independent, and the other as dependent is not a statistical problem, of course, but rather a matter of judgment and circumstance. The dependent variable may be construed as the effect of the independent variable, or as the outcome to be predicted from the predictor variable. In many instances, there may be no clear causal dependency at all, because both variables may be found to be the consequences of an unidentified third factor.

Next, scales are established on the axes in such a manner as to accommodate, with a margin to spare, the observed ranges of the respective variables. Thus, the horizontal scale covers the distance from $200 to $1,200, while the vertical scale extends from 10 to 20. Needless to say, enough markers are set up on each axis to ensure accurate and effortless plotting. Unlike the

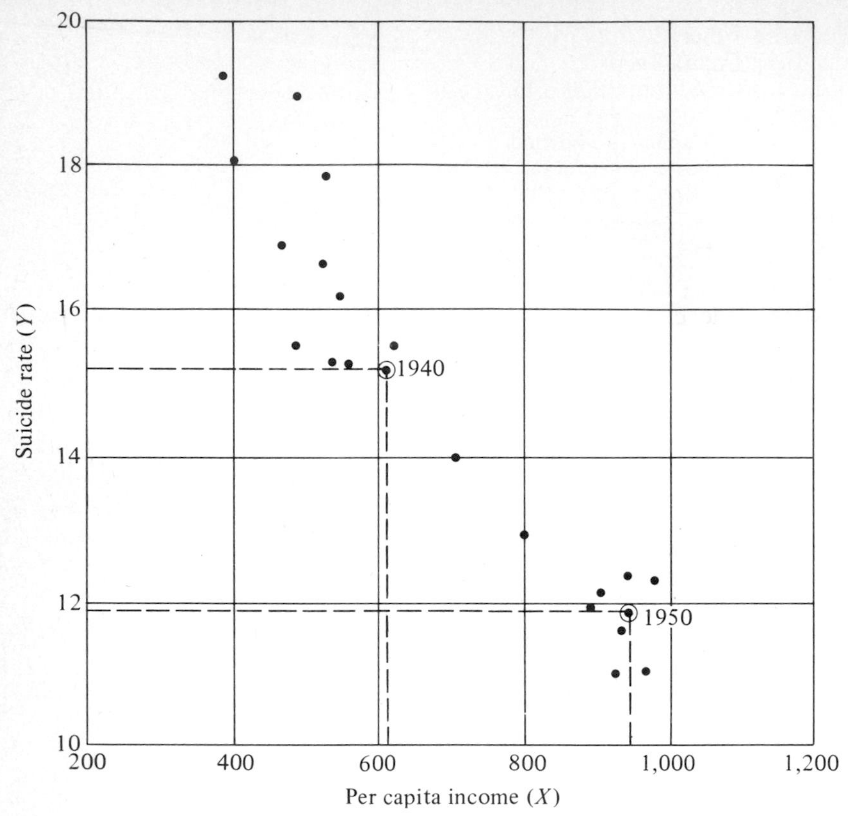

Figure 9.2.1 *Scatter Diagram, Suicide Rate by Annual Per Capita Income, Central United States, 1929–1950*

histogram, the vertical scale in a bivariate plot need not begin with zero, because the focus of attention is on the contour of the scatter rather than on relative frequency as gauged by the height of the curve.

Having drawn and scaled the axes, we are ready to plot each pair of values by a double-duty point. For any given case, the Y-value fixes the height of the point above the base line and the X-value fixes its horizontal distance from the vertical axis. Thus, 1940 is represented in Figure 9.2.1 by a point located at the intersection of guide lines extending perpendicularly from $Y = 15.2$ and $X = 602$; similarly, 1950 is represented by a point at the intersection of 11.9 and 947. The swarm of all such points constitutes the scatter diagram.

Types of Scatter It is the pattern of this swarm that enables us to judge the nature of the relationship between two variables—and such a judgment is a preliminary essential to the proper measurement of that relationship. Thus, it appears that a fixed increase in income is accompanied by a fixed decrease in the suicide rate—that is, the suicide rate changes by a constant amount per unit income. Such a relation is termed *rectilinear*, because the trend of scatter conforms to the track of a straight line.

Any such trend line, whether freehand or mathematically fitted, is technically termed a *line of regression*. This concept was coined in 1877 by Galton, who used it in connection with his correlational studies of the characteristics of parents and their offspring. He perceived, for example, that the sons of very tall fathers were, on the average, shorter than their fathers, while the sons of very short fathers were, on the average, taller than their fathers. Galton termed this phenomenon *regression toward the mean* and observed that in a scatter diagram representing the heights of fathers and sons, the slope of the straight line connecting the mean heights of sons measured this regression tendency. Accordingly, the line itself was called the regression line, a designation that has survived and enjoys wide usage in statistics, although it is no longer restricted to its original connotation.

The relation in Figure 9.2.1 is said to be negative because the slope of the regression line is downward, that is, as per capita income increases the suicide rate decreases. A relationship is said to be positive when the slope of the regression line is upward; as one variable increases the other increases also. A positive rectilinear relationship is displayed in Figure 9.2.2; family size increases as farm acreage increases.

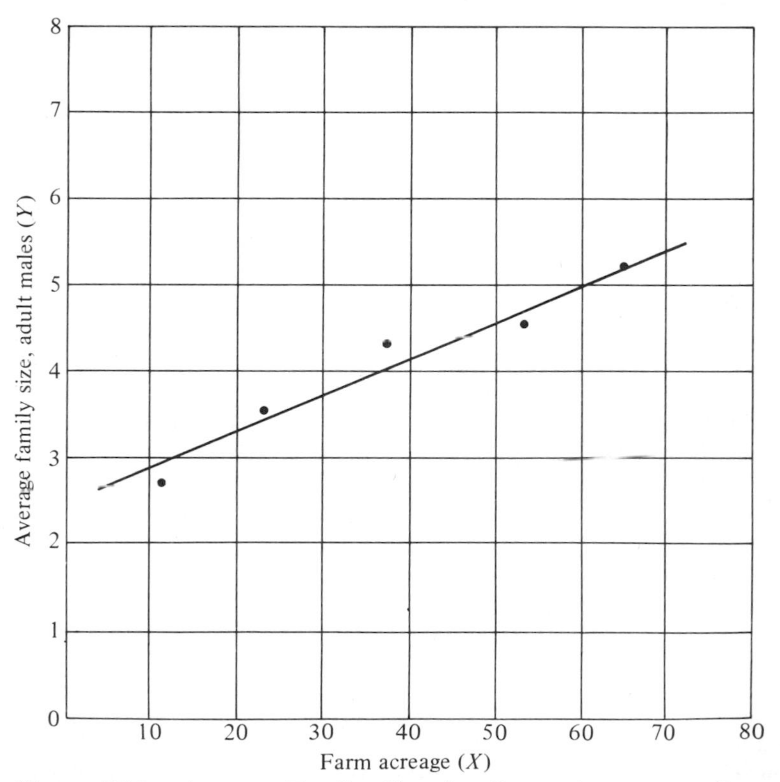

Figure 9.2.2 *Average Family Size by Farm Acreage in Pre-Communist China*

Source: John L. Buck, *Chinese Farm Economy*. Chicago: University of Chicago, 1930. Copyright © 1930 by The University of Chicago Press. Reproduced by permission.

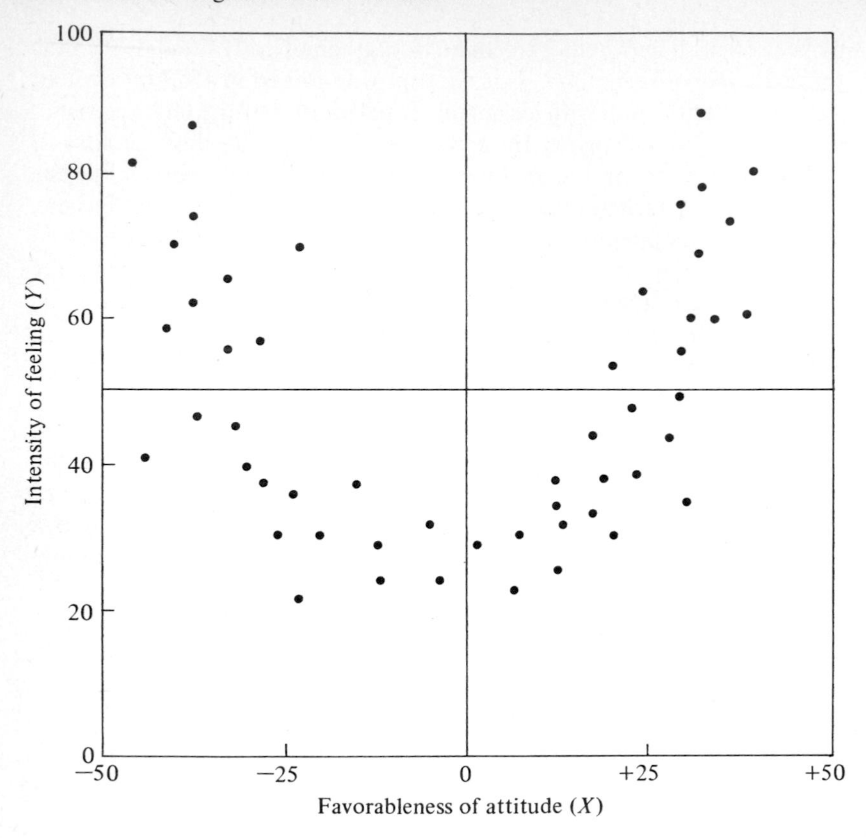

Figure 9.2.3 *Scatter Diagram, Intensity of Feeling by Favorableness of Attitude (Hypothetical Data)*

The trend of the scatter will not always be rectilinear; rather it may be *curvilinear* and take on any one of innumerable curve patterns. A simple example is provided in Figure 9.2.3, which portrays the relation between attitude toward a minority people and intensity of feeling. As might be anticipated, a decided opinion—whether pro or con—is held with considerable intensity of feeling, while a less decided or neutral opinion arouses no very strong feelings. Whether this pattern—extreme opinion, strong effect; neutral opinion, weak affect—holds under all conditions is not our concern here. That would be a matter for empirical investigation. Here we are interested only in exhibiting a type of relationship that may appear in sociological studies.

The three scatter diagrams presented thus far have distinctive trends and one would have little hesitancy in describing them as exclusively rectilinear or curvilinear. But scatters of empirical observations are seldom so clean and unambiguous; more often both linear and curvilinear tendencies combine in the same data and thereby complicate the problem of representing correlation

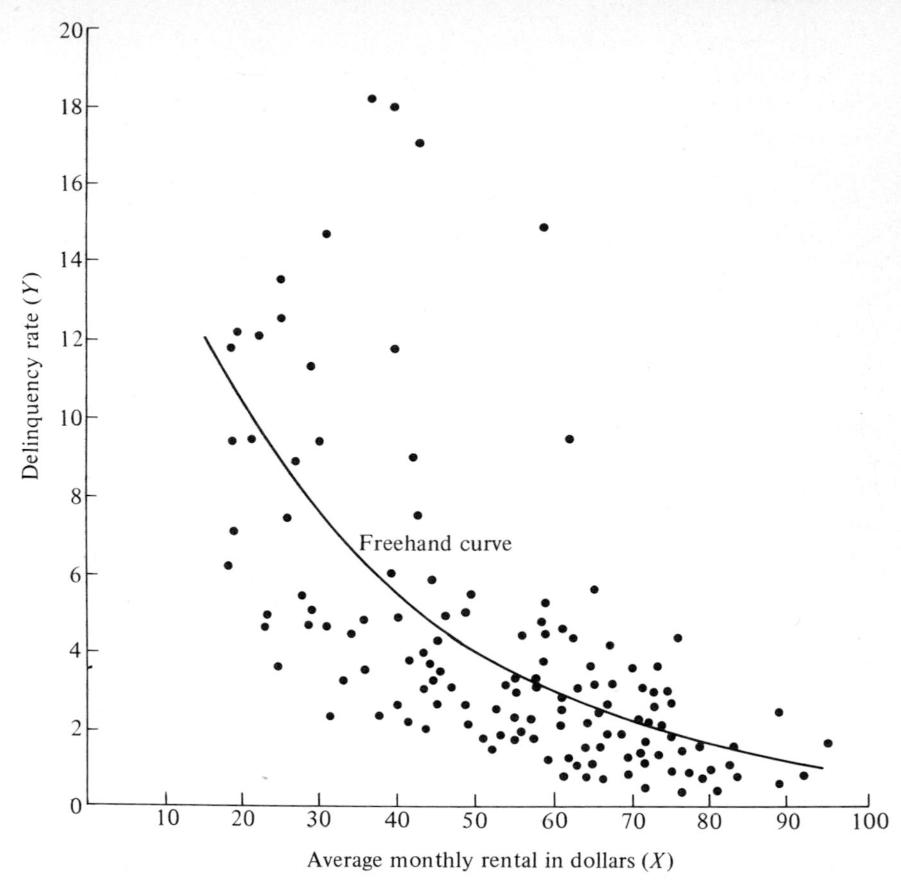

Figure 9.2.4 *Scatter Diagram and Freehand Regression Line, Delinquency Rate by Monthly Rental, 142 Local Areas, Chicago, 1930*

by an overall measure. For example, the scatter of delinquency rates and average monthly rentals for 140 small census tracts in Chicago (Figure 9.2.4) is marked by some linearity, yet it appears that a curved trend line would fit the entire scatter better. From left to right, as rentals increase, the delinquency rate responds by decreasing, but at a progressively slower rate. This is evidenced by the straightening of the swarm. Of course there are several striking exceptions to the foregoing generalization—a few extremely high delinquency rates occur with above average rentals. These *outliers* would require special analysis, since they represent a breakdown in the rule. Yet, in the main, the relationship holds fairly well, affording some predictability of one variable from the other. For example, if we knew the rental to be $60, we would predict the delinquency rate to be approximately 3.0, which is the height of the freehand regression curve at that point. To be sure, such a prediction would not be free of error, for the obvious reason that none of the observed values fall precisely on the curve at that point—all deviate to a greater or lesser extent. Evidently, the accuracy of any such prediction would vary according to the tendency of the points to hug the line of relationship between the two series. When the points move within a narrow band, predictive accuracy, and therefore correlation, would be high; when the points are widely scattered, predictive accuracy would be correspondingly low. Only when all points fall exactly on the regression line would prediction and correlation be perfect. At the other extreme, when the scatter is purely random, then we may just as well ignore the so-called predictor variable. For any or all rentals, our best guess would be the overall mean of the delinquency rates.

Scedasticity From Figure 9.2.4, even knowing a rental to be \$30, we still could not accurately forecast the corresponding level of delinquency, because the delinquency values are widely scattered for that rental value; on the other hand, for a rental of \$70, we could predict with much greater accuracy, owing to the bunching of delinquency rates around the regression line. This scatter of Y-values for the respective X-values is known as *scedasticity*. If the degree of variation in delinquency rates— the width of the scatter band—had been uniform for all values of X, then we could have spoken of Y as being *homoscedastic* in respect to X. Actually, the degree of scatter in Y diminishes as X changes, so that Y is *heteroscedastic* in respect to X. Heteroscedasticity implies that the degree of predictability is not uniform throughout the entire series; hence, its presence reduces the representativeness of a single overall measure of correlation, which, after all, is an average. Just as we hesitate to compute the mean of heterogeneous bimodal data, similarly we hesitate to calculate an average measure of correlation of a heteroscedastic scatter.

Although the concept of the regression line commonly leads us to perceive the relationship between variables as conforming to some unbroken graph line, whether rectilinear or curvilinear, this perceptual tendency can be usefully discarded for certain heteroscedastic patterns in scatter diagrams. In the scatter diagram shown in Figure 9.2.5, for example, the pattern *might* be described as a monotonic curvilinear relationship with heteroscedasticity; the mean number of hierarchical levels at headquarters increases at a decreasing rate as the size of the agency increases. But the scatter of points may also be described as fitting a triangular pattern since all points fall within a clearly discernible triangle. The triangular pattern carries the suggestion that the smaller agencies have wide flexibility in the number of organizational levels while the larger organizations are increasingly constrained to have a relatively large number of levels. The tendency to perceive a line pattern in heteroscedastic scatter may obscure the nature of the social processes that have given rise to the heteroscedasticity.

This section has stressed the important role of the scatter diagram in the

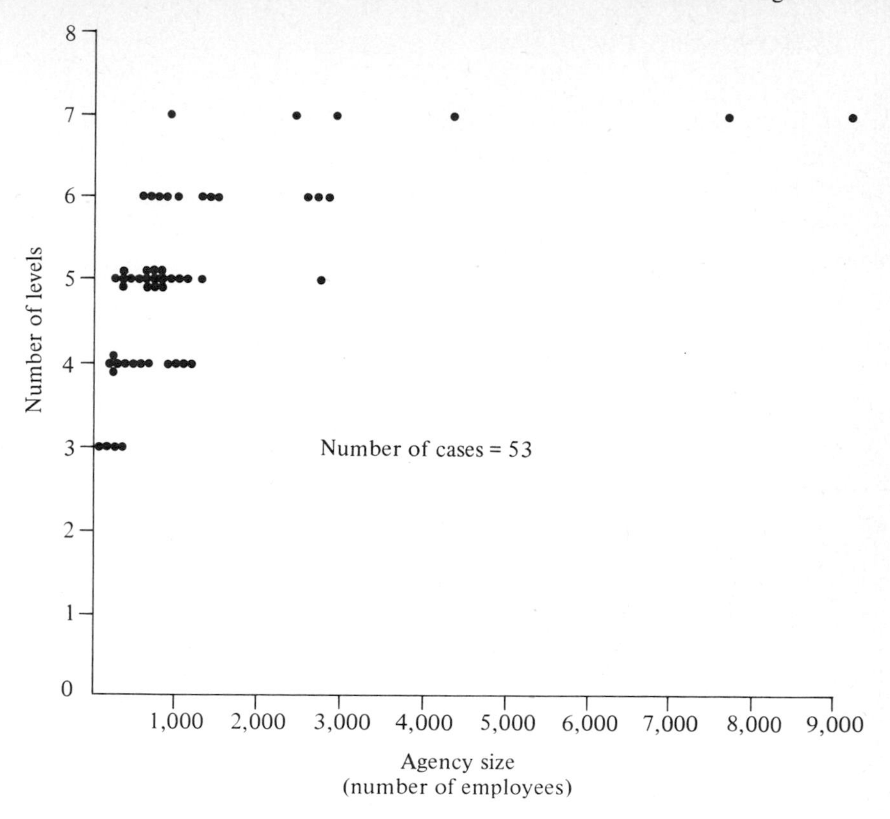

Figure 9.2.5 *Size of Agency and Number of Hierarchical Levels at Headquarters*

Source: Peter M. Blau, "A Formal Theory of Differentiation in Organizations," *American Sociological Review* 35 (1970), Figure 1 on p. 205. Reproduced by permission.

preliminary analysis of covariation between two quantitative variables. From these charts it is possible to determine whether the relationship is (a) rectilinear or curvilinear; (b) positive or negative; (c) weak or strong; (d) homoscedastic or heteroscedastic. Scatter diagrams also permit the analyst to locate cases that depart strikingly from the general pattern and to determine whether the pattern is well described by any kind of line pattern. Because of these important disclosures, one is well advised to construct and study the scatter diagram before attempting to measure the correlation between quantitative variables, and to take into account the insights provided by the scatter diagram in deciding how best to measure association and interpret the results. Without these visual aids, we run the risk of arriving at misleading conclusions.

QUESTIONS AND PROBLEMS

1. Define the following concepts:
   scatter diagram — homoscedasticity
   bivariate data — heteroscedasticity
   rectilinearity — curvilinearity
   scedasticity

2. When the regression is curvilinear, why is homoscedasticity not likely to be uniform in both directions? Show graphically.

3. (a) Plot a scatter diagram from Table 9.2.2.
   (b) What does the scatter of this graph suggest concerning the "pull" of economic conditions?

Table 9.2.2 *Income and Geographic Mobility by State, 1970*

State	Index of Income[a]	Percent Mobile[b]	State	Index of Income	Percent Mobile
Alabama	74	14.7	Montana	86	23.9
Alaska	119	44.0	Nebraska	90	20.2
Arizona	94	27.4	Nevada	114	30.6
Arkansas	69	18.7	New Hampshire	96	20.4
California	116	22.6	New Jersey	118	18.2
Colorado	100	31.2	New Mexico	78	22.4
Connecticut	125	14.9	New York	116	15.9
Delaware	105	16.4	North Carolina	79	16.6
Florida	98	28.0	North Dakota	79	20.5
Georgia	85	21.6	Ohio	103	14.3
Hawaii	108	24.8	Oklahoma	86	23.4
Idaho	85	25.0	Oregon	101	25.5
Illinois	112	14.0	Pennsylvania	98	11.1
Indiana	98	16.2	Rhode Island	100	15.5
Iowa	92	17.3	South Carolina	74	16.4
Kansas	94	22.8	South Dakota	77	19.8
Kentucky	78	15.1	Tennessee	79	15.1
Louisiana	75	15.7	Texas	90	22.0
Maine	82	15.4	Utah	86	19.4
Maryland	113	22.2	Vermont	89	19.8
Massachusetts	109	14.8	Virginia	96	26.3
Michigan	108	16.4	Washington	108	25.3
Minnesota	97	18.9	West Virginia	75	13.1
Mississippi	62	15.7	Wisconsin	97	15.5
Missouri	95	19.3	Wyoming	93	26.3

[a] Ratio of 1969 state per capita income to 1969 national per capita income.
[b] Percent of 1970 population who were living in a different county or abroad in 1965.
Source: U.S. Bureau of the Census. *Statistical Abstract of the United States:* 1974. Tables 45 and 627. Washington, D.C.: U.S. Government Printing Office. 1974.

3 LINEAR REGRESSION AND CORRELATION

When the basic pattern of the relationship between two variables has been determined with the aid of the scatter diagram, a succinct and precise expression of that pattern may be formulated in an equation for the regression line. This equation will serve as a rule by which values on one variable may be predicted from values on the other. The regression line for a rectilinear relationship may be algebraically expressed by a relatively simple equation, while curvilinear relationships call for more complex equations or for some transformation of the data to *linearize* the regression. In this section we direct our attention to the regression equation for rectilinear bivariate relationships and to the measurement of the degree of association in terms of the relative reduction in prediction error.

The equation for a straight line constituting the regression of Y on X has the general form

$$\hat{Y} = a_{yx} + b_{yx}X . \tag{9.3.1}$$

The change in $\hat{Y}$ per unit change in X is the slope of the regression line, which is represented in the equation by b_{yx}. When $X = 0$, it is evident that $\hat{Y} = a_{yx}$; hence, a_{yx} in Formula 9.3.1 is the point at which the regression line intercepts the vertical axis at $X = 0$, and this point is known as the "Y-intercept." Our problem is to find the numerical values of a_{yx} and b_{yx} such that the plot of Formula 9.3.1 will best fit the scatter. The standard specification for "best fit" is known as the *least squares criterion* and may be formulated as follows: The line is located so as to minimize the sum of the squares of the vertical deviations of the points from the line. We seek to minimize the quantity

$$\Sigma(Y - \hat{Y})^2$$

where Y = observed value and $\hat{Y}$ = predicted value. If we have the numerical values of a_{yx} and b_{yx} that satisfy this criterion, then the plot of Formula 9.3.1 will fit best by the criterion of least squares. Such a regression line is analogous to the mean of a univariate distribution, since the mean is the point that minimizes the sum of squared deviations. Furthermore, like deviations from the mean, the algebraic sum of the vertical deviations around the regression line will equal zero. Thus it is appropriate to refer to the regression line as the central tendency of the scatter.

The values of a_{yx} and b_{yx} for the line of best fit may be computed by the following formulas, presented here without an explanation of their derivation:

$$b_{yx} = \frac{\Sigma(X - \bar{X})(Y - \bar{Y})}{\Sigma(X - \bar{X})^2} \tag{9.3.2}$$

$$= \frac{N\Sigma XY - (\Sigma X)(\Sigma Y)}{N\Sigma X^2 - (\Sigma \bar{X})^2} . \tag{9.3.3}$$

$$a_{yx} = \bar{Y} - b_{yx}(\bar{X}) . \tag{9.3.4}$$

We note that the numerator for the slope (Formula 9.3.2) is a sum of *cross-products*, that is, the sum of the products obtained by multiplying the deviation of X from $\bar{X}$ by the deviation of the Y for that same case from $\bar{Y}$. The mean of the cross-products is known as the *covariance* of X and Y, symbolized by σ_{xy}. The defining formula for the covariance is thus

$$\sigma_{xy} = \frac{\Sigma(X - \bar{X})(Y - Y)}{N}. \qquad (9.3.5)$$

We note also that the denominator in the formula for the slope is the sum of squared deviations from the mean of the predictor variable, X. The mean of these squared deviations is the variance of X. Expressed in symbols:

$$\sigma_x^2 = \frac{\Sigma(X - \bar{X})^2}{N}. \qquad (9.3.6)$$

Thus the slope may be expressed as the ratio of the covariance of the two variables to the variance of the predictor

$$b_{yx} = \frac{\sigma_{xy}}{\sigma_x^2}. \qquad (9.3.7)$$

When variates have been expressed in the form of standard measures (that is, when deviations from the respective means are expressed in standard deviation units), σ_x^2 necessarily implies unity and the slope is the covariance of the standard measures. The slope of the regression line for variables expressed in standard measures is conventionally symbolized by r and we therefore have

$$r_{yx} = \sigma_{x'y'} \qquad (9.3.8)$$

where $x' = \dfrac{X - \bar{X}}{\sigma_x}$,

$y' = \dfrac{Y - \bar{Y}}{\sigma_y}$.

Since $\sigma_{x'y'} = \sigma_{y'x'}$, it follows that $r_{yx} = r_{xy}$. Stated in words, when variates have been expressed in the form of standard measures, the slope of the regression of Y on X is equal to the slope of the regression of X on Y. The student should note, however, that this identity of slopes holds only for variates expressed in the form of standard measures; as indicated by Formula 9.3.2 and again by Formula 9.3.7, the slope for variates in their "raw" (that is, unstandardized) form will differ depending on which variable is designated as the independent or predictor variable.

We illustrate the computation of the parameters of the regression equation, for variates in both unstandardized and standardized form, with the data of Table 9.3.1. In Table 9.3.2, we present the computations of a_{yx} and b_{yx} for the data of Table 9.3.1.

Table 9.3.1 *Husband's Occupational Prestige and Family Social Status Score*

Family	Husband's Occupational Prestige (X)	Family Social Status Score (Y)
A	75	80
B	70	87
C	60	91
D	55	44
E	50	22
F	40	58
G	25	52
H	20	10
I	15	38
J	10	18

Source: Hypothetical data contrived to approximate the correlation between husband's occupational prestige and family status for black families judged by white respondents, as reported in William A. Sampson and Peter H. Rossi, "Race and Family Social Standing," *American Sociological Review* 40 (April 1975): 201–214.

Table 9.3.2 *Calculation of* a_{yx} and b_{yx} for the *Data of Table 9.3.1*

X	Y	X^2	Y^2	XY
75	80	5,625	6,400	6,000
70	87	4,900	7,569	6,090
60	91	3,600	8,281	5,460
55	44	3,025	1,936	2,420
50	22	2,500	484	1,100
40	58	1,600	3,364	2,320
25	52	625	2,704	1,300
20	10	400	100	200
15	38	225	1,444	570
10	18	100	324	180
420	500	22,600	32,606	25,640

$$b_{yx} = \frac{N\Sigma XY - (\Sigma X)(\Sigma Y)}{N\Sigma X^2 - (\Sigma X)^2} = \frac{256{,}400 - 210{,}000}{226{,}000 - 176{,}400}$$

$$= \frac{46{,}400}{49{,}600}$$

$$= .935$$

$$a_{yx} = \bar{Y} - b_{yx}\bar{X} = 50 - .935\,(42)$$

$$= 10.73$$

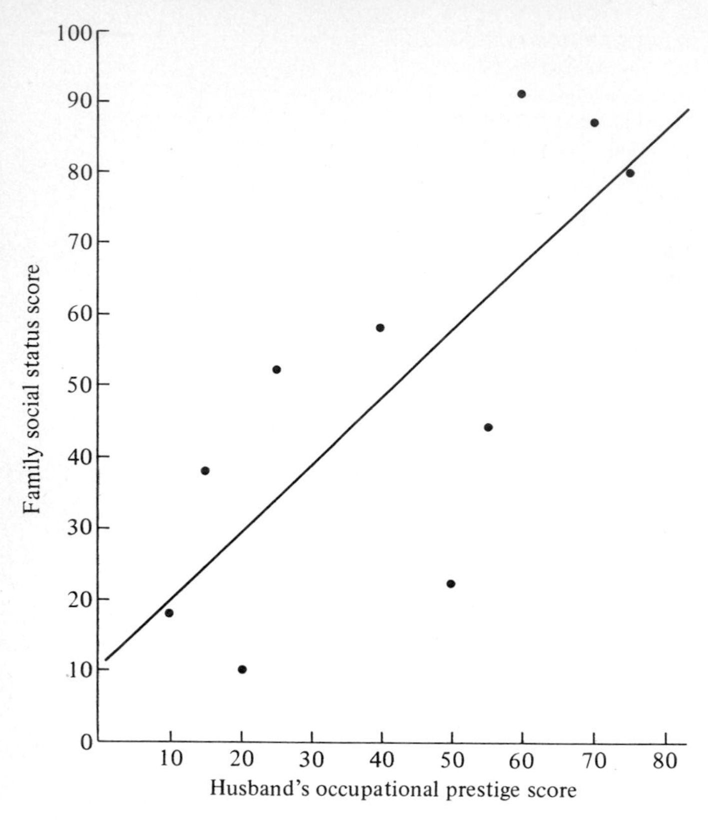

Figure 9.3.1 *Scatter Diagram and Regression Line, Husband's Occupational Prestige and Family Social Status Score*

The regression equation, $\hat{Y} = 10.73 + .935X$, may be regarded as a rule for predicting Y from X. It is drawn on the scatter diagram in Figure 9.3.1. It requires that Y increase by .935 units for every unit increase in X; hence for every unit increase in the husband's occupational prestige (Table 9.3.1) there is an increase of .935 status score points for the family. Other sources of influence on the family social status score, for example, the wife's occupation and the educational levels of both husband and wife, are responsible for the deviations of status scores from the regression line. A given family's status score will be above or below the score expected on the basis of the husband's occupation (the regression line prediction) depending on whether these additional factors enhance or depress the family social status. Since social variables are generally influenced by a number of factors, we do not expect the points for real cases to fall exactly on the regression line representing the dependence of such a variable on a single predictor.

The computation of the regression slope for the variates expressed in the form of standard measures is illustrated in Table 9.3.3. Although there is a

Table 9.3.3 *Deviations as Standard Measures (from Table 9.3.1) and the Regression Slope* r *as the Covariance of Standard Measures*

Family	Deviations		Standard Measures		Cross-Products
	$x = X - \bar{X}$	$y = Y - \bar{Y}$	$x' = \frac{x}{\sigma_x}$ $\sigma_x = 22.3$	$y' = \frac{y}{\sigma_y}$ $\sigma_y = 27.6$	$x'y'$
A	33	30	1.48	1.09	1.6132
B	28	37	1.26	1.34	1.6884
C	18	41	.81	1.49	1.2069
D	13	− 6	.58	− .22	− .1276
E	8	−28	.36	−1.01	− .3636
F	− 2	8	− .09	.29	− .0261
G	−17	2	− .76	.07	− .0532
H	−22	−40	− .99	−1.45	1.4355
I	−27	−12	−1.21	− .43	.5203
J	−32	−32	−1.43	−1.16	1.6588
					8.1231
					− .5705
					7.5526

$$r = \frac{\Sigma x'y'}{N} = \frac{7.5526}{10} = .755$$

more convenient computing formula for r (to be described subsequently), we have here computed the regression slope for standard measures by taking the deviation of each variate from its respective mean, dividing each such deviation by the standard deviation of the appropriate distribution, computing the cross-products of the resulting standard measures, and taking the mean of these cross-products to obtain the covariance of the standard measures, which is r (Formula 9.3.8).

A scatter diagram of the standard measures is shown in Figure 9.3.2. The configuration of the points is the same, of course, whether the scatter diagram is based on standard measures or unstandardized scores; the only alteration is in the unit of measurement. In Figure 9.3.2, two regression lines have been drawn to emphasize that, although $r_{yx} = r_{xy}$, the two regression lines are not the same line except where $r = 1.0$. The two regression lines are most divergent when $r = 0$ and converge toward each other as the magnitude of r increases. Thus it is possible to judge the magnitude of r by the angular distance between the two regression lines. As indicated in the figure, y' increases by .76 standard deviation for every increase of 1.0 standard deviation in x'. Similarly, x' increases by .76 standard deviation for every increase of 1.0 standard deviation in y'. The regression line of y' on x' is located so as to minimize the sum of the squared *vertical* deviations of

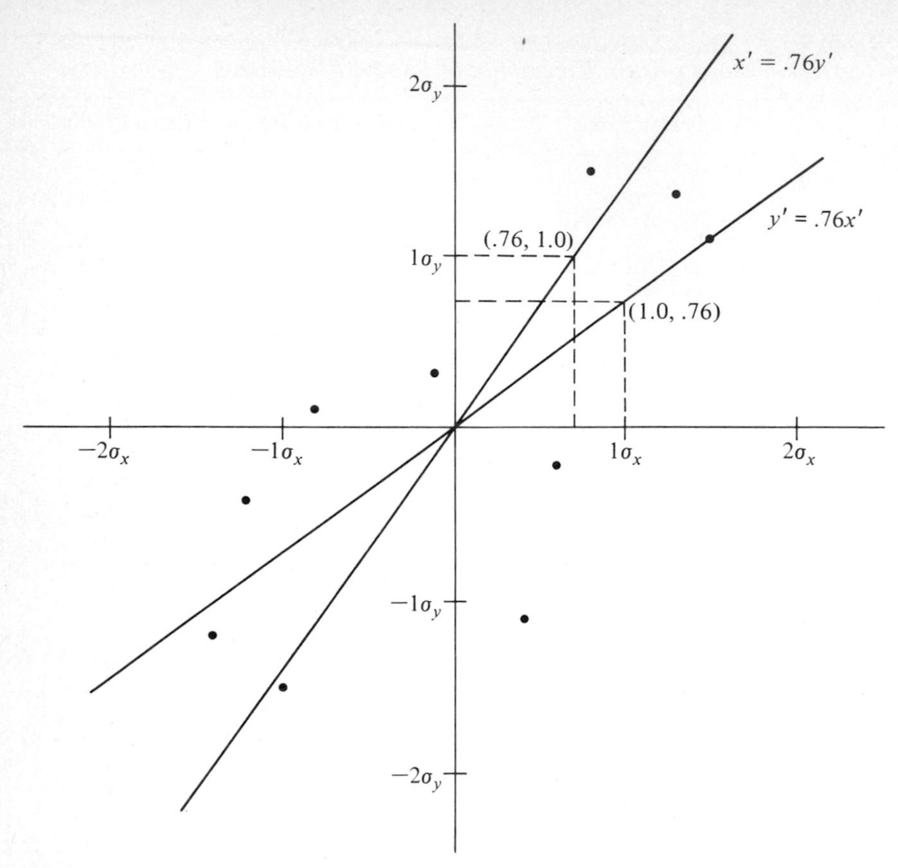

Figure 9.3.2 *Scatter Diagram of Standard Measures, with Regression of* y′ *on* x′ *and Regression of* x′ *on* y′

points from the regression line, whereas the regression line of x' on y' is located so as to minimize the sum of the squared *horizontal* deviations. Since the best fitting regression lines necessarily pass through the point at the intersection of guide lines extending perpendicularly from the two means, for standardized variates both regression lines pass through the point 0,0. Hence the y' intercept and the x' intercept are at zero. The regression equations for standard measures therefore assume the general forms:

$$\hat{y}' = rx' \tag{9.3.9}$$

$$\hat{x}' = ry' \tag{9.3.10}$$

where $x' = \dfrac{X - \overline{X}}{\sigma_x}$

$$y' = \frac{Y - \overline{Y}}{\sigma_y}.$$

(It should be noted that, while $\hat{y}'$ in Equation 9.3.9 is the product of a standard measure, x', and a constant, r, it is not itself a standard measure. It has a mean of zero but a standard deviation smaller than 1.00. The same point holds for $\hat{x}'$ in Equation 9.3.10.)

Since r is the slope of the regression line for variates expressed as standard measures, it may be construed as the mean change in Y for every unit change in X, or the reverse, for standard measures. Thus, if a given husband's occupational prestige deviates by 1 sigma (σ) from its mean, we predict the associated family status score will deviate from its mean by .76 sigma (σ).

The slope of the regression line for variates expressed as standard measures, r, is commonly referred to as the *correlation coefficient*, the *product-moment correlation coefficient*, or *Pearsonian r*. The measure was devised by Karl Pearson and hence bears his name. In statistics, the mean of a variate is sometimes referred to as a *moment coefficient*; if the variate is a product, then we refer to the *product-moment coefficient*. Because his measure was the mean of the cross-products of standard measures, Pearson referred to his measure of correlation as the product-moment correlation coefficient. His name for this coefficient has been shortened in common usage to correlation coefficient.

The Relative Reduction in Prediction Error Having found the equation for the best fitting regression line, we may now proceed to determine how much improvement in predictive accuracy it affords. We therefore seek a measure with the general form of a relative reduction in error measure:

$$\frac{E_1 - E_2}{E_1}.$$

Prediction error in this instance is measured by the sum of squared deviations, as in the case of the correlation ratio. Prediction error in predicting Y from knowledge of its own distribution is therefore the sum of squared deviations of the Y values around $\overline{Y}$; thus, $E_1 = \Sigma(Y - \overline{Y})^2$. Prediction error in predicting Y from knowledge of X is measured by the sum of squared deviations around the regression line; thus $E_2 = \Sigma(Y - \hat{Y})^2$. We therefore express the relative reduction in prediction error as follows:

$$\frac{E_1 - E_2}{E_1} = \frac{\Sigma(Y - \overline{Y})^2 - \Sigma(Y - \hat{Y})^2}{\Sigma(Y - \overline{Y})^2}. \tag{9.3.11}$$

The computation of these sums of squared deviations and of the relative reduction in prediction error is illustrated in Table 9.3.4. A third sum of squared deviations is also displayed in Table 9.3.4; the sum of squared deviations of the predicted values from the mean of Y, $\Sigma(\hat{Y} - \overline{Y})^2$. This computation is included to illustrate once again the additive nature of the sums of squared deviations. As in the case of the correlation ratio, the total variation, $E_1 = \Sigma(Y - \overline{Y})^2$, may be decomposed into two components: the

Table 9.3.4 *Illustrative Computation of Total Variation, Error Variation, and Explained Variation (from the Data of Table 9.3.1)*

X	Y	$\overline{Y}$	$\hat{Y} = 10.73 + .935\,X$	Total Variation E_1 $(Y - \overline{Y})^2$	Error Variation E_2 $(Y - \hat{Y})^2$	Explained Variation $(\hat{Y} - \overline{Y})^2$
75	80	50	80.855	900	.7310	952.0310
70	87	50	76.180	1369	117.0724	685.3924
60	91	50	66.830	1681	584.1889	283.2489
55	44	50	62.155	36	329.6040	147.7440
50	22	50	57.480	784	1258.8304	55.9504
40	58	50	48.130	64	97.4169	3.4969
25	52	50	34.105	4	320.2310	252.6510
20	10	50	29.430	1600	377.5249	423.1249
15	38	50	24.755	144	175.4300	637.3100
10	18	50	20.080	1024	4.3264	895.2064
420	500	500	500.000	7606	3265.3559	4336.1559

Total variation = *error variation* + *explained variation*

$$7606 = 3265 + 4336$$

$$= 7601^{a}$$

$$\frac{E_1 - E_2}{E_1} = \frac{4336}{7606} = .57$$

[a] Equality not exact because of accumulated rounding error.

error variation, $E_2 = \Sigma(Y - \hat{Y})^2$, and the explained variation, $\Sigma(\hat{Y} - \overline{Y})^2$. Since $E_1 - E_2$ = explained variation, we may write the relative reduction in prediction error in the following form:

$$r^2_{yx} = \frac{E_1 - E_2}{E_1}$$

$$= \frac{\Sigma(\hat{Y} - \overline{Y})^2}{\Sigma(Y - \overline{Y})^2}$$

$$= \frac{\textit{Explained variation}}{\textit{Total variation}}. \qquad (9.3.12)$$

In Table 9.3.4, the figures show that 57 percent of the total variation is explained variation. Alternatively stated, the regression line, $\hat{Y} = 10.73 + .935\,X$, predicts family social status score with 57 percent less prediction error than the overall mean of 50.

Although the computations are not shown here, the same relative reduction in prediction error would be obtained if we made the computations on standardized measures or if we computed the relative reduction in prediction error for predicting X from Y instead of Y from X.

Computing Formulae The above computations for r and for the relative reduction in prediction error have been carried out in an illustrative manner designed to reveal the nature of these measures, rather than in the most convenient way. The computation of r does not require that each variate first be transformed into a standard measure; r may be computed from the raw measures as follows:

$$r = \frac{N\Sigma XY - (\Sigma X)(\Sigma Y)}{\sqrt{[N\Sigma X^2 - (\Sigma X)^2][N\Sigma Y^2 - (\Sigma Y)^2]}}. \tag{9.3.13}$$

The requisite sums appear in Table 9.3.2, and by making the appropriate substitutions in Formula 9.3.13 we have

$$\begin{aligned} r &= \frac{256{,}400 - 210{,}000}{\sqrt{[226{,}000 - 176{,}400][326{,}060 - 250{,}000]}} \\ &= \frac{46{,}400}{\sqrt{(49{,}600)(76{,}060)}} \\ &= \frac{46{,}400}{\sqrt{3{,}772{,}576{,}000}} \\ &= \frac{46{,}400}{61{,}421} \\ &= .755\,. \end{aligned}$$

This result is identical to that obtained by finding the covariance of the standard measures (Table 9.3.3) although rounding errors will sometimes make for minor differences in the results of the two computing procedures. The student should recall that even though r may be computed directly from the raw measures without first transforming them into standard measures, as illustrated above, the essential meaning of r as a regression slope applies only to standard measures.

The computation of the relative reduction in prediction error does not require that the regression predictions for all cases be computed, as illustrated in the computations in Table 9.3.4 above. The meaning of the relative reduction in prediction error is best conveyed by finding the errors of prediction around the overall mean and the errors of prediction around the regression line, but this quantity may be computed more directly. The algebraic equivalence is not demonstrated here, but the relative reduction in prediction error can be most readily computed by squaring r. Thus

$$\frac{E_1 - E_2}{E_1} = \frac{\Sigma(Y - \bar{Y})^2 - \Sigma(Y - \hat{Y})^2}{\Sigma(Y - \bar{Y})^2} = \frac{\textit{Explained variation}}{\textit{Total variation}} = r^2 \quad (9.3.14)$$

Having obtained r by Formula 9.3.13 above, we may proceed to compute the relative reduction in prediction error:

$$r^2 = (.755)^2 = .570$$

We may also note that the regression slopes for unstandardized measures, b_{yx} and b_{xy} can be computed from r and the standard deviations of X and Y. The appropriate computing formulae are:

$$b_{yx} = r\frac{\sigma_y}{\sigma_x} \quad (9.3.15)$$

$$b_{xy} = r\frac{\sigma_x}{\sigma_y}. \quad (9.3.16)$$

Using Formula 9.3.15, we compute

$$b_{yx} = .755\frac{27.58}{22.27} = .935.$$

We may also compute r from the regression slopes for unstandardized measures. From Equations 9.3.15 and 9.3.16, it follows that

$$r = b_{yx}\frac{\sigma_x}{\sigma_y} = b_{xy}\frac{\sigma_y}{\sigma_x}. \quad (9.3.17)$$

Rank Order Correlation When cases have been ranked from high to low on a given variable, the rank of each case is sometimes represented by an ordinal number. If there are 15 cases, and the highest is represented by "1" the lowest would be represented by "15." Since these numbers show order alone, they are labeled with ordinal numbers. Unlike cardinal numbers, ordinal numbers do not measure intervals between cases. For example, the first case may be barely above the second case, and both may be far above the third, but such inequalities among intervals are disregarded in the assignment of ranks.

The study of social phenomena sometimes calls for a measure of the degree of correspondence or agreement between two rankings. For example, one may wish to know the degree of agreement between the seriousness rankings of crimes by blacks and whites, males and females, or urban and rural residents. Because such rankings are ordinal measures, the association between ranks should, strictly speaking, be measured by a measure of association for ordinal data (for example, gamma (γ), Chapter 8). Nonetheless, it has long been common in research to measure the correlation between ranks by the product-moment correlation coefficient. Because of this long precedent, we discuss briefly this application of the correlation coefficient, which entails treating ordinal numbers as if they were cardinal numbers.

There is no urgent necessity for a special formula for the computation of the correlation between ranks; applying Formula 9.3.13 to the two series of ranks will yield the correlation between them, just as it yields the correlation between any two series of paired quantitative variates. However, the mean of a set of ranks (without ties) depends only on the number of ranked items, N:

$$\textit{Mean rank} = \bar{R} = \frac{N+1}{2} .$$

Also, provided there are no tied rankings, the sum of squared deviations of each rank from the mean rank depends only on N:

$$\textit{Sum of squared deviations} = \Sigma(R - \bar{R})^2 = \frac{N(N^2 - 1)}{12} .$$

These specialized expressions permit the derivation (not shown here) of a simplified computing formula for the product-moment correlation for ranks, which is commonly referred to as the *rank order correlation coefficient*, here symbolized by ρ_s.[2] The computing formula is as follows:

$$\rho_s = 1 - \frac{6\Sigma D^2}{N(N^2 - 1)} \tag{9.3.18}$$

where D = difference between paired ranks
N = number of items ranked.

We illustrate the application of this computing formula with the data of Table 9.3.5, showing the ranking of occupations according to their prestige, average income, and psychosis rate. The differences in ranks for income and occupational prestige, and the squares of these differences, are shown in the table. The sum of these squared differences and the number of items ranked are the only quantities required to compute the rank order correlation between occupational income and occupational prestige. For these variables

$$\begin{aligned} \rho_s &= 1 - \frac{6(126)}{17(289 - 1)} \\ &= 1 - \frac{756}{4896} \\ &= 1 - .15 \\ &= .85 . \end{aligned}$$

[2] The rank order correlation coefficient is traditionally symbolized by ρ (rho). This use of a Greek letter symbol for a statistic violates the convention that reserves Greek letters for population parameters. To preserve the symbol ρ for the parameter corresponding to the statistic r, while also abiding by the well-established tradition of symbolizing the rank order correlation by ρ, we have here attached to ρ the subscript s, for Spearman, the statistician who originally developed the computing formula given in Equation 9.3.18. Thus ρ_s may be read "Spearman's rho."

Table 9.3.5 *Occupational Groupings Ranked by Income, Prestige, and Psychosis Rate*

(1)	(2)	(3)	(4)	(5)	(6)	(7)	(8)
Occupational Grouping	Income Rank	Prestige Rank	D (2)–(3)	D^2	Psychosis Rate Rank	D (3)–(6)	D^2
Peddlers	17	17	0	0	2	15	225
Waiters	16	16	0	0	1	15	225
Domestics	15	14	1	1	4	10	100
Barbers, beauticians	14	13	1	1	7	6	36
Semiskilled and unskilled	13	15	−2	4	3	12	144
Sales people	12	9	3	9	8	1	1
Skilled workers	11	12	−1	1	6	6	36
Office employees	10	8	2	4	14	−6	36
Semiprofessional (e.g., druggists)	9	4	5	25	9	−5	25
Small tradespeople	8	5	3	9	15	−10	100
Subexecutives	7	6	1	1	10	−4	16
Police officers, fire fighters	6	10	−4	16	13	−3	9
Major salespeople	5	7	−2	4	16	−9	81
Minor government employees	4	11	−7	49	5	6	36
Clergy, teachers, social workers	3	2	1	1	12	−10	100
Technical engineers	2	3	−1	1	11	−8	64
Large owners, doctors, lawyers, dentists	1	1	0	0	17	−16	256
				126			1490

Source: Robert C. Clark, "Psychoses, Income and Occupational Prestige," *American Journal of Sociology* 54, 1949, p. 433.

This result indicates a relatively high degree of agreement between the rankings of these occupations on income and prestige. Approximately 72 percent ($\rho_s^2 = .85^2 = .72$) of the variation in one set of ranks is predicted by variation in the other.

For the rank order correlation between occupational prestige and psychosis rate we have

$$\rho_s = 1 - \frac{6(1490)}{17(289 - 1)}$$

$$= 1 - \frac{8940}{4896}$$

$$= 1 - 1.83$$

$$= -.83 .$$

This result indicates a relatively high *negative* correlation between the rankings of these occupations on prestige and psychosis rate; for instance, an occupational grouping ranked high in prestige is likely to be ranked low in psychosis rate, and vice versa. Approximately 69 percent ($\rho_s^2 = -.83^2 = .69$) of the variation in one set of ranks is predicted by variation in the other, but the covariance of the ranks is negative.

If two or more cases seem to share the same rank they are said to have *tied ranks*. Since the number of ranks and the number of items must coincide, it simply will not be possible for two items to share the same rank, so they must be given adjoining ranks. In such instances, both items will be assigned the arithmetic mean of the adjoining ranks. Thus, if the third and fourth items are tied, each is given the rank of 3.5. If three or more items are tied, the same rule of averaging applies.

Formula 9.3.18 is based on the assumption of no tied ranks, and if ties appear, that equation will yield an inflated value for ρ_s. Alternatively stated, with tied ranks, the results given by Formulas 9.3.18 and 9.3.13 will not be identical, and Formula 9.3.18 will yield a value that is too large in absolute value. If the number of tied ranks relative to the number of items ranked is small, the effect will be slight and may be disregarded. If the number of ties is large, Formula 9.3.18 will yield a heavily inflated value of ρ_s and that equation should therefore not be used. In this circumstance, the investigator should reconsider the question of whether a product-moment correlation between the heavily tied ranks is appropriate. If so, Formula 9.3.13 may be used to obtain an uninflated value of ρ_s, and if not, some alternative measure (for example, gamma (γ)) may be computed.

Point Biserial Correlation (r_{pb}) In Section 1 of this chapter we computed the correlation ratio between sex and salary for librarians. That example illustrates the general class of problems in which the independent variable is a dichotomy and the dependent variable is a quantitative variable. In such a circumstance, the two classes of the dichotomy may be coded 0 and 1 and the product-moment correlation between this coded dichotomy (X) and the quantitative variable (Y) may be computed by Formula 9.3.13. Such a correlation is referred to as a *point biserial correlation coefficient*,[3] and may be symbolized r_{pb}. The student may wish to compute r_{pb} for the data of Table 9.1.1 and verify the following:

$$r_{pb}^2 = \eta^2 = \text{the correlation ratio}$$

$$r_{pb}\frac{\sigma_y}{\sigma_x} = b_{yx} = \text{the difference between means}$$

$$a_{yx} = \text{the mean } Y \text{ of the } X \text{ category coded 0.}$$

[3] The point biserial correlation coefficient is not to be confused with the *biserial correlation coefficient*, which might be computed for the same data. The biserial correlation coefficient, which is not discussed in this book, is an estimate of what the correlation coefficient between the two variables would be if both were measured as continuous variables with a bivariate normal distribution. The point biserial correlation coefficient should not be construed as such an estimate.

The Correlation between Two Dichotomies (ϕ) The two classes of any dichotomized variable may be coded 0 and 1, and the correlation between two such coded dichotomies may be computed by Formula 9.3.13. This correlation coefficient, commonly referred to as ϕ (phi), may be computed as follows:

$$\phi = \frac{ad - bc}{\sqrt{(a + b)(c + d)(a + c)(b + d)}}. \tag{9.3.19}$$

The letters represent frequencies in the cross-classification of the dichotomies as follows:

		X 0	X 1
Y	0	a	b
Y	1	c	d

As an exercise, the student may wish to code the double dichotomy shown in Table 8.2.5 into two 0,1 variables and verify that r computed by Formula 9.3.13 is the same as ϕ (Formula 9.3.19). The student may then proceed to verify that $\phi^2 = \tau$ (see Chapter 8, Section 3).

Comparison of r and r^2 On the surface, the difference between r and r^2 appears trivial: a simple detail of exponent, with each value easily convertible into the other, after one has been computed. Nevertheless, these two measures respectively focus on two distinct but interrelated aspects of co-variation—a distinction which the design of this chapter has deliberately sought to portray.

Although it is computationally convenient to obtain r^2 from r, an r^2 derived from r does not convey the meaning of the relative reduction in prediction error or the proportion of total variation that is explained, which r^2 measures. The "square of the slope" suggests nothing about the relative magnitude of either prediction error or explained variation. Neither does an r, converted arithmetically from r^2, convey the meaning of the slope. The "square root of the proportion of variation explained" cannot be visualized as slope. Although r and r^2 are interdependent, r^2 measures the proportion of the total variation in one variable that is associated with, or "explained" by the other. On the other hand, r measures the rate of change in one variable relative to the other, where both have been represented in standard form. Because of this distinction, r is primarily a predictive device whereas r^2 measures the relative accuracy of the predictions.

Since r is slope, it manifestly must have a direction, predominantly up or down, according to whether the variables are positively or negatively related.

It follows that the direction of the slope reflects the type of relation, which is then symbolized by a plus or minus sign. Since r^2 is a proportion of the total variation, it necessarily carries no sign.

It should be clear that r and r^2 are not interchangeable; nor should they be mechanically derived from one another until their structural meanings are thoroughly understood. Since r is always larger than r^2, it could be used deliberately or unwittingly to exaggerate the strength of the association and thereby mislead the reader; for example, $r = .5$ may seem to signify a reasonably strong association, but $r^2 = .25$ indicates that only 25 percent of the variation in either variable is accounted for by the other, or, alternatively stated, that the regression line affords a 25 percent reduction in prediction error. When the emphasis is on the strength of the overall relationship between two variables, as is frequently the case in sociological studies, r^2 is the pertinent statistic.

QUESTIONS AND PROBLEMS

1. Define the following concepts:

error variation	line of least squares
explained variation	product-moment correlation coefficient
total variation	slope
unexplained variation	Y-intercept
regression line	

2. If $r_{yx} = .3$, how much of the variation in Y is linearly associated with variation in X? If $r_{xy} = .3$?

3. Both X and Y are quantitative variables. Dividing X into a series of equal intervals, η^2_{yx} is computed. Under what conditions will η^2_{yx} be equal to r^2_{yx}?

4. Both X and Y are quantitative variables. Show graphically the conditions under which r_{yx} would be approximately 0 and η^2_{yx} would be 1.00.

5. Compute r and r^2 between state income and percent of the population mobile (Table 9.2.2). Interpret your results.

6. Compute b_{xy} for the data of Table 9.3.1, that is, find the regression slope of the husband's occupational prestige on the family social status score. Compare with b_{yx}. Compute a_{yx} and compare with a_{yx}.

7. For every rise or fall of one percentage point in the unemployment rate of whites in the United States, would you anticipate that the unemployment rate of nonwhites would rise or fall, on the average, by more than one percentage point or by less than one percentage point? Restate your preliminary answer as a hypothesis about the magnitude of the regression coefficient of nonwhite rates on white rates and use the data of Table 9.3.6

***Table 9.3.6** Unemployment Rates for White and Nonwhite Males, 16 Years and Over, in the United States, 1948–1974*

Year	Unemployment Rate	
	White	Nonwhite
1948	3.4	5.8
1949	5.6	9.6
1950	4.7	9.4
1951	2.6	4.9
1952	2.5	5.2
1953	2.5	4.8
1954	4.8	10.3
1955	3.7	8.8
1956	3.4	7.9
1957	3.6	8.3
1958	6.1	13.8
1959	4.6	11.5
1960	4.8	10.7
1961	5.7	12.8
1962	4.6	10.9
1963	4.7	10.5
1964	4.1	8.9
1965	3.6	7.4
1966	2.8	6.3
1967	2.7	6.0
1968	2.6	5.6
1969	2.5	5.3
1970	4.0	7.3
1971	4.9	9.1
1972	4.5	8.9
1973	3.7	7.6
1974	4.3	9.1

Source: U.S. Department of Labor, *1975 Manpower Report of the President.* Table A–20. Washington, D.C.: U.S. Government Printing Office. April, 1975.

to check your hypothesis. What proportion of the variance in the unemployment rates of whites and nonwhites over this 27-year period is common variance (that is, what is r^2)?

8. School segregation indexes for 1967 are given for 60 cities in Table 9.3.7. Residential segregation indexes for 1960 are given for the same 60 cities in Table 9.1.2. Construct a scatter diagram for the relationship between residential and school segregation. Compute r and r^2 and interpret. Is the regression slope of school segregation on residential segregation greater or smaller for southern cities as compared to cities outside the South? Interpret your results.

Table 9.3.7 *School Segregation Indexes for 60 United States Cities, 1967*

City (by Region)	School Segregation Index	City (by Region)	School Segregation Index
Northeast		South (continued)	
Boston	74	Birmingham	94
Buffalo	80	Charlotte	77
Jersey City	57	Corpus Christi	77
Newark	68	El Paso	51
New York City	52	Fort Worth	93
Philadelphia	76	Houston	92
Pittsburgh	85	Jacksonville	92
Rochester	61	Louisville	76
Yonkers	60	Memphis	95
		Miami	92
North Central		Nashville	85
Akron	70	New Orleans	87
Chicago	92	Norfolk	90
Cincinnati	77	Oklahoma City	97
Cleveland	90	Richmond	95
Columbus	81	Saint Petersburg	91
Dayton	90	San Antonio	88
Des Moines	76	Tampa	88
Detroit	79	Tulsa	97
Indianapolis	85	Washington, D.C.	77
Kansas City	79		
Milwaukee	88	West	
Minneapolis	74	Denver	82
Omaha	88	Long Beach	78
Saint Louis	91	Los Angeles	89
Saint Paul	62	Oakland	64
Toledo	80	Portland, Oregon	74
Wichita	86	Sacramento	39
		San Diego	78
South		San Francisco	67
Atlanta	95	San Jose	49
Austin	86	Seattle	65
Baltimore	87	Tucson	68

Source: Reynolds Farley and Alma F. Taeuber, "Racial Segregation in the Public Schools," *American Journal of Sociology* 79 (January 1974): 888–905.

4 SOME GENERAL GUIDES TO THE INTERPRETATION OF CORRELATION

The skillful computation of an abstract index of correlation does not exhaust the sociologist's responsibility, nor is the calculated result—often obtained more or less routinely—a guarantee that the sociological significance has been grasped. It is quite possible to be thoroughly familiar with the purely statistical character of a correlational index and still be completely unaware of its sociological import.

The measurement of a correlation may be reduced to a mere computing routine, but the interpretation of that correlation cannot be similarly routinized. Although some statistical procedures (to be discussed in Chapter 10) are undoubtedly useful in interpreting correlations, no statistical procedure can substitute completely for an intimate familiarity with the subject matter or for creative intelligence in solving the riddle of how variables come to be related to one another. The following statements concerning the possible sources of statistical linkage between variables may be helpful in analyzing observed associations.

(1) Correlation as Evidence of Cause and Effect In lay language, as well as in statistical usage, there is no concept so commonly employed, and so useful in the understanding of the world about us, as the concept of *causation*. We speak of the causes of death, the causes of accidents, the causes of divorce, of delinquency, and of crime. It is thought that, if the causes of delinquency, disease, and war could only be uncovered, we could control their occurrence. Statistical researches in biology are built around correlations that are designed to reveal, for example, the effect of nutrition on plant and animal growth.

The causal influence may flow unilaterally in one direction, or bilaterally in both directions simultaneously. For example, long hours of study may cause good grades, and good grades also may encourage more study; higher wages may cause higher prices, and higher prices bring higher wages in their famous spiral effect. Marriage may increase length of life, and healthy, long-lived people tend to marry.

In spite of the wide prevalence of the concept of causation and its quite obvious utility, in some intellectual circles there is a fashionable objection to it. The notion that one variable may exercise a force upon another seems to some thinkers too mystical to be credible. According to one school of thought, "causation" is merely empirical association and sequence. As Hume's argument runs, we cannot perceive causation, only statistical contingency.

According to this view, therefore, we may state only that correlations reveal a mathematical relation, in the sense that values vary with one another. Consequently, a rugged empiricist will leave the question open as to whether there is such a thing as causation in the first place, and if so, whether its direction can be ascertained by statistical means.

Although these sophisticated questions are of undoubted fascination, common sense will not renounce the concept of cause. No amount of disputation will raise doubts in the mind of a gardener that *for all practical purposes* rain "causes" the plants to grow and not the other way around. An intellectual who may reject the concept of cause will nevertheless, in real life, act as though it exists. The applied scientist, who is called on to manipulate and control, will use the concept of causative factors as an indispensable guide to action. Statistical regression is an aid in discovering, if only by inference, where causation operates, and how strong it is.

Such causal inferences are essentially quite similar to inferences drawn from carefully controlled experimental studies. If an experimentally induced change in one variable is followed by a change in a second while all other variables remain unchanged, variation in the first variable is identified as a cause of variation in the second. The second variable may have still other causes as well, for example, a change in the other variables also may lead to a change in it. In fact, it is common in laboratory studies to vary two experimental conditions to appraise the causal impact on the dependent variable of each individually and both jointly.

The existence of a correlation between two variables is not an unfailing clue to the existence of a cause-and-effect relationship between them. Although a correlation based on nonexperimental data indicates that Y changes as X changes, it carries no assurance that all other variables have remained fast. Thus, the observation of a positive correlation between shoe size and reading ability among elementary school children would not be interpreted as an indication that variation in shoe size leads to variation in reading ability. The correlation arises because with increasing age both shoe size and reading ability increase—older children have larger feet and have learned to read better than younger children. The covariation of two variables, X and Y, *may* arise because X has an effect on Y, Y has an effect on X, or each has an effect on the other. A causal relationship is only one among several possible sources of covariation.

That the effect of one variable on another is *mediated* through an intervening variable does not change our conception of cause-and-effect relationships in any fundamental way. Practically all causal relations between variables are mediated by intervening variables, even though these may remain unspecified. Thus, the effect of rainfall on corn yield is achieved through processes of plant chemistry that could be represented as intervening variables. An increase in wages leads to an increase in prices through the intervening variable of increased production costs. Therefore, when we affirm a causal relation between two variables, we do not deny the presence of an intervening variable mediating the effects of one on the other. In Figure 9.4.1 we show schematic causal diagrams in the form of arrows to represent effects in a specified direction. In Figure 9.4.1a, we have classified as causal correlations those that arise from the effect of one variable on another, whether or not the intervening variable mediating that relationship is explicitly identified.

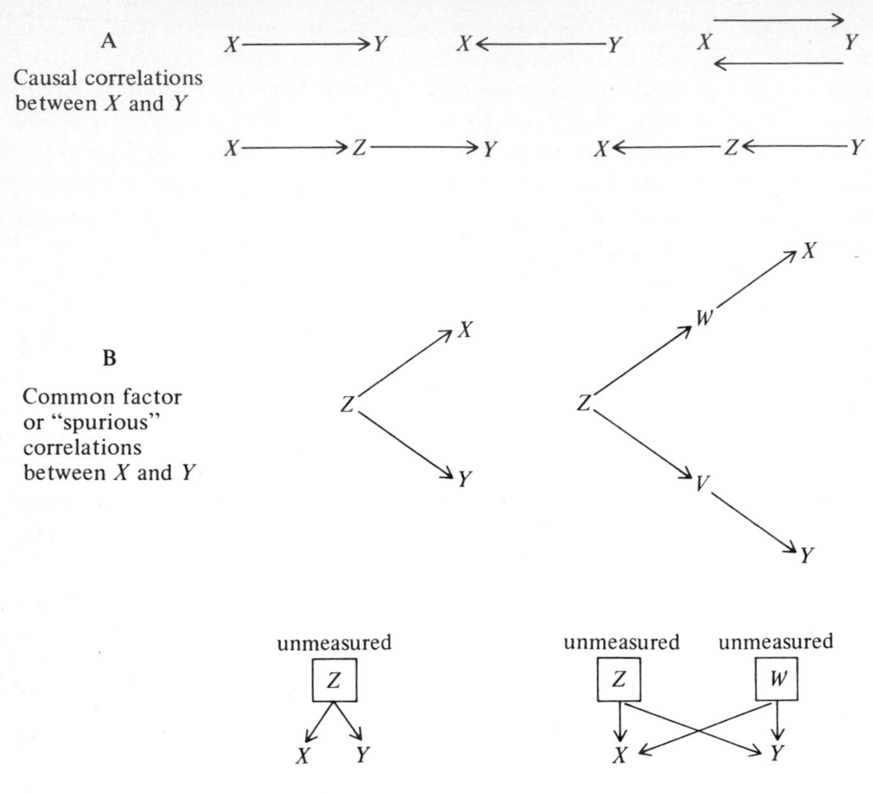

Figure 9.4.1 *Schematic Causal Diagrams*

(2) Common Factor and "Spurious" Correlations The high correlation expected between grades in economics and sociology cannot be explained primarily by a causal relation between the two, but rather by the common factors which make for academic achievement in these two related subjects. The grades achieved in economics and sociology are merely slightly different manifestations of almost exactly the same phenomenon, the components of which are intelligence, study habits, grading system, similarity of subject matter, common motivation, and interest. High delinquency and high truancy similarly may be associated, not because one causes the other, but because they too, like sociology and economics marks, are affected by the same underlying factors. In the field of heredity, the characteristics of siblings, which show many similarities, likewise are the results of common antecedent hereditary factors. In fact, this was the original conception of correlation as propounded by Francis Galton, who first gave currency to the concept through his studies in heredity during the last quarter of the nineteenth century.

The distinction between causal correlations on the one hand, and *common factor* or *spurious* correlations, on the other, is shown schematically in

Figure 9.4.1b. The diagrams represent the circumstance in which a correlation between X and Y occurs because both are affected by common factors, which may or may not have been measured. In this circumstance it is somewhat misleading to refer to r^2 as the proportion of variance in one variable that is explained by the other, even though that terminology is well established as a description for the meaning of r^2. For the diagrams in Figure 9.4.1b it is appropriate to regard r^2 as the proportion of variance in X and Y that is due to the common factor present in both, rather than as the proportion of variance in Y that is explained by X.

The bivariate correlation formula accommodates only two variables, but by restricting our computations to two variables we do not thereby immobilize or exclude from operation, within the field before us, the numerous other factors that continue to produce their effects on our measures whether or not we happen to be looking at them. For example, when we measure a person's height or weight, we also are measuring the effects of that person's age, sex, race, lifelong nutrition, and way of life, and even the time of day, because height normally shrinks during the day, and weight increases. These factors that intrude in our measurements are called *concealed factors*, because the surface information carries the effect of such hidden factors without identifying them. If ministers' salaries are positively correlated with the amount of alcohol consumed by the members of their congregations, we reasonably doubt that such a correlation arises because of a cause-and-effect relationship between these two variables. We are inclined instead to assume that the correlation is spurious, that it arises from the effects of such concealed factors as the size of the city and the socioeconomic status of the church members, which factors affect both the ministers' salaries and alcoholic consumption among congregations. The intrusion of concealed factors into an observed correlation is not always so evident as in this example, and we therefore run the risk of crediting to one variable the effects of its concealed correlates.

A *spurious correlation* may be defined as a correlation in which hidden factors exert an effect on Y which is erroneously credited to X. The correlation between X and Y is not unreal as a consequence, but the assumption of a cause-and-effect relationship between X and Y is false. Clearly, it is the interpretation that is spurious, rather than the statistical correlation as such.

Techniques of statistical control permit us to remove the effect of a third variable in exploring the correlation between two variables of special interest. Standardization was presented in Chapter 6 as one such technique, and the elaboration of association (subclassification) was discussed in Chapter 8 as another procedure that may be useful in ferreting out the substantive meaning of statistical associations. The techniques of partial correlation and multiple regression to be discussed in Chapter 10 offer additional aids for the interpretation of correlation. But even with such aids as these, it remains difficult in some instances to distinguish between the effect of an intervening variable and a common antecedent variable.

(3) Artifactual Correlations Correlations that arise because of the effects of variables that remain concealed in the analysis are *real correlations*. They exist in nature, so to speak, although we misinterpret them and infer from them a cause-and-effect relationship when none exists. *Artifactual correlations*, in contrast, do not exist in nature, but emerge from the data because they have been "built into the data" by the investigator. In this sense, they are the artifacts of the social analyst and may, accordingly, be called artifactual correlations. One type of artifactual correlation is sometimes labeled a *part-whole correlation*, indicating that the same observations are duplicated, in part, in the two series of measures. In a correlation of freshman and all-university grades in American colleges and universities, the freshman grades appear as duplicates within the total university measures. To the extent that freshmen are a smaller or larger component of the respective student bodies, the resulting correlation automatically will be lower or higher, for the very evident reason that correlation between near-identities is necessarily high. The result will be tautological rather than substantive, and it would be more informative to correlate the separate components rather than correlating the part with the whole.

The data for correlations in social science are frequently in the form of synoptic measures, such as means, rates, and percentages. Special care must be exercised in interpreting correlations based on such synoptic measures. In the correlation between the death rate and the size of the population, for example, population size appears as a component of both variables. If we choose pairs of numbers at random, call one of each pair X and the other Y, we would expect no correlation between them. But if we create a synoptic variable, X/Y and correlate this with Y, we expect a negative correlation. This is analogous to correlating the death rate with the size of the population, where the death rate has population size as its denominator. Because the composition of the synoptic measure in this instance makes the correlation "lean" in the direction of a negative correlation, the observation of a small negative correlation may be an artifact. If the correlation between the death rate and population size is positive, on the other hand, this cannot be artifactual because of the appearance of population size as a component in both variables, and the investigator may reasonably proceed to seek other interpretations.

(4) Chance Elements in Observed Correlations Most observed correlations are based on a sample of measures, each subject to observational error. Not all cases can be tabulated because of practical limitations, so it is necessary to resort to a sample. Our measures may be the best available but still may be subject to sizeable errors in observation. Two variables may appear to be associated simply because the measures in the sample happen to display some association by chance, even though the more complete set of cases would show no association at all. From a scatter diagram representing zero correlation, we might, by chance, select a subset of cases showing either positive or negative correlation, and for a few possible subsets, a high degree of correlation. Sampling variation is discussed in greater detail in Chapters 12 through 15. At this point it is enough to say that the observed correlation always may be expected to differ from the true correlation to some degree, and the smaller the number of cases, the less reliable will be the observed correlation as an estimate of the true correlation.

An investigator must never lose sight of the obvious fact that a correlation formula usually reflects only a few of the variables in the context under study, although memory and understanding may accommodate a very large number of variables, depending upon the investigator's experience, intelligence, and opportunities for observation. Therefore, it is highly probable that on many occasions an experienced worker will, in a certain sense, know much more than a given formula may show. An analyst can attend to a larger number of variables, but probably not quite so accurately as can mathematical tools. A good worker, therefore, always will combine and blend the potentialities of mathematical tools with the subtlety of imagination.

QUESTIONS AND PROBLEMS

1. Discuss how knowledge of subject matter will influence the interpretation of obtained correlation measures.
2. Give several illustrations of spurious correlation, and show how the spurious element may be isolated. (Hint: Review section on Elaboration, Chapter 8.)
3. Define and illustrate a concealed factor, and show how it may deceive an observer.
4. Why are there so many coefficients of association and correlation?
5. What special problems are posed by correlating group means (cf. Robinson, 1950)?
6. Explain why zero correlations are unsatisfactory from the standpoint of the objectives of science.

SELECTED REFERENCES

Robinson, William S.
1950 "Ecological correlations and the behavior of individuals." American Sociological Review 15: 351–357.

Snedecor, George
1956 Statistical Methods. Fifth edition. Ames: Iowa State College Press. Chapters 6 and 7.

10 Partial Correlation, Multiple Regression, and Path Analysis

1 AN OVERVIEW OF MULTIVARIATE ANALYSIS

Purposes and Uses of Multivariate Analysis In social research, it is much more common to pose questions that require multivariate analysis (that is, the simultaneous analysis of three or more variables) than it is to be satisfied with the measure of correlation or association between pairs of variables. The attempt to interpret or explain bivariate correlations commonly gives rise to multivariate questions. For example, in a study of complex organizations, Hall[1] found a positive correlation between the perceived degree of control exercised by the authority hierarchy and the degree to which impersonality was perceived as expected and practiced in interpersonal relationships. What gives rise to such a correlation? Are these two variables positively correlated because both are consequences of the degree of specialization in work roles? Or does a high degree of authority diminish the personal involvement of employees in their work, which leads, in turn, to the development of impersonal relationships with fellow workers? Or are both of these variables manifestations of a more basic feature of organizations, for example, the degree to which they are bureaucratized? Still other mechanisms can conceivably give rise to the positive correlation between perceived control and perceived impersonality, but most such interpretations, like those suggested here, will incorporate additional variables into the analysis; checking the

[1] Richard H. Hall, "The Concept of Bureaucracy: An Empirical Assessment." *American Journal of Sociology* 69 (July, 1963): 32–40.

tenability of these interpretations will then entail some form of multivariate analysis.

Multivariate analysis techniques permit the exploration of the relationship between two variables with selected additional variables controlled or held constant. A variety of such techniques are available, each appropriate for particular purposes. One might ask, for example, if the positive correlation between poverty and the frequency of interracial violence in cities would disappear if city size were statistically controlled, upon the reasonable assumption that larger cities have higher concentrations of poor persons and also provide, for reasons unrelated to poverty, a more favorable setting for incidents of interracial violence. To explore the tenability of the contention that such a bivariate correlation is spurious, the *partial correlation coefficient* may be used. In a similar vein, wc may ask whether the average change in earned income for each additional year of schooling is the same for men and women after controlling for occupational prestige and other variables. *Unstandardized partial regression coefficients* are responsive to this research need and permit an investigator to describe such an average change in one variable per unit change in another variable, with still other variables held constant. The comparison of such coefficients for men and women will then serve to indicate whether an additional year of schooling is worth the same amount of income for both sexes, after "partialling out" the effects of other variables such as occupational prestige.

We may also be interested in comparing the relative effects of two different variables in the same population as well as the relative effects of the same variable in two populations. For example, we may be interested in determining for males whether income changes more as education changes or as IQ changes. Even though the unstandardized partial regression coefficients will indicate the average change in income per unit change in education with IQ controlled and the average change in income per unit change in IQ with education controlled, these results are not comparable to each other because the units of measure for education and IQ are not comparable. To make them comparable, all variates may be expressed as standard deviates (deviations from their respective means measured in their respective standard deviation units). Regression coefficients based on variables measured in standard deviate form are known as *standardized partial regression coefficients* and permit an informative comparison to be made of the effects of variables measured in different units.

In still other circumstances, an investigator can readily conceive of variables affecting each other in a well-defined causal sequence, and one may wish to measure separately the direct effects and the indirect effects (those effects mediated by a specified intervening variable). For example, one may assume that association with deviant persons has an effect on attitudes toward the norms that prohibit such deviation, and that these attitudes, in turn, affect the propensity to engage in such deviation. But one may also assume that association with deviant persons has an effect on the propensity to engage in such deviation through mechanisms other than the attitudes that such association fosters. Thus, one may wish to measure separately the effect of association with deviant others as mediated through attitudes (the indirect effect) and compare its magnitude with the direct effect of such association (for example, the effects mediated by other unnamed variables or mechanisms). One's conclusions about the processes of becoming deviant will differ depending on whether the direct effect or the indirect effect constitutes the predominant part of the observed correlation between deviant associates and deviant behavior. For such purposes as this, *path analysis* or *path regression analysis* will be useful.

Multivariate analysis techniques may be viewed as a statistical adaptation to meet the impracticality of conducting large scale social experiments on certain topics. In an experiment the investigator systematically varies an independent variable and observes the effect of that variation on a dependent variable, while literally holding constant or equalizing the effects of other variables that might otherwise confound the conclusions from the experiment. An investigator utilizing the multivariate analysis techniques discussed in this chapter does not ordinarily manipulate a given variable in order to observe its effects. Instead, the investigator takes advantage of existing variations that occur naturally. But since these "natural" variations are ordinarily accompanied by (or covary with) variations in other dimensions that confound the conclusions that might be drawn from the description of bivariate correlations, the effects of these other extraneous variables need to be removed. These extraneous variables are not literally held constant in multivariate analysis, but statistical adjustments are made to remove their estimated effects. In this way, the statistical findings are uncontaminated by the intrusion of the extraneous variables controlled. The major shortcomings of these techniques stem from (a) imperfections in measurement—statistical adjustment for an imperfectly measured variable will control for that variable imperfectly; (b) the difficulty of including all relevant variables in the analysis—only those variables that are measured can be included, and measures may not be available for all potentially relevant extraneous variables; and (c) uncertainties about causal direction—although we may assume that X affects Y, the correlation between the two may be due, in part, to the effect of Y on X.

In spite of these shortcomings commonly encountered in applying multivariate techniques and interpreting their results, these techniques facilitate the understanding of social phenomena in greater depth than the examination of bivariate relationships affords. They permit the exploration of hypotheses concerning the causal structure surrounding large scale, multifaceted, macrosociological events that cannot feasibly be explored in a laboratory setting. And they afford an opportunity to investigate behavior in response to natural conditions, without imposing the artificial conditions that occasionally limit the applicability of the conclusions from laboratory experiments to nonlaboratory settings. Multivariate analysis techniques remain less than ideal for drawing inferences about cause and effect, but they provide a firmer basis for such inferences than bivariate modes of analysis, and they constitute the preferred mode for drawing such inferences when experimentation is not feasible. They are most useful, of course, when combined with a well-articulated theory that provides initial leads to the causal structure underlying the events being explored. Because of their utility, and in spite of their shortcomings (which should not be overlooked in drawing conclusions from their results), multivariate analysis techniques are now widely used in social investigations in an attempt to clarify the causal connections between variables, and multivariate analysis seems to be well on the way to replacing simple bivariate analysis as the core technique of quantitative social research.

The utility of certain multivariate analysis techniques, especially multiple regression analysis and the multiple correlation coefficient, is not limited to checking the tenability of hypotheses about the causal structure of events. These techniques may also be useful in certain kinds of applied prediction problems. For example, colleges and universities are commonly interested in anticipating the academic performance of prospective students, and they find information such as high school grade point averages and scores on achievement tests useful in making predictions about performance levels in college courses. Multiple regression analysis provides a procedure for weighting each such predictor so as to maximize the predictive accuracy of the entire set of predictor variables in a linear regression equation. And regression techniques will aid in selecting the most efficient combination of predictors from a large pool of potential predictors. The multiple correlation coefficient may be used to measure the predictive efficiency of such efforts in terms of the relative reduction in prediction error, just as other measures of association and correlation discussed in previous chapters measure predictive efficiency. These applications of multivariate analysis may proceed without concern for the causal structure responsible for the observed correlations. In current social science research multivariate analysis techniques are commonly used to "fish" for such predictor variables as well as to check the tenability of theoretical propositions about cause-and-effect relationships.

The Logic of Controlling by Statistical Adjustment In Chapter 8, the student was introduced to the concept of exploring the association between two variables while controlling for a third variable. The technique of *subclassification* is one way of holding an extraneous factor constant. Subclassification entails subdividing a set of cases into subsets that are homogeneous on the control variable, and thus it means actually holding the control variable constant (or approximately constant) within such homogeneous subsets. Controlling by statistical adjustment, the mode of control for extraneous factors that is implicit in the multivariate analysis techniques discussed in this chapter, is designed to accomplish the same purpose without the subclassification of cases. Although subclassification is a potentially useful and informative procedure for controlling extraneous variables, it is a procedure that is costly in the number of cases required. The finer the subclassification, and the more variables to be controlled by subclassification, the larger the number of cases required in the total sample in order to have a sufficient number of cases in each subset to permit reliable conclusions to be drawn. By utilizing statistical adjustment to control for extraneous variables, we avoid spreading our cases into many subclasses for separate analysis.

But how can an extraneous variable be controlled without examining cases that are homogeneous on that variable, that is, without subclassifying into homogeneous subsets? The simplest form of statistical adjustment controls for an extraneous variable by making a simple linear estimate of the effect of that extraneous variable, and then removing the variation attributed to that effect. To illustrate, we consider the positive correlation over the years 1930 to 1970 between the mean annual salary of American public school teachers and the proportion of persons 18 to 22 years of age enrolled in colleges and universities. Before concluding from this bivariate correlation that an increased investment in teachers' salaries results in a higher level of college enrollment, we should explore some other possible reasons for the correlation. We might consider first the possibility that there is a long-term trend of increasing demand for higher education (for whatever reason) and a simultaneous long-term inflationary trend in teachers' salaries. The coincidence of the two long-term trends would suffice to yield a correlation between salaries and enrollments even if there were no other connections between these two variables. To guard against the possibility that this coincidence alone is responsible for the correlation between salaries and enrollments, we should express teachers' salaries in "constant dollars," that is, dollars with constant purchasing power. This represents one kind of statistical adjustment designed to guard against drawing erroneous conclusions. Such an adjustment insures that the unit of measure for teachers' salaries is comparable from year to year. But this adjustment may also be conceived as an operation that removes the variance in teachers' salaries that is attributable simply to the fluctuating value of the dollar. Thus this adjustment, like many others encountered in multivariate analysis, consists in removing a part of the variation that threatens to confound the conclusions drawn.

But suppose there remains a positive correlation between the proportion of the college age population enrolled in colleges and universities and teachers' salaries expressed in "constant dollars"? We may entertain the possibility that the correlation is spurious because of the dependence of both salaries and enrollments on the general level of prosperity. When times are good, the argument runs, teachers are paid more and high school graduates are more likely to be able to afford college, while hard times depress teachers' salaries and decrease the likelihood that high school graduates will continue their education in college. If, as suggested, both the level of teachers' salaries and the level of college enrollment are manifestations of the level of general prosperity, then the correlation between salaries and enrollments may be due to this common dependence rather than to an effect of one variable on the other. We may guard against this possible source of spuriousness in the correlation by statistical adjustment, assuming we have an appropriate measure of general prosperity for these same years, for example, the per capita gross national product. With such a measure of prosperity, we may estimate the variation in salaries and in enrollments that is attributable to variations in prosperity, remove that part of the variation, and proceed to examine the relationship between salaries and enrollments with the effects of prosperity removed.

The multivariate statistical techniques discussed in this chapter control for variables extraneous to our immediate interest, not by actually holding them constant, but by making statistical adjustments for their effects. When the effects have thus been "partialled out," we may draw conclusions from results which are no longer confounded by the variables controlled. The general strategies of statistical adjustment entailed in multivariate analysis will be discussed in greater detail as we proceed to consider partial correlation, multiple regression analysis, and path analysis in subsequent sections of this chapter.

QUESTIONS AND PROBLEMS

1. Define the following concepts:
   multivariate analysis
   control by statistical adjustment
   spurious correlation

2. If an investigator hypothesizes that A is positively correlated with B, that C is negatively correlated with D, and that E is positively correlated with F, six variables are entailed in these hypotheses. Does that mean that this investigator's hypotheses are multivariate hypotheses calling for multivariate analysis in testing them against data? Explain your answer.

2 THE PARTIAL CORRELATION COEFFICIENT

In a study of student disorders in New York City high schools during the 1968–1969 academic year,[2] the investigators reasoned that the variation among the high schools—the units of analysis in the inquiry—in the rate of such disorders might be an effect of variation in school and student characteristics such as student achievement level. When they examined the bivariate correlation between achievement level and the frequency of "political" student disorders (involving more than five students and of such seriousness

[2] Paul Ritterband and Richard Silberstein, "Group Disorders in the Public Schools," *American Sociological Review* 38 (August, 1973): 461–467.

that the police were notified), they found this correlation to be in the expected direction and notably different from zero ($r = -.36$). This finding seemed to give empirical support to the supposition that a high concentration of students with low achievement levels makes a high school disorder-prone. However, this is not what the investigators finally concluded.

Reasoning that student political disorders might not only reflect the characteristics of schools and the academic characteristics of students, but might also be affected by events in the larger community as they impinge on each school area with its distinctive ethnic and racial composition, the investigators examined the correlation between the racial composition of the schools and the frequency of student disorders. They found that the percentage of students who are black in the schools was even more highly correlated with the frequency of disorders ($r = .54$) than were any of the other school characteristics, and they then examined the correlation between the frequency of disorders and the other characteristics with the percentage black statistically controlled. For this purpose, the investigators employed the partial correlation coefficient. After examining the partial correlation between achievement level and frequency of student disorders, controlling for the percentage of students who are black, the investigators concluded that the original bivariate correlation was spurious. Whereas the original correlation was $-.36$, the partial correlation was close to zero. In the light of this notable reduction in the correlation when the percentage black was controlled, the investigators concluded that, instead of achievement level having an effect on the frequency of student disorders, these two variables were correlated only because both had been affected by the percentage of blacks in the student body, along with the propensity, at that time and place, for black students to follow the example of "politicizers" who set a model for participation in political disorders. This application of the partial correlation coefficient illustrates how conclusions based on bivariate correlations may be strikingly altered by statistical control for additional variables, and how the partial correlation coefficient may be used to check an investigator's hypothesis that a bivariate correlation is spurious. We proceed now to examine the statistical character of the partial correlation coefficient and the principles governing its use and interpretation.

The *partial correlation coefficient* is defined as the product-moment correlation between the regression residuals of two variables, each regressed on a common set of additional variables. Since the set of additional variables may include one or several, a partial correlation coefficient may entail the statistical control of one or several variables. A partial correlation between X and Y based on the regression residuals on a single additional variable, Z, is symbolized $r_{XY.Z}$ (Read: r sub XY dot Z, or the partial r between X and Y controlling for Z). This is known as a first-order partial correlation (that is, one variable has been controlled). A partial correlation based on the regression residuals with two variables controlled is symbolized $r_{XY.ZW}$ and it is known as a second-order partial correlation since there are two variables controlled. A third-order partial ($r_{XY.ZWV}$) would refer to a correlation in which three variables have been controlled, and so on. By a simple extension of this same terminological convention, a simple bivariate correlation (r_{XY}) is known as a zero-order correlation (that is, one having no variables controlled). Although a tenth-order partial correlation, or even a hundredth-order partial, is possible, in practice partial correlations of higher order than the third (having three variables controlled) are rarely utilized, primarily because higher order partial correlations tend to be cumulatively affected by measurement error.

The First-Order Partial Correlation Coefficient Although the computational formula makes it unnecessary to actually compute the regression residuals and the correlation between them in order to obtain a first-order partial correlation coefficient, the understanding of partial correlation will be enhanced by describing the computation of a first-order partial by the more tedious route before presenting the convenient computational formula. In this description we will focus on the computation of the first-order partial between mean achievement level and frequency of student disorders, controlling for the percentage of each school's student body that is black.

We begin by finding the regression of both student achievement level and frequency of student disorders on the control variable, percentage black in the student body. Hypothetical data representing these three variables for ten schools are displayed in Table 10.2.1. The student can readily verify that the product moment correlation (r) between mean achievement level and number of disorders is $-.36$, as reported in the study of New York City schools. The correlation between percentage black and student disorders for the data in Table 10.2.1 is .54, also as reported. But we should emphasize that these data are hypothetical and for illustrative purposes only; these are not

Table 10.2.1 *Mean Achievement Level, Percentage Black, and Number of Disorders for 10 Hypothetical Schools, 1968–1969*

School	Mean Achievement Level Y	Blacks in Student Body (Percent) Z	Number of Disorders X
A	65	72	5
B	72	55	2
C	90	60	8
D	76	92	4
E	97	38	1
F	105	59	5
G	84	93	5
H	93	12	3
I	121	24	2
J	109	34	0

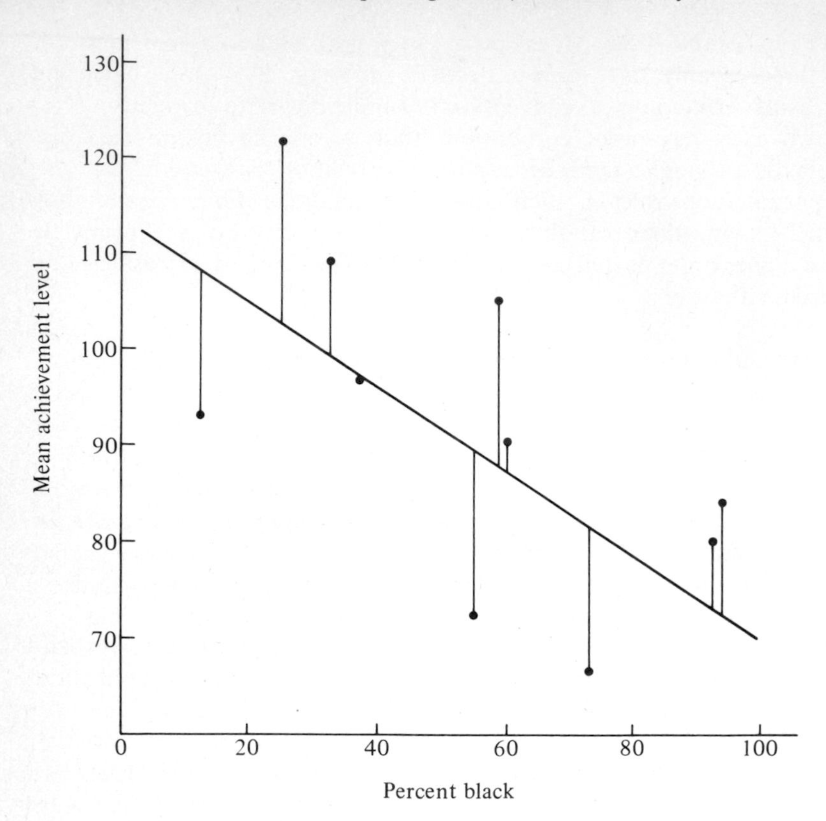

Figure 10.2.1 *Regression of Mean Achievement Level of Percentage Black, Showing Regression Errors (Residuals) for Each School*

the actual data on which Ritterband and Silberstein based their conclusions discussed above.

Following the reasoning of the investigators who explored the correlates of student disorders in New York City schools, we suspect that the correlation between school achievement level and frequency of disorders is a spurious correlation, that is, a correlation that appears because both mean achievement level and the number of disorders have been affected by the percentage black in the student body. Hence we proceed to find the regression residuals of each on percentage black, which residuals are graphically shown in Figures 10.2.1 and 10.2.2.

It is evident from Figure 10.2.1 that a portion of the variation in mean achievement level is associated with variation in percentage black ($r = -.63$, $r^2 = .40$), but for each case there is a regression error, or residual. These residuals represent the effects on mean achievement levels of factors other than percentage black, and we shall presume that these factors continue to operate in a similar manner—making the mean achievement level for a given

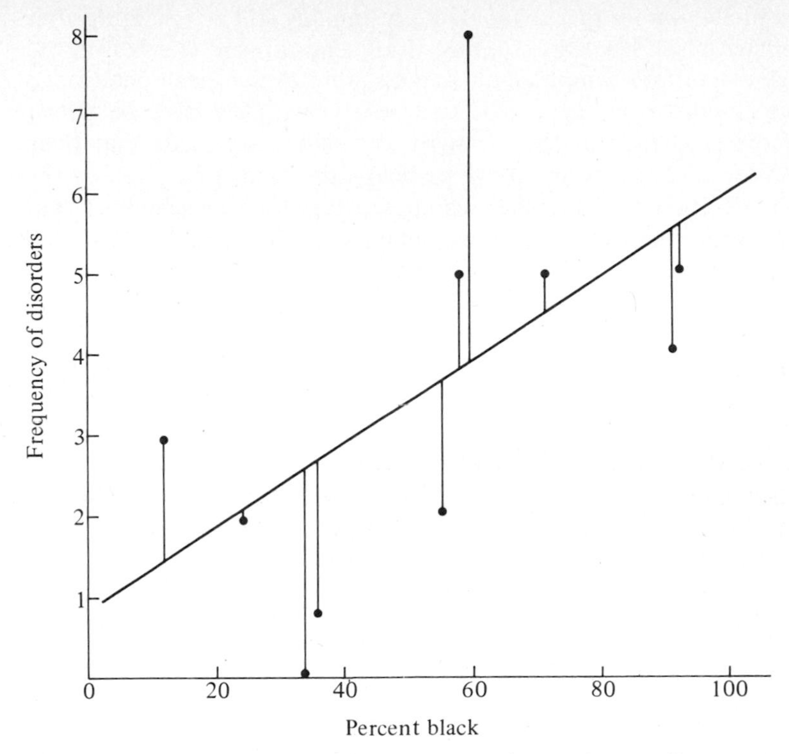

Figure 10.2.2 *Regression of Frequency of Disorders on Percentage Black, Showing Regression Errors (Residuals) for Each School*

school higher or lower than the score predicted on the basis of the percentage black—even as the percentage black changes.

The residuals are symbolized by the e_i in the following equation:

$$Y_i = a_{YZ} + b_{YZ}Z_i + e_i \,. \tag{10.2.1}$$

The residuals are obtained by subtracting the regression prediction ($\hat{Y}_i = 113.3 - .41Z_i$) from the observed achievement level for each school, taking care to retain the sign which indicates whether the observed achievement level is above the regression line (positive difference) or below the regression line (negative difference).

Now we may ask how the mean achievement levels of these same schools would vary if they all had the same percentage black among their students. As this percentage increases, we expect a decrease, on the average, in the mean achievement level, because the correlation is negative. This does not mean that we should anticipate that the mean achievement level will fall exactly at the predicted level for a hypothetical new percentage black; the same factors that make the achievement level higher or lower than the regression prediction for that school will presumably still be operating after the percentage black has shifted. Thus, if all schools were identical in the percentage black (for example, if all schools were at the mean percentage black), we could expect their respective achievement levels to be above or below the prediction at the mean by the same magnitude that their present achievement levels are above or below the prediction given by the regression line. By substituting the mean percentage for each school's actual percentage black in Formula 10.2.1, we obtain an achievement level for each school adjusted to the mean percentage black. In symbols:

$$\begin{aligned} Y'_i &= 113.3 - (.41)(53.9) + e_i \\ &= 91.2 + e_i \\ &= \bar{Y} + e_i \, . \end{aligned} \qquad (10.2.2)$$

The observed school achievement levels, their respective deviations from the regression prediction (based on Formula 10.2.1), the adjusted school achievement levels (based on Formula 10.2.2), and their respective deviations from the mean of these adjusted levels are shown in Table 10.2.2.

Table 10.2.2 *Observed Mean Achievement Levels, Regression Predicted Mean Achievement Levels, Adjusted Mean Achievement Levels, and Residuals for the 10 Schools Shown in Table 10.2.1*

	(a)	(b)	(c)	(d)	(e)
School	Observed Mean Ach Level	Ach Level Predicted by % Black[a]	Regression Residual[b]	Ach Level Adjusted to Mean % Black[c]	Deviation of Adjusted Ach Level from Mean
A	65	83.8	−18.8	72.4	−18.8
B	72	90.7	−18.7	72.5	−18.7
C	90	88.7	+ 1.3	92.5	+ 1.3
D	76	75.6	+ 0.4	91.6	+ 0.4
E	97	97.7	− 0.7	90.5	− 0.7
F	105	89.1	+15.9	107.1	+15.9
G	84	75.2	+ 8.8	100.0	+ 8.8
H	93	108.4	−15.4	75.8	−15.4
I	121	103.5	+17.5	108.7	+17.5
J	109	99.4	+ 9.6	100.8	+ 9.6
Mean	91.2	91.2	0.0	91.2	0.0
Variance	277.0	110.0	167.0	167.0	167.0
Proportion of total variance	1.00	.40	.60	.60	.60

[a] $Y_i = 113.3 - .41Z_i$.
[b] $e_i = Y_i - \hat{Y}_i$.
[c] $Y_i' = 113.3 - (.41)(53.9) + e_i$.

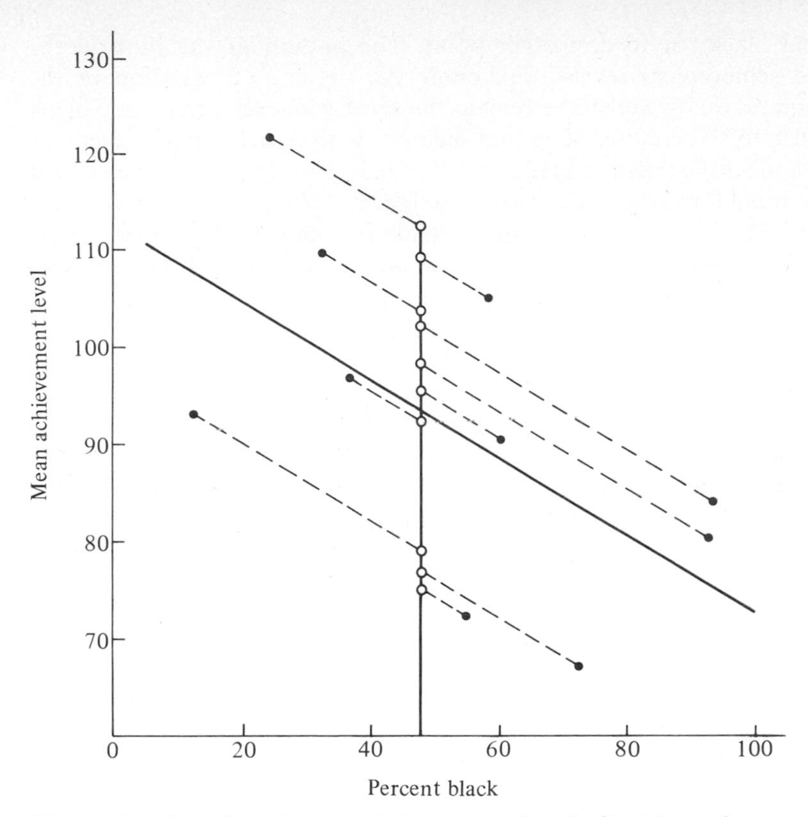

Figure 10.2.3 *Actual Mean Achievement Levels (Solid Dots) and Mean Achievement Levels Adjusted to Identical Percentage Black in Each School (Hollow Dots)*

The adjustment of school achievement levels to the levels expected if all schools had the mean percentage black may be visualized as in Figure 10.2.3, showing the preservation of the pattern of residuals as each school's achievement level is increased or decreased to the mean for all schools. The adjustment proceeds as if the achievement levels were moved from their actual positions along a straight line parallel to the regression line. Since the adjustment modifies the achievement level by a change parallel to the regression line, the original regression error for each school is precisely preserved in the adjustment and becomes the deviation of that school's adjusted level from the mean for all schools.

It should be intuitively evident from Equation 10.2.2 and from Figure 10.2.3 that the pattern of variation in the adjusted scores does not depend on the percentage black to which the achievement levels are adjusted. The expected pattern of variation in adjusted achievement levels is the same whether we adjust to the assumption that all schools have the mean percentage black, that they all have 100 percent black, that they all have

0 percent black, or to any other point. The pattern of variation in the adjusted achievement levels is precisely the pattern of variation in the regression residuals, and these remain the same whatever percentage black we adjust to. Therefore, it is not necessary to specify the percentage to which the adjustment is made; we may refer simply to the achievement levels adjusted for variation in percentage black.

Having obtained the regression residuals for school achievement levels regressed on the percentage students who are black in ten schools, we now proceed to find the regression residuals for the number of student disorders regressed on the percentage black. These residuals are displayed in Table 10.2.3. Again, the pattern of variation in the residuals is presumed to represent the variation among these schools that would obtain if all had the same percentage black in their respective student bodies.

Since we have adjusted out of each of the variables that part of the variation that is attributable to variation in percentage black, we may proceed to correlate the residuals to obtain the correlation between school achievement level and number of student disorders, with percentage black held constant.

Table 10.2.3 *Observed Frequency of Disorders, Regression Predicted Frequency of Disorders, Adjusted Frequency of Disorders, and Residuals for the 10 Schools Shown in Table 10.2.1*

	(a)	(b)	(c)	(d)	(e)
School	Observed Frequency of Disorders	Frequency of Disorders Predicted by % Black[a]	Regression Residuals[b]	Frequency of Disorders Adjusted to Mean % Black[c]	Deviation of Adjusted Frequency of Disorders from Mean
A	5	4.35	+0.65	4.15	+0.65
B	2	3.55	−1.55	1.95	−1.55
C	8	3.79	+4.21	7.71	+4.21
D	4	5.29	−1.29	2.21	−1.29
E	1	2.76	−1.76	1.74	−1.76
F	5	3.74	+1.26	4.76	+1.26
G	5	5.34	−0.34	3.16	−0.34
H	3	1.53	+1.47	4.97	+1.47
I	2	2.10	−0.10	3.40	−0.10
J	0	2.57	−2.57	0.93	−2.57
Mean	3.50	3.50	0.00	3.50	0.00
Variance	5.05	1.48	3.58	3.58	3.58
Proportion of total variance	1.00	.29	.71	.71	.71

[a] $\hat{X}_i = .97 + .047Z_i$.

[b] $e_i = X_i - \hat{X}_i$.

[c] $X_i' = .97 + .047(53.9) + e_i$.

Table 10.2.4 *Correlation of Residuals from Tables 10.2.2 and 10.2.3 to Obtain the Partial Correlation Coefficient*

Regression Residuals of Achievement Level on Percentage Black	Regression Residuals of Number of Disorders on Percentage Black			
$Y - \hat{Y}$	$X - \hat{X}$	$(Y - \hat{Y})^2$	$(X - \hat{X})^2$	$(Y - \hat{Y})(X - \hat{X})$
−18.8	0.65	353.44	0.4225	−12.220
−18.7	−1.55	349.69	2.4025	28.985
1.3	4.21	1.69	17.7241	5.473
0.4	−1.29	0.16	1.6641	− 0.516
− 0.7	−1.76	0.49	3.0976	1.232
15.9	1.26	252.81	1.5876	20.034
8.8	−0.34	77.44	0.1156	− 2.992
−15.4	1.47	237.16	2.1609	−22.638
17.5	−0.10	306.25	0.0100	− 1.750
9.6	−2.57	92.16	6.6049	−24.672
0.0	0.0	1671.29	35.7898	− 9.064

$$r = \frac{-9.064}{\sqrt{(1671.29)(35.7898)}} = -.037$$

This computation is displayed in Table 10.2.4. The first-order partial correlation coefficient here, as in the original study, is very close to zero, and we conclude that there is essentially no correlation between mean achievement level and number of student disorders in these schools after the percentage black has been held constant.

While the computation of a first-order partial correlation coefficient by the long route described above is evidently possible, it is typically very tedious, and such tedium is unnecessary. It is more convenient to obtain the first-order partial correlation by the following short computational formula:

$$r_{XY.Z} = \frac{r_{XY} - r_{XZ}r_{YZ}}{\sqrt{(1 - r_{XZ}^2)(1 - r_{YZ}^2)}} . \qquad (10.2.3)$$

For the data of Table 10.2.1 we have

$$\begin{aligned} r_{XY.Z} &= \frac{-.36 - (.54)(-.63)}{\sqrt{(1 - .54^2)(1 - .63^2)}} \\ &= \frac{-.02}{.65} \\ &= -.03 . \end{aligned}$$

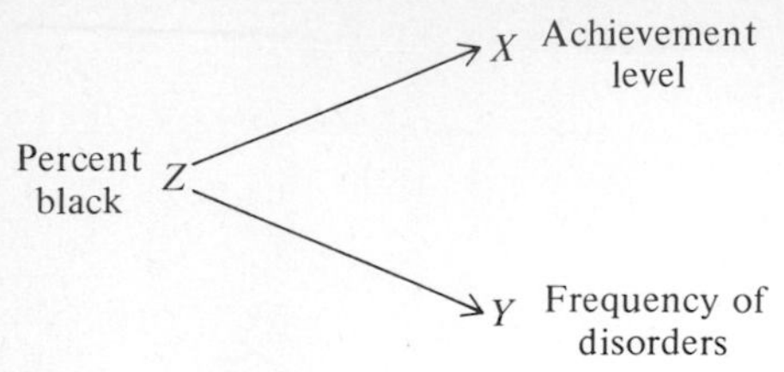

Figure 10.2.4

This result agrees, of course, with the result of computing the correlation between regression residuals, but the computation based on the three zero-order correlations is much less trouble to carry out.

The conclusion that the bivariate correlation between student achievement level and the frequency of disorders is spurious is based not simply on the fact that the partial correlation controlling for percentage black is approximately zero, but also on assumptions about the causal structure linking these three variables. The causal structure assumed in drawing this conclusion of spuriousness is represented in Figure 10.2.4. This structure may also be succinctly expressed in words: It is assumed that the percentage black in each school affects the average achievement level of that school and that the percentage black also affects the frequency of student disorders in that school. Given these causal assumptions, and the fact that the correlation essentially disappears when the percentage black is controlled (that is, the partial correlation is very close to zero), it is appropriate to conclude that the bivariate correlation between achievement level and frequency of disorders is spurious.

There are, however, causal assumptions other than those represented in Figure 10.2.4 that would also be compatible with a zero partial correlation. Such alternatives are represented in Figure 10.2.5, where the percentage black is represented as an intervening or mediating variable between achievement level and frequency of student disorders. Since each of these alternative causal structures also implies a zero partial (as does the causal structure represented in Figure 10.2.4), finding a zero partial is not helpful in deciding which assumption about causal order best represents how things work. Therefore, finding a zero partial does not demonstrate the validity of the conclusion that the bivariate correlation is spurious, because the bivariate correlation between achievement level and frequency of disorders is not spurious if the causal structure represented in either Figure 10.2.5a or b is correct.

To substantiate the conclusion that the bivariate correlation between achievement level and frequency of student disorders is spurious, the alternative causal structures (Figures 10.2.5a and b) must be judged implausible on grounds other than the observed bivariate correlations and the partial correlation of zero. These alternative causal structures *are* implausible on the basis of common-sense reasoning about the events they portray. It is contrary to our common-sense reasoning to propose that the

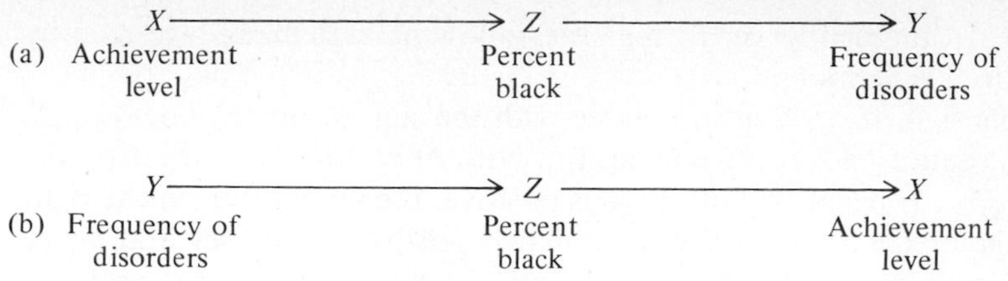

(a) $X \longrightarrow Z \longrightarrow Y$
Achievement level → Percent black → Frequency of disorders

(b) $Y \longrightarrow Z \longrightarrow X$
Frequency of disorders → Percent black → Achievement level

Figure 10.2.5

percentage of students who are black enrolled in New York City high schools in 1968–1969 was affected by the achievement level of the school; this would be plausible only if students were relatively free to choose the high school they attend, and if black students were notably more inclined than white students to attend high schools with a reputation for low achievement. It would be unrealistic to assume that such conditions prevailed in New York City in 1968–1969; hence the causal structure represented in Figure 10.2.5a is not a plausible alternative. It is also contrary to our common-sense reasoning to propose that the frequency of student disorders in each school had an effect on the percentage black in that school; the racial composition of the New York City high schools in 1968–1969 antedated the upsurge of student political disorders that occurred in the late 1960s. Hence the causal structure portrayed in Figure 10.2.5b is not a plausible alternative. This leaves us with one plausible alternative that implies the obtained zero partial (Figure 10.2.4), and because this structure explains the observed correlation between achievement level and frequency of student disorders by the dependence of both on the percentage black, it is appropriate to conclude (since the partial correlation actually did turn out to be approximately zero) that the original bivariate correlation is spurious. The general point to be emphasized is that a zero partial correlation does not, in itself, demonstrate that the original correlation was spurious. The interpretation of partial correlations must always take account of the underlying causal structures that are plausible in the particular circumstances being considered.

Rules of Thumb for Inferring Partials† It is useful to note certain conditions under which a first-order partial correlation will be zero and certain conditions under which the first-order partial will not approach zero. Reference to these relatively simple conditions will often permit one to tell from a quick examination of the zero-order correlations what will happen in the partial, without making tedious calculations. For this discussion we repeat Formula 10.2.3 here for easy reference:

$$r_{XY.Z} = \frac{r_{XY} - r_{XZ}r_{YZ}}{\sqrt{(1 - r_{XZ}^2)(1 - r_{YZ}^2)}}.$$

†This topic may be omitted without loss of continuity.

The student can readily verify that the following statements are true by reference to this formula.

(1) If the sign of r_{XY} does not agree with the sign of the product $r_{XZ}r_{YZ}$, then $r_{XY.Z}$ cannot be zero. If one but not both of r_{XZ} or r_{YZ} is negative, the product $r_{XZ}r_{YZ}$ is negative. But if r_{XY} is positive, the numerator on the right of Formula 10.2.3 becomes the sum of two positive quantities and hence cannot be zero. Similarly, if both r_{XZ} and r_{YZ} are negative, or if both are positive, the product $r_{XZ}r_{YZ}$ is positive. But if r_{XY} is negative, the numerator on the right of Formula 10.2.3 becomes the sum of two negative numbers and hence cannot be zero. A simple examination of the signs of the zero-order correlations may suffice to determine in some circumstances that the first-order partial correlation cannot be zero.

(2) If $r_{XZ}r_{YZ} = r_{XY}$, then $r_{XY.Z} = 0$.
Under the conditions stated, the numerator on the right of Formula 10.2.3 is zero; hence the entire right-hand side of the equation is zero and the partial is zero.

One practical implication of this is that a quick mental calculation is often sufficient to indicate that the first order partial will be approximately zero; if the product $r_{XZ}r_{YZ}$ is approximately equal to r_{XY}, then one can anticipate that the first-order partial will be approximately zero. In applying this rule of thumb, one should recognize that the difference $r_{XY} - r_{XZ}r_{YZ}$ is the numerator but not the whole of the computing formula for the first-order partial. Since the denominator is almost always less than 1.0, the numerator alone yields an undersestimate of the first-order partial (except where both r_{XZ} and r_{YZ} are zero). It may also be noted that the numerator alone underestimates the first-order partial more when the correlations with the control variable are strong than when these correlations are weak.

(3) If $r_{XZ} = 0$, or if $r_{YZ} = 0$, then $r_{XY.Z} \geq r_{XY}$. Note that if either r_{XZ} or r_{YZ} (or both) is zero, the numerator on the right of equation 10.2.3 is just r_{XY}. Furthermore, if one but not both of these correlations between the control variable and the two variables of primary interest is zero, one of the terms in the denominator is 1.0 and the other will be less than 1.0. If, then, we divide r_{XY} by a quantity less than 1.0, we will obtain a quotient that exceeds r_{XY}. If both r_{XZ} and r_{YZ} are zero, the denominator is exactly 1.0 and the partial is just r_{XY}.

The useful implication of this is that the introduction of a control variable that is uncorrelated or only weakly related to one or both of the two variables of primary concern cannot be expected to yield a first-order partial that is very different from the original bivariate correlation. Since controlling for an unrelated variable has no effect, we need not be concerned to control for unrelated variables.

Second-Order and Higher-Order Partial Correlations The second-order partial correlation is based on the same principles as the first-order partial, but the second-order partial, $r_{XY.ZW}$, is defined as the product-moment

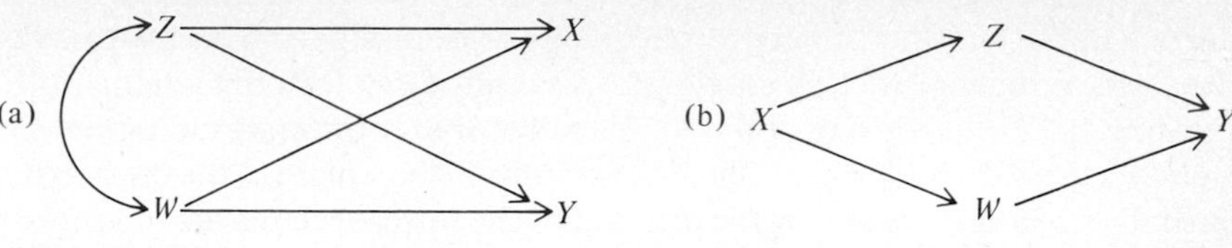

Figure 10.2.6

correlation between the regression residuals of X on both Z and W simultaneously, and of Y on both Z and W simultaneously. Assumed causal structures that would make a second-order partial correlation appropriate are represented graphically in Figures 10.2.6a and b. In Figure 10.2.6a, the hypothesis is that r_{XY} is spurious, not because of the common dependencc of these variables on Z alone, nor because of the common dependence of these variables on W alone, but because of their common dependence on both Z and W. If that hypothesis is true, then the second-order partial should be approximately zero. In Figure 10.2.6b, the hypothesis is that Z and W are both intervening between X and Y, with each of these intervening variables mediating part of the effect of X, which has no direct effect on Y. Again, if this hypothesis is true, the second-order partial should be approximately zero.

As in the case of the first-order partial, the second-order partial could be computed the long way, entailing the actual computation of the appropriate multiple regression residuals (see Section 3 of this chapter) and the correlation between those residuals. But it may also be computed by a short computational formula, which is much less tedious and which typically entails less rounding error. The second-order partial correlation may be computed from appropriate first-order partials as follows:

$$r_{XY.ZW} = \frac{r_{XY.W} - r_{XZ.W}r_{YZ.W}}{\sqrt{(1 - 4^2_{XZ.W})(1 - r^2_{YZ.W})}}$$

$$= \frac{r_{XY.Z} - r_{XW.Z}r_{YW.Z}}{\sqrt{(1 - r^2_{XW.Z})(1 - r^2_{YW.Z})}} . \qquad (10.2.4)$$

The square of the second-order partial may also be obtained by an alternative computing formula based on the standardized partial regression coefficients (symbolized by b*) or the unstandardized partial regression coefficients (symbolized by b) as follows:

$$r^2_{XY.ZW} = b^*_{XY.ZW}b^*_{YX.ZW} = b_{XY.ZW}b_{YX.ZW} \cdot \qquad (10.2.5)$$

The computation of these regression coefficients is discussed in Section 3 of this chapter.

Higher-order partial correlations can always be obtained from the partials of the next lower order in a manner analogous to Formula 10.2.4. However, the computational effort becomes tedious beyond the second-order partial, and it is ordinarily more efficacious to obtain partial correlations of the third or higher order by an alternative computing formula analogous to Formula 10.2.3, using one of the standard computer programs to obtain the partial regression coefficients. Thus, for example, the square of the third-order partial correlation is given by the product of the third-order partial regression coefficients, as follows:

$$r^2_{XY.ZWV} = b^*_{XY.ZWV} b^*_{YX.ZWV} = b_{XY.ZWV} b_{YX.ZWV} \,. \tag{10.2.6}$$

While this procedure has the disadvantage of requiring an additional step to determine the sign of the partial correlation, that step is easily accomplished. The two partial regression coefficients will have the same sign; one extracts either the positive root or the negative root of the square of the partial correlation depending on whether the signs of the partial regression coefficients are positive or negative.

QUESTIONS AND PROBLEMS

1. Define the following terms:

zero-order correlation	partial correlation coefficient
regression residual	first-order partial correlation
linear regression adjustment	second-order partial correlation

2. The diagrams below hypothesize certain causal structures linking variables X, Y, and Z. For each such hypothesized structure, indicate whether or not the partial correlation $r_{XY.Z}$ would be useful in assessing the tenability of the hypothesis. Describe your reasoning in each instance.

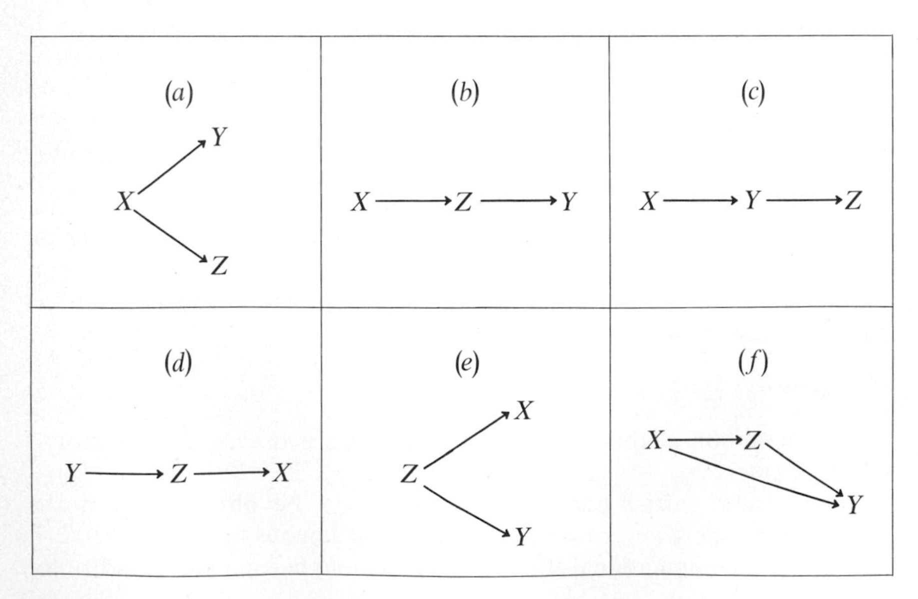

3. An investigator reasons that the zero-order correlation between X and Y is probably spurious because of Z. Without actually computing the first-order partial correlation (instead use the rules of thumb given on pp. 289–290), indicate for each of the sets of findings listed below whether the investigator's reasoning is tenable. Describe your reasoning.
   (a) $r_{XY} = .37; \; r_{XZ} = 0; \; r_{YZ} = .56$
   (b) $r_{XY} = .24; \; r_{XZ} = -.60; \; r_{YZ} = -.40$
   (c) $r_{XY} = .66; \; r_{XZ} = .35; \; r_{YZ} = .30$
   (d) $r_{XY} = .35; \; r_{XZ} = -.70; \; r_{YZ} = .50$
   (e) $r_{XY} = .40; \; r_{XZ} = .20; \; r_{YZ} = .20$
   (f) $r_{XY} = .42; \; r_{XZ} = .60; \; r_{YZ} = 0$
   (g) $r_{XY} = -.32; \; r_{XZ} = .80; \; r_{YZ} = -.40$
4. For each of the sets of findings listed in the exercise above, compute $r_{XY.Z}$ and check the accuracy of your application of the rules of thumb in Problem 10.2.3.

3 MULTIPLE REGRESSION AND MULTIPLE CORRELATION

Multiple regression is analogous to bivariate regression (Chapter 9); but in bivariate regression a dependent variable is regressed on a single independent variable, while in multiple regression a dependent variable is regressed on two or more independent variables simultaneously, with each independent variable weighted so as to yield predicted values with minimum prediction error by the criterion of least squares. The weight for each of the independent variables is known as a *partial regression coefficient*, and such coefficients are geometrically interpreted as slopes, just as in bivariate regression. For a given independent variable, the partial regression coefficient indicates how the dependent variable would regress on that independent variable after the effects of all of the other independent variables included in the analysis have been statistically eliminated. The *multiple correlation coefficient* is the product-moment correlation between the observed values on the dependent variable and the values predicted by the weighted combination of independent variables.

To illustrate the use of multiple regression and multiple correlation in sociological research, we consider briefly a series of studies in which this mode of analysis has been employed.

(1) In a study of coup d'état activity in 35 independent nations in sub-Sahara Africa, an investigator developed an index to reflect the degree to which each nation-state had experienced coups d'états (whether attempted or completed), and then regressed this index of coup activity on a number of demographic, socioeconomic, and military variables describing each nation.

The results of the analysis indicate that, after adjusting for the effects of each of the other independent variables, six predictors were found to account for slightly less than one-half of the variation in the index of coup activity. The predictors with positive effects were (a) population growth rate; (b) the degree to which economic and political activity concentrated in a single city; (c) the ratio of the national defense budget to the gross national product; and (d) the percentage of men of military age on active service in the armed forces. The two predictors with negative effects were (e) access of the population to radio communications, and (f) the per capita gross national product. Using all predictors, the correlation between the predicted level of coup activity and the actually measured level of coup activity (that is, the multiple correlation coefficient) was .75. After adjustment, certain variables that might be assumed to be predictive of coup d'état activity were found to have almost no effect. For example, the degree of urbanization, the size of the military forces, and the amount of United States aid made a negligible contribution to the predictability of coup activity.[3]

(2) What most influenced the educational attainment of American males in the 1960s? Social class background, mental test scores, high school grades, occupational aspirations expressed in high school, and perceived educational plans of high school friends were among the variables positively correlated with the educational level actually attained, as measured in a study based on a follow-up of a group originally studied as high school seniors. But in the regression analysis of these and other variables as predictors of educational attainment, the investigator found that for both blacks and whites, the largest standardized partial regression coefficient (indicating the greatest contribution to the predictability of educational attainment) was found for measured intelligence, although several additional variables retained some predictive utility even after adjustment control for the other predictors in the analysis. The multiple correlation achieved with these predictors was .62 for whites and .55 for blacks.[4]

(3) Although it has been contended that the rapidity of urbanization, rather than the urban setting per se, increases the incidence of crimes against property in urbanizing areas and periods, the findings in a multiple regression analysis based on nineteenth-century data describing French *départements* (political subdivisions) casts doubt on that contention. The investigators regressed the number of crimes against property in each French *département* on the percentage of the population urban, the recent rate of urbanization, and other variables. Results for a series of years consistently show that, after adjustment for each of the independent variables in the analysis, the percentage urban had the largest standardized regression coefficient, while the

[3] Alan Wells, "The Coup d'État in Theory and Practice: Independent Black Africa in the 1960's," *American Journal of Sociology* 79 (January, 1974): 871–887.

[4] James N. Porter, "Race, Socialization and Mobility in Educational and Early Occupational Attainment," *American Sociological Review* 38 (June, 1974): 303–316.

rate of urbanization had a negligible effect. In these analyses, the multiple correlation coefficient was consistently close to .90.[5]

(4) A syndrome of values presumed to represent an adaptation to the complexities of urban life has been called "modernity" and has been explored by a number of sociologists. The correlates of modernity explored in a Latin American sample included personal characteristics of respondents such as education, occupation, and length of urban residence as well as more distant background features such as parental education, parental place of birth, and father's occupation. In a multiple regression analysis, only the personal characteristics emerged with notable regression coefficients; the more distant background features had very small partial regression coefficients after adjusting for the personal characteristics. The multiple correlation was .69 when all the predictors were included, but dropped only to .68 when the predictors were limited to the three personal characteristics alone.[6]

(5) Racial disorders were frequent in some American cities in the late 1960s, while other American cities had few such disorders during the same period. What characteristics of cities were associated with the frequency of such disorders? From a multiple regression analysis of the *frequency* of racial disorders, Seymour Spilerman[7] concluded that the variables usually suspected—low incomes among blacks, poor housing, police discrimination, and so forth—did not contribute to the prediction of variation in the frequency of disorders after controlling for the size of the black population. However, in a multiple regression analysis of the *severity* of racial disorders in American cities, William Morgan and Terry Clark[8] found that the degree of housing and job inequality for blacks did contribute substantially to the prediction of variation in severity, even after the size of the black population was statistically controlled. Taking the two studies together, the appropriate conclusion seems to be that a large black population increases the likelihood that racial incidents will develop, but does not increase the likelihood that such incidents will escalate into a severe riot. Such escalation is likely, however, if the grievance level of the minority population is high, regardless of the size of the minority population.

These illustrative studies drawn from sociological journals suggest the variety of substantive content usefully analyzed by multiple regression techniques. This very limited sample does not do full justice to the utility of multiple regression, which has been used in the analysis of data pertaining to an almost limitless variety of topics. With appropriate adaptation, multiple

[5] Abdul Qaiyum Lodhi and Charles Tilly, "Urbanization, Crime, and Collective Violence in 19th Century France," *American Journal of Sociology* 79 (September, 1973): 296–318.
[6] Alejandro Portes, "The Factorial Structure of Modernity: Empirical Replications and a Critique," *American Journal of Sociology* 79 (July, 1973): 15–44.
[7] Seymour Spilerman, "The Causes of Racial Disturbances: A Comparison of Alternative Explanations." *American Sociological Review* 35 (August, 1970): 627–649.
[8] William R. Morgan and Terry Nichols Clark, "The Causes of Racial Disorders: A Grievance-Level Explanation," *American Sociological Review* 38 (October, 1973): 611–624.

regression can be used with nonquantitative variables ("dummy variable" analysis). The analysis can proceed in a sequence so as to include only those variables that give maximum predictability (step-wise multiple regression) instead of including all potential predictors simultaneously as in the above illustrations. The basic logic of the technique can be adapted to provide for the possibility that the effect of one variable may depend on the level of another (interaction effects). Multiple regression analysis provides the core and the computational procedure for estimating parameters in path analysis (to be discussed in Section 4), and given appropriate conditions and suitable modifications, multiple regression may even be used to estimate the effects of unmeasured variables, although the presentation of these procedures is beyond the scope of this book. This versatile and useful analytic tool deserves careful study.

The Special Case of Two Predictors The simplest multiple regression problem entails the prediction of a dependent variable on the basis of two independent variables or predictors. Since many of the basic features of multiple regression analysis can be well illustrated in the relatively simple two predictor case, we limit our discussion here to two predictors.

One assumes in multiple regression analysis that each of the independent variables may make a unique contribution to the prediction of the dependent variable. In the two predictor case, this assumption can be represented in the following equation:[9]

$$\hat{Y} = a_{Y.12} + b_{Y1.2}X_1 + b_{Y2.1}X_2 \,. \qquad (10.3.1)$$

Formula 10.3.1 is known as a *multiple regression equation*; it describes the regression plane for predicting Y from knowledge of the two independent variables, X_1 and X_2. Such a plane has been graphically portrayed in Figure 10.3.1. As suggested in the figure, we can locate each case on the base of the cube utilizing the X_1 and X_2 values for that case. Corresponding to each such location on the base, there is a height to the regression plane, that is, a predicted value. The observed values will typically fall above or below the plane so that there will be regression errors or residuals. This is evidently parallel to the case of simple bivariate regression in which each case is located along the axis representing the independent variable, and corresponding to each such location on that axis there is a height to the regression line, which is the predicted value. Again, the observed values typically fall above or below the line and hence give rise to prediction errors or residuals.

There is a statistical adjustment inherent in multiple regression such that the multiple regression coefficient of Y on X_2 controlling for X_1 is the simple

[9] To simplify notation, we represent the Xs in subscripts by their numbers only. Thus, for example, $b_{Y1.2}$ is short notation for $b_{YX_1 \cdot X_2}$. This simplification in subscript notation is especially helpful with a large number of predictors.

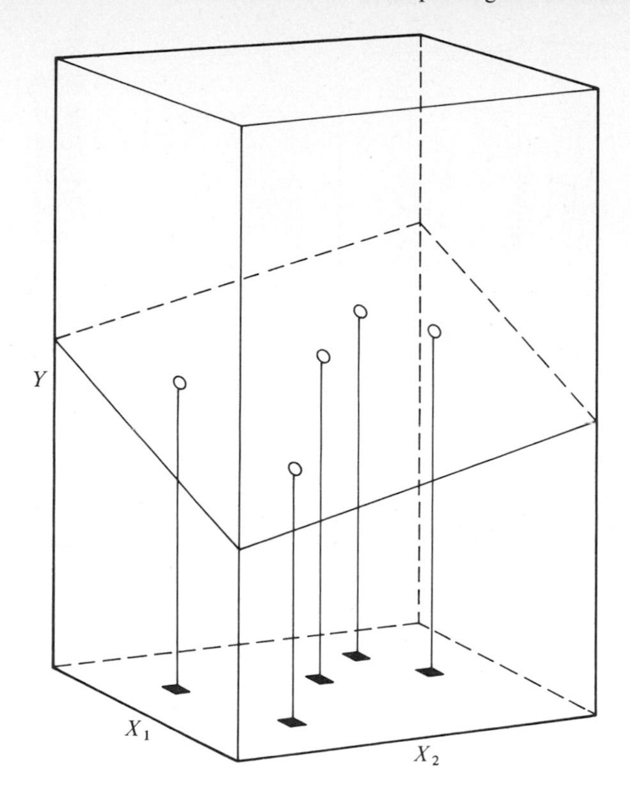

Figure 10.3.1 *Graphic Portrayal of a Regression Plane for Making Predictions on a Dependent Variable,* Y, *from Two Independent Variables,* X_1 *and* X_2

linear regression of Y on X_2 with X_1 at a constant value for all cases. This adjustment is diagrammatically portrayed in Figure 10.3.2.

The adjustment proceeds as if the observed Y values for each case were moved along a straight line parallel to the regression plane to a common point on the control variable. In Figure 10.3.2, each value has been moved in this way to a point where all have a common X_1 value. As in the linear regression adjustment inherent in partial correlation, the residuals are precisely preserved in this adjustment process. Expressed in the form of an equation, this adjustment entails the substitution of a single value for X_1 for all cases. If we let that value be C, the equation for each adjusted score becomes:

$$\hat{Y}_{i\,adjust} = a_{Y.12} + b_{Y1.2}C + b_{Y2.1}X_2 + e_i \,. \tag{10.3.2}$$

The variable X_1 has now been held constant (at C) and the resulting adjusted scores are a function of a constant term ($a_{Y.12} + b_{Y1.2}C$), of the variable X_2, and of the error term characteristic of each individual case.

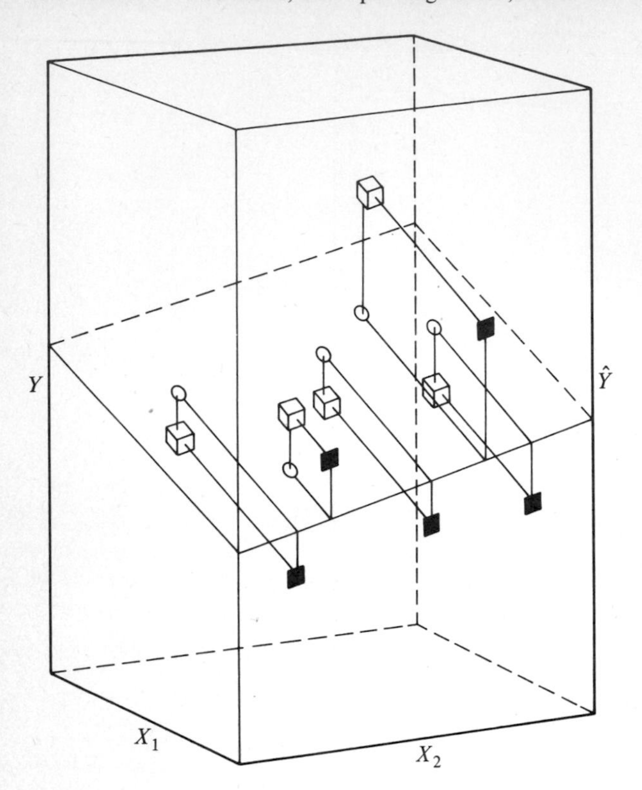

Figure 10.3.2 *Graphic Portrayal of Multiple Regression Adjustment of* Y *to a Common Value of* X_1. [*Adjustment "moves" observed value of* Y *(small cubes) to adjusted* Y *value (solid squares) along line parallel to the regression plane slope with respect to* X_1 ($b_{YX_1.X_2}$), *preserving the magnitude of the residual* (e_i).]

Using these adjusted scores, we might then proceed to find the regression of the *Y* values, adjusted to a constant X_1, on the other independent variable, X_2. But if we proceeded to make such a computation, we would find that the regression we obtain is precisely the same as the slope of the regression plane with respect to X_2. Hence, obtaining the slopes of the regression plane is the crux of the problem of multivariate adjustment. In finding the *adjustment* slopes, we have found the *adjusted* slopes.

While most students find it helpful to be able to visualize geometrically a multiple regression prediction based on two predictors, such an image does not provide a solution to the fundamental computational problem in multiple regression, which is to find the appropriate values for the intercept ($a_{Y.12}$) and the slopes ($b_{Y1.2}$ and $b_{Y2.1}$) in the multiple regression equation. In seeking this solution, we seek those values which will satisfy the least squares criterion, that is, those which will minimize the sum of squared prediction errors. Expressed in symbols, we seek to find those coefficients which will minimize $\Sigma(Y - \hat{Y})^2$. These partial regression coefficients, once found, will (a) provide us with the weights that will minimize prediction error by the criterion of least squares, (b) indicate for each of the independent variables the appropriate slope to use in adjusting out its effects, and (c) indicate the regression slope of the dependent variable on each of the independent variables after the effects of the other independent variable have been appropriately adjusted out. Clearly, the partial regression slopes provide the fundamental information that a multiple regression analysis can yield.

Computing Partial Regression Coefficients: Two Predictors Partial regression coefficients (slopes) may be expressed in two forms: *standardized* and *unstandardized.* Standardized partial regression coefficients are the slopes of the regression plane when all variables have been translated into standard measures, that is, when all deviations from the respective means have been divided by the standard deviations. Such coefficients typically range from -1.0 to $+1.0$ like r, which is the slope for a bivariate regression when both variables have been translated into standard measures. We will symbolize standardized partial regression coefficients by b^* with appropriate subscripts.[10] For two predictors, the two standardized partial regression coefficients may be computed as follows:

$$b^*_{Y1.2} = \frac{r_{Y1} - r_{Y2}r_{12}}{1 - r^2_{12}}; \tag{10.3.3}$$

$$b^*_{Y2.1} = \frac{r_{Y2} - r_{Y1}r_{12}}{1 - r^2_{12}}. \tag{10.3.4}$$

The first subscript for a partial regression coefficient symbolizes the dependent or predicted variable. The second subscript symbolizes the independent or predictor variable, while additional subscripts following the "dot" symbolize the variable(s) controlled. Thus, $b^*_{Y1 \cdot 2}$ is read "bee star sub Y one dot two" and stands for the partial regression of the dependent variable Y on the independent variable X_1 controlling for the other predictor, X_2.

Unstandardized partial regression coefficients are the slopes of the regression plane for variables in their raw or unstandardized form. Like the unstandardized regression slope in a bivariate regression, these values are not restricted to the range from -1.0 to $+1.0$, and their magnitudes are, in part, dependent on the units of measure. For example, the regression coefficient of income on education controlling for IQ will be different depending on whether income is measured in dollars, francs, or yen, and on whether

[10] In many discussions of multiple regression, the symbol β (beta) is used to represent the standardized regression coefficients, instead of b^*, which we selected to use. The use of the Greek β (beta) in this context does not comply with the convention well-established in statistics that Greek letters be used to represent population parameters while Latin letters are used to represent sample statistics. In any case, the student should recognize that what other discussions refer to as *beta* or *beta weight* is simply the regression coefficient, here symbolized by b^*.

education is measured in years of school completed or in months of school completed. In contrast, the standardized partial regression coefficient will be the same whatever the units of measure. We will symbolize unstandardized partial regression coefficients by b, with appropriate subscripts. The unstandardized partial regression coefficients and the intercept necessary to complete the regression equation for variables in unstandardized form may be obtained from the standardized coefficients and the means and standard deviations, as follows:

$$b_{Y1.2} = b^*_{Y1.2}\frac{s_Y}{s_1}; \tag{10.3.5}$$

$$b_{Y2.1} = b^*_{Y2.1}\frac{s_Y}{s_2}; \tag{10.3.6}$$

$$a_{Y.12} = \overline{Y} - b_{Y1.2}\overline{X}_1 - b_{Y2.1}\overline{X}_2. \tag{10.3.7}$$

The computation of partial regression coefficients for more than two predictors is tedious by hand, although computer programs are widely available to accomplish this task in milliseconds. The computational procedure for three or more predictors is even tedious to describe, and such procedures are therefore not presented here.

Rationale for the Computational Formulae for Partial Regression Coefficients†
Although the mathematical derivation of the computing formulae for partial regression coefficients is beyond the scope of this book, the interested student may find the following abbreviated exposition helpful in understanding multiple regression. We assume here that all variables have been translated into standardized form, here symbolized by y' for Y in standardized form, x'_1 for X_1 in standardized form, and so on. With each variable thus transformed into standard measures, the multiple regression equation may be written as follows:

$$\hat{y}' = b^*_{Y1.2}x'_1 + b^*_{Y2.1}x'_2. \tag{10.3.8}$$

Note that the intercept no longer appears in the multiple regression equation for standard measures. The intercept for standard measures is always zero in the multiple regression case as it is in the bivariate regression case.

Although we will not describe the algebraic derivation in detail here, we summarize briefly the reasoning that leads to computational solutions for the regression coefficients. The least squares criterion implies that the covariation between the prediction errors and either of the predictor variables should be zero. If this were not so, then some modification of the regression coefficients would yield a smaller sum of squared prediction errors, and the least squares criterion would not have been satisfied. But this condition (that

†This part may be omitted without loss of continuity.

the covariation between prediction errors and predictor variables should be zero) is equivalent to the condition that all of the covariation between each predictor and the dependent variable should be included in (that is, be equal to) the covariation between each predictor and the regression predictions for the dependent variable. Hence, for each predictor, we may write the following equations continuing to express each variable in standard form:

$$cov\,(x_1', y') = cov\,(x_1', \hat{y}') = cov\,[x_1', (b^*_{Y1.2}x_1' + b^*_{Y2.1}x_2')]; \qquad (10.3.9)$$

$$cov\,(x_2', y') = cov\,(x_2', \hat{y}') = cov\,[x_2', (b^*_{Y1.2}x_1' + b^*_{Y2.1}x_2')]. \qquad (10.3.10)$$

Since the covariance between two standardized variables is the zero-order correlation between those variables, the left-hand expression in Formula 10.3.9 is r_{Y1} and the left-hand expression in Formula 10.3.10 is r_{Y2}. Furthermore, the expression farthest to the right in each equation may be rewritten in the form of a sum of covariances, to yield the following:

$$r_{Y1} = b^*_{Y1.2}\,cov\,(x_1', x_1') + b^*_{Y2.1}\,cov\,(x_1', x_2'); \qquad (10.3.11)$$

$$r_{Y2} = b^*_{Y1.2}\,cov\,(x_2', x_1') + b^*_{Y2.1}\,cov\,(x_2', x_2'). \qquad (10.3.12)$$

Again, we have the covariance of two standardized variables which may be replaced by the appropriate r. Since the correlation of a variable with itself

$$r_{x_1'x_1'}$$

is 1.0, we may rewrite Formulas 10.3.11 and 10.3.12 in the following form:

$$r_{Y1} = b^*_{Y1.2} + b^*_{Y2.1}r_{12}; \qquad (10.3.13)$$

$$r_{Y2} = b^*_{Y1.2}r_{21} + b^*_{Y2.1}. \qquad (10.3.14)$$

Formulas 10.3.13 and 10.3.14 are known as the *normal equations* for the two predictor multiple regression problem. In general (for standardized variables) there will always be as many normal equations as there are predictors, and since there are also as many unknown regression coefficients (the b^*s) as predictors, we will always have a set of k simultaneous equations in k unknowns, where k indicates the number of predictors. In this instance of two simultaneous equations in two unknowns, the student proficient in elementary algebra should be able to verify that the solutions for b^* are precisely those given in Formulas 10.3.3 and 10.3.4 above.

The Multiple Correlation Coefficient The *multiple correlation coefficient*, symbolized by R, is the product-moment correlation between the observed values on the dependent variable and the values predicted by the multiple regression equation. In symbols:

$$R = r_{Y\hat{Y}}\,. \qquad (10.3.15)$$

The square of the multiple correlation coefficient (R^2) measures the relative reduction in prediction error achieved by shifting from the mean as the prediction for all cases to the regression equation (Formula 10.3.1) as the prediction for each case.

Formula 10.3.15 does not provide the most convenient computational formula for obtaining R; such a computation would require the prior computation of a predicted value for each case before computing the multiple correlation coefficient. A more convenient computational formula for the two predictor case is given by:

$$R^2 = b^*_{Y1.2} r_{Y1} + b^*_{Y2.1} r_{Y2} \,. \qquad (10.3.16)$$

Using this computational formula, the square of the multiple correlation coefficient may be obtained from the standardized regression coefficients and the zero-order correlations, and the tedious computation of a predicted value for each case is unnecessary. To obtain R from R^2, the positive root is taken, since the correlation between observed values and values predicted from the multiple regression equation is never negative.

An Illustration of Multiple Regression: Two Predictors To illustrate the computation of regression coefficients and the multiple correlation coefficient in the two predictor case, we take three variables from a study of educational and occupational attainment among American males. The zero-order correlations, means, and standard deviations for the white males are displayed in Table 10.3.1. The dependent variable (Y) is the status index of the occupation held five years after graduation from high school. The two

Table 10.3.1 *Zero-Order Correlations, Means, and Standard Deviations for Occupational Attainment* (Y), *Parental SES* (X_1), *and High School Grades* (X_2) *for 14,891 White American Males Who Graduated from High School* c. *1960*

	X_1	X_2	Mean	Standard Deviation
Y	.197	.236	46.4	22.3
X_1		.122	36.4	23.2
X_2			2.29	.69

Source: James N. Porter, "Race, Socialization and Mobility in Educational and Early Occupational Attainment," *American Sociological Review* 39 (June, 1974): 303–316; adapted from Table 1, p. 307. Reprinted with permission.

predictors are parental socioeconomic status (X_1, measured by the status index of the occupation of the head of the household in which the boy lived while in grade 12) and high school grades (X_2, measured by the average of self-reported high school grades in five academic subjects). Using the zero-order correlations displayed in Table 10.3.1, we compute the standardized regression coefficients using equations 10.3.3 and 10.3.4:

$$b^*_{Y1.2} = \frac{.197 - (.236)(.122)}{1 - (.122)^2} = \frac{.168}{.985} = .171;$$

$$b^*_{Y2.1} = \frac{.236 - (.197)(.122)}{1 - (.122)^2} = \frac{.212}{.985} = .215.$$

The standardized partial regression coefficients may be used to compare the relative contributions of the independent variables to the overall predictability achieved by the two variables in combination. In this particular instance, the difference between the two is slight, suggesting that the two independent variables (grades and parental socioeconomic status) contribute about equally to the variation in occupational attainment. The student will note that in this instance the partial regression coefficients are very similar to the corresponding zero-order correlations (.236 as compared to .215, and .197 as compared to .171). This is because the two predictors are very weakly correlated with each other; in the special instance of two predictors completely uncorrelated with each other, it will be evident from Formulas 10.3.9 and 10.3.10 that the partial regression coefficients would be identical to the corresponding zero-order correlations. Even in this illustration of very weakly correlated predictors, however, each partial regression coefficient is slightly less than the corresponding zero-order correlation, indicating that a small part of the zero-order correlation between occupational attainment and each of the predictors was attributable to the other predictor.

With the solutions for the standardized partial regression coefficients, we may proceed to find the multiple correlation coefficient, using equation 10.3.16:

$$R^2 = (.171)(.197) + (.215)(.236) = .034 + .051 = .085;$$

$$R = .29.$$

The calculations indicate that the product-moment correlation between predicted and observed levels of occupational attainment is .29. Alternatively stated, slightly over 8 percent of the variation in observed levels of occupational attainment is removed by statistical adjustment for the two predictor variables used in this analysis.

Using the standardized regression coefficients obtained above and the means and variances given in Table 10.3.1, we may now proceed to compute the unstandardized regression coefficients and the intercept, using equations 10.3.5, 10.3.6, and 10.3.7:

$$b_{Y1.2} = .171 \frac{22.3}{23.2} = .164;$$

$$b_{Y2.1} = .215 \frac{22.3}{.69} = 6.95;$$

$$\begin{aligned} a_{Y.12} &= 46.4 - (.171)(36.4) - (.215)(2.29) \\ &= 39.7. \end{aligned}$$

With these computations, we may write the equation for predicting occupational attainment from the high school grades and parental socioeconomic status scores for each case. The prediction equation is as follows:

$$\hat{Y} = 39.7 + .164X_1 + 6.95X_2 .$$

The unstandardized regression coefficients represent the average change in occupational attainment for each unit change in high school grades, after the effects of parental socioeconomic status have been adjusted out ($b_{Y2.1}$), and the average change in occupational attainment for each unit change in parental socioeconomic status, after the effects of high school grades have been adjusted out ($b_{Y1.2}$). Since the unstandardized regression coefficients represent these effects in the units of the respective variables rather than in standard deviation units, they are not directly comparable with each other. A comparison of their magnitudes is not informative about the relative magnitudes of the contribution of each predictor to the predictability of the dependent variable; we concluded above that the relative contribution of each of these two predictors was about equal, but this conclusion is not at all evident in the unstandardized regression coefficients. Although it is not appropriate to compare the unstandardized regression coefficients for different predictors, these unstandardized coefficients may be used to summarize how the dependent variable changes with changes in each independent variable, after controlling for the other independent variable. Since $b_{Y2.1} = 6.95$, we may state, for example, that one unit increase in high school grades is worth about seven status points; for every increase of one high school grade unit, the young adult males in this sample experienced, on the average, an increase of about seven status index units in the occupations they held five years after graduation. And since $b_{Y1.2} = .164$, each unit of parental socioeconomic status is worth about .16 status index units in the early occupational positions of sons.

Unstandardized regression coefficients are also useful in comparing the effects of the same predictor in different populations. Whereas the comparison of the *standardized* coefficients between populations will be confounded by differences in variances between those populations, the *unstandardized* coefficients will indicate the unit change in the dependent

***Table 10.3.2** Zero-Order Correlations, Means, and Standard Deviations for Occupational Attainment* (Y), *Parental SES* (X_1) *and High School Grades* (X_2) *for 435 Black American Males Who Graduated from High School* c. *1960.*

	X_1	X_2	Mean	Standard Deviation
Y	.136	.134	32.1	20.8
X_1		.054	21.8	17.3
X_2			2.30	.64

Source: James N. Porter, "Race, Socialization and Mobility in Educational and Early Occupational Attainment," *American Sociological Review* 39 (June, 1974): 303–316; adapted from Table 1, p. 307. Reprinted with permission.

variable per unit change in the independent variable in each population, and hence represent the kind of comparison ordinarily desired in determining the similarity or difference of effect in different populations. To illustrate this point, we consider the regression coefficients for black males and compare these to the regression coefficients for white males reported above. The zero-order correlations, means, and standard deviations for black males are displayed in Table 10.3.2, and we utilize this information to obtain the standard regression coefficients for black males:

$$b^*_{Y1.2} = \frac{.136 - (.134)(.054)}{1 - (.054)^2} = \frac{.129}{.997} = .129;$$

$$b^*_{Y2.1} = \frac{.134 - (.136)(.054)}{1 - (.054)^2} = \frac{.127}{.997} = .127.$$

If we stop at this point and compare the standardized regression coefficients for blacks and whites, we are comparing coefficients that reflect the influence of the differences in variance between the two populations as well as the influence of any difference in the effects of the independent variables. In examining the effects of parental socioeconomic status, for example, ($b^*_{Y2.1}$) we note that the regression coefficient for whites is .171 while the regression coefficient for blacks is .129. The comparison suggests that the effect of parental status is greater for whites than for blacks. Before drawing this conclusion, however, we should examine the unstandardized regression coefficients and compute them for blacks:

$$b_{Y1.2} = .129\,\frac{20.8}{17.3} = .155;$$

$$b_{Y2.1} = .127\,\frac{20.8}{.64} = 4.13.$$

It is now evident that the unstandardized regression coefficient representing the effect of parental status on occupational attainment, controlling for high school grades, is .164 for whites and .155 for blacks. It seems appropriate, therefore, to conclude that the effect of parental status is approximately the same for blacks and whites, even though a comparison of the standardized regression coefficients initially suggested otherwise. For both populations, a one unit increase in parental status is worth approximately .16 of a unit of occupational attainment; the standardized regression coefficients differ primarily because of the difference in variances between blacks and whites, and not because of any difference in the unstandardized slope of the regression. On the other hand, the unstandardized regression coefficient for high-school grades is 6.95 for whites and 4.13 for blacks; hence each unit increase in high school grades has more effect on early occupational attainment for whites than for blacks.

As a final point in this comparison, we compute the multiple correlation for blacks, using Formula 10.3.16, and compare it to the multiple correlation for whites.

$$R^2 = (.129)(.136) + (.127)(.134) = .035;$$

$$R = .186.$$

The variation in early occupational attainment is not very completely accounted for by variation in these two predictors for either blacks or whites, but the predictors are especially poor for blacks. Among blacks approximately 3 percent of the variation is accounted for by these two predictors, whereas among whites these predictors accounted for approximately 8 percent of the variation.

Elaborations on Multiple Regression In this chapter, we do not presume to present an exhaustive treatment of multiple regression analysis, and many features of this mode of analyzing data must necessarily be omitted. It seems useful, nonetheless, to comment briefly on selected extensions of multiple regression analysis so that the student will have some minimum familiarity with these extensions which are likely to be encountered in social research.

(1) Step-wise Regression As indicated in the discussion above, any number of predictors may be simultaneously incorporated into a multiple regression analysis. For certain special purposes, predictors may be entered sequentially into the regression equation rather than all being entered simultaneously. One such purpose is to determine the increase in explained variance that is achieved with the addition of a single predictor or set of predictors to an already existing set. If the increase is judged trivial, the added predictors may be judged unnecessary or unimportant, even though their regression coefficients are nonzero. Another such purpose is to find the best set of predictors from a large pool of potential predictors. If this is the

purpose, the step-wise regression procedure is an empirical search procedure in which the predictor with the highest zero-order correlation with the dependent variable is entered first, the additional predictor that adds most to the explained variance achieved by the first is added second, the additional predictor that adds most to the explained variance achieved by the first two is added third, and so on until the best additional predictor fails to add more than a trivial amount to the explained variance already achieved. Considerable caution should be exercised in the interpretation of the results of such a step-wise regression procedure because the conclusions drawn from the simultaneous inclusion of predictors and from the sequential inclusion of predictors are not always the same.

(2) *Dummy Variable Analysis* Although multiple regression analysis was originally designed for use with quantitative variables, nominal variables may be incorporated into regression analysis with readily interpretable results. A simple dichotomy (for example, male–female, urban–rural) may be included as a predictor variable by coding one category 0 and the other category 1. Dummy variable analysis is a procedure for including into a multiple regression analysis nominal variables with more than two classes. This is accomplished by translating such a classification variable into a series of dichotomies, wherein each case is classified as being included in (coded 1) or not included in (coded 0) each of the several categories. To clarify, assume we wish to enter a three-category variable for religious denomination (Protestant, Catholic, Jew) as a predictor in a multiple regression analysis. We translate this three-category nominal variable into three "dummy" variables, X_1, X_2, and X_3, as follows:

$X_1 = 1$ or 0; If Protestant, $X_1 = 1$; otherwise $X_1 = 0$.

$X_2 = 1$ or 0; If Catholic, $X_2 = 1$; otherwise $X_2 = 0$

$X_3 = 1$ or 0; If Jew, $X_3 = 1$; otherwise $X_3 = 0$.

For reasons not explained here, we then proceed (without loss of information) to include any two of these dummy variables as predictors in the multiple regression equation. Although the number of dummy variables entered is always one less than the number of categories in the nominal variable, the effect of the omitted category as well as the effects of those included can be recovered from the regression analysis results, and controls for each of the categories (including the one excluded) are incorporated into the regression coefficients for any other predictors included in the analysis. Dummy variable analysis thus permits greater flexibility in the application of regression analysis than would be feasible if predictors were limited to quantitative variables.

(3) *Multiplicative Terms or Interaction Effects* A basic assumption in multiple regression analysis, as discussed so far in this section, is that the separate linear effects of each predictor may be additively combined to yield appropriate predictions of the values on the dependent variable. It is sometimes the case, however, that one's substantive reasoning suggests that predictors should be multiplicatively rather than additively combined, or multiplicatively as well as additively combined, to yield the most accurate predictions. A substantive interpretation of the multiplicative combination of predictors is that the effect of one variable on the dependent variable depends on the level of a second variable, that is, that the two variables have an interaction effect on the dependent variable. Terms may be included in a multiple regression equation to represent multiples of the separate predictors; the regression coefficients for such multiplicative terms are then estimates of the interaction effect for that combination of variables.

An investigator may utilize step-wise regression procedures to decide whether the interaction terms contribute to the prediction of the dependent variable beyond the predictive accuracy achieved by the additive effects alone. This is accomplished by entering the additive terms alone, then entering the additive and multiplicative terms, and then noting the increase in explained variance achieved. Alternatively, if theoretical reasoning leads to the anticipation of interactive effects rather than additive effects, one may enter the multiplicative terms first and then note the increase in explained variance achieved by the inclusion of the additive terms along with the multiplicative terms.

The interpretation of multiplicative effects when the multiplicative term in question is the product of two or more quantitative variables is hazardous and should be approached with caution. When the multiplicative term in regression analysis is the product of two dummy variables, the interpretation is statistically identical to the interpretation of interaction effects in the analysis of variance (Chapter 15).

These elaborations on multiple regression analysis—the step-wise procedure, dummy variable analysis, and the inclusion of multiplicative terms in the regression equation—makes multiple regression analysis a highly flexible and very general system for data analysis, with a wide variety of potential applications in the analysis of social data.

Special Problems in Determining the Relative Importance of Variables in Multiple Regression Analysis It has become very common practice in social science research for investigators to compare the magnitudes of standardized partial regression coefficients and attribute a certain "importance" to variables according to such magnitudes. Among the examples introducing this section, for example, we cited a study by Porter in which the largest standardized partial regression coefficient in predicting educational attainment was for intelligence. This finding may readily lead to the conclusion that the most important determinant of educational attainment is intelligence. Such a conclusion should be made with caution, and with due consideration for the following possible hazards in drawing such a conclusion:

(1) Partial regression coefficients are attenuated (pulled closer to zero) by random measurement error or measurement unreliability. If two predictors really have exactly equal effects on a dependent variable, but one of these predictors is measured with high reliability and the other with low reliability, the partial regression coefficients for the two predictors will not be equal. The predictor measured with high reliability will have the larger partial regression coefficient. Hence, if intelligence is measured with higher reliability than parental socioeconomic status, intelligence may have a larger standardized partial regression coefficient than parental socioeconomic status, partly or wholly because of differing reliabilities, rather than because of differences in real effects. Techniques for correcting for random measurement error in multiple regression analysis are beyond the scope of this book.

(2) If multiple indicators for the same underlying concept are included as predictors in a multiple regression analysis, the real effects of that underlying dimension will be split among the indicators. Hence, if a single score represents intelligence among the predictors, but parental socioeconomic status is represented in the analysis by three different predictors, the regression coefficient for parental socioeconomic status will be divided among the three predictors, and the partial regression coefficient for each may be smaller than the partial regression coefficient for intelligence simply because of this subdividing of the effect of a single dimension among its several indicators. If predictors include several measures of the same dimension it is ordinarily advisable to combine them into a single index in the regression analysis so as to avoid misleading conclusions. Alternative techniques are available for estimating the regression coefficients for unmeasured dimensions represented by several indicators, but those techniques are too complex to include here.

(3) Closely correlated predictors make the partial regression coefficients unreliable. The correlation among predictors is known as *multicolinearity* and modest correlations among predictors are common and unproblematic in multiple regression analysis. As the correlations among predictors approach 1.0, however, the results of the multiple regression analysis become increasingly unreliable; in the extreme case of two perfectly correlated predictors, their partial regression coefficients are indeterminate. It is difficult to state a general and simple rule of thumb for alerting consumers of statistics to the hazards of multicolinearity because the hazards depend on the number of predictors and their respective correlations with the dependent variable as well as on the correlations among the predictors themselves. Techniques are available for estimating the *standard errors* of partial regression coefficients (and this is a common feature of the print-out from computer programs for multiple regression analysis) and extreme multicolinearity will be reflected in high standard errors. The concept of standard errors is discussed in Chapters 13 and 14.

(4) The relative importance of variables depends not only on the relative magnitude of the standardized regression coefficients, interpreted with due

caution for the points made above, but also on their relative importance in the underlying causal structure. In multiple regression analysis, no assumptions need to be made about the causal structure linking the predictor variables. However, interpretation of the results of a multiple regression analysis in terms of the relative importance of variables does require some assumptions about the causal structure linking the predictors (assuming they *are* linked and hence correlated with each other). Comparing the magnitudes of standardized regression coefficients entails the comparison of the direct effects of the predictors on the dependent variable, but conclusions about the relative importance of variables must also take into account the indirect effects of the predictors, that is, effects that are mediated through other predictors. Such decisions are best handled by making causal assumptions explicit as in path analysis, the topic of the next section.

QUESTIONS AND PROBLEMS

1. Define the following concepts:
   multiple regression equation
   standardized partial regression coefficient
   unstandardized partial regression coefficient
   normal equations
   intercept
   multiple correlation coefficient
   step-wise regression
   dummy variable
   interaction effects
   multiplicative terms in regression equations
   additive terms in regression equations
   attentuation
   multicolinearity

2. Using the data of Table 10.3.3, answer the following:
   (a) In which country, the United States or Great Britain, do education (school leaving age) and occupational status account for a larger proportion of the variation in annual income, or is R^2 about the same for both countries? (Note: the inclusion of father's occupational status as a predictor would add little to the explained variance in either country.)
   (b) Considering education and occupational status as predictors of annual income, is an additional year of schooling worth more in annual income in the United States or in Great Britain? Is an additional unit of occupational status worth more in the United States or in Great Britain?
   (c) On the basis of your answers to (a) and (b) above, do you conclude that a higher R^2 in one population as compared to another implies that the effect of at least one predictor is greater in that population, or is such a conclusion inappropriate? What difference in the income distributions of

Table 10.3.3 *Product-Moment Correlations, Means, and Standard Deviations for Selected Variables Pertaining to Status Attainment of White Males Aged 25–64 in the United States (1963) and Great Britain (1963)*

	School Leaving Age	Occupational Status	Annual Income	Mean	Standard Deviation
United States					
Father's occupational status	.27	.26	.16	41.0	10.3
School leaving age		.53	.42	11.1	3.4
Occupational status			.39	43.1	12.1
Annual income				6,397	3,659
Great Britain					
Father's occupational status	.25	.35	.31	35.3	10.5
School leaving age		.45	.49	14.5	1.3
Occupational status			.62	38.8	11.5
Annual income				2,541	1,238

Source: Adapted from Donald J. Treiman and Kermit Terrell, "The Process of Status Attainment in the United States and Great Britain," *American Journal of Sociology* 81 (November, 1975): 563–583. Table 3, p. 573.

the United States and Great Britain is helpful in understanding your findings in (a) and (b) above?

(d) In which country do education and father's occupational status account for a larger proportion of the variation in occupational status, or is R^2 about the same for both countries?

(e) Which of these two variables (education or father's occupational status) is the better predictor of occupational status in the United States? Which is the better predictor of occupational status in Great Britain? What do you conclude about the impact of education and parental status on occupational achievement in these two countries?

3. (a) Using the data of Table 10.3.4, determine what proportion of the county variation in intensity of rebellion is explained by the additive effects of "traditionalism" and "commercialization."

(b) (For students with access to a computer and a computer program for multiple regression.) Include three predictors of the intensity of rebellion

Table 10.3.4 *Product-Moment Correlations between Selected Variables Describing County Characteristics during the Romanian Peasant Rebellion of 1907*

	Traditionalism	Commercialization	CT[d]
Intensity of Rebellion[a]	.45	.71	.78
Traditionalism[b]		.29	.77
Commercialization[c]			.79

[a] An index based on the proportion of villages in which incidents of rebellion occurred and the estimated number of deaths attributed to the rebellion in each county.
[b] An index based on the literacy rate, i.e., 1.0 − proportion literate.
[c] An index based on the proportion of land devoted to the main cash crop (wheat). Thus the commercialization index is simply the percentage of cultivated land in each county devoted to wheat over a five-year period (1900–1904).
[d] The product of the traditionalism and commercialization indexes, in other words, the *interaction term.*
Source: Adapted from Daniel Chirot and Charles Ragin, "The Market, Tradition and Peasant Rebellion: The Case of Romania in 1907," *American Sociological Review* 40 (August, 1975): 428–444, Table 2, p. 439. Reprinted with permission.

in a multiple regression equation: "traditionalism," "commercialization," and the product of these two variables (CT). Compare the relative magnitudes of the standardized regression coefficients. Explain why the inclusion of the interaction (product) term increases the explained variance and the coefficient for the interaction term is larger than for either of the two components of it. (For clues, consult the article from which the table was adapted.)

4 PATH ANALYSIS†

Path analysis is a method that permits the social analyst to use explicit causal assumptions in the analysis of data. In path analysis causal assumptions are incorporated into a set of multiple regression equations and the coefficients are estimated in the usual way. These coefficients give numerical values to the direct effects of selected causal variables on each of a series of dependent variables, and indirect effects can be readily obtained from the complete set of direct effects. In interpreting the results of path analysis it is important to recognize that no path analysis can be carried out until the causal order assumptions have been stated since these assumptions are an integral part of this method and are not conclusions drawn from the results.

The explicit statement of assumptions about the causal structure underlying a set of observed correlations is known as a *causal model*; and the term *path model* is usually restricted to causal models in which no feedback effects are postulated. The nature of path models and of the corresponding path

† This section may be omitted without loss of continuity.

Table 10.4.1 *Zero-Order Correlation Coefficients, Means, and Standard Deviations for Parental Socioeconomic Status (X_1), High School Grades (X_2), Educational Attainment (X_3), and Occupational Attainment (X_4) for 14,891 White American Males Who Graduated from High School* c. *1960*

	X_1 SES	X_2 GRADES	X_3 EDUC	X_4 JOB	Mean	Standard Deviation
X_1 SES		.122	.303	.197	36.40	23.20
X_2 Grades			.377	.236	2.29	.69
X_3 Education				.489	6.24[a]	.79
X_4 Job					46.40	22.30

[a] Educational attainment is not expressed in grades of school completed. Highest grade in school completed was coded into the standard census education categories.

Source: James N. Porter, "Race, Socialization and Mobility in Educational and Early Occupational Attainment," *American Sociological Review* 39 (June, 1974): 303–316, adapted from Table 1, p. 307. Reprinted with permission.

analysis can be most readily presented by considering a simple example. For this purpose, we expand on the illustration used in the previous section.

In the previous section, we considered parental socioeconomic status and high school grades as predictors of occupational attainment among American males. We now expand this illustration to include educational attainment as a third variable presumed to have an effect on occupational attainment. The zero-order correlations between all pairs of these variables, and the means and standard deviations for each, are shown in Table 10.4.1. These entries provide a convenient point of departure for our discussion of path analysis, but it will be evident that we begin this discussion with the data in an advanced stage of processing, that is, the table presents the results of prior computations on the raw data rather than the raw data themselves.

To formulate the causal assumptions constituting the path model, we may utilize prior information and common-sense reasoning (a) to designate the causal ordering of the variables, and (b) to propose the path or paths by which each antecedent variable has its presumed effects on variables that follow it in the proposed causal ordering. For the four variables in Table 10.4.1, the causal ordering is largely unproblematic because each variable describes a chronologically ordered experience in the individual biographies; parental socioeconomic status is causally prior to educational and occupational attainment of the son, for example, as are the son's high school grades. The causal ordering of parental socioeconomic status and high school grades may be considered more problematic, and one might want to leave open the question of whether the correlation between these two variables results from the effect of one on the other or from their common dependence on antecedent variables not included in this analysis. Path analysis can accommodate this kind of uncertainty, as indicated in greater detail below. The path model states not only the causal ordering of the variables, but also the analyst's assumptions about the direct and indirect paths by which one variable has an effect on another. For example, the path model states whether parental socioeconomic status affects occupational attainment of the son directly (that is, without this effect being mediated by another variable in the model), indirectly through educational attainment (as when parental socioeconomic status affects educational attainment which, in turn, affects occupational attainment), indirectly through some other mediating variable, or in all of these manners. In short, the path model makes explicit the analyst's best judgment about the causal structure by which the variables are interconnected.

The assumptions about causal structure—the path model—may be expressed either in the form of a set of equations or in the form of a corresponding *path diagram*. Three possible path models for the variables of Table 10.4.1 are presented in Figures 10.4.1, 10.4.2 and 10.4.3 where each model is expressed in both diagrammatic and equational form. Although these equivalent forms convey similar kinds of information, both may be advantageously used. The diagrammatic form is more readily comprehended at a glance, but the equations are basic to the analysis and make the assumption of linear dependence explicit.

A path model, in either diagrammatic or equational form, may be read by considering the postulated effects of each variable on every other variable.

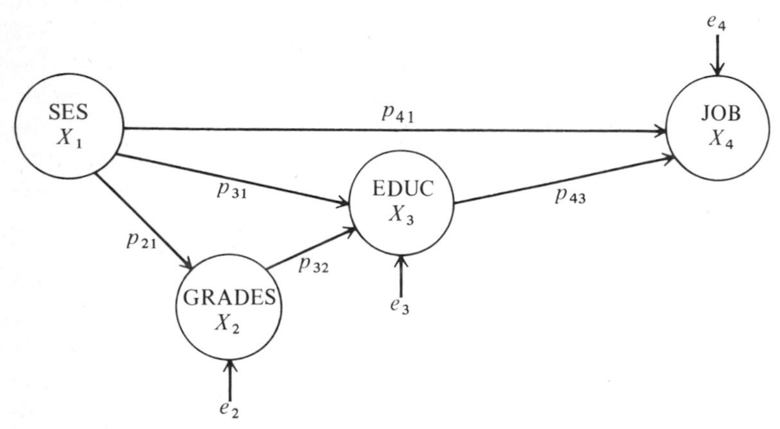

Corresponding equations:

$$\text{SES} = p_{1e}e_1 \qquad + (0)(\text{GRADES}) + (0)(\text{EDUC}) + (0)(\text{JOB})$$

$$\text{GRADES} = p_{2e}e_2 + p_{21} \qquad + (0)(\text{EDUC}) + (0)(\text{JOB})$$

$$\text{EDUC} = p_{3e}e_3 + p_{31}(\text{SES}) + p_{32}(\text{GRADES}) \qquad + (0)(\text{JOB})$$

$$\text{JOB} = p_{4e}e_4 + p_{41}(\text{SES}) + (0)(\text{GRADES}) + p_{43}(\text{EDUC})$$

$$r_{e_i e_j} = 0 \text{ for all } i \neq j$$

$$r_{e_i X_j} = 0 \text{ for all } i \neq j$$

Figure 10.4.1 *Path Model for Variables Affecting Occupational Prestige—(1)*

For example, two of the assumptions expressed in Figure 10.4.1 are: (1) that parental socioeconomic status (SES) has an effect on son's grades (GRADES), and (2) that GRADES has no effect on SES. These two assumptions are represented in the diagram by an arrow from SES to GRADES and by the absence of an arrow from GRADES to SES. The same assumptions are represented in the equations by the presence of a nonzero coefficient for the effect of SES on GRADES (see the equation in which GRADES appears on the left-hand side of the equation) and by a zero coefficient for the effect of GRADES on SES (see the equation in which SES appears on the left-hand side of the equation).

Although this model assumes that SES has an effect on GRADES, it does not assume that SES completely determines GRADES; for this reason the effects of all other unspecified variables on GRADES are represented by a *residual* or *error* term, symbolized by e_2 in both the diagrammatic and the equational form of the model.[11] In the diagram, the absence of arrows connecting the error terms with each other or with more than a single variable in the model represents the assumption that error terms are uncorrelated with these other elements. In the equational form of the model, this assumption is made explicit by presenting, in addition to the "dependence" equations, statements of noncorrelated residuals. By assuming uncorrelated residuals we may use the regression techniques described in the previous section to estimate the path coefficients. Such estimates will be inaccurate if the assumption of uncorrelated residuals is contrary to fact. The model shows, for example, a zero correlation between SES and e_2, which implies that the only reason for covariation between SES and GRADES is the effect of SES on GRADES; if this assumption is not true, the model is said to be *misspecified.* A model is misspecified if it fails to represent completely the connections (causal or otherwise) between all elements in the model. When the model has been misspecified, the estimated effects will be in error to a greater or lesser degree.

In the model represented in Figure 10.4.1, SES is not causally dependent on any other variables included in the model. SES is therefore said to be an *exogenous* variable in this model. All other variables in this model are *endogenous* because the model proposes that they are dependent on at least one other variable in the model. In presenting the equations for the model, we have included an equation in which SES appears on the left-hand side in

[11] The terminology for referring to the sources of variation other than the variables explicitly included in the model is not standardized, and the symbols vary with the terminology. Thus the terms *residual* (symbolized by r or R), *error* (symbolized by e) and *unique factors* (symbolized by u) are synonymous. Because r and R might be confused with the bivariate and multiple correlation coefficients and because the term *unique factors* seems more appropriate in the context of factor analysis, where it was originally used, we shall use the symbol e and refer either to the error term or the residual term. We should note explicitly that the error term includes but is not limited to random measurement error as a source of unexplained variation. Whether e is called an error term or a residual, it will refer, in the context of path analysis, to *all* sources of unexplained variation or error variation in a given variable.

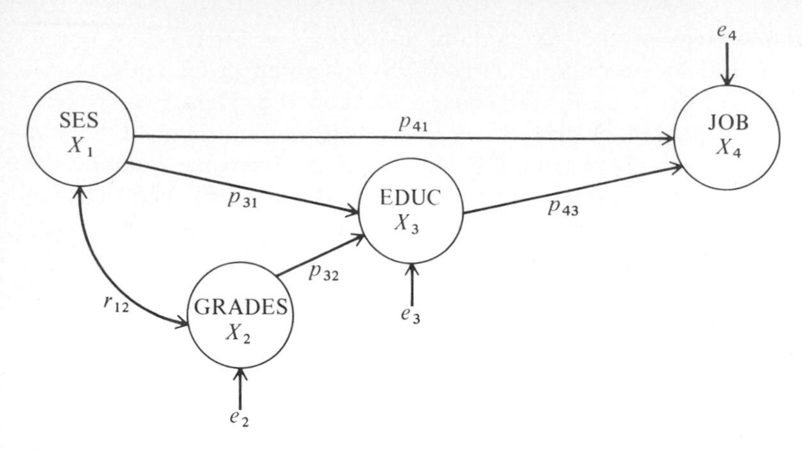

Corresponding equations:

$$\begin{aligned}
\text{SES} &= p_{1e}e_1 \qquad\qquad\qquad + (0)(\text{GRADES}) + (0)(\text{EDUC}) + (0)(\text{JOB}) \\
\text{GRADES} &= p_{2e}e_2 + (0)(\text{SES}) \qquad\qquad\qquad + (0)(\text{EDUC}) + (0)(\text{JOB}) \\
\text{EDUC} &= p_{3e}e_3 + p_{31}(\text{SES}) + p_{32}(\text{GRADES}) \qquad\qquad + (0)(\text{JOB}) \\
\text{JOB} &= p_{4e}e_4 + p_{41}(\text{SES}) + (0)(\text{GRADES}) + p_{43}(\text{EDUC})
\end{aligned}$$

$$r_{e_ie_j} = 0 \text{ for all } i \neq j \text{ except } r_{e_1e_2} \text{ may be nonzero}$$

$$r_{e_iX_j} = 0 \text{ for all } i \neq j \text{ except } r_{e_1X_2} \text{ and } r_{e_2X_1} \text{ may be nonzero}$$

Figure 10.4.2 *Path Model for Variables Affecting Occupational Prestige—(2)*

order to make explicit the assumption that it is *not* dependent on any of the variables appearing on the right-hand side of that equation. It is, of course, necessary to include in that equation something on which SES *is* dependent, and we have therefore designated it as dependent on e_1, an error term analogous to the other error terms, which represents all sources of unexplained variation not otherwise included in the model. For an exogenous variable, such as SES, the error term represents the source of all variation, since none of the variation in an exogenous variable is explained by other variables in the model. The residuals for exogenous variables have been omitted from the diagrams since they are superfluous there.

The model represented in Figure 10.4.2 is identical to that in Figure 10.4.1 except that the covariation of SES and GRADES is left unanalyzed in the model. In the model of Figure 10.4.2, no assumption is made that SES has an effect on GRADES, or the reverse. Although the correlation between these two variables, if any, will be taken into account in estimating the other coefficients of the model, no claim is made as to how this correlation is generated. Such an unanalyzed correlation between variables in a causal model is represented diagrammatically by a curved, double-headed arrow, in

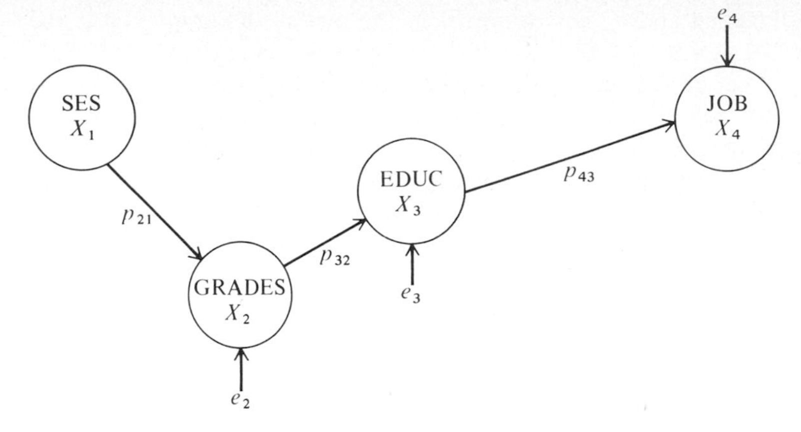

Corresponding equations:

$$\begin{aligned}
\text{SES} &= p_{1e}e_1 + (0)(\text{GRADES}) + (0)(\text{EDUC}) + (0)(\text{JOB}) \\
\text{GRADES} &= p_{2e}e_2 + p_{21}(\text{SES}) + (0)(\text{EDUC}) + (0)(\text{JOB}) \\
\text{EDUC} &= p_{3e}e_3 + (0)(\text{SES}) + p_{32}(\text{GRADES}) + (0)(\text{JOB}) \\
\text{JOB} &= p_{4e}e_4 + (0)(\text{SES}) + (0)(\text{GRADES}) + p_{43}(\text{EDUC})
\end{aligned}$$

$$r_{e_i e_j} = 0 \text{ for all } i \neq j$$

$$r_{e_i X_j} = 0 \text{ for all } i \neq j$$

Figure 10.4.3 *Path Model for Variables Affecting Occupational Prestige—(3)*

contrast to the straight, single-headed arrows used to represent postulated effects. When the model is stated in equational form, this unanalyzed correlation is represented by statements which permit r_{12} to be nonzero.

It should be evident that there are two exogenous variables in Figure 10.4.2; both SES and GRADES are exogenous because the model does not propose that either is dependent on any other variable in the model. Although it would be possible to propose a model in which all variables but one are exogenous, the term *path analysis* is ordinarily used to refer to models which have two or more endogenous variables.

Figure 10.4.3 differs from the two preceding path models in that it postulates a simple *causal chain*. The model proposes that SES affects GRADES which affect EDUC which affects JOB, and the effect of each antecedent variable is assumed to be mediated entirely by intermediate, or intervening, variables in the chain. Stated otherwise, the model postulates that there is no direct effect of SES on JOB, but that there is an indirect effect mediated by GRADES and EDUC. The model also postulates no direct effect of GRADES on JOB but an indirect effect on JOB mediated by EDUC. The student will recognize that causal chains are also postulated in the two models discussed above. For example, in the model of Figure 10.4.1, the effect of GRADES on JOB is assumed to be mediated entirely by EDUC, while the effect of SES on JOB is assumed to be both direct and indirect. A path analysis will provide numerical estimates for both direct and indirect effects, and these results will either confirm or fail to confirm the assumption that certain direct effects are absent. In this sense, a path model may be "tested," but it should be noted that such a test pertains to the presence or absence of direct effects rather than constituting a test of the causal ordering of the variables.

The three models represented in Figures 10.4.1, 10.4.2, and 10.4.3 are all alike in postulating no feedback effects, and all are therefore called *recursive* models. A causal model is said to be recursive if it proposes that no causal variable is affected, directly or indirectly, by any of its effects. Thus in a recursive model, a given variable "recurs" in its subsequent effects, and in the effects of its effects, but an effect does not feed back on its cause. The dependence equations for a recursive model can always be arranged so that the coefficients above the diagonal running from upper left to lower right are

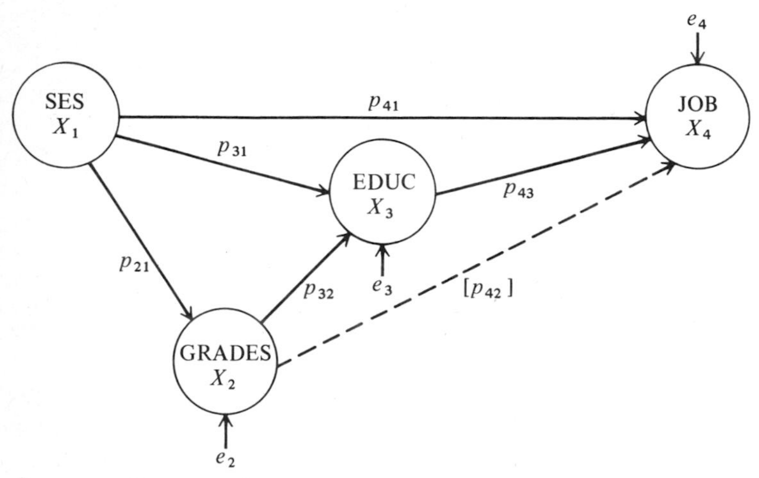

Corresponding equations:

$$\text{SES} = p_{1e}e_1$$

$$\text{GRADES} = p_{2e}e_2 + p_{21}(\text{SES})$$

$$\text{EDUC} = p_{3e}e_3 + p_{31}(\text{SES}) + p_{32}(\text{GRADES})$$

$$\text{JOB} = p_{4e}e_4 + p_{41}(\text{SES}) + [p_{42}](\text{GRADES}) + p_{43}(\text{EDUC})$$

$$r_{e_i e_j} = 0 \text{ for all } i \neq j$$

$$r_{e_i X_j} = 0 \text{ for all } i \neq j$$

Figure 10.4.4 *Fully Recursive Path Model Corresponding to the Model of Figure 10.4.1*

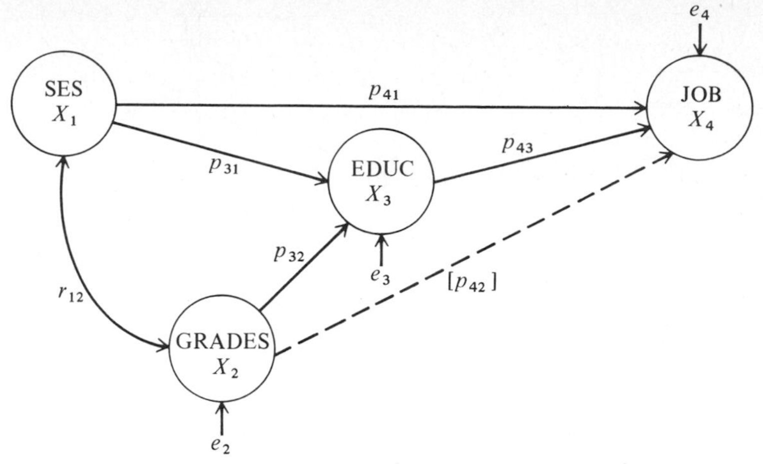

Corresponding equations:

$$\text{SES} = p_{1e}e_1$$

$$\text{GRADES} = p_{2e}e_2$$

$$\text{EDUC} = p_{3e}e_3 + p_{31}(\text{SES}) + p_{32}(\text{GRADES})$$

$$\text{JOB} = p_{4e}e_4 + p_{41}(\text{SES}) + [p_{42}](\text{GRADES}) + p_{43}(\text{EDUC})$$

$$r_{e_i e_j} = 0 \text{ for all } i \neq j \text{ except } r_{e_1 e_2} \text{ may be nonzero}$$

$$r_{e_i r_j} = 0 \text{ for all } i \neq j \text{ except } r_{e_1 X_2} \text{ and } r_{e_2 X_1} \text{ may be nonzero}$$

Figure 10.4.5 *Fully Recursive Path Model Corresponding to the Model of Figure* 10.4.2

uniformly zero, as illustrated in the dependence equations for each of the models represented above. If the model is *nonrecursive*, that is, if there *are* direct or indirect feedback effects, the equations cannot be arranged to meet this condition. The estimation of nonrecursive models is beyond the scope of this book and our attention here is limited to recursive models. It should be clear that a nonrecursive model is implausible for the variables represented in Table 10.4.1. For example, it is implausible to assume that JOB (as an adult) affects GRADES (in high school). Thus, in this as in many other problems of interest in social research, recursive models provide a reasonable basis for the analysis and interpretation of social data.

Before proceeding to compute estimates for the coefficients in the models presented above, we can transform each into a *fully recursive* model, as shown in Figures 10.4.4, 10.4.5, and 10.4.6. A model is said to be fully recursive if each variable is directly influenced by all variables antecedent to it in the postulated causal order. The additional direct effects necessary to make each of the models in Figures 10.4.1, 10.4.2, and 10.4.3 fully recursive have been represented in Figures 10.4.4, 10.4.5, and 10.4.6 by arrows with broken

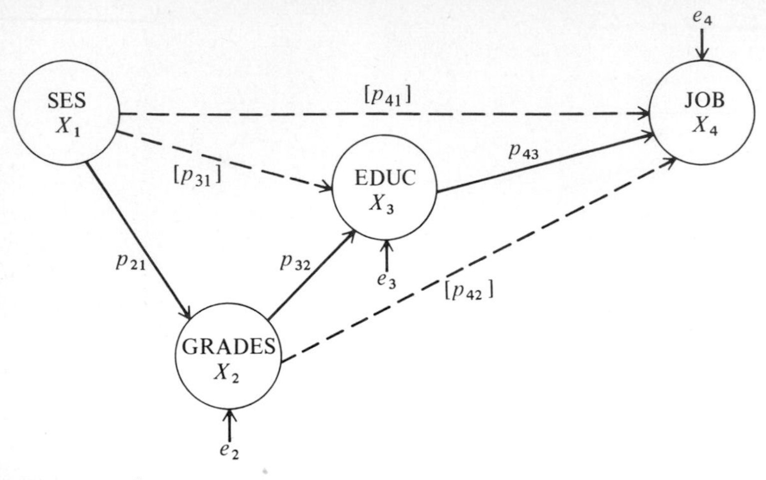

Corresponding equations:

$$\text{SES} = p_{1e}e_1$$

$$\text{GRADES} = p_{2e}e_2 + p_{21}(\text{SES})$$

$$\text{EDUC} = p_{3e}e_3 + [p_{31}](\text{SES}) + p_{32}(\text{GRADES})$$

$$\text{JOB} = p_{4e}e_4 + [p_{41}](\text{SES}) + [p_{42}](\text{GRADES}) + p_{43}(\text{EDUC})$$

$$r_{e_i e_j} = 0 \text{ for all } i \neq j$$

$$r_{e_i X_j} = 0 \text{ for all } i \neq j$$

Figure 10.4.6 *Fully Recursive Path Model Corresponding to the Model of Figure 10.4.3*

lines, and the corresponding coefficients have been enclosed in brackets. The models proposed earlier postulate that these bracketed coefficients will be zero. When the fully recursive model is estimated, the tenability of these postulates can be determined.

It is evident from the above figures that, after translation into fully recursive form, models 10.4.4 and 10.4.6 consist of an identical set of equations, and the estimates for the coefficients based on these fully recursive models must therefore be identical. The difference between these two models lies entirely in which of these estimates are hypothesized to be zero. Model 10.4.5 in fully recursive form differs from the others in only one respect: the variable GRADES is permitted to be correlated with SES but is not proposed as being dependent on SES—a difference that will not affect the estimates for the other coefficients in the model. The equivalence of the estimates for the models in fully recursive form—whatever direct effects may have been omitted in the investigator's postulated model—provides the rationale for making the estimates on the basis of a fully recursive model. Such estimates will not be affected by an investigator having erroneously omitted a direct effect that should have been included.

To estimate the coefficients in the fully recursive models, we note that the equations constituting the model have the form of regression equations; they represent the dependence of one variable on one or more causally antecedent variables and require a determination of the best fitting coefficients to describe that dependence. Hence, we may rewrite the equations using appropriate regression coefficients in place of path coefficients, as shown below. In making the transformation to regression equations we also remove the error terms, and in recognition of this removal, put a "hat" on the left-hand side to indicate that the equation yields a predicted rather than an actual value. The resulting equations for the model of Figure 10.4.4 are:

$$\hat{X}_2 = b^*_{21} X_1; \tag{10.4.1}$$

$$\hat{X}_3 = b^*_{31.2} X_1 + b^*_{32.1} X_2; \tag{10.4.2}$$

$$\hat{X}_4 = b^*_{41.23} X_1 + b^*_{42.13} X_2 + b^*_{43.12} X_3. \tag{10.4.3}$$

The standardized regression coefficients, which are the same as the standardized path coefficients, may be estimated by the procedures described in the preceding section. It will be evident that the estimating equations above are also the estimating equations for the model of Figure 10.4.6. For the model of Figure 10.4.5, the equations are as follows:

$$\hat{X}_3 = b^*_{31.2} X_1 + b^*_{32.1} X_2; \tag{10.4.4}$$

$$\hat{X}_4 = b^*_{41.23} X_1 + b^*_{42.13} X_2 + b^*_{43.12} X_3. \tag{10.4.5}$$

The student will note that these equations are identical to Formulas 10.4.2 and 10.4.3 above.

The standardized path coefficients computed from the data of Table 10.4.1 using Formulas 10.4.1 to 10.4.3 above are displayed in Figure 10.4.7a. The unstandardized path coefficients (sometimes called *path regression coefficients*) may be readily obtained from the standardized coefficients by multiplying the latter by the appropriate ratio of standard deviations. Thus

$$b_{21} = b^*_{21} \frac{\sigma_2}{\sigma_1};$$

$$b_{31.2} = b^*_{31.2} \frac{\sigma_3}{\sigma_1};$$

$$b_{32.1} = b^*_{32.1} \frac{\sigma_3}{\sigma_2};$$

and so forth.

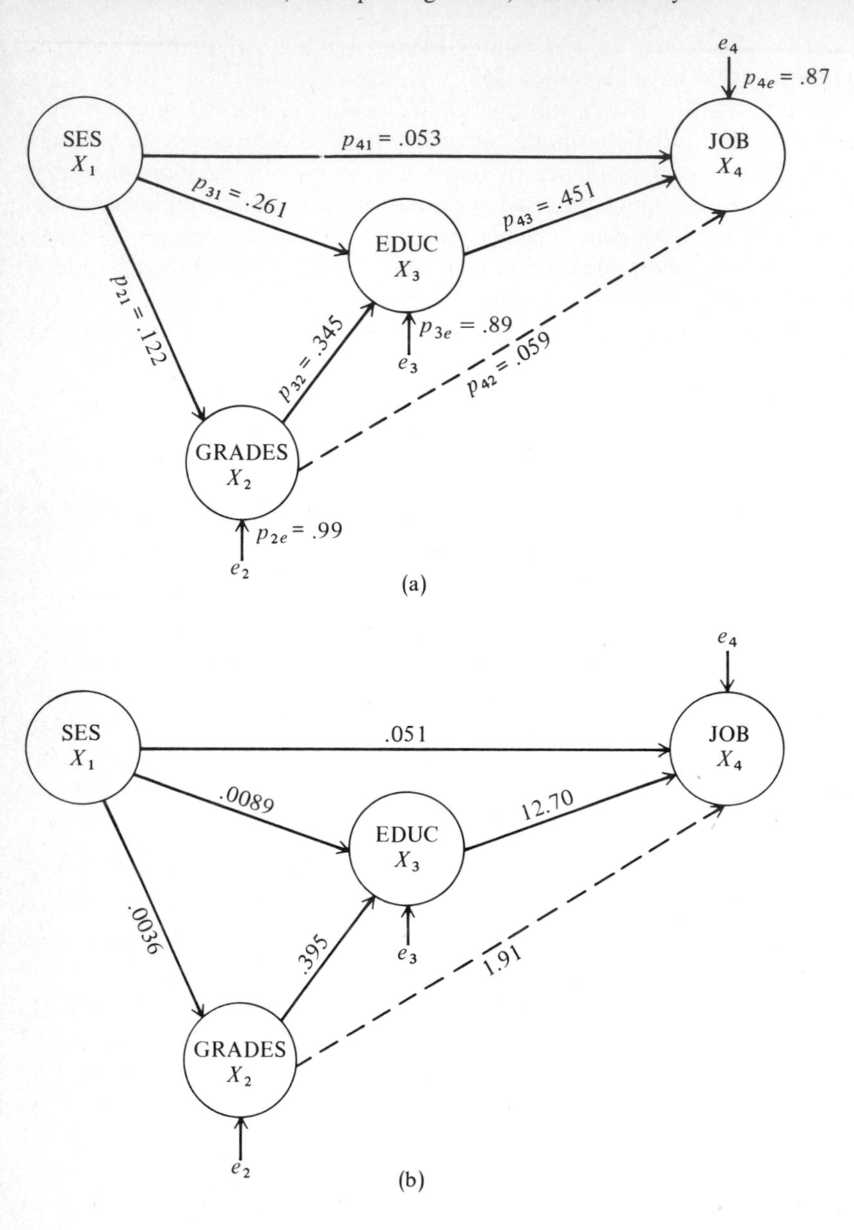

Figure 10.4.7 *Path Coefficients (Standardized and Unstandardized) for the Model of Figure 10.4.4*

The unstandardized path coefficients are displayed in Figure 10.4.7b. (As an exercise, the student may wish to verify the coefficients for X_3 as a dependent variable using the formulae given in the previous section.)

It may be shown algebraically that the standardized path coefficient for the residual or error term is equal to the square root of $1 - R^2$, where R^2 is the square of the multiple correlation between the dependent variable in question and the variables causally antecedent to it as predictors. Thus

$$p_{4e} = \sqrt{1 - R^2_{4.123}};$$

$$p_{3e} = \sqrt{1 - R^2_{3.12}};$$

$$p_{2e} = \sqrt{1 - R^2_{2.1}} = \sqrt{1 - r^2_{21}}\ .$$

It follows from the equations above that the square of the coefficient for the residual or error term indicates the proportion of the variation in that variable that is *not* explained by the variables in the path model. Thus, $(p_{4e})^2 = (.87)^2 = .76$, and this indicates that the variables in the model account for approximately 24 percent of the variance in JOB (occupational prestige), while the remaining 76 percent of that variation has sources outside this model.

The unstandardized path coefficients depend on the units of measure and the magnitudes of the standard deviations of each variable, and a comparison of one with another in the same path diagram is difficult to interpret. For example, the unstandardized path coefficient for JOB on GRADES (1.91) is approximately 37 times as large as the unstandardized path coefficient for JOB on SES (.051). This does *not* mean, however, that the direct effect of GRADES accounts for 37 times more of the variation in JOB than does SES. As the comparison of their standardized coefficients indicates, the direct effects of GRADES and SES are about equal. Unstandardized path coefficients (path regression coefficients) indicate the unit change in the dependent variable per unit change in the antecedent variable, when all other variables antecedent to the dependent variable in the model have been held constant. Thus, JOB (occupational prestige) increases by 1.91 units for every unit increase in GRADES, when SES and EDUC are held constant. In addition to such descriptive statements, unstandardized path coefficients are useful in making comparisons between the direct (or indirect) effects in different populations. For example, we might wish to compare the direct effect of GRADES on JOB for blacks and whites, or for males and females; in such a comparison, the unstandardized path coefficients are the appropriate parameters to compare.

Decomposing the Zero-Order Correlations The basic theorem of path analysis pertains to the components of the zero-order correlations between pairs of variables. That theorem may be abstractly stated as follows:

$$r_{ij} = p_{ij} + \sum_k p_{ik} r_{jk} \qquad (10.4.6)$$

where i and j denote two variables in the model and k denotes the complete set of variables in the model with paths leading directly to the ith variable. The first term on the right-hand side of the equation is the direct causal effect of X_j on X_i. Each of the elements in the summation term may be further decomposed since each r_{jk} may be decomposed by succeeding applications of the same decomposition formula. In this presentation, we forego a formal analysis of this theorem and present instead a discussion that depends on the student's intuitive grasp of the logic of path models.

We will express each of the zero-order correlations as the sum of three components, each of which has a different substantive meaning. The direct causal effect of X_2 on X_4, for example, is the effect that is not mediated by any other variable included in this model. This direct effect may be read directly from Figure 10.4.7: $p_{42} = .059$. In addition to this direct effect, there are as many indirect effects as there are paths from X_2 to X_4 through other variables in the model. Thus, the indirect causal effect of X_2 on X_4 is constituted by effects mediated by other variables. In this instance, there is only one indirect effect, and it is through X_3. The magnitude of this indirect effect may be obtained by taking the product of the corresponding path coefficients; thus the indirect effect of GRADES on JOB through EDUC is

$$p_{43}p_{32} = (.451)(.345) = .156.$$

Finally, the model attributes some part of the correlation between X_2 and X_4, not to the effects of X_2 on X_4 (whether direct or indirect), but to a common antecedent variable: SES. We shall refer to that part as the spurious component. To find the spurious component, we decompose the correlation into its component parts using Formula 10.4.6 and find those parts that are neither direct effects nor indirect effects. Thus, by Formula 10.4.6, we have

$$r_{42} = p_{42} + p_{41}r_{21} + p_{43}r_{32}.$$

Applying Formula 10.4.6 again to decompose r_{32} we obtain

$$r_{32} = p_{32} + p_{31}r_{12}.$$

Substituting, we have the complete decomposition of r_{42}:

$$r_{42} = p_{42} + p_{41}r_{21} + p_{43}p_{32} + p_{43}p_{31}r_{12}.$$

The direct effect has already been identified as p_{42}. The indirect effect has been identified as $p_{43}p_{32}$. The remaining parts constitute the spurious component of r_{42}. Hence the spurious component is $p_{41}r_{12} + p_{43}p_{31}r_{12}$. Substituting magnitudes from Figure 10.4.7a, we have:

$$\begin{aligned} p_{41}r_{12} + p_{43}p_{31}r_{12} &= (.053)(.122) + (.451)(.261)(.122) \\ &= .006 + .013 \\ &= .019. \end{aligned}$$

We may now express the observed correlation between X_2 and X_4 as the sum of the direct effect, indirect effect, and spurious component.

$$
\begin{aligned}
r_{42} &= \text{direct effect} + \text{indirect effect} + \text{spurious component}\\
&= p_{42} \qquad + p_{43}p_{32} \qquad + p_{41}r_{12} + p_{43}p_{31}r_{12}\\
&= .059 \qquad + .156 \qquad + .019\\
&= .234.
\end{aligned}
$$

The result differs from r_{42} computed directly (Table 10.4.1: $r_{42} = .236$) because of rounding errors. The results indicate that GRADES and JOB are correlated primarily because of the indirect effect of GRADES on JOB through EDUC.

In decomposing the correlation between SES and JOB, there will be no spurious component since there is no variable in the model antecedent to SES. Hence this correlation will be decomposed into the direct effect and a set of indirect effects. The student will note that in Figure 10.4.7, the indirect effects of SES on JOB are represented by the following paths:
(1) from SES to EDUC to JOB;
(2) from SES to GRADES to EDUC to JOB;
(3) from SES to GRADES to JOB.
Hence, we have

$$
\begin{aligned}
r_{41} &= \text{direct effect} + \text{indirect effects}\\
&= p_{41} \qquad + p_{43}p_{31} + p_{43}p_{32}r_{12} + p_{42}r_{12}\\
&= .053 \qquad + .118 \quad + .019 \qquad + .007\\
&= .053 \qquad + .144\\
&= .197.
\end{aligned}
$$

The results here indicate that the correlation between parental socioeconomic status and the prestige level of the son's first job is largely the consequence of educational attainment which is influenced by SES and which, in turn, influences occupational prestige.

The student should be able to verify the following equations, using Formula 10.4.6, the path diagram of Figure 10.4.7, and the principles illustrated in the discussion above. Using Formula 10.4.6:

$$r_{43} = p_{43} + p_{41}r_{31} + p_{42}r_{32};$$

$$r_{31} = p_{31} + p_{32}r_{12};$$

$$r_{32} = p_{32} + p_{31}r_{12};$$

$$r_{43} = p_{43} + p_{41}p_{31} + p_{41}p_{32}r_{12} + p_{42}p_{32} + p_{42}p_{31}r_{12}.$$

Inspection of the figure reveals that there is no indirect effect. Hence we have:

$$
\begin{aligned}
r_{43} &= \text{direct effect} + \text{spurious component} \\
&= p_{43} \qquad + p_{41}p_{31} + p_{41}p_{32}r_{12} + p_{42}p_{32} + p_{42}p_{31}r_{12} \\
&= .451 \qquad + .014 \quad + .002 \qquad + .020 \quad + .002 \\
&= .451 \qquad + .038 \\
&= .489.
\end{aligned}
$$

$$
\begin{aligned}
r_{31} &= \textit{direct effect} + \textit{indirect effect} \\
&= p_{31} \qquad + p_{21}p_{32} \\
&= .261 \qquad + (.122)(.345) \\
&= .261 \qquad + .042 \\
&= .303.
\end{aligned}
$$

$$
\begin{aligned}
r_{32} &= \textit{direct effect} + \textit{spurious component} \\
&= p_{32} \qquad + p_{31}p_{21} \\
&= .345 \qquad + (.261)(.122) \\
&= .345 \qquad + .032 \\
&= .377.
\end{aligned}
$$

$$
\begin{aligned}
r_{21} &= \textit{direct effect} \\
&= p_{21} \\
&= .122.
\end{aligned}
$$

The substantive interest in these decompositions of the zero-order correlations is primarily the light they shed on the reasons for the correlations between early occupational prestige, on the one hand, and parental socioeconomic status and high school grades on the other. The results indicate that the effects of both parental SES and high school grades are largely mediated by educational attainment. We may also note that the direct effect of high school grades on early occupational attainment is about the same as the direct effect of parental socioeconomic status, and that the indirect effects of these two variables are also roughly similar. These results are therefore incompatible with two rather popular conceptions of the determinants of early occupational attainment; early occupational attainment is neither a simple effect of the advantage of family position (parental SES) nor a simple effect of ability as measured by high school grades. Both of these conceptions are partially—but only partially—supported by the results of the path analysis.

The decomposition of zero-order correlations in models with correlated exogenous variables (for example, the model represented in Figure 10.4.5) proceeds in a manner analogous to that illustrated above, but the interpretation of the components is sometimes complicated in such a model by the lack of an explicit claim about the reasons for the correlations between the exogenous variables. The zero-order correlations can always be decomposed into direct effects and "not direct" effects but it is not always possible in such models to distinguish between indirect effects and spurious components. For example, for the model of Figure 10.4.5:

$$\begin{aligned} r_{41} &= \textit{direct effect} + \textit{indirect effect} + \textit{``not direct'' effect component} \\ &= p_{41} \qquad + p_{31}p_{43} \qquad + p_{43}p_{32}\gamma_{21} + p_{42}\gamma_{21} \\ &= .053 \qquad + .118 \qquad + .026 \\ &= .197. \end{aligned}$$

This decomposition of r_{41} does not differ numerically from the decomposition previously made for the model of Figure 10.4.5. But in this decomposition, part (in this particular instance, a very small part) of what was previously identified as an indirect effect cannot be unambiguously so labelled; this ambiguous component (the "*not direct*" *effect component*) may be either an indirect effect or a spurious component, and the nature of the model does not permit us to readily identify it as either.

Selecting the Final Path Model Having estimated the coefficients of the fully recursive models, we may return to the path models originally presented in Figures 10.4.1 through 10.4.3 and ask which, if any, is correct. We may imagine that each of these models represents the thinking of a different investigator, and the results of the path analysis will serve as the basis for deciding which investigator has the "best" model. We note first that model 10.4.2 differs from the other two models in not postulating a causal order between SES and GRADES. Assumptions about causal order are inputs into the path analysis and the results of that analysis cannot determine the validity of such assumptions. Hence, the results do not permit a decision about whether model 10.4.1 or model 10.4.2 is "better." Model 10.4.1 and model 10.4.3, on the other hand, differ in claiming which direct effects will be zero and which will not. The results of the path analysis *are* relevant to these competing claims, and we may consider the tenability of these two models in the light of these results. Two types of criteria, expressed here in the form of questions, may be used to assess the tenability of path-model claims that certain direct effects are zero: (1) Is the magnitude of the path coefficient such that it is plausible to assume that it differs from zero only because of random sampling variation? Sampling variation is discussed in Chapters 13, 14, and 15, and we will not consider it further at this point except to note that very large samples ordinarily make it implausible to assume that even very small path coefficients differ from zero only because of sampling variation. Since the illustrative data above were based on an unusually *large* sample of almost 15,000 cases, we would conclude by this criterion that none of the models originally presented is adequate, since they postulate zero direct effects that are not zero. (2) Can the zero-order correlations be reasonably well predicted from the path coefficients if certain paths hypothesized to be zero are treated as if they had a coefficient of zero? This latter criterion is ordinarily applied following the exclusion of very small path coefficients in the fully recursive model by the first criterion.

In applying the second criterion, the model is ordinarily re-estimated using only those paths retained after the exclusions. The resulting estimates are then entered into equations analogous to the decomposition equations illustrated above to yield predicted zero-order correlations comparable to the observed zero-order correlation between each pair of variables. For example, if p_{41} and p_{42} were excluded, by hypothesis, the student can readily verify that the estimates for this reconstituted model would be as follows:

$$p_{21} = b^*_{21} = r_{21} = .122;$$

$$p_{31} = b^*_{31.2} = .261;$$

$$p_{32} = b^*_{32.1} = .345;$$

$$p_{43} = b^*_{43} = r_{43} = .489;$$

$$p_{41} = 0 \text{ (by hypothesis)};$$

$$p_{42} = 0 \text{ (by hypothesis)}.$$

Returning now to the decomposition equation for r_{41} above and omitting all terms which include the hypothesized zero coefficients we have

$$\begin{aligned}\hat{r}_{41} &= p_{43}p_{31} + p_{43}p_{32}p_{21}\\ &= (.489)(.261) + (.489)(.345)(.122)\\ &= .128 = .021\\ &= .149.\end{aligned}$$

We previously observed that $r_{41} = .197$, so the reconstituted correlation does not reproduce it closely. The "goodness of fit" of the new model must be assessed by considering the discrepancies between each observed and each reconstructed correlation. By the application of this second criterion, the student will find that with any of the direct effects excluded (treated as zero), the discrepancies are too large to meet the usual criteria, that is, a maximum discrepancy of .02 or, in some instances, .05. We therefore conclude that, by either of the criteria described above, none of the path models originally described provides an adequate fit for the data. The fully recursive model, which necessarily reconstructs the zero-order correlations without error and which is, in that sense, not falsifiable, is therefore tentatively retained as the superior model.

Parental socioeconomic status, high school grades, and educational attainment do not provide a very complete explanation of the variation in early occupational attainment, as indicated by the fact that these three variables in combination explain only about 24 percent of the variation in the prestige of first jobs ($R^2_{r.123} = .24$). This indicates, not that the fully recursive path model is "wrong" but that it is incomplete, that other variables not included in the model have effects on early occupational attainment. Despite such incompleteness, the path model is helpful in illuminating features that might be overlooked or misinterpreted by other modes of analysis. The zero-order correlations between JOB and the three other variables permit the analyst to order the other variables according to the strength of their association with early occupational attainment, but fail to inform us about how each of these relationships changes when the other variables are statistically controlled. A multiple regression analysis with JOB as the dependent variable and SES, GRADES, and EDUC as predictors imposes statistical controls but fails to alert us to the implications of the causal structure among the predictor variables themselves. Such a multiple regression analysis would yield as regression coefficients the direct effects on JOB shown in Figure 10.4.7, and we might erroneously interpret these to mean that SES and GRADES are inconsequential in regard to early occupational attainment. In contrast, the path analysis, with its explicit assumptions about the causal structure among the predictors, suggests a different conclusion: SES and GRADES are of considerable importance in early occupational attainment, but their effects are largely indirect through their effects on the amount of education completed which, in turn, influences early occupational attainment. Such a conclusion is of noteworthy substantive importance and it suggests the general utility of path analysis in drawing informed conclusions from social data.

QUESTIONS AND PROBLEMS

1. Define the following concepts:

path analysis	endogenous variable
direct effects	misspecified model
indirect effects	causal chain
causal model	feedback
path model	recursive model
path diagram	nonrecursive model
residual or error term	fully recursive model
exogenous variable	

2. An investigator reasons that the individual's "sense of powerlessness" in today's world will be minimized by community participation and by active social involvement with friends. The investigator reasons further

that community participation and active social involvement will be positively affected by income level and education, and that both income and education will have negative effects on the "sense of powerlessness." Represent these assumptions in the form of a path model. Write the multiple regression equations that would need to be estimated to obtain the path coefficients.[12]

SELECTED REFERENCES

Blalock, Hubert M., Jr.
1969 Theory Construction. Englewood Cliffs, N.J.: Prentice-Hall.

Cohen, Jacob
1968 "Multiple regression as a general data analytic system." Psychological Bulletin 70:426–443.

Duncan, O. D., Jr.,
1966 "Path analysis: sociological examples." American Journal of Sociology 72:1–16.

Fennessey, James
1968 "The general linear model: A new perspective on some familiar topics." American Journal of Sociology 74:1–27.

Gordon, Robert
1968 "Issues in multiple regression." American Journal of Sociology 73:592–616.

Land, K. C.
1969 "Principles of path analysis." In E. Borgatta (ed.), Sociological Methodology 1969. San Francisco: Jossey-Bass.

Suits, Daniel
1957 "The use of dummy variables in regression equations." Journal of the American Statistical Association 52:548–551.

[12] For a more elaborate model which includes additional variables and estimates for path coefficients, see Luther B. Otto and David L. Featherman, "Social, Structural and Psychological Antecedents of Self-Estrangement and Powerlessness," *American Sociological Review* 40 (December, 1975): 701–719.

Probability 11

1 NATURE OF PROBABILITY

Probabilities both originate in uncertainty and provide a basis for coping with it. People are willing to travel by air because they believe that the chances of a fatal crash are slim; an applicant is admitted to graduate school because that applicant's estimated chances of completing the program appear to be excellent. We say rain is very probable and advise carrying an umbrella (or not) according to our confidence in such a forecast; we may gamble on heads or tails; we speculate on the sex of an expected baby. Sociologists try to measure the probability of job success, of a juvenile delinquent's becoming an adult criminal, or of a marriage terminating in divorce. The layperson and the actuary, in their own ways, estimate the probability of death before 70, of a wife's survival, and even the probability of twins.

Once we leave the realm of *descriptive statistics* and enter into the realm of *inferential statistics*—estimation and hypothesis-testing—we discover that very little is certain and that practically everything is uncertain. Since probability provides a basis for coping with that uncertainty, we must now explore its meaning.

Basis of Probability Probability measures may be based on subjective ideas, theoretical principles, empirical frequencies, or a mixture of all three. Some consist of more or less strongly held subjective feelings and hopes. Thus, the expectation of rain, depression, or war may stem from a combination of limited personal observations, wishful thinking, and pessimistic fears. But such beliefs, however confidently entertained, do not necessarily provide a reliable view of the future. To be sure, persons rich in clinical experience or possessed of insight and balanced judgment often are amazingly accurate in their forecasts; but such predictions are not subject matter for the field of statistics. To come within the scope of statistics, such subjective expectations must be quantified. In some cases, it may suffice to declare that the probability of one event is higher than the probability of another; in other cases, it may be necessary to give every probability a precise numerical value. In any case, for purposes of statistical analysis, some measure of probability is required, even though that measure is quite tentative and subject to revision in the light of subsequent experience and observation.

Some probability measures are based primarily on theoretical principles. These are not always readily distinguishable from so-called empirical probabilities, next to be taken up, because of the close interplay between fact and theory. A theoretically-deduced probability measure may be adjusted after a few trials, and an empirical probability may be revised as its underlying theory comes to be better understood.

In tossing coins, the 50 percent chance of getting heads is usually cited as a theoretical probability, because its formulation seemingly entails no facts, only reasoning from theoretical principles. We reason that, if the coin is perfectly symmetrical and if the tossing is perfectly mechanical, the probability of heads is just equal to the probability of tails. In this case, since we have no reason to think otherwise, we conclude that heads is as likely as tails and that each therefore has a probability of .50. In analogous fashion, we could reason out the probability of getting two heads on two trials. In this case, there are four possible outcomes instead of two: HH, HT, TH, TT. Since heads and tails are equally likely on a single trial, there is no reason to believe that one combination is more likely than another and we conclude that the probability of any one combination is 1 in 4, or .25.

Although the deductive method of fixing probabilities finds many applications in science generally, it rarely appears in sociology as such. The infrequency of its use is attributable not to the limitations of the method but rather to the nature of sociological theory as presently constituted. At present, there is little if any sociological theory tied to a set of numerical probabilities as a clear corollary.

If weights are assigned to possibilities on the basis of empirical observations of previous outcomes, these possibilities become known as *empirical probabilities*. As a simple practical example of the calculation of empirical probabilities we present an abridged life table of the United States (1970) and eighteenth-century Berlin (Table 11.1.1). The purpose of this record is to set up the probabilities of death for various age groups. The elementary outcomes are, of course, life and death in a given year. The weights of these two possibilities are based on the average number of age-specific deaths over a reasonable period of observation. The average of the past thereby becomes the best estimate for the probability of the future. Thus, if, out of a cohort of

Table 11.1.1 *Abridged Life Table, United States* 1970 *and Berlin c.* 1750

Age	Number Surviving at Specified Ages	
	United States	Berlin
0	100,000	100,000
10	97,722	54,000
20	96,980	49,600
50	89,474	31,300
60	80,347	22,600

Sources: Johann P. Süssmilch, *Die Göttliche Ordnung*, second edition, Berlin, 1762, Vol. II, pp. 319–322 (adapted); and U.S. Department of Health, Education, and Welfare, *Vital Statistics of the United States:1973*. Vol. II, Section 5. "Life Tables." Table 5.1. Washington, D.C.: U.S. Government Printing Office. 1975.

100,000 United States births, approximately 89,000 persons have survived to the age of 50, we may conclude that in the future, as well, anyone in a similar cohort would have an 89 percent probability of surviving to that age. In eighteenth-century Berlin, the corresponding person would have had only a 31 percent probability of surviving to age 50. A 20-year-old in the United States would have an 83 percent probability (80,347 ÷ 96,980) of reaching his or her sixtieth birthday. Of course, this is the type of empirical actuarial data on which life insurance premiums are computed.

Probability Defined In giving a pat definition of probability, our purpose is not to start a philosophical debate, but rather to reduce the scope and complexity of the discussion. With that object in mind, we offer the following definition: Probability is the *proportion* of *successes* on a *large number* of *trials.* It is not a prediction for a given trial, rather it is the proportion of successes (however defined) on innumerable, conceptually repeatable trials. We may base a prediction on a probability, but that probability as such is not a prediction. We may predict rain for tomorrow if the probability of rain is .80, but that probability itself is not a prediction. The probability gives the chances of rain—it is more conservative than a categorical forecast.

Probabilities range in value between 0 and 1. Very strictly speaking, they cannot take values of 0 and 1, since these values stand for certain outcomes. Being proportions, probabilities based on the same total add up to 1.00 (provided that none is omitted). The probability of a male birth plus the probability of a female birth equals 1.00; the probability of drawing a black card from an ordinary deck, added to the probability of drawing a red card necessarily adds up to 1.00, since no other colors are possible.

In dealing with probabilities, a "successful" event, or *success*, is wholly a matter of definition. If only two outcomes are possible on any trial, we arbitrarily label one a success, the other a failure. If a girl (female birth) is considered a success, a boy (male birth) would be considered a failure. If it created no confusion, we could define a failing grade as a success, and a passing grade as a failure. In discussions of probability, the term *success* carries no evaluative connotation—a successful event is not cheered, nor is a failure jeered. It is simply a convenient way of grouping all possible outcomes into two classes: the successes and the failures.

The term *possibility set* is sometimes applied to the outcomes that are possible on any trial; and the probabilities corresponding to those possibilities are then labeled the *probability set*. The possibility set must include at least two outcomes, but beyond that minimum number there is no upper limit. If the possibility set consists of three or more outcomes, and we require the probability of each one, then each outcome in turn would be regarded as a success. For example, instead of the probability of a batter getting a hit, we may require the probability of that batter getting an extra-base hit. In that case, we would redefine success as a double, a triple, or a home run, and take the proportion of these extra-base hits in a large number of trials as the probability of success.

With a quantitative variable, there is no limit theoretically to the number of outcomes in the possibility set. We may ask for the probability that a randomly drawn measure will be larger than the median; or we may ask for the probability that a randomly drawn measure will exceed the 90th centile point. We have already anticipated such questions in our analysis of the normal curve (p. 170). In that analysis, we obtained the proportion of cases under any segment of the normal curve from the table of normal areas. Any one of these proportions may be construed as the probability that a measure randomly drawn from a normal population will fall in that interval. From this perspective, for example, the probability that a randomly drawn measure will exceed the mean by 1.96 standard deviations is .025; by 2.33 standard deviations, .010; and by 2.58 standard deviations, .005. In setting up confidence intervals (Chapter 13) and in testing statistical hypotheses (Chapter 14), we will have frequent occasion to make use of such normal probabilities.

A *trial* is whatever process results in a success or failure. It is not necessarily a contrived act such as the tossing of a coin. It could be the birth of a boy or a girl; it could be the process of a jury reaching a verdict; it could be the process of manufacturing an acceptable product. It could be life itself, since life may expire at any instant. For purposes of setting up probabilities, any one of these processes may be considered to be a trial.

A *large number of trials* here means an *unlimited* number of trials. We require an unlimited number of trials, since the stated probability of success will realize itself only in the limit. In a limited number of trials, the observed proportion may and usually does deviate from the expected proportion or probability. For instance, in 100 tosses of a penny, we may observe 53 heads, instead of the expected 50, for a deviation of 3 in 100; in another 100 tosses, we may observe 48 heads, for a deviation of 2 in 100; and so on. Although the magnitude of the deviation between observed and expected proportions will decrease as the number of trials increases—in accordance with the law of large numbers—it will reach zero only in the case of an infinity of trials. Therefore, we must place no bound on the number of trials.

Since we can never conduct an endless number of trials, we require only that trials be *conceptually repeatable*. We may conduct a large number of trials and set our probability equal to the observed proportion of successes in that limited sequence of trials; however, that observed proportion is conceptually different from the probability itself. The probability of success is a claim that a given numerical proportion will realize itself after an endless number of imaginary trials. Probability statements thus may be made even though no actual trials are ever run. The probability of getting a high correlation in a random sample, when the correlation in the sampled population is zero, may be readily calculated before any sample is drawn (p. 450). Similarly, the probability of a relatively large difference between sample means, when there is no difference between population means, may be established without making a single empirical comparison. The odds for or against a horse winning the Kentucky Derby are meaningful only before the race is run; in this case, all trials are necessarily conceptual.

The Validity of the Probability Measure Probabilities are of no use unless they are borne out by subsequent events—that is, unless they forecast the future with reasonable accuracy. It is a matter of common knowledge that the actual outcomes hardly ever correspond perfectly to the probability statements. These statements must be regarded as conceptual models idealized on infinity. The subsequently observed finite ratios can agree only occasionally with these models. Only rarely will we observe exactly 50 heads on 100 tosses of even a perfect penny, although the probability of heads is quoted as 50–50. Such discrepancies between expected and observed frequencies will result either from (1) continued operation of chance factors that introduce an indefinite amount of variation, or from (2) the operation of changing conditions that make the initial statement of probability invalid. These discrepancies ultimately lead to one of the central issues of hypothesis-testing (Chapters 14 and 15), which at this time may be given only preliminary mention: How may the discrepancy between expected and observed values be interpreted—as chance variation around the stated probability value, or as evidence against the initial statement of probability?

Let us suppose that, in a large number of tosses of the same coin, a gambler observes heads only 40 percent of the time, instead of the a priori 50 percent. At this point in the game, how should the gambler react on the next bet? The gambler must decide whether to accept the discrepancy as mere chance variation to be ignored or reject the initial probability statement as false and revise his or her expectations.

The gambler's first line of reasoning might be: The *law of large numbers* prescribes that with an increasing number of fair tosses, the approximation to the hypothetical true proportion will tend to become correspondingly closer. The "deficit" of heads must have resulted from excessive *runs* of tails. Hence, sooner or later the heads will have to "catch up" in order to make up the deficit. Therefore, "I will now bet on heads." However, such a judgment is inconsistent with the usual assumption in coin-tossing that each outcome is independent of all previous ones. The coin has no "memory" of previous outcomes, and therefore heads and tails are still *equiprobable* on the next toss. After all, the 50–50 division of events holds good only for an infinitely large number of throws, or trials. Any finite number of trials is not enough to equalize the outcomes. Actually, the deficit need never be made up in a finite number of trials. Therefore, in spite of the unusual runs of tails, the bettor should not change the pattern of strategy, but accept the discrepancy as due to the play of chance factors.

On the other hand, if the run of tails persists beyond the limits of reasonable tolerance, the player might well raise the question whether the orthodox hypothesis of equiprobability was valid in the first place. Perhaps the coin was not symmetrical after all; perhaps the coin favors the tail, as the empirical evidence seems to indicate. In other words, if the discrepancy between expected and observed frequencies becomes too great, we should question the hypothesis and reject it accordingly. Although there is no fixed rule which could guide the observer in determining at what point the length of the run becomes suspicious, sooner or later it would become profitable to bet consistently on tails and be right more than half the time. In this case, our gambler rejects the initial probability statement and establishes a new one consistent with the observed frequencies. Such rejection is known as a *statistical decision*.

The same testing procedure is applied to empirical probabilities. In the case of life tables, a certain chance variation is permitted without casting doubt on the reliability of the empirically predicted death rates. But when the variation becomes too great, the actuarial tables are revised and a new norm or model is set up for subsequent observations.

From the foregoing discussion, it is obvious that the determination of the possibilities and their respective weightings is the core of probability calculations. The most elementary rules of such calculations are given next.

QUESTIONS AND PROBLEMS

1. Define the following terms:

probability	trial
subjective probability	conceptually repeatable trial
theoretical probability	possibility set
empirical probability	probability set
success	law of large numbers

2. By January 1969, 8 of 35 former Presidents of the United States had died in office. Comment on the validity of the statement that former Vice President Spiro Agnew had, at that time, 8 chances in 35 to succeed former President Richard Nixon in office during his first term.
3. From Table 11.1.2, calculate the chances of men remarrying by the end of the second year of divorce. Calculate the proportion (probability) of women remarried by the end of the fourth year of divorce.
4. On a true–false test of 50 items, a student marks 25 correctly. Did the student know the answers to those 25 questions, or was he or she guessing on all 50? How would you decide?
5. According to the International List, there are 200 causes of death, of which cancer is one; hence, a person has 1/200 probability of dying of that disease. Comment.

Table 11.1.2 *Average Annual Probabilities of First Remarriage in Each Year of Divorce for Persons Whose First Marriage Was Ended by Divorce, by Sex, for the United States:* 1960–1966

Year of Divorce	Persons Divorced at Beginning of Year of Divorce, 1960–1966			
	Men		Women	
	Number divorced (numbers in thousands)	Remarried per 1,000 divorced	Number divorced (numbers in thousands)	Remarried per 1,000 divorced
Total	9,776	180	13,349	139
1st year	1,546	318	1,779	250
2nd year	1,001	231	1,224	197
3rd year	775	213	1,022	179
4th year	646	187	858	157
5th year	517	164	751	116
6th year	434	187	727	129
7th year	402	198	624	161
8th year	388	56	521	83
9th year	412	75	523	99
10th year	405	146	474	44
11th to 15th year	1,594	111	2,135	96
16th year and over	1,657	130	2,713	90

Source: U.S. Department of Commerce, Bureau of the Census. *Current Population Reports*, Special Studies, Series P–23, No. 32. Table A. "Probabilities of Marriage, Divorce, and Remarriage." Washington, D.C.: U.S. Government Printing Office. July 29, 1970.

2 PROBABILITY CALCULATIONS

To ask, "What is the probability of death from cancer?" is for all practical purposes the same as asking, "What fraction of all deaths have been caused by cancer?" Although a competent demographer would wish to specify the population cohort to which the question referred, this would not change the fundamental nature of probability as a frequency ratio. Hence, to answer the question we would have to express the frequency of death by cancer as a proportion of deaths by all possible causes, including cancer. About 19 percent of all deaths now occurring in the United States are attributed to cancer; therefore, the probability of ascribing death to cancer among subsequent deaths would likewise be 19 percent. This experience ratio of approximately 1 in 5 trials is used to forecast the events of the long run, on the assumption that the conditions of death remain constant.

Similarly, to ask, "What is the probability of a multiple birth?" is practically the same as asking, "What fraction of all births are multiple births?" In answering that question, we would count the total number of births in a given year, and divide that total into the number of multiple births in that year. For example, in 1968 there were about 3,500,000 live births in the United States, of which approximately 70,000 were live multiple births. Dividing 7 by 350 gives the probability of a multiple live birth:

$$\frac{7}{350} = \frac{1}{50} = 0.02.$$

It is to be understood that this estimate of 0.02 is provisional and subject to change; an estimate based on five years of experience probably would differ from 1 in 50 and carry more confidence than an estimate based on a single year.

While all probabilities may be expressed in the general form

$$Pr(A) = \frac{N_A}{N} \qquad (11.2.1)$$

where A stands for a success (however defined), N = the total number of trials, and N_A = the number of successful trials, it is not always necessary to establish N_A and N by actual counting. In many problems, instead of actually enumerating successes, we take advantage of what are essentially computing formulas. We give a few of the simpler ones here.

Equally Likely Outcomes With equally likely events, the general formula (11.2.1) specializes to

$$Pr(A) = \frac{N_A}{kN_A}$$

$$= \frac{1}{k} \qquad (11.2.2)$$

where k is the total number of possible outcomes (all equally likely). In this case, to get the probability of a successful outcome (A), we simply divide 1 by k. For example, in tossing a coin we assume that heads and tails are equally likely and that the probability of heads (or tails) is

$$\begin{aligned} Pr(H) &= \frac{1}{2} \\ &= .50. \end{aligned}$$

Similarly, since the six sides of a die are assumed to be equally likely, the probability of throwing any one of them, say a deuce, is

$$\begin{aligned} Pr(2) &= \frac{1}{6} \\ &= .167. \end{aligned}$$

This method of computing probabilities holds only for equally likely events, and is of little practical importance once we leave the realm of games of chance.

Alternative Events As previously noted, the definition of success is arbitrary. We may narrow it or broaden it according to our purposes. We may broaden it from the ace of spades to any black ace; we may narrow it from any ace to a red ace. In broadening our definition of success, we necessarily increase the probability of success, since the proportion of failures will be smaller; in narrowing our definition, we decrease the probability of success, since the proportion of failures will be larger. If we bet on a horse to win, we get a success only if the horse finishes first; but if we bet on a horse to show, we get a success if the horse finishes third or better. The odds of a horse showing are better than the odds of the horse winning.

With two or more mutually exclusive events—if one occurs the other cannot—the probability that any one of them will occur is simply the sum of the individual probabilities. Instead of enumerating successes, we simply add individual probabilities (assuming they are given). As a formal rule: The probability that one of two or more *mutually exclusive events* will occur is the sum of the probabilities of the individual events. By this rule the probability of either A or B is the probability of A added to the probability of B. In symbols:

$$Pr(A \text{ or } B) = \frac{N_A}{N} + \frac{N_B}{N}. \tag{11.2.3}$$

Entering the probability of the ace of spades and the probability of the ace of clubs in (11.2.3) gives the probability of a black ace:

$$Pr(\text{Black Ace}) = \frac{1}{52} + \frac{1}{52}$$
$$= \frac{2}{52}.$$

Entering the probability of a black ace and the probability of a red ace in the same formula gives the probability of either a black ace or a red ace:

$$Pr(\text{Black Ace or Red Ace}) = \frac{2}{52} + \frac{2}{52}$$
$$= \frac{4}{52}.$$

Caution: We must be careful not to apply the foregoing addition rule to alternative outcomes that may happen together. If for example, we wish to establish the probability of drawing either an ace *or* a black card, then we must take into account the probability that these two outcomes may occur together. We take up this complication later.

Compound Event Instead of the probability of alternative events, we may be required to find the probability of a combination of events, or a *compound event*. For example, instead of the probability of drawing an ace or a black card from a full deck we may require the probability of drawing a card that is both black and an ace. Similarly, we may require the probability of heads on two trials, instead of the probability of heads on a single trial; instead of the probability of throwing a six on a single trial, we may seek the probability of throwing a six on two consecutive trials.

With independent events—where one event does not affect another—the probability of their occurring together is simply the product of their individual probabilities. By this rule, the probability of A and B occurring together is the probability of A multiplied by the probability of B. In symbols:

$$Pr(A \text{ and } B) = Pr(A)Pr(B). \tag{11.2.4}$$

To be more specific: If, in throwing dice, the probability of a 5 on any die is 1/6, then the probability of a 5 *and* a 5 with two dice would be (according to Formula 11.2.4) 1/36. This result may be clarified visually by means of the accompanying diagram, which indicates the 36 possible ways in which the two dice may fall together—that is, the 36 joint outcomes. While there is only one way in which the two fives may fall together to give a probability of 1/36, a 2 and 3 may also occur as 3 and 2—in two ways—for a probability of 2/36. The probabilities of other joint outcomes may similarly be read from this diagram:

Die 1	Die 2	Die 1	Die 2	Die 1	Die 2
	1		1		1
	2	3 ⟷	2		2
1	3		3		3
	4		4		4
	5		5	5 ⟷	5
	6		6		6
	1		1		1
	2		2		2
2 ⟷	3	4	3	6	3
	4		4		4
	5		5		5
	6		6		6

The cumbersome procedure of listing every possible combination is obviated in practice by the aforesaid multiplication rule (Formula 11.2.4). Instead of counting every possible combination and dividing that number into the total number of successful combinations, we merely find the probabilities of individual outcomes and multiply these together for the required answer.

Independent and Dependent Outcomes Formula 11.2.4 holds for independent events—for events having no effect on one another. For example, the probability of a head on a given trial is independent of the previous fall, whether head or tail; the probability of the next outcome has not changed because of the previous event.

However, events are not always independent; one event may be dependent on another, and that dependency must be taken into account before applying the multiplication rule. For example, before applying the multiplication rule to get the probability of A and B, given that B is dependent on A, we must adjust the probability of B for its dependency on A. In symbols:

$$Pr(A \text{ and } B) = Pr(A)Pr(B|A) \qquad (11.2.5)$$

where $Pr(B|A)$ is the probability of B subject to the condition that A has occurred. Because the probability of B is conditioned by the occurrence of A, we speak of it as a *conditional probability*. To be more concrete: In computing the probability of drawing two aces in succession from a bridge deck of 52 cards, the probability of drawing the first ace is, of course, 4/52. But the drawing of one ace (without replacing it) obviously affects the probability of drawing another ace on the next trial. There being only 51 cards remaining (including three aces), the probability of the next ace is now 3/51. The joint probability is the product of the individual probabilities:

$$\frac{4}{52} \times \frac{3}{51} = \frac{12}{2{,}652} \cdot$$

This means that, in the long run, out of every 2,652 trials consisting of two random draws without replacement, one would expect 12 outcomes of double aces on the average.

This illustration, which is convenient because of its easy manipulation, is a formal, routine demonstration of a familiar principle. However, in the practical affairs of social life—to say nothing of pure science—it is not always so easy to detect the presence of dependence, and still more difficult to measure its degree. Dependence is often very complex—as are all social events—and accurate calculation is often impossible. Consequently, we frequently are obliged to disregard the presence of dependence and forego its calculation, since the needed empirical data are inaccessible.

For example, actuaries calculate the joint (compound) probability of spouses surviving to celebrate their golden wedding. If, at marriage, the bride is 21 and the groom 26, they would have to live at least to the ages of 71 and 76, respectively. Now, we know from the general life tables for the United States (1973) that the 26-year-old male has slightly fewer than 2 in 5 chances of surviving to that age; the 21-year-old female has slightly more than 7 out of 10 chances to survive for 50 years after marriage. By applying the product rule, we find that the probability of both surviving to the required ages would be

$$\frac{2}{5} \times \frac{7}{10} = \frac{14}{50} = .28.$$

But this probability is computed on the assumption that the survival ratios of the two spouses are mutually independent. It seems incredible, however, that marriage partners are chosen at random. Since it is plausible to believe that healthy spouses tend to marry healthy mates, in that respect selective factors are in operation, as well as chance factors. Furthermore, since husband and wife share a common way of life, their longevity is likely to be correlated. Hence, their survival ratios are not as completely independent as the above probabilistic calculation assumes. Strictly speaking, the joint probability of .28 is an overstatement or understatement, depending upon the health of the cohort. Only if men and women of the respective ages 26 and 21 were randomly paired would the joint probability of .28 be valid.

Outcomes Not Mutually Exclusive Formula 11.2.3 holds only for events that are *mutually exclusive*—that is, for outcomes that cannot occur together on a single trial. Thus, a head and a tail cannot both occur when a single coin is tossed; a birth cannot be single and multiple at the same time. Yet, in many instances, a given object may be classified or perceived in more than one way: a playing card may be read simultaneously as an ace and a spade; an individual may be characterized as both single and Catholic. Such traits are not at all mutually exclusive since they represent different characteristics of the same event. Therefore, it would be erroneous to apply the simple addition theorem, which holds only for mutually exclusive events, to events which may occur together. Thus, to calculate the probability of an ace *or* a spade, we must not add the probabilities of these simple events and let it go at that. The probability of an ace or a spade is *not*:

$$\frac{4}{52} + \frac{13}{52} = \frac{17}{52}. \qquad \text{(Wrong!)}$$

This probability is too high, because we are erroneously counting the ace of spades twice instead of once. To obtain the correct probability, we must deduct from the inflated value the relative number of times the ace and the spade would occur together. But this is their compound probability, or the product of their individual probabilities. The foregoing addition rule therefore must be amended by subtracting the duplicate occurrence, which is measured by its joint probability, or 1/52. As a rule: The probability that one of two alternative events, not mutually exclusive, will occur is the sum of the individual probabilities, minus their compound (joint) probability. In symbols:

$$Pr(A \text{ or } B) = Pr(A) + Pr(B) - Pr(A)Pr(B). \tag{11.2.6}$$

According to Formula 11.2.6 the probability of an ace *or* spade would be:

$$\frac{4}{52} + \frac{13}{52} - \frac{1}{52} = \frac{16}{52}.$$

Fourfold Table The foregoing distinctions among probabilities may be advantageously displayed by means of a fourfold table of cell and marginal frequencies (including the total frequency):

	B	$\bar{B}$	
A	N_{11}	N_{12}	$N_{1.}$
$\bar{A}$	N_{21}	N_{22}	$N_{2.}$
	$N_{.1}$	$N_{.2}$	N

Dividing marginal frequencies by the total frequency gives the marginal, or unconditional, probabilities. For example:

$$Pr(A) = \frac{N_{1.}}{N}. \tag{11.2.7}$$

Dividing cell frequencies by corresponding marginal frequencies gives the conditional probabilities. For example:

$$Pr(B|A) = \frac{N_{11}}{N_{1.}} \tag{11.2.8}$$

where $Pr(B|A)$ = probability that a case is B given that it is A. Dividing cell frequencies by the total frequency gives the joint (compound) probabilities, one for each cell in the table. For example, the probability of A and B occurring together is symbolized

$$Pr(AB) = \frac{N_{11}}{N}. \tag{11.2.9}$$

Multiplying any conditional probability by its corresponding marginal probability gives the joint probability:

$$Pr(AB) = Pr(A)Pr(B|A). \tag{11.2.10}$$

Dividing both sides of Formula 11.2.10 by $Pr(A)$ gives the conditional probability as the ratio of the joint probability to the marginal probability:

$$Pr(B|A) = Pr(AB)/Pr(A). \tag{11.2.11}$$

When A and B are statistically independent,

$$Pr(B|A) = Pr(B)$$

and

$$Pr(AB) = Pr(A)Pr(B). \tag{11.2.12}$$

This agrees with Formula 11.2.4.

The fourfold table thus serves to give visibility to relationships that otherwise might remain obscure. Consider the above example of drawing two aces (without replacement) from a full deck. The probability of this is

$$\begin{aligned}\frac{4}{52} \times \frac{3}{51} &= \frac{1}{13} \times \frac{1}{17} \\ &= \frac{1}{221}.\end{aligned}$$

These probabilities may be easily obtained from the entries of the fourfold table. After 1, 17, and 221 have been entered (encircled), we get the remaining entries by subtraction:

	B	$\bar{B}$	
A	(1)	16	(17)
$\bar{A}$	16	188	204
	(17)	204	(221)

A glance at this table reveals that the probability of two aces (successes) is 1/221; a second glance reveals that the probability of no successes (neither card an ace) is 188/221; a third glance reveals that *A* and *B* are not statistically independent, since 17/221 = 1/13 is not equal to 1/17. The ratios of entries from a fourfold table thus serve to bring out the distinctions and relations among probabilities which may be a trifle puzzling when presented separately one by one.

The Probability of a Single Event Since the concept of probabilities assumes innumerable repeated trials, it is conventionally asserted that we cannot assign a probability to the outcome of a single trial. This apparent truism is not as simple at it appears. There are three possible reactions to this ambiguous statement.

In the first place, we do act on probabilities, and we commonly do apply them to single trials. Thus, "in the clutch," a pinch hitter is selected on the differential probability of his getting a hit. The manager will reason as follows: "Jones has a batting average of .275 (empirical probability); if I send him up to bat, he is more likely to succeed than Smith, who has only a .225 average." Analogously, a person who is ill calculates his or her own private chances for recovery and makes plans accordingly, however indifferent the insurance company may be to the identity of the individual case in its actuarial mass statistics. Thus, each person continually codifies experiences and probabilistically prepares for the alternative outcomes in each single instance.

In the second place, a single case may be thought of as a wholly unique event. With this interpretation, it would seem doubly clear that no probability statement could be made, since there can be no long run of unique events—for that matter, no run at all. Being unique, the present event could not have occurred in the past; and obviously could not occur again. Therefore, a probability statement would bc both impossible and useless. For example, if Julius Caesar had tossed one of his coins, we could state, properly and reliably, the probability of his having turned up his image thereon. But we cannot estimate with the same precision the numerical probability of his having been in England. Nevertheless, historians constantly speculate intelligently on the "unique" actions of Caesar and many other personages in history. How can the statistical principle forbidding such speculations be reconciled with the familiar and useful deductions of thinking people?

The deception lies in the fact that, sociologically speaking, a unique event is never totally unique. Although there was only one Julius Caesar, he still possessed traits in common with other leaders of men, not only in his broad characteristics, but also in the potentialities of a trip across the English Channel at that time. With all his singularities, he was also a member of a class (a subset) representing a large number of similar historical instances. Therefore, the question should be rephrased to read: What is the probability of persons like Caesar, under similar circumstances, having gone to Britain? Although these forces do not lend themselves to quantitative measurement, an intuitive but profitable approximation of probability can be obtained. Many important human decisions, including the adoption of scientific hypotheses and legal judgments, rest on such qualitative bases rather than on methodical quantitative procedures. Some statisticians would reserve the concept "likely" for such circumstantial judgments, and employ the concept "probable" only when estimates are found on rigorous quantitative observations.

A third possible interpretation of the foregoing statement holds that a single outcome never can set up, modify, or test a probability statement. The reason for this is simply that probabilities are established or confirmed only by the results of a larger number of trials. In that sense, probability does fail to tell us anything about the single case. A single parole violation would not be sufficient to disprove the correctness of a probability of 80 percent success established by previous empirical observations, since 20 percent of the cases are confidently expected to be failures anyway. It is only after a large number of subsequent observations, when the constant factors have had an opportunity to manifest their force, that the validity of a probability statement can be made plausible, although never proved. We would question the hypothesis of a fair coin after a run of 3 or 5 heads and certainly reject it after a run of perhaps 25.

QUESTIONS AND PROBLEMS

1. Define the following terms:

alternative events	conditional probability
compound event	marginal probability
mutually exclusive events	joint probability
independent event	

2. Does the following table show that *A* and *B* are independent or dependent? Explain your answer.

	B	$\bar{B}$	
A	19	21	40
$\bar{A}$	8	32	40
	27	53	80

3. In a group of 6 persons, there are 4 boys and 2 girls. If we select two individuals by chance, what would be the probability of drawing a boy first and a girl second? a girl first and a boy second? two girls? two boys?

4. For the following distribution of Jewish and Protestant males and females, calculate the probability of mixed marriages and homogamous marriages,

assuming that all marry and that religion of husband and wife are independent.

	Female	Male
Protestant	20	30
Jewish	60	50
	80	80

5. Assume a group of 100 persons, of whom 20 are Republicans and 80 Democrats. If 5 different persons are selected by chance, what would be the probability that all would be Republican?

6. In a large number of throws of two true dice, how often would you expect a total of 7?

7. If two dice were repeatedly thrown, which event would surprise you more: a double-six on the first throw, or on any one of the subsequent throws?

8. Comment on the statement: "All individuals in the group are by definition subject to the same probability statement; hence, the individual, in the cohort for which the prediction is made, cannot justifiably exempt himself from the operation of the quoted probability."

9. If the Internal Revenue Service selects a 25 percent random sample of the income tax returns for checking, what would be the probability of a person being selected for checking two years in succession? Comment. (Assume yearly samples are independent.)

3 COMBINATIONS AND PERMUTATIONS

By definition, a compound event is a grouping of individual events. It could comprise the letters in a word, a bridge foursome, or a group of musical notes. But groups usually can be assembled in more than one way. Thus, there are 36 ways in which two dice may fall; a 3 and a 5 may occur in two ways since each figure can appear on either one of two dice, as has been shown previously. Similarly, letters can be scrabbled—arm, ram, and mar; four persons may be seated in various orders around a table; the same musical tones may take various positions in a chord or in a melodic sequence. A facility in such regrouping is of prime importance in the calculation of probabilities, because the magnitude of a joint probability is necessarily increased in proportion to the number of ways in which a specified joint event can occur.

This count will depend upon whether we focus on the membership of the group or on the order in which the component members are placed—that is, upon whether we are dealing with a *combination* or a *permutation*. Statistically speaking, combination denotes the identity of the individuals that compose a group, while permutation denotes the specific order, or arrangement, in which the members may be placed.

To determine the number of combinations or permutations in any given set, we could laboriously enumerate all possible ways, as has been done for the tosses of two dice (Section 2). But, in practice, there are more efficient methods to achieve this result. We now turn to these counting procedures.

Combinations Any set of distinct objects is called a combination, symbolized *C*. Such a joint event may consist of a list of digits, a handful of coins, a group of nine baseball players, a series of births, a set of postage stamps, a bridge hand, a list of true–false questions, or a sample of a human population. The substitution, addition, or subtraction of even a single item in the given set produces a different combination. Thus, the two sets of digits, 2698 and 2697, constitute two different combinations. The component items must be distinct but need not be distinguishable, as in 4444 and 44444, which are different combinations. For a given problem, a set of five pennies may be considered as indistinguishable, although each bears a different date—a fact which may be important to a numismatist, however.

Permutations The above illustrations make it evident that in many situations we are interested not only in the identity of the elements in the combination, but also in the order of their arrangement. A telephone number of 2698 is not identical with 2986; the same 9 baseball players may be put forth in different batting orders; 5 classmates may be chosen in different orders of sociometric preference; the tumblers in a Yale lock may be placed in different orders so that only the corresponding key will fit it. In the above cases the number and identity of the items remain unaltered; they remain the same *combination*, but they take on different *permutations* because two or more items in the arrangement are reordered. When the focus of attention is on the order of things, the series is a permutation, symbolized by *P*.

Popular language does not always correctly distinguish these two terms. To open a bank vault lock consisting of gears and pins which must be actuated in a specific sequence, the operator must know, strictly speaking, not merely the combination, but also the permutation of the dial numbers. A rearrangement of any two (or more) of the pins would produce another permutation and make the vault impossible to open.

The Calculation of Permutations Any or all of the objects in a given combination may be permuted, provided the objects are distinguishable. This proviso is quite logical since a reordering of indistinguishable objects is meaningless: A rearrangement cannot be recognized insofar as the objects are considered to be absolutely identical.

If the number of objects in a combination is small enough, the permutations can be formed by inspection and counted. Thus, the three digits, 2, 5, and 6, can be permuted in six ways as follows:

256	562	625
265	526	652

The addition of one digit to the set (2568) would not increase the number of combinations, it would increase the number of possible permutations from 6 to 24, as the *tree diagram* in Figure 11.3.1 demonstrates. But the same result could have been achieved more quickly without such extensive and detailed figuration by simple multiplication according to the following logic.

Examination of the foregoing 24 permutations reveals that the first digit in a permutation could be any one of the four—that is, the first digit could occur in four ways. After each of these four possibilities, the next digit could fall in any one of three remaining ways to join each of the preceding four, yielding so far twelve different permutations, two digits at a time. Thus, the pattern of calculation could continue until all the available digits are used up as above. Summarizing the arithmetical procedure to count the number of ways, we have: $4 \times 3 \times 2 \times 1 = 24$ ways. If all ten digits (0–9) were available, the computation would be $10 \times 9 \times 8 \times 7 \times 6 \times 5 \times 4 \times 3 \times 2 \times 1 = 3{,}628{,}800$. The principle here involved is called the *multiplication theorem*, the generalized formulation of which is:

If an event can occur in N_1 ways, and thereafter in N_2 ways, and so on, these successive events can occur in that order in $N_1 \times N_2 \times \cdots N_N$ ways.

This theorem is used in counting both complete and partial permutations.

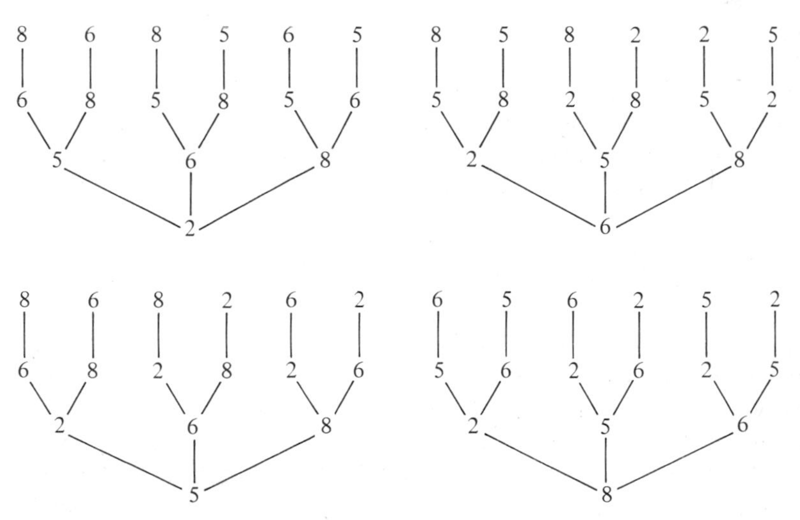

Figure 11.3.1 *Tree Diagram of Permutations*

Complete Permutation A complete permutation consists of permuting all objects in a set, all at a time: that is, N objects, N at a time. The resulting number of permutations is called the *factorial of the number*, and is symbolized:

$$P_N^N = N! \qquad (11.3.1)$$

In permuting four distinguishable objects, 4! would cumulate to 24 different orders, as has already been illustrated. Therefore a factorial of a number (N) is defined as the product of all integers N to 1, inclusive.

Partial Permutation A permutation of N distinguishable objects, less than N at a time, may be called partial permutation, symbolized by P_r^N, to be read "the number of permutations of N objects, r at a time." This formula is a directive to carry out the permutations only through r stages, instead of through the total N.

$$P_r^N = N(N - 1)(N - 2) \cdots (N - r + 1). \qquad (11.3.2)$$

For example, the number of four-digit telephone numbers, when all ten digits are available (but when the same digit is allowed to appear only once in any given telephone number), would be derived as follows:

$$P_4^{10} = 10 \times 9 \times 8 \times 7 = 5{,}040.$$

Recurring Items Combinations which contain indistinguishable items are amenable to permutation only in proportion to the number of distinguishable objects. Nevertheless, there are certain combinations in which the recurrence of items is not only permitted, but even required, as, for instance, in telephone numbers and computer punch cards.

In the case of discrete physical objects, such recurrences obviously are impossible. In permuting nine children on a baseball field, physical circumstance precludes their recurrence in successive positions on the same trial. A child cannot play first base and center field at the same time. The same principle holds for sociometric arrangements. However, numerical symbols and letters of the alphabet do not suffer from this limitation; hence, if one wishes to determine the total number of all possible arrangements, with no restriction on recurrences, one must allow the multiplication theorem to take over from the factorial principle. Whereas all factorials are based on the multiplication theorem, not all cases in which this theorem is employed need be factorials, as in the instance of telephone numbers. The number of possible four-digit telephone numbers, allowing all ten digits to recur, would be $10 \times 10 \times 10 \times 10 = 10{,}000$. The symbol for such an operation is N^r, where r is the number of positions to be filled, in this instance, 10^4. If the N objects are to be serialized in N positions, the symbol would be N^N.

Combining *N* Objects, *r* at a Time There is no difficulty in the calculation of the number of combinations, when *N* objects are taken *N* at a time, for there can be only one combination. A problem ensues only when fewer than the available *N* objects are used to form the combinations—when we take them *r* at a time.

As in the case of permutations, when *N* is small, the subsidiary combinations can be found by inspection. A group of four children, A, B, C, D, taken two at a time, could combine in the following six ways:

AB	BC	CD
AC	BD	
AD		

The partial permutations of the same group, two at a time, would total 12. Clearly, there will always be fewer combinations than permutations, for any given *N* and *r*.

With larger *N*s, it is tedious to determine the number of combinations by means of a tree diagram. Instead, it is more convenient to compute first the number of permutations of *N* objects, taken *r* at a time. The second step consists in clearing the result of these permutations by dividing by $r!$, which is the number of ways in which each combination can be permuted. Since each combination can have $r!$ permutations, we may obtain the number of combinations by dividing the number of permutations by $r!$ This division obviously reduces the permutations to the corresponding number of combinations. The formula is as follows:[1]

$$C_r^N = \frac{P_r^N}{r!}, \tag{11.3.3}$$

where N = total number of items, and
r = number of items in combination.

According to this formula, the number of different ways in which we may select 4 items from a pool of 10 would be:

$$C_4^{10} = \frac{10 \times 9 \times 8 \times 7}{4 \times 3 \times 2 \times 1}$$

$$= 210 \text{ combinations.}$$

These combinatorial methods are of special significance in the calculation of the binomial probabilities; these are the subject of the following section.

[1] In many texts in college algebra, as well as in statistics, the formula is written:

$$C_r^N = \frac{N!}{r!\,(N-r)!}$$

This formulation is obtained by multiplying both members of the fraction by $(N - r)!$, thus producing complete permutations in both the numerator and denominator, which is considered by some to be a simpler procedure.

QUESTIONS AND PROBLEMS

1. Define the following terms:
   combination
   permutation
   multiplication theorem
   recurring items
   tree diagram

2. A population can be classified by race in 3 ways, by religion in 3 ways, and by nativity in 2 ways. Prepare a tree diagram to show in how many ways persons may be classified by all three together.

3. How many telephone numbers of 4 digits each can be made from all digits 0 to 9 if:
   (a) no telephone number begins with zero?
   (b) no duplicate digits are permitted?
   (c) all digits are permitted to recur?

4. In the columns of an IBM card, each labeled 0 to 9 inclusive, how many different code numbers can be made using 1, 2, or 3 columns?

5. It is usually assumed that additional culture traits (discoveries or inventions) increase the possibilities of new inventions, since new traits may combine with any one of a large number of already existing traits. In principle, this is a statistical approach to the problem of culture growth. Assuming that all traits may freely combine, how many permutations could be formed with 2, 3, 4, 5, or 6 traits? What generalization does this suggest?

6. What would be the more probable bridge hand: any specified set of 13 cards, or a complete suit? Explain.

7. In a toss of 10 coins, why are 10 heads (tails) so much less probable than any one of the other combinations? (Hint: Calculate C_r^N for $N = 10$, and all possible values of r.)

4 BINOMIAL PROBABILITIES: INDEPENDENT EVENTS

The Binomial Every possibility set obviously may be reduced to two mutually exclusive possibilities and the resulting outcomes labeled "success" and "failure." The outcome termed success may be an elementary event, such as a male birth, or it may be one of a group of elementary events, such as four aces in a deck of playing cards. In these instances, the probabilities of success would be 1/2 and 4/52, respectively. But however defined, whenever the possibility set consists of only two alternatives, there can be only two corresponding probabilities: the probability of success, p, and the probability of failure, q. Hence, the probabilities of the twofold set may always be written:

$$(p + q) = 1.00.$$

Since the expression on the left is the algebraic sum of two terms, it is a *binomial.*[1]

While the layperson's interest almost always is centered on the probability of success on a single trial, the statistician's interest usually lies in the probability of r success on N trials. The statistician is interested in the general quantitative laws that govern the behavior of events on repeated trials and enable the likelihood of their occurrence to be predicted. For example, an analyst may desire the probability of two heads on successive throws, or the probability of a 5 and 5 when two dice are thrown together. We have shown that the probability of a 5 and 5 is 1/36, since a 5 and 5 may occur in only one way. On the other hand, the probability of a 3 and a 5 is 2/36, since a 3 and 5 may occur in two ways. Analogously, the probability that the faces of two dice will sum to 7 is 6/36, and the probability that they will sum to 8 is 5/36.

Clearly, a specified outcome may occur in more than one way. Hence, to calculate the probability of that outcome, we always must determine the number of ways in which it can occur; otherwise we cannot compute its probability. As before, we pursue a very simple example: the probability of parents having a boy and a girl (ignoring multiple births). We first calculate the probability of a male birth and a female birth *in that order*. According to the product theorem for compound events,

$$\begin{aligned} pq &= \frac{1}{2} \times \frac{1}{2} \\ &= \frac{1}{4}, \end{aligned}$$

where $p =$ the probability of a male, and
$q =$ the probability of a female.

But the children may come in reverse order—the girl on the first trial and the boy on the second trial—making a total of two ways in which the specified compound event can occur. Therefore, disregarding birth order, the probability of a male birth and a female birth would be

$$\begin{aligned} Pr(m, f) &= \frac{1}{4} + \frac{1}{4} \\ &= 2(\tfrac{1}{4}) \\ &= \frac{1}{2}. \end{aligned}$$

[1] In the following analysis, outcomes are assumed to be independent from trial to trial—that is, the probabilities of success and failure are constant from trial to trial.

Similarly, we may establish the probability of guessing 3 right and 2 wrong on a true–false test of five items. First we find the probability of 3 right and 2 wrong in that order:

$$\begin{aligned} pppqq &= p^3q^2 \\ &= (\tfrac{1}{2})^3(\tfrac{1}{2})^2 \\ &= \tfrac{1}{32}, \end{aligned}$$

where p = probability of guessing right, and
q = probability of guessing wrong.

Next, we determine the number of ways in which 3 right (R) and 2 wrong (W) could occur on five trials. By listing, we discover a total of 10 ways:

RRRWW WRWRR
RRWRW RRWWR
RWRRW RWWRR
WRRRW RWRWR
WRRWR WWRRR

Combining the two results, we arrive at a probability of 10/32 for the specified joint outcome, 3 rights and 2 wrongs.

Full Set of Binomial Probabilities By identical logic, we could obtain the probabilities of 2 and 4 rights, respectively. The probability of all 5 right by guessing would be 1/32, because there is only one way of obtaining 5 successes on five trials. Likewise, the probability of all wrong on five trials would be 1/32, there being only one possible way of getting every item wrong. All these results may be usefully assembled in a single display which suggests the law of their formation (Table 11.4.1).

From an inspection of this complete set of binomial probabilities, we may discern the general rule: To find the probability of r successes in N independent trials, find the probability of r successes and $N - r$ failures in that order, then weight that probability by the number of ways in which r successes can occur. But we may permute N outcomes, of which r are successes and $N - r$ are failures, in

$$\frac{N!}{r!(N-r)!}$$

ways. Hence, the probability of r successes in N trials is:

$$Pr(r \text{ successes}) = C_r^N p^r q^{N-r}. \tag{11.4.1}$$

The Expanded Binomial Were we to express each probability in the above notation and arrange them in order from five successes to none at all, we would have the following sequence:

$$C_5^5p^5q^0,\ C_4^5p^4q^1,\ C_3^5p^3q^2,\ C_2^5p^2q^3,\ C_1^5p^1q^4,\ C_0^5p^0q^5.$$

Table 11.4.1 *Combinations of Rights and Wrongs, Five True–False Items*

	Number Right						Total
	5	4	3	2	1	0	
	RRRRR	RRRRW RRRWR RRWRR RWRRR WRRRR	RRRWW RRWWR RWWRR WWRRR WRWRR WRRWR WRRRW RWRWR RWRRW RRWRW	WWWRR WWRRW WRRWW RRWWW RWRWW RWWRW RWWWR WRWRW WRWWR WWRWR	WWWWR WWWRW WWRWW WRWWW RWWWW	WWWWW	
C_r^N	1	5	10	10	5	1	32
$Pr(R)$	$\frac{1}{32}$	$\frac{5}{32}$	$\frac{10}{32}$	$\frac{10}{32}$	$\frac{5}{32}$	$\frac{1}{32}$	1.00

And, were we to sum these terms, we would have an instance of the *expanded binomial*, since the algebraic results correspond to raising the binomial $(p + q)$ to the Nth power, $(p + q)^N$; in our example,

$$\left(\frac{1}{2} + \frac{1}{2}\right)^5:$$

$$\frac{1}{32} + \frac{5}{32} + \frac{10}{32} + \frac{10}{32} + \frac{5}{32} + \frac{1}{32} = 1.$$

Naturally, this probability series sums to unity, since every possible joint outcome is accommodated herein.

Binomial Probability Distribution Since probability is identical with relative frequency, the binomial probabilities may be viewed as the frequency distribution of the variable "r successes on N trials," r taking all integral values from zero through N. Such a distribution is appropriately entitled a *binomial probability distribution* and may be displayed in either tabular or graphic form. Thus, where

$$p = \frac{1}{2},$$

$$q = \frac{1}{2},$$

and $N = 5$,

Table 11.4.2 Binomial Frequency Table, N = 5, p = .5

r	$Pr(r)$
5	$\frac{1}{32}$
4	$\frac{5}{32}$
3	$\frac{10}{32}$
2	$\frac{10}{32}$
1	$\frac{5}{32}$
0	$\frac{1}{32}$
	$\frac{32}{32} = 1.00$

as in the foregoing example, the binomial probability table would be constituted as shown in Table 11.4.2. The corresponding histogram is shown in Figure 11.4.1.

Mean and SD of Binomial Distribution By definition, the mean of a binomial distribution is the average number of successes on all trials; therefore, it is the expected number per trial. The *SD* is, of course, an average of chance variation around that mean expectation. Both quantities may be computed directly in the usual manner. Thus, to find the mean, we (1) weight each value of r by its corresponding probability (frequency); (2) sum these weighted

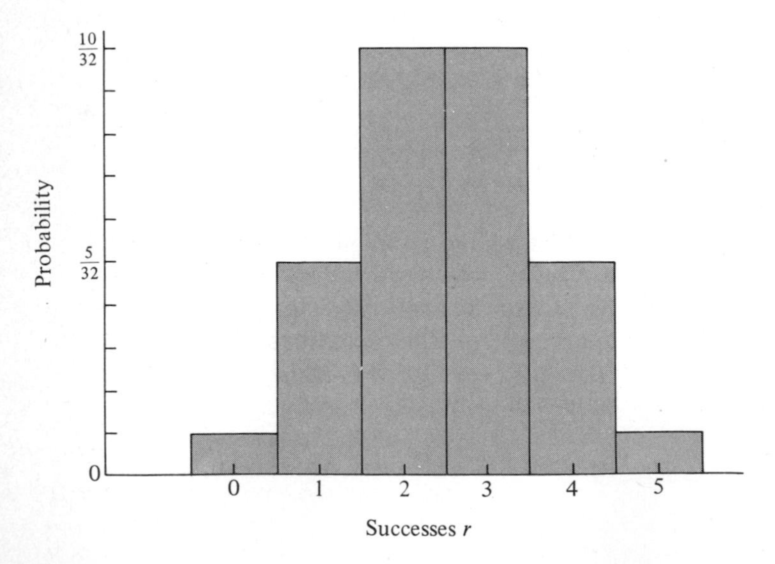

Figure 11.4.1 Histogram, Binomial Probability Distribution (N = 5, p = .5)

Table 11.4.3 *Computation of Mean, Binomial Frequency Distribution*

r	$Pr(r)$	$Pr(r) \times r$
5	$\frac{1}{32}$	$\frac{5}{32}$
4	$\frac{5}{32}$	$\frac{20}{32}$
3	$\frac{10}{32}$	$\frac{30}{32}$
2	$\frac{10}{32}$	$\frac{20}{32}$
1	$\frac{5}{32}$	$\frac{5}{32}$
0	$\frac{1}{32}$	0
	1.00	$\frac{80}{32} = 2.5$
$\overline{X} = \frac{2.5}{1.00} = 2.5$		

values; and (3) divide that sum by the total probability (frequency). Since the sum of the probabilities is always unity, the mean of the binomial probability distribution is simply the sum of the values of r weighted by their respective probabilities. The *SD* is analogously computed. In Table 11.4.3, we illustrate the computation of the mean of the binomial frequency distribution of Table 11.4.2.

Practically, such cumbersome calculations are unnecessary, since the mean of the binomial distribution of r successes on N trials is always p of N, usually written Np, and the *SD* is $\sqrt{pq}$ of N, usually written $\sqrt{Npq}$. Applying these formulas to the above data we obtain:

$$\overline{X} = Np. \tag{11.4.2}$$

$$\overline{X} = 5(\tfrac{1}{2}) = 2.5.$$

$$\sigma = \sqrt{Npq}. \tag{11.4.3}$$

$$\sigma = \sqrt{5(\tfrac{1}{2})(\tfrac{1}{2})} = \sqrt{1.25} = 1.12.$$

Empirical Probabilities The binomial expansion need not result in a symmetrical distribution, nor need it refer only to theoretical probabilities. For instance, over a period of time, if 5 percent of the freshmen in English I fail to pass, 95 percent would succeed in passing. Presumably, future students would conform to those probabilities. The corresponding binomial therefore would be: (.05 + .95) = 1. Accordingly, we may ask illustratively with what frequency the various combinations of successes and failures would occur by chance in the long run in classes of 10 students. The solution is provided by the successive terms of the expanded binomial shown in Table 11.4.4.

Table 11.4.4 *Binomial Probabilities,* N = *10,* p = *.05*

Number Failing (r)	$Pr(r) = C_r^N p^r q^{N-r}$ (rounded to the nearest thousandth)
10	.000
9	.000
8	.000
7	.000
6	.000
5	.000
4	.001
3	.010
2	.075
1	.315
0	.599

The probability that the whole class would fail is represented by the first term of the expansion:

$$(.05)^{10} = \frac{1}{10{,}240{,}000{,}000{,}000}.$$

Obviously, this is an extremely improbable event. However, a freak event does occur occasionally, however miraculous it may seem when it happens. This particular event is only 15 times as unlikely as a specific combination of cards in any given hand of bridge.

Binomial and Normal Distributions By the binomial expansion, it is a simple matter to determine, for example, the probabilities of 3, 2, 1, and 0 boys in families of three children. However, this method of calculating probabilities becomes cumbersome when the number of permutations is large—for example, when we wish to calculate the probability of 90 or more heads in 100 throws, or of 60 or more right by guessing on a true–false test of 100 items. Confronted by such problems, statisticians posed the question of whether the binomial distribution approaches some fixed pattern as N gets larger, which in turn might be used to provide approximately the desired probabilities.

This interesting and fruitful possibility was explored and resolved by De Moivre over 200 years ago in his discovery that the binomial distribution more and more nearly resembles the normal curve as N increases without limit. In fact, N has only to exceed 20 to produce a very good fit between binomial and normal distributions, provided p is not too divergent from .5. This convergence of the discrete binomial on the continuous normal distribution is illustrated in Figures 11.4.2 and 11.4.3, in which $p = .5$ and $N = 5$ and 10, respectively.

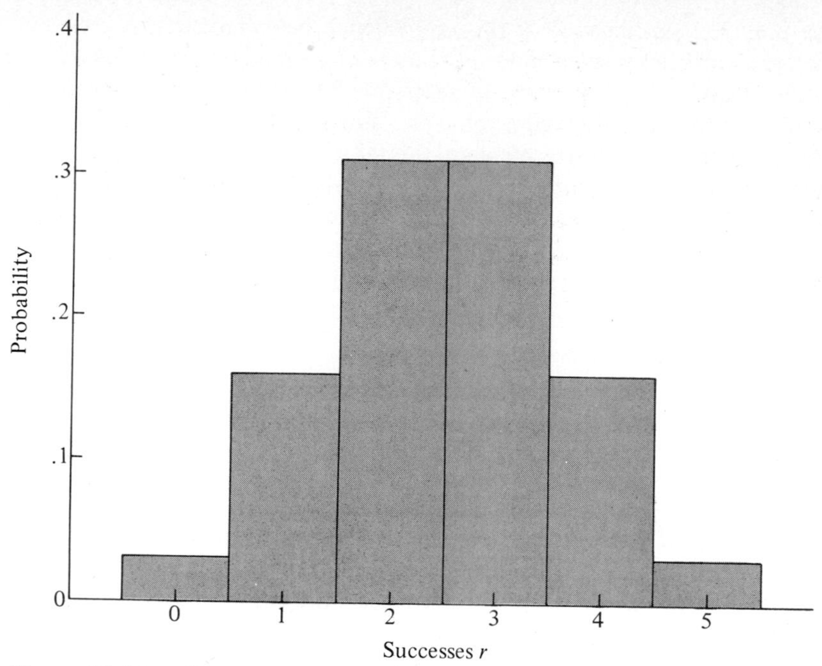

Figure 11.4.2 *Histogram, Probability Distribution* (N = 5, p = .5)

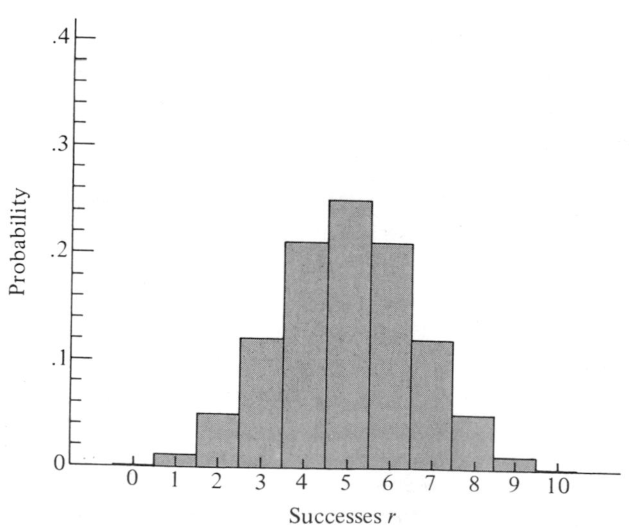

Figure 11.4.3 *Histogram, Probability Distribution* (N = *10*, p = .5)

The practical significance of this equivalence between the binomial and normal probabilities resides in the fact that the binomial probabilities may be obtained handily from the usually accessible Table of Normal Areas. We need only express an observed r-value as a standard measure and then evaluate that standard measure by means of this table. Thus, to find the probability of guessing 60 items *or more* right on a true–false test of 100 items, we first express 60 as a standard measure and then determine the frequency with which more exceptional values are expected to occur. To convert 60 to standard form, we require the mean and *SD* of the distribution:

$$\begin{aligned} \bar{X} &= Np \\ &= 100(\tfrac{1}{2}) \\ &= 50. \end{aligned} \qquad \begin{aligned} \sigma &= \sqrt{Npq} \\ &= \sqrt{100(\tfrac{1}{2})(\tfrac{1}{2})} \\ &= \sqrt{25} \\ &= 5. \end{aligned}$$

By means of these values, we convert 60 to a standard measure in the usual manner, with one minor adjustment. Since the normal curve is based on a continuous variable, we treat discrete 60 as the midpoint of an interval extending from 59.5 to 60.5. Since our problem is to determine the probability of 60 *or more* right by guessing, not *more than* 60, our standard measure is calculated on 59.5 instead of 60.5.

$$\begin{aligned} \frac{r - Np}{\sqrt{Npq}} &= \frac{59.5 - 50}{5} \\ &= 1.9. \end{aligned}$$

From the table we find that approximately 3 percent of all items in a normal distribution lie beyond 1.9 sigmas; hence, the probability of obtaining 60 or more by guessing is approximately 3 percent. In other words, in a class of guessers, 3 percent could expect to score 60 or higher.

Although 3 percent of a large class of students could be expected to answer 60 or more out of 100 true–false questions correctly by sheer chance, we cannot infer from such an observed score that they actually did *guess*. An obtained score of 60, for example, could mean that the student was correctly informed about 60 percent of the items, was misinformed on the remaining 40 percent, and guessed at none. More plausibly, perhaps, an obtained score of 60 could mean that the student was correctly informed about something less than 60 percent of the items, was misinformed on an additional fraction, and guessed at the remainder. But such a score also might have been obtained by guessing all answers. How, then, is the professor, or any other observer, to determine the correct explanation of such a score? The answer is that on the basis of the statistical evidence, an observer simply cannot know for sure. The central issue of what is known in statistical interpretation as *decision making* is now joined: whether to accept the hypothesis of chance, or to explain the score of 60 by the presence of some determining factor such as studiously acquired knowledge or possibly even cheating. There is no formula for adjudicating between these two alternatives, although, as the probability of the chance event becomes smaller and smaller, the chance hypothesis of guessing becomes correspondingly less and less acceptable. Such questions will be treated in Chapters 14 and 15 on hypothesis-testing.

QUESTIONS AND PROBLEMS

1. Define the following terms:
   binomial
   binomial probabilities
   binomial probability distribution
   independent outcomes
2. It has been contended that the excess of male over female births in the United States is due to the tendency of families to have no children after a male birth. Would such a parental bias affect the sex ratio?
3. A person claims to be able to discriminate between the tastes of two cigarettes. That person is given Brands A and B in pairs, 5 times. Assuming the subject merely guesses, calculate the probability of 5 correct choices; 4 correct choices; at least 4 correct choices. What chance factors would interfere with the subject's judgment? How many rights would you require to be convinced that the subject has the power to discriminate?
4. Are the following probabilities identical: 4 heads in 4 tosses of 1 coin; 4 heads in 1 toss of 4 coins? State the rule.
5. In tossing a coin, how long could one throw only heads without arousing suspicion of a biased coin? Explain your reasoning.
6. In a throw of two coins, what is the probability of obtaining exactly 50 percent heads and 50 percent tails? Also calculate for 4, 6, 8, and 10 coins. What happens to the probabilities of an even split as the number of coins is increased from 2 to 10?
7. If a true–false examination consists of 6 questions, what is the probability that the student will mark by chance:
   (a) all correctly?
   (b) half correctly?
   (c) all incorrectly?
8. Does the expanded binomial always result in a symmetrical distribution? Calculate the distribution of fives and no fives with 2, 4, and 5 dice. What generalization can you offer?
9. The Old Testament states that Jacob had 12 sons. We may hypothesize either (a) that there were actually no daughters among the children, or (b)

that the daughters were not recorded. Determine the probability of the first alternative and state which assumption you favor. What other factors, besides statistical evidence, might enter into your decision?

SELECTED REFERENCES

David, F. N.
1962 Games, Gods and Gambling. London: Charles Griffin.

Gnedenko, B. V. and A. Ya. Khinchin
1962 An Elementary Introduction to the Theory of Probability. Translated from fifth Russian edition by Leo F. Boron. New York: Dover Publications, Inc.

Kemeny, John G., J. Laurie Snell, and Gerald L. Thompson
1956 Introduction to Finite Mathematics. Englewood Cliffs: Prentice-Hall. Chapters 3 and 4.

Levinson, Horace C.
1963 Chance, Luck and Statistics. New York: Dover.

Marschak, Jacob
1954 "Probability in the social sciences." Pp. 166–215 in Mathematical Thinking in the Social Sciences, edited by Paul F. Lazarsfeld. Glencoe, Illinois: Free Press.

Polya, George
1954 Mathematics and Plausible Reasoning, Vol. II, Patterns of Plausible Inference. Princeton: Princeton University Press. Chapters 14 and 15.

Sampling 12

The Purpose of Sampling In drawing conclusions from data the social scientist's case generally rests on partial information. In a survey, it usually is impractical to question all possible respondents; in an experiment, it is not feasible to test the hypothesis on all possible subjects. But conclusions based on a fragment of the total aggregate may be astonishingly accurate. Well-selected fragments may effectively reflect the characteristics of the whole: a small snippet from a bolt of cloth; a few drops of blood from the patient's total supply; a few thousand survey votes, by which we describe the political intentions of millions of voters. Such procedure is standard practice in everyday social and economic life, as well as in the branches of scientific activity. In such instances, when the data are partial rather than complete, and when they are used to characterize the entire set, we call the fragment a *sample* and the total aggregate a *universe*, or *population*. We name a specified value of the universe, such as the mean, a *parameter*, and its counterpart in the sample we term a *statistic*. The objective of sampling, therefore, is to draw an inference about the parameter, which is unknown, from the sample statistic which is observed. This process of generalizing in a prescribed manner from sample to universe has come to be known as *statistical inference*.

Advantages of Sampling The collection of sample data naturally requires less time and effort than does the compilation of complete data. Hence, surveyors of American public opinion universally avail themselves of samples of respondents, since results based on time-consuming total enumerations would be obsolete before they could be tabulated and published. Similarly, since 1940, sampling has been extensively employed by the United States Bureau of the Census in its decennial enumerations to provide more promptly detailed descriptions of various characteristics of the American population.

Because sampling is also less costly, it may be quite feasible when the financial burden of full coverage is prohibitive. For example, although a single-handed sociologist may believe it desirable to interview every divorced couple in a large city, he or she may find such a program impracticable to execute because of limited financial resources. So the analyst must be content with a carefully selected sample. But this is not an unusual restriction. In general, research workers expect to arrive at valid generalizations on the basis of sample materials presumed to be representative of some wider domain. Thus, the anthropologist draws conclusions about the predominant values in a culture on the basis of a sample of native informants; the United States Children's Bureau relies on a sample of juvenile courts to establish national trends in delinquency.

Sample materials are the stuff of which scientific generalizations are made, but the sampling method involves more than mere recognition of that principle. Modern sampling practice is distinguished by (1) its emphasis on the well-defined universe, (2) the random selection of cases, and (3) the estimate of the reliability of the sample statistic—that is, how closely it probably conforms to the unknown parameter. These concepts form the basis of the discussions in this and the following chapters.

1 UNIVERSE AND SAMPLE

Definition of Universe In colloquial speech, the term "universe" suggests the entire Creation. But in statistical language, *universe* refers merely to a complete set of elements and their characteristics, about which a conclusion is to be drawn on the basis of a sample. Such sets are also termed statistical *populations*. We may cite as universes: the incomes of all American families in 1970, the opinions of all college students on the subject of war, or the social-status ratings of all residents in Yankee City. From these examples we discern that the statistical universe may be conceived conveniently as having two aspects: (1) the cases which are the units actually sampled and which, therefore, are called the *sampling units*; and (2) the *sampling trait* (for example, income) possessed by the sampling units, measures of which are subsequently manipulated statistically. Clearly the same population of cases or sampling units has numerous sampling traits, one or more of which may be the center of attention in any given set of calculations.

Our sampling may attempt to describe the distribution of one trait or the joint distribution of two or more traits. In brief, our problem may be to describe a *univariate*, *bivariate*, or *multivariate* population distribution. If our task is to draw conclusions about a univariate distribution, we employ such measures as the mean, median, range, and standard deviation; if our focus is on a bivariate population, we utilize an appropriate measure of simple correlation. When attention is focused on three or more traits, we manipulate our data to yield a conclusion about the correlation of one variable with a composite of the others (a multiple correlation), or about the correlation between any two variables with the others held constant or controlled (a partial correlation). The purposes of sampling are by no means limited to making estimates of a univariate parameter. The cautions and guidelines that govern good sampling practice are always relevant. They do not become trivial side issues because the intent of a study is to explore the association, for example, between income and vote intention rather than to estimate the number intending to vote for a particular candidate. But the sources of sampling bias may be different in these two instances. In estimating the number of voters intending to vote for a candidate, the oversampling or undersampling of persons with characteristics related to vote intention will bias the sample estimate, whereas in estimating the association between income and vote intention, it is the oversampling or undersampling of groupings among whom this association is especially high or especially low (that is, subgroupings that exhibit interaction effects; see Chapter 15) that will bias the sample estimate.

Sampling units are physically selected: families with incomes, farms with acres, students with opinions, workers with occupations. Further, the well-defined criterion that determines eligibility for such selection here applies to the sampling units rather than to the traits. Thus, a population of families consists of all human groups that satisfy the working definition of a family.

On the other hand, it is the variable properties of the sampling units that command our ultimate interest. If families are sampled, our interest will lie in one or more of their relevant traits: income, nationality, size, social status, religion, and so on. Variables such as these are subjected to statistical measurement after the sampling units have been drawn and their characteristics determined.

When, for any reason, it is unnecessary to identify explicitly both aspects of the universe, we may quite properly employ elliptical phrases such as "the population of college students" or "the universe of attitudes." But such abbreviated statements omit the sampling trait of attitude (of the college students) in the first instance, and the sampling unit (the college student whose attitudes are polled) in the second instance. It should be evident that the sampling unit or the case in social research need not be the individual person; it may be a collectivity (for example, families, gangs, work crews, cities, and so forth), a time period (hours, days, months, years), or some other relevant unit (for example, newspaper editorials, television programs, works of art, segments of conversation, episodes of delinquency, and so on).

Finite and Infinite Universes The universe may be *finite* or *infinite*, depending upon whether the sampling units are finite or infinite in supply. By definition, a finite universe contains a countable number of elements. It may be relatively small, as, for example, all students enrolled in a particular college in a given year; or it may be relatively large, as all college enrollees in the United States. But the infinite universe consists of an endless number of elements, such as an unlimited number of penny tosses or other experimental trials. The infinite universe is thus purely conceptual and may even seem metaphysical to the finite mind. And yet it is often heuristically postulated in statistical inference. For one reason, an infinite population permits the reliability of the sample findings to be more simply evaluated by formulae than does a finite population. Consequently, we resort to the assumption of an infinite population whenever the size of a finite population is large enough to justify it. Additionally, the infinite universe may be invoked when it is not reasonable to limit the size of the universe at all, as in the case of infinitely repeatable experimental trials.

The Target Universe and the Sampled Universe The universe which is actually sampled (the *sampled universe*) will not always coincide with the universe on which our sights are fixed (the *target universe*). The target universe represents ideally the territory we intend our generalizations to cover—the domain to which we eventually apply our sample knowledge. Our ultimate interest may lie, for instance, in the patterns of adjustment of all married students on campuses in the United States, but for practical reasons it may be necessary to restrict the sampling to the available couples on a particular campus, and these couples then become the sampled universe. When a mailed questionnaire is returned by less than 100 percent of the sample, as is usually the case, we may conceive of the target universe as the complete mailing list from which the sample was selected, and the sampled universe as all persons on that list who theoretically would return that questionnaire if given an opportunity to do so. Thus, the concept *target universe* may be applied broadly to an idealized extension of the sampled universe, or narrowly to a universe in which a fraction of the units, for one reason or another (such as refusals or not-at-homes), are inaccessible to measurement.

Strictly speaking, statistical inference should be rigorously limited to the sampled universe; and yet the social analyst can scarcely refrain from speculating about the target universe. No research study in the social sciences would ever be made if its findings could not be extrapolated imaginatively beyond the limited universe from which the sample has been derived. Such speculations are both justifiable and desirable, provided their tentative nature is recognized and understood. A sampling study of mental health and social class in a New England community is of significance primarily for the light that it sheds on the relation between social structure and personality deterioration in other American communities. And so the investigator quite naturally probes sample materials for their wider generality and makes the most of costly data. In projecting the findings of the study onto the vaguely defined target universe, the analyst necessarily proceeds without benefit of strict reliability procedures. Yet the analyst may be, and usually is, engaged in fruitful and necessary scientific activity. Nevertheless, such a liberal statistical morality, which is practiced by even the most thoughtful and productive social scientists, is by no means license for irresponsible and sloppy statistical generalizations.

It may be useful to introduce some terminology to facilitate the distinction between generalizations from a sample to the sampled universe based on a statistical formula and less rigorous generalizations from a sample to an ill-defined target universe based on seasoned judgment and broader knowledge. The former may be conveniently referred to as *statistical inferences*; procedures for making such inferences will be discussed in Chapters 13–15. The latter may be termed *general inductions*. Whereas statistical inferences are made on the basis of well-defined procedures designed to guard against inappropriate conclusions, they apply to the sampled universe and not necessarily to the target universe. No such simple procedural guidelines are available for making general inductions about the target universe, and a judicious conclusion about the target universe depends primarily upon the investigator's knowledge and experience.

Problems of Sampling The process of sampling is in its scientific sense a technical operation which must be conducted according to standard prescriptions in order to secure all its benefits. In fact, sometimes costly social investigations have been severely blemished because the sampling tactics were crude and inadequate. Because of the admitted difficulties in sampling human populations, the discussion of the theory and practice of sampling always must occupy an important place in the domain of social statistics.

By way of illustration, consider the task of sampling a set of college students—a common assignment for majors in journalism or sociology—with a view to generalizing about the entire student body. Disregarding momentarily the kind of data sought—attitude toward communism, number of dates per week—how may we obtain a sample which will do justice to the student body? This is no simple task. The sampler might give way to the first impulse to take all persons enrolled in elementary sociology—a rather attractive possibility since such classes include a wide variety of students and, in addition, are easily reached for study or survey purposes. However, on second thought, the sampler will realize that college courses are almost sure to exert some selective influence among students. Thus, sociology is likely to attract persons whose primary interest is in social issues but may hold no appeal for those principally interested in the physical world. In restricting the sampling to sociology students, therefore, there is a danger of excluding certain types of individuals who are negatively selected by this subject. To forestall this outcome, we might propose using persons enrolled in English Composition, as this subject may be required of all. Though reasonable enough on first thought, this alternative will seem less satisfactory when we realize that only underclassmen are enrolled; mature upperclassmen would not be included in the sample.

Many other possibilities will suggest themselves in the search for representative coverage. Thus, we might consider canvassing the men's dormitories. But this scheme rather glaringly omits women, and so it would have to be modified to include women's residence halls as well. Even with this modification, the plan is rather obviously unsound: it makes no provision for students housed outside the residence halls. Apparently, other plans will have to be tried until an adequate one is developed.

The general shortcoming of the aforementioned alternatives will be recognized by even the casual reader: they do not afford each sampling unit an equal opportunity of being selected. Ideally, of course, our sampling procedure would exclude no unit in the population from the possibility of being drawn into the sample, and each unit would have a known probability of inclusion. Our inferences from sample to sampled universe are greatly simplified if each case is known to have a probability of inclusion equal to that for any other case and if all combinations of cases of a given size are equally probable. But strict inferences are possible even when the probability of selection varies from case to case, provided such probabilities are known. Sampling procedures in which each case has a known probability of inclusion (whether equal for all cases or not) produce a *probability* or *random sample*. When the probabilities of inclusion are unknown or disregarded, the sample is referred to as a *nonprobability* or a *nonrandom sample*.

Nonprobability sampling procedures are such that the representativeness of the sample cannot be estimated and, consequently, it is impossible to generalize confidently from the sample to the sampled universe. Therefore, nonprobability procedures are seldom considered to be ideal; nevertheless, they are often justifiably resorted to in social research because of practical necessity. Beginning with the least useful, several nonprobability procedures are presented in the following section in ascending order of utility so the student will be familiar with their respective merits and shortcomings.

QUESTIONS AND PROBLEMS

1. Define the following terms:

universe	sampled population
sample	target population
infinite population	statistical inference
finite population	general induction

2 SAMPLING PROCEDURES

Haphazard Sampling The acceptance of whatever cases one fortuitously happens to encounter, without any consideration whatsoever for their degree of representativeness, may be termed *haphazard sampling*.[1] This practice

[1] Synonymous concepts that appear in statistical writing include *accidental sampling* and *convenience sampling*.

is exemplified by the old-fashioned "straw vote" in which citizens are accosted on the street to ascertain their voting intentions. From these "straws in the wind," the election forecast is made. The obvious objection to such casual procedure is that the man-in-the-street simply is not representative of the total electorate. Similarly, in a survey of student opinion on the propriety of final examinations, those nighthawks whom we happen to find breakfasting in the coffee shop at 11 a.m. will be somewhat less than a fair cross-section of the entire student body. Such crude chunks of data are samples only in the loosest sense of the term. When seriously used, they constitute an unflattering reflection on the sophistication of those who resort to, and accept, such data.

Availability Sampling Although fully aware of the limitations of nonrandom sampling, sooner or later the experienced social scientist will realize that some form of it is often the only alternative to abandoning the inquiry. In many instances, an analyst may be required to seize whatever opportunity is available. The Kinsey survey of sexual behavior in the United States male population was severely criticized because, to a large extent, it was based on data provided by solicited subjects who made themselves available. It was plausibly contended that persons who volunteer to provide information about their sexual behavior are likely to differ significantly from persons who decline to be interviewed. But the investigators considered the acceptance of volunteer respondents as a pragmatic solution, rather than the preferred procedure. Their reasonable assumption was that many random selectees could not be induced to relate their sex histories, and that any possibility of obtaining a completely random sample was precluded.

For similar reasons, other types of sociological studies often must rely on available opportunities. For example, it may bc impossible for the behavioral scientist to sample all workers in a given industry, yet possible to observe workers in a local plant; impossible to poll a group selected from all school children in the state, but possible to poll those attending various local schools. Social psychologists are likely to find little humor in the sarcastic remark that their broad "universal" generalizations are founded on experiments on college sophomores. Above all, they would prefer to study all kinds of individuals—old as well as young, noncollege as well as college—in formulating the laws of social learning. But often it is a question of sampling the available students or otherwise abandoning the project. Hence, social psychologists must convert a captive audience of students into a valid sample. Similarly, research on delinquency typically has been based on readily available delinquents—boys on probation, boys in court, or boys in a correctional training school. Recent studies of delinquency have moved toward a more representative sampling of delinquents, the unapprehended as well as the apprehended, both through observations on the streets and through the use of questionnaires and interviews designed to obtain an anonymous self report of delinquent activity.

It would be pedantic to deny the uses of available opportunities, even though they do not yield ideal data. Social scientists, like everyone else, often must content themselves with compromises. Notwithstanding their shortcomings, availability samples do yield significant information and insights. For example, the Burgess and Cottrell sample of 526 married couples produced many revealing propositions on the positive and negative factors in marital happiness; Znaniecki's renowned sample of letters written by Polish peasants provided suggestive hypotheses on basic human motives. Moreover, inexpensive captive samples—college classes, prison inmates, members of the armed forces—often can be utilized to develop research techniques and to supply experience in applying them. No one would seriously protest against studying hospital patients because they do not perfectly represent all diseased persons; few would abandon the study of prison inmates merely because they are not an exact replica of the criminal population. Availability samples are quite legitimate, so long as the inferences drawn from them are accompanied by reservations which are made necessary by the ill-defined relation between universe and sample.

Judgment (Purposive) Sampling When the composition of the universe is known to the sampler, and when a small sample of only one or a few items is required, the sampler may elect to rely on his or her mature, sound judgment in choosing the sample. Thus, judgment sampling will be most effective in the hands of an expert who knows the population, and who can readily spot the typical case. Why should such an expert choose a sample by lot and thereby run the risk of drawing an extreme case, when it is possible to obtain a representative item by deliberate choice? The expert may legitimately pit his or her personal judgment against the operation of the laws of probability. The Lynds did not leave the choice of "Middletown" to blind chance; if they had done so, a highly atypical community might have presented itself for study and might have defeated the very purpose of the investigation. These sociologists were interested in a single modal community, as indicated by their choice of pseudonym. Therefore, judgment sampling is especially well-suited to the *case study*, in which many aspects of a single representative case are observed and analyzed.

While judgment sampling is considered efficient for selecting only one or two items, it has been applied on occasion to procure samples of considerable bulk. The *Time* magazine sample of United States college graduates (1952), consisting of individuals whose last names began with "Fa," was justified by the surveyors on the ground that such names probably were scattered randomly in the various ethnic, religious, and socioeconomic groups of the population. And since a genuine randomized selection would have required a much more cumbersome sampling apparatus, along with trained hands to operate it, the "Fa" type of sample was more practicable, and hence, in this instance, the preferred alternative.

However, the seductive plausibility of judgment sampling should not be permitted to conceal its hazards. Even the expert's judgment is subject to human bias and error. And even when the universe is in plain sight, it is probable that the observer will misjudge to some extent the representativeness of a sample. To demonstrate such bias, the English statistician Yates once requested 12 persons to select 3 samples of 20 items each from a collection of 1,200 stones, each sample to represent as accurately as possible the size distribution of that experimental universe. Although the observers were free to view the collection at their leisure, still there was a consistent tendency to exaggerate the average size of the stones and to minimize their variation. In short, there was a constant error in judgment. We cannot be sure that analogous errors did not bias the "Fa" sample. Because of the practical certainty of human bias, judgment sampling must be applied with great caution. But whatever the limitations of nonrandom samples may be, the techniques of descriptive statistics obviously must be applied competently to assure their maximum utility.

Definition of Random Sampling If available samples are fallible, and expert judgment is not to be trusted, what are the factors that should determine the composition of the sample? The answer is: chance factors. However, chance factors are usually held in low regard as guides to action; therefore, it is surprising that we should so willingly lay aside our cumulated knowledge and experience and go to the other extreme, permitting blind chance to determine the choice of the sample. In most human situations, we wish to eliminate chance, since it disturbs our predictions. Nevertheless, the ideal sampling procedure is one in which the drawings are affected by impartial chance factors alone, with the result that one item in the universe is as likely to be included in the sample as another. No item is accorded a preferential advantage. In fact, *simple random sampling* is defined as a procedure that provides an equal opportunity of selection to each unit in the population, and provides an equal opportunity for all combinations of units of the given sample size. *Equal opportunity* or *same probability* refers to the expectation that each unit and each combination of units would be drawn an equal number of times if identical drawings were conducted indefinitely. Of course, an empirical demonstration of innumerable drawings is not undertaken; nevertheless, it is tacitly assumed that the a priori expectation of equal occurrences would be confirmed eventually.

Of course, equal probability of selection cannot be assumed if some cases are less accessible than others or if the selection mechanism functions imperfectly. Either of these related contingencies will defeat the aim of random sampling and will result in *biased sampling*. Thus, in selecting a sample of 10 slips from a receptacle containing 200, randomness would be precluded if the slips were unequal in size, shape, or weight; or if they were carelessly mixed so that the last names dropped into the hat would have the best opportunity of being drawn. But the outward appearance of the extracted sample never would reveal such procedural flaws, or biasing factors. The composition of a nonrandom sample is visibly no different from that of a sample of the same size selected by random procedures; the sampling operation leaves no telltale mark. Thus, if we were to come upon two different scatters of 10 coins each strewn on two tables, the one all heads and the other showing 6 heads and 4 tails, no amount of visual inspection would tell us whether either or both of them had been carefully laid down or whether they had been tossed at random. Nevertheless, the observer will conclude intuitively that the first probably had been laid down and the second had been tossed arbitrarily. Why? Because the probability of 10 heads together is so small, and the division of 6 and 4, much greater. In fact, the probabilities compare as 1 to 210. Still, we never can be certain.

A suspicion of biased sampling will be aroused by the subsequent discovery of a marked discrepancy between the sample value and the true value of the universe, when and if the latter becomes known. The now classic example of such a melancholy outcome is the notorious *Literary Digest* presidential pre-election poll of 1936. While Roosevelt obtained approximately 60 percent of the popular vote in the actual election, his percentage in the *Literary Digest* sample of over 2,000,000 respondents was only 40 percent—a difference of 20 percentage points! This discrepancy in such a large sample was symptomatic of a gross defect in the sampling procedure, which statisticians easily established upon later review and analysis. Although such a discrepancy could have occurred by chance, it would have been extremely unlikely. Hence the deduction that the method of sampling was biased. But such appraisals are always retrospective. In advance of the sampling, one can only provide for adequate machinery which will be reasonably certain to yield a random selection.

Simple Random Sampling If called upon to devise a do-it-yourself sampling technique, the inexperienced analyst probably would procure as many slips as there are items in the population to be sampled. Slips of paper are more easily shuffled than people. We may surmise that the sampler would then number these slips consecutively from 1 through N, corresponding to the numbered units of the universe. Next, he or she would place the slips in a suitable receptacle and mix them until the set was thoroughly scrambled. Finally, the analyst would reach in and take out as many cases as desired. For a sample of 10 cases, 10 different slips would be removed. Statisticians recognize this as the simplest type of random sampling and therefore have dubbed it *simple random sampling*.

They do not describe it as "reaching in and drawing out n different items." Rather, they define it as that procedure in which every distinct sample of n items has an equal probability of selection from a finite population of N items. This definition expresses the long-run consequence of "reaching in and taking out n different items." For, if that procedure were applied indefinitely to a given population (restoring the entire sample after each trial), each different sample would tend to reappear an equal number of times. In its procedural aspect, simple random sampling is the apparatus that guarantees the fulfilment of this criterion of equal probability. In its substantive aspect, it is the very criterion itself.

To unfold further the meaning of simple random sampling, let us consider the number of ways in which samples of 2 items can be selected from a miniature population of 5, whose members we shall designate: a, b, c, d, e. By manipulation, we discover that there are 10 different possible combinations, or samples:

ab	bc	cd	de
ac	bd	ce	
ad	be		
ae			

Hence, the probability of each combination in simple random sampling must be 1 in 10; each sample is expected to occur once in every 10 trials on the average.

For finite universes and samples of any size, this probability may be expressed in terms of the now familiar combinatorial formula:

$$Pr\text{ (Given Sample)} = \frac{1}{C_n^N}.$$

Applying the formula to the illustrative data given above, we have

$$C_n^N\text{ (Total Possible Samples)} = \frac{5 \times 4}{2 \times 1} = 10$$

$$Pr\text{ (Given Sample)} = \frac{1}{10},$$

which agrees with the previous result.

Sampling by Random Digits A rudimentary technique for carrying out simple random sampling has already been set forth: (1) represent units on slips; (2) thoroughly scramble; and (3) draw the required number of slips. However, the drawing need not be carried out laboriously in this manual fashion. Usually it will be inefficient to do so, especially when the population is large. It is much more practicable to substitute a table of randomly ordered digits for the shuffled numbered slips—a common procedure of research workers. This technique necessarily requires that we number serially all the units in the population from 1 through N and then draw from the corresponding table of random digits as many different numbers (combinations of digits) as there are cases to be included in the sample. The cases whose serial numbers correspond to those drawn from the table constitute the sample. Table VII of the Appendix presents such a list of random numbers.

Simple random sampling is the most primitive, least complex selection procedure and most clearly exposes the essential operation of randomness. Being free of procedural modifications which are often made necessary by practical circumstances, it is conceptually the simplest of all sampling routines and therefore is so labeled. Since it is random sampling in its most uncomplicated form, it serves as a standard of sampling efficiency against which other types are compared and evaluated. These other types are made necessary by the fact that simple random sampling in its pure form almost never can be employed in large-scale social research. It is far too impractical and costly and often even impossible. Nevertheless, it is essential that the student clearly comprehend its basic characteristics in order to be able to recognize and appreciate the degree to which the alternative methods depart from this standard model. Three of the most prevalent alternative types are set forth here as a brief introduction to the subject: (1) *stratified*; (2) *cluster*; and (3) *interval* sampling. These sampling devices themselves are largely an outgrowth of the sheer practical problems that have arisen in the sample surveys of large human aggregates, which accounts for their wide currency in social science.

Stratified Sampling "Do the citizens of Brownville favor busing to achieve racial integration of schools?" No competent surveyor of public opinion would attempt to answer that question without canvassing both white and black residents of the community. Public sentiment on this issue would not be portrayed reliably by a sample that slighted either group, since opinions on this issue are so divergent. Yet such an imbalance between groups might occur under simple random sampling unless the sample were made large enough to forestall that eventuality. A more economical alternative would be to sample each subgroup separately and combine the results, thereby avoiding a costly inflation of sample size. Such an operation, which first separates the entire population into relevant *strata* before randomly drawing the sample, is known as *stratified sampling*.

From a procedural standpoint, therefore, stratified sampling consists of the following stages: (1) division of the total universe into subclasses, or strata; (2) the selection of a random sample from each stratum; and (3) the consolidation of the subsample statistics into a combined statistic weighted for size of strata. In this context, the term *stratification* of course does not connote a hierarchy, as the ranks of an army, or the geological seams of the earth crust; rather, it signifies the categories of a statistical variable, such as race, sex, or religion, into which the total population is conveniently divided.

Nor does stratification imply a relaxation of the requirement of randomized selection, although sometimes that inference has been drawn mistakenly. Possibly this misconception may reflect a failure to distinguish clearly between stratified random sampling and so-called *quota sampling*, which has been used widely in opinion polling. In quota sampling, quotas are preassigned to strata, but the final selection of cases is left to the interviewer's discretion. However, if the benefits of random sampling are to be attained, subsamples from strata must be chosen randomly. The resort to convenience or judgment sampling is no more warranted within a given stratum than it is in the whole, unstratified population, and it is almost certain to lead to biased results.

Since stratified sampling is more complex than simple random sampling, we may rightly ask what its compensating advantages are. Briefly put, it is a labor-saving device for securing equivalent accuracy with fewer cases than is likely under simple random sampling. It is essentially for this reason that national public opinion polls typically employ some form of stratified sampling to keep the size of the sample down to manageable proportions. Thus, subsamples usually are selected from each of the broad geographical regions of the United States, it being plausibly assumed that public opinion on many issues varies from one sector to another: for example, the Midwest has been shown to be internationally more isolationistic than the East. Similarly, subsamples commonly are drawn from various age and economic levels, as these factors are also known to exert an influence on the content and intensity of public opinion. Older persons generally are more conservative than younger; political opinions tend to parallel economic interests. On the other hand, national opinion polls would never stratify the population by hair color, since there is probably no genuine correlation between hair color and political opinion. In general, stratification serves no purpose when the stratifying factor is uncorrelated with the sampling trait being measured.

Comparison of Strata While the principal justification of stratified sampling is equal accuracy with a smaller sample, the comparative data which are its natural by-product provide an additional inducement for using it. Thus, the decision to stratify by age in a survey of opinion on student participation in college administration may be prompted as much by the wish to compare the characteristics of the various age groupings as by the need to economize on sample size. The differences among age groupings (Table 12.2.1) on the

Table 12.2.1 *Opinion Concerning Student Participation in College Administration by Age and Percentage Distribution, United States,* 1969

"Do you think college students should or should not have a greater say in the running of colleges?"

Age	Should	Should Not	No Opinion	Total
21–29	45%	53%	2%	100%
30–49	26	71	3	100
50 and over	16	76	8	100
Total	25	70	5	100

Source: *Gallup Opinion Index*, Report Number 46, April, 1969, p. 10. (Gallup national survey results are based on interviews with a minimum of 1,500 adults.) Reprinted with permission.

question "Do you think college students should or should not have a greater say in the running of colleges?" may, in fact, be even more pertinent and revealing than the overall weighted average of 25 percent, which necessarily conceals such differences. Since strata often are treated individually, it has been suggested that the term *domains of study* be applied to strata when they are being analyzed in this segregated manner.

Cluster Sampling Simple random sampling implies not only a random selection procedure but also a complete list, so that every unit is accessible to the draw. When the universe to be sampled is vast and extends over a wide area, the compilation of such an indispensable list is a laborious undertaking; and even when such a list is available (for example, a city directory), usually it is not up-to-date even on the date of issue because of the mutability of the population.

Therefore, if an adequate list of elementary sampling units is not available, we may turn instead to more or less permanent groupings into which the population is naturally divided and which can be listed conveniently. Human beings are usually found in prevailingly standard groups: they are clustered geographically by states, counties, municipalities, neighborhoods, blocks, precincts, and dwelling units; people work together in factories, offices, and stores; they are organized in innumerable clubs, lodges, schools, and miscellaneous associations. These groups, or *clusters*, may be serially utilized as sampling units through which we reach the ultimate elementary unit (for example, the person or household) which is the objective of our survey. Therefore, such a procedure is termed *cluster sampling*. We reach the elementary unit through a shorter or longer chain of samplings of the more easily listed clusters. Since sampling is carried out in successive stages before reaching our destination, this type of sampling is also called *multistage sampling*. When the clusters at any stage consists of territorial units, we may describe that stage as *area sampling*.

Let us suppose that we wish to survey the occupational ambitions of high school pupils, ages 15–17, in Chicago. It would be exceedingly laborious and financially prohibitive, as well as hazardous in accuracy, to compile a list of all specified pupils in that metropolis. However, a permanent list of high schools is easy to obtain. These schools could be sampled randomly, and a list of students in the desired age categories then would be obtained from the comparatively few high schools in the sample. From that list, the sample of students finally would be drawn. Similarly, if the households of the city were to be surveyed, we might first draw a sample of blocks, then dwelling units, and finally households. Even though such a multistage procedure still would require a source list at each level, the lists would be smaller and more current. In such economy lies the first advantage of the cluster approach.

However, the execution of any social survey is not fulfilled by the mere drawing of the sampling units. We still must make personal contact with

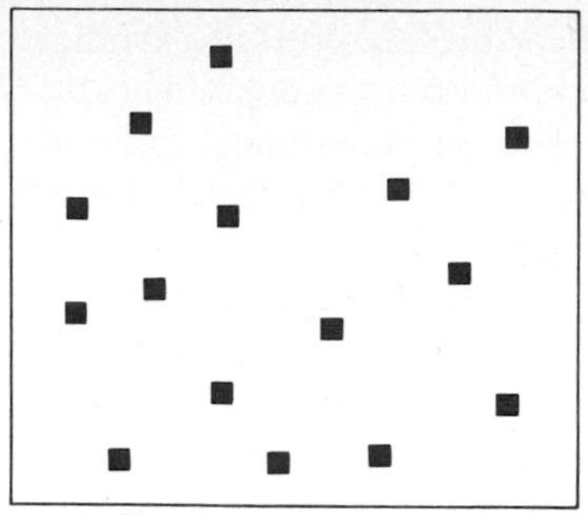

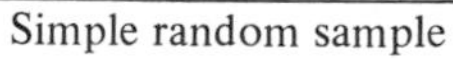

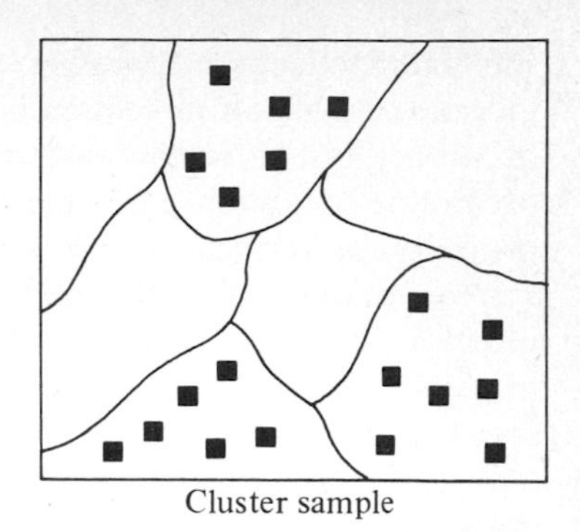

Figure 12.2.1 *Schematic Diagram, Simple Random and Cluster Sampling*

the selected elementary units to obtain the information that was the objective of the study in the first place. This problem of personal contact develops into a major obstacle in simple random sampling, which may yield a sample whose elements are widely dispersed over a wide area, requiring a prohibitive expenditure of time and energy to reach. In such a plan for personal interviews cluster sampling displays another technical advantage. Under this procedure, the elementary units are concentrated territorially and therefore are more easily accessible, with less wastage in transportation. This condition is shown in Figure 12.2.1.

The physical and mechanical convenience of cluster sampling is purchased at a price in quality. Generally, to a degree, cluster sampling lacks the very characteristic that is the objective of good sampling: typicality and representativeness. Its reduced representativeness is due to the fact that the elementary sampling units within clusters, particularly human clusters, are likely to be closely similar in regard to their social characteristics; consequently, sampling within these clusters understates the dispersion and provides unnecessary duplication. For example, the residents of a given city block are likely to belong to the same socioeconomic class, and therefore may hold the same political and social opinions. By the same token, they are not likely to be representative of any diversity of public opinion in the large aggregate of unsampled neighborhoods; a cluster sample is thus *less* likely to represent the variety of opinions than a simple random sample of the same size scattered over the community. This lack of heterogeneity within clusters reduces the comparative effectiveness of cluster samples. Because any one cluster is likely to be more homogeneous than the total population and may differ strikingly from it, a sample consisting of a single cluster does not allow a confident generalization to the universe. If the sample is to contain the diversity characteristic of the universe, many small clusters which differ widely from one another, even though each may be homogeneous within itself, would be preferable to a few large clusters lacking diversity both among and within themselves.

Interval (Systematic) Sampling Whenever the sampling units are arranged in some kind of natural sequence, such as consecutive admissions to a hospital or library books in the card catalogue, it may be economical and even preferable to obtain a sample by taking cases at a fixed interval. Such a procedure is termed *interval sampling*, or more commonly but less aptly, *systematic sampling*. The selection of every tenth name from the telephone directory, after a random start among the first ten names, illustrates the process of interval sampling. Such a procedure has obvious utility for the social scientist who frequently has occasion to study a series of events such as a file of newspapers, the characteristics of dwelling units in a given ecological area, the cases on docket in a criminal court, or a card catalogue of welfare case records.

To establish the width of the sampling interval (k) in any given problem, we merely find the ratio of population size (N) to desired sample size (n):

$$k = \frac{N}{n}.$$

Thus, if the sample is to contain 5 percent of the universe, or 1 out of every 20 cases, obviously the sampling interval would be 20; and we would draw every 20th item, randomly starting with any number within the first interval of 20. Such calculation presupposes, of course, that sample size has been fixed in advance of the sampling. But if n has not been set, and an arbitrary interval is employed, it is still necessary to pass through the entire sequence, even though we may seem to have an ample number of items after we have proceeded only part of the way. If we discontinued the drawings before completing the entire circuit, we would deprive the units in the omitted segment of the opportunity of being chosen and thereby we would destroy the randomness of the operation. For example, by skipping names L to Z in an alphabetical listing, almost certainly we would produce a biased sample.

Since each interval contributes one and only one item to each sample, it follows that there can be no more different samples than there are items within the interval. Thus, if the sampling interval is equal to 10, there can be only 10 possible samples, regardless of the size of the universe, be it 1,000 or 1,000,000. But the number of different samples would be almost incalculable in simple unrestricted random sampling. This restriction in the number of possible samples serves to distinguish interval sampling from simple random sampling, since the latter furnishes an equal opportunity of selection to every distinct combinatorial sample of n items.

In spite of this severe limitation, interval sampling often will produce results that compare favorably in representativeness to those yielded by simple random sampling. In particular, whenever the values may be presumed to be randomly ordered, simple random and interval sampling will yield identically accurate results. Such a presumption often is reasonable when items have been alphabetically listed, since there is usually no correlation between the alphabetical order of names and the traits the named objects possess. Thus, we have seen previously that the alphabetical listing of large American cities orders the respective suicide rates in a sequence which is seemingly purely random. Similarly, an alphabetical listing of students may be expected to result in a sequence of grade averages that is wholly random and, therefore, free of trends and cycles. In such cases, it makes no ultimate difference whether we select random digits or draw every *k*th unit, except that interval sampling is usually more simple to execute. The long-run results would be virtually identical for any given size sample.

However, there is one notable circumstance that constitutes a special hazard for interval sampling and may easily lead to erroneous conclusions. When the universe values form a *cyclical progression*, the sampling interval may coincide with the phase of the cycle, causing interval sampling to yield an unrepresentative set of identical values. Let us consider a fictitious sequence whose phase is four: 1, 2, 3, 2; 1, 2, 3, 2; 1, 2, 3, 2. Now, if the sampling interval is set equal to 4, any sample necessarily will consist of a set of identical values. It will consist of all 1s, all 2s, or all 3s. In no way will such samples do justice to the variation in the universe.

This type of pitfall is illustrated in a sampling study of June issues of the Sunday *New York Times*, 1932–1942, which disclosed that only Protestant marriages were featured on the society page of the sampled issues. From this finding, the conclusion was drawn that the upper social class of New York City was preponderantly Protestant in religious background. But this inference was challenged immediately on the ground that, for ceremonial reasons, few Jewish marriages are performed in June, and therefore few notices of Jewish marriages could have appeared in the sampled issues of the *Times*.[1] A check sampling, undisturbed by daily and monthly cycles, revealed that Jewish marriages in fact were regularly featured by the society editors during the appropriate seasons. By that criterion, Jews were represented proportionately in the upper social strata. In this instance, the sample interval led to an overrepresentation of Protestants, an error compounded by the unfortunate judgmental selection of June as a point of origin.

But interval sampling also carries its intrinsic advantages. In fact, when the numerical values form an *arithmetic progression*, interval sampling will be even more effective than simple random sampling. For, in that event, the sample will necessarily distribute itself evenly over the entire range of values and thereby will provide a reliable miniature of the population distribution. For example, if we select every 5th girl from a lineup according to height, the resulting sample necessarily will be representative of the distribution of girls' heights. Analogously, if we sample every 25th dwelling unit along a metropolitan avenue after a random start, probably we would obtain an accurate

[1] David and Mary Hatch, "Criteria of Social Status as Derived from Marriage Announcements in the *New York Times*," American Sociological Review 12(1947): 396–403; and W. J. Cahnman, "A Note on Marriage Announcements in the *New York Times*," American Sociological Review 13(1948): 96–97.

cross-section of the various socioeconomic groupings on that street, since dwelling units are segregated and ordered according to social status. In this way interval sampling may supply its own stratification.

From this, it is evident that the principal advantage of interval sampling lies in the mechanical ease with which it can be applied to such natural sequences as rows of dwelling units, card files, city directories, and so on. Its special hazard is the cyclical sequence, and we must maneuver to circumvent that danger when it is thought to exist.

Interrelatedness of Random Sampling Procedures Quite obviously, the foregoing random procedures are not mutually exclusive. They may be—and usually are—combined in a variety of ways. Cluster sampling may be used within broad strata, and interval sampling may be used within these clusters. No single *sample design* is best for all purposes. In all sampling, an attempt is made to attain the desired degree of representativeness as economically as possible, which is the guiding criterion of modern sampling design. We should recognize that the designing of a sample is a form of statistical engineering and accounting, requiring appropriate skills and knowledge. In the foregoing statement, we have merely hinted at the technical aspects of sampling, which of course are fully developed in treatises on that subject.

However, effective sampling requires much more than mere technique. If the assets of a given sampling procedure are to be fully realized, it is essential that all necessary discretionary as well as mechanical steps be expertly performed. Thus, the anticipated benefits of stratified sampling will not be attained unless the strata into which the population was judgmentally divided before the sample was drawn actually differed among themselves on the sampled trait. In the aforementioned instance, stratification by race would be profitless if blacks and whites shared the same opinions on busing. On the other hand, if blacks and whites differed widely, then, statistically speaking, there would have been considerable variation between strata, but relatively little within each stratum, thereby validating the original decision to stratify by race. It is more efficient in terms of sample size to sample from two homogeneous strata than from one very mixed, heterogeneous stratum. The knowledgeable worker must anticipate the validity of the stratifying criterion before the sampling begins.

Analogously, the effectiveness of cluster sampling will be enhanced by expertly composing and recomposing clusters before the sampling so that each cluster is as representative of the entire population as possible. Insofar as that effort can achieve success, given the heterogeneity of the grand universe itself, the respective clusters will tend to resemble one another, while there will be considerable statistical variation within each cluster. Thus, in cluster sampling we invert the specifications of stratified sampling: instead of homogenizing strata we diversify the elements within clusters. For example, we may combine precincts into larger geographic districts to increase the diversification within clusters and thereby raise their representativeness of the entire electorate and fulfill their function as samples.

In sum, no sampling technique is completely automatic; all involve subject-matter decisions. Hence, firsthand practical experience with the concrete subject matter contributes quite as much to fulfillment of a sampling project as does dexterity in the mechanical routines of applied statistics.

QUESTIONS AND PROBLEMS

1. Define the following terms:

probability sampling	simple random sample
availability sample	stratified sample
haphazard sample	cluster sample
judgment sample	interval sample
random sample	

2. Distinguish clearly between the concepts *cluster* and *stratum* in method of selection. What are the desirable characteristics of a cluster? a stratum?

3. (a) Since random sampling gives the "deviant" case the same opportunity to be selected as the "typical," what is its justification?
   (b) How do unrepresentative cases contribute to the representativeness of the sample?
   (c) Would a sample, carefully and randomly drawn, necessarily be representative?

4. If you had to select and interview a random sample of residents (adult) in the community, what procedural difficulties would you almost certainly encounter?

5. In a community survey, what would be the disadvantages of selecting a sample of families from only one neighborhood? Under what circumstances might such sampling procedure be acceptable?

6. Explain: "Stratified sampling is possible only when there is some previous information on the population."

7. How would you proceed to sample the following?
   (a) the students on a campus for occupation of fathers
   (b) households in a city for number of gainfully employed workers
   (c) a concert audience for socioeconomic membership
   (d) a theatre audience for reaction to play
   (e) library reference room readers for length of stay

8. Distinguish stratified random sampling and quota sampling.

9. A sample of all persons 65 and over is to be drawn in a given community. Suggest possible standard clusters that might be used to expedite such a plan.

10. "An area cluster sample could be reused for successive studies of a metropolitan community over a period of one year." Comment.

11. From a class of 25 pupils, how many distinct triads (groups of 3) can be formed?

12. (a) If a person were dealt a bridge hand (sample) of 13 black cards, would that prove the deck (population) consisted of all black cards?
(b) Does it prove the drawing (deal) was biased rather than random?

13. Consider a universe of 12 elements: A to L
(a) How many simple random samples of Size 4 can be formed?
(b) How many interval samples of Size 4?
(c) How many stratified samples of Size 4, when Stratum 1 consists of elements A–F, and Stratum 2 consists of elements G–L, and 2 elements are to be randomly selected in simple manner from each stratum?

14. (a) Consult the table of random digits in the Appendix (Table VII). Write a directive for selecting by random digits a simple random sample of 100 (n) items from a universe of 13,300 (N).
(b) In this instance, what is the sampling ratio or fraction?
(c) How many more cases would have to be added if the sampling ratio were fixed at 1/100?

SELECTED REFERENCES

Cochran, William G.
1963 Sampling Techniques. Second edition. New York: Wiley.

Lazerwitz, Bernard
1968 "Sampling theory and procedures." Pp. 278–328 in Hubert M. Blalock, Jr. and Ann B. Blalock (eds.), Methodology in Social Research. New York: McGraw-Hill.

Sharp, Harry and Allan Feldt
1959 "Some factors in a probability sample survey of a metropolitan community." American Sociological Review 24 (October): 650–661.

Parameter Estimation 13

1 CONFIDENCE INTERVALS

Sampling Error It should be unnecessary to repeat that a sample is useful only for the information it supplies on the characteristics of the universe. Thus, the sampling process finds its ultimate consummation in a description of the population from which the sample is drawn. But this description can be only an approximation: an average monthly expenditure of $152.75 in a random sample of college students will not correspond exactly to the average expenditure of the whole student body which the sample is designed to represent. Nor, obviously, will a second sample's mean of $143.62 necessarily correspond more closely to the unknown parameter. Therefore, owing to the vagaries of chance sampling, if no two samples are alike and all are in error, with how much confidence can we speak of the value of the universe? Clearly, we cannot merely project the value of the sample onto the universe, and let it go at that. This procedure of adopting the sample value in lieu of the parameter, or universe value, is hedged about with as many regulations as is the procedure of sampling itself. We call this set of prescribed procedures *statistical inference*.

Since, in the practical affairs of life, it is so rare that we are able to examine a whole universe, and since sampling is so frequently the resort, the procedures of statistical inference loom quite large in the repertory of statisticians. In fact, some actually would identify the science of statistics with the problem of sampling and decision-making—an extreme emphasis which seems unrealistic to the authors, since descriptive statistics have a validity and importance in their own right.

The problems of statistical inference begin with the legitimate assumption of a discrepancy between the variable sample estimate and the universe parameter, which is constant. When this discrepancy is the result of random sampling, we call it a *sampling error*. If the average monthly expenditure of all students is \$155 and the sample mean is \$152, the sampling error is \$3.00. In general, the sample rarely provides us with a perfect description of the universe, because few if any sample values are entirely free of sampling error.

Sampling Distribution Even though such errors cling to almost every sample statistic, there is no reason to be dismayed by such a natural consequence of random sampling. These errors, when taken in the mass, do not behave chaotically; on the contrary, they behave systematically. They exhibit a characteristic distribution, and therefore they are conveniently amenable to analysis. There is not only a definite zone surrounding the parent value beyond which the samples rarely stray, but in addition the variety of sample values distribute themselves in a recognizable pattern which permits statistical inferences to be drawn confidently. For example, we can be certain that the average age of a sample of high school students will not fall below 10, nor exceed 21; and we also can be reasonably certain that it will not fall short of 14 nor surpass 17. We can assert this because we are so thoroughly familiar with the ages of high school students. But whether or not we are intimately familiar with the universe, we still can be assured that the sample value will always bear some resemblance to the corresponding universe value. We can assert with confidence that the statistic will always adhere, to a greater or lesser extent, to its parameter.

To bring this very important fact home to us, we shall illustratively take a known universe, extract a large number of samples from it, calculate the mean for each, and then compare these means with the population mean. Then we will have demonstrated to what extent the sample means actually digress from, or converge on, the parameter itself. Of course, in an actual problem, we will not have knowledge of the universe mean, nor will we take a large number of samples. Instead we will be ignorant of the true mean and will take only one sample, which we will have to make do. But with the background afforded by this experiment, we will be fortified with knowledge of how samples in general behave, all of which will help us to infer how the particular sample, which we have drawn, is probably behaving.

For our working universe, we shall take the 1970 suicide rates of 229 Standard Metropolitan Statistical Areas given in Table 13.1.1. This set of rates constitutes the *sampled* universe. If the *target* universe were the rates for the same cities at another time (for example, 1980 or 1960) or the rates at the same time for a larger set of cities (for example, United States cities with 25,000 population and over) some judgment of changes over time or of differences in rates between cities of different sizes would be required to draw

Table 13.1.1 *Array of Suicide Rates for 229 United States Standard Metropolitan Statistical Areas, 1970*

2.7	7.3	8.7	9.6	10.7	11.6	12.7	14.3	16.9
3.3	7.4	8.8	9.7	10.7	11.7	12.8	14.3	17.0
3.8	7.4	8.8	9.7	10.7	11.7	12.8	14.5	17.2
4.6	7.4	8.8	9.7	10.8	11.7	12.8	14.6	17.5
5.0	7.4	8.9	9.8	10.9	11.8	12.8	14.7	17.8
5.2	7.5	8.9	9.8	10.9	11.8	12.8	14.9	17.9
5.2	7.5	8.9	9.8	11.0	11.8	12.8	15.1	18.3
5.5	7.6	9.0	9.9	11.0	11.9	12.9	15.1	18.4
6.0	7.6	9.0	9.9	11.1	11.9	12.9	15.2	18.6
6.3	7.7	9.1	9.9	11.2	12.0	13.0	15.2	18.7
6.3	7.7	9.1	9.9	11.2	12.0	13.0	15.4	19.0
6.4	7.7	9.2	10.0	11.2	12.0	13.1	15.5	19.4
6.5	7.8	9.2	10.0	11.3	12.0	13.2	15.6	20.0
6.5	7.9	9.2	10.0	11.3	12.1	13.2	16.0	20.1
6.6	7.9	9.3	10.0	11.3	12.2	13.2	16.0	20.6
6.6	8.0	9.3	10.1	11.4	12.2	13.5	16.1	20.9
6.7	8.2	9.4	10.2	11.4	12.2	13.6	16.1	21.0
6.7	8.4	9.4	10.3	11.5	12.3	13.6	16.1	21.8
6.7	8.4	9.4	10.3	11.5	12.3	13.7	16.1	22.0
6.9	8.5	9.4	10.4	11.5	12.4	14.0	16.2	22.1
6.9	8.5	9.4	10.5	11.6	12.4	14.0	16.3	22.5
7.1	8.5	9.5	10.5	11.6	12.5	14.0	16.4	22.5
7.2	8.6	9.5	10.5	11.6	12.7	14.0	16.5	24.8
7.2	8.6	9.5	10.6	11.6	12.7	14.1	16.7	24.9
7.3	8.7	9.6	10.6	11.6	12.7	14.1	16.9	25.0
7.3	8.7	9.6	10.7					

Sources: U.S. Bureau of the Census, *County and City Data Book, 1972.* Table 3. Washington, D.C.: U.S. Government Printing Office. 1973. U.S. Department of Health, Education and Welfare, *Vital Statistics of the United States, 1970,* Vol. II: Mortality, Part B. Washington, D.C.: U.S. Government Printing Office. 1974.

a conclusion about that target universe. The purpose of this illustration is to demonstrate the manner in which samples behave in relation to the parameters of the sampled universe.

From this universe we shall (1) draw 100 random samples—restoring each sample to the universe before drawing another—and (2) compute the mean of each sample. This experiment ought to be sufficient to reveal something about the regularity in the behavior of the seemingly wayward samples. These experimental means are arrayed in Table 13.1.2 and grouped in Table 13.1.3.

Inspecting this lot, we see that only 6 of the 100 samples coincide exactly with the universe mean, known to be 11.7. All others contain a sampling error; in fact, every one of them probably would have shown a sampling error if we had insisted on greater decimal accuracy. However, we can see that the universe mean is nested snugly in the middle of the whole array of sample means, and that most of the sample means are huddled compactly around the true mean of 11.7, which is practically at the 50 percent division.

Table 13.1.2 *Array of 100 Sample Means,* n = *30*

10.3	10.8	11.5	11.8	12.3
10.4	10.8	11.5	11.8	12.3
10.4	10.9	11.5	11.9	12.4
10.4	10.9	11.5	11.9	12.4
10.4	10.9	11.5	11.9	12.4
10.5	11.0	11.5	11.9	12.4
10.6	11.0	11.6	11.9	12.4
10.6	11.0	11.7	11.9	12.4
10.6	11.1	11.7	11.9	12.5
10.6	11.2	11.7	11.9	12.5
10.6	11.2	11.7	11.9	12.5
10.6	11.2	11.7	12.1	12.9
10.6	11.3	11.7	12.1	12.9
10.8	11.3	11.8	12.1	13.0
10.8	11.3	11.8	12.2	13.0
10.8	11.3	11.8	12.2	13.0
10.8	11.3	11.8	12.2	13.0
10.8	11.3	11.8	12.2	13.0
10.8	11.3	11.8	12.2	13.2
10.8	11.4	11.8	12.2	13.2

Table 13.1.3 *Frequency Tally, Empirical Sampling Distribution, 100 Sample Means,* n = *30*

Class Interval	Tally	Frequency (f)
10.0–10.9	卌 卌 卌 卌 卌	25
11.0–11.9	卌 卌 卌 卌 卌 卌 卌 卌 卌 I	46
12.0–12.9	卌 卌 卌 卌 II	22
13.0–13.9	卌 II	7
		100

Almost four-fifths of the 100 means fall between 10.7 and 12.7, or within one point of the true mean. Characteristically, the sample means gravitate toward the parent mean, or the center of the universe distribution.

It now becomes evident how improbable it is that a sample mean will deviate seriously from the universe mean. Furthermore, not only do sample means cluster around the true mean, but their pattern of distribution takes on a shape unmistakably approaching the normal curve (Figure 13.1.1). If we had drawn, processed, and tabulated all possible samples,[1] instead of

[1] $C^{N}_{n} = C^{229}_{30} = \dfrac{P^{229}_{30}}{30!}$

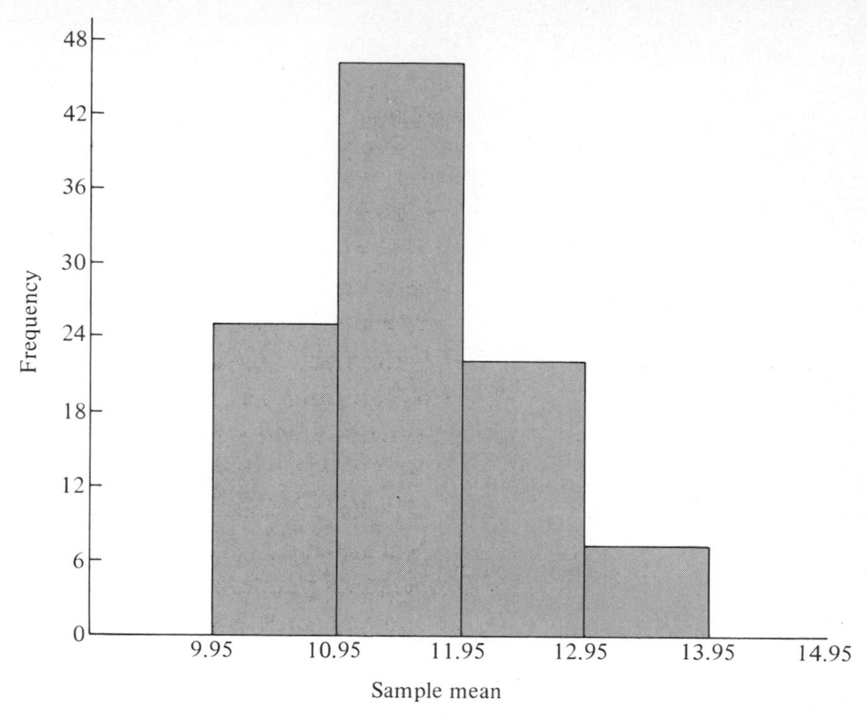

Figure 13.1.1 *Histogram of 100 Sample Means,* n = *30*

merely 100 of them, we would have a *pattern* of distribution which would display still greater conformity to the smooth normal curve. This hypothetical frequency curve of all possible sample means is labeled the *theoretical sampling distribution,* and has, as its mathematical model, the ideal normal curve. Any experimental distribution of a limited number of samples is called an *empirical sampling distribution.* Of course, the theoretical sampling distribution would be attained only by an infinite number of samples, since it would require that many to assure the distribution of sample means in their proper, expected proportions. However, the significant procedural point in this type of problem is that we usually can treat the theoretical sampling distribution of means as if it were a perfect, smooth, ideal normal curve.

Distribution of Sampling Errors By subtracting the universe mean from each of the 100 sample means, we obtain the 100 sampling errors (Table 13.1.4), which of course have the same curve pattern as the means themselves (Table 13.1.3). All we have done is move the zero origin to the universe mean (11.7).

The distribution of sampling errors (Table 13.1.5) again demonstrates that the small deviations are very numerous; the larger discrepancies are few in number. If we had drawn a single sample only, we could be practically

Table 13.1.4 *Array of Sampling Errors, 100 Samples,* n = *30*

−1.4	−.9	−.2	.1	.6
−1.3	−.9	−.2	.1	.6
−1.3	−.8	−.2	.2	.7
−1.3	−.8	−.2	.2	.7
−1.3	−.8	−.2	.2	.7
−1.2	−.7	−.2	.2	.7
−1.1	−.7	−.1	.2	.7
−1.1	−.7	0	.2	.7
−1.1	−.6	0	.2	.8
−1.1	−.5	0	.2	.8
−1.1	−.5	0	.2	.8
−1.1	−.5	0	.4	1.2
−1.1	−.4	0	.4	1.2
−.9	−.4	.1	.4	1.3
−.9	−.4	.1	.5	1.3
−.9	−.4	.1	.5	1.3
−.9	−.4	.1	.5	1.3
−.9	−.4	.1	.5	1.3
−.9	−.4	.1	.5	1.5
−.9	−.3	.1	.5	1.5

Table 13.1.5 *Frequency Distribution of Sampling Errors*

Class Interval	f
−1.7 to −1.1	13
−1.0 to −0.4	26
−0.3 to 0.3	32
0.4 to 1.0	20
1.1 to 1.7	9
	100

certain that our sample mean would not have missed the true mean by more than two points, plus or minus; indeed, 78 percent of the samples are in error by less than plus or minus one.

The presence of sampling errors obviously complicates the task of drawing an inference about the parameter; but the neat pattern of these errors renders the uncertainties of such inference less formidable. Our dissection of the empirical sampling distribution has given us considerable reassurance of the reasonableness of the behavior of samples, which will permit us to estimate the precision of our sample statistic.

Parameter Estimates with Allowance for Sampling Error We now take up the technique which will make appropriate allowance for the ever present sampling error, and thereby enable us to make reliable estimates of the parameter. For, unless we can rely with a certain degree of confidence on our inferences, there is no point in making them at all. The entire purpose of sampling is to substitute a sample of given reliability for the prohibitively expensive enumeration of the unknown universe.

But we cannot measure the sampling error of our single sample directly since we are never given the true mean from which to compute it. That would suppose that we already have what we have set out to find in the first place. However, the single sample does provide two items of information: (1) a single value that serves as an estimate of the parameter (*point estimate*), and (2) an estimate of the error to which such point estimates are subject (*standard error*). On the basis of these two items of information we may derive a third: (3) a range of values carrying a specified probability of including the parameter, or a *confidence interval.*

The procedures for estimating the magnitude of sampling error are numerous, since they differ according to the sample statistic (for example, whether mean or percentage) and according to the makeup of the sample (whether large or small, simple or stratified). Here we present only two of the simplest techniques for estimating sampling error and for creating interval estimates for parameters; these apply to means and percentages of *large random samples*, respectively. They will suffice for our purposes, since other techniques produce results which in principle carry the same interpretation as the method here unfolded.

Interval Estimate of the Mean Let us suppose that, in taking a single sample ($n = 30$) from the 229 suicide rates, we obtain a mean of 12.5. In such a realistic situation, we do not know the value of the true mean and will never know it. At this stage of the inquiry, 12.5 is the only estimate we have of the true mean, and therefore it is our best estimate. We call this a *point estimate*. No one would claim that this estimate is any more than approximately correct. The weakness of the point estimate is that it provides no indication of the sampling error, and consequently we do not know how much confidence to accord it. This disadvantage is overcome by setting up an interval which probably contains the true mean, and which we therefore label an *interval estimate*. To construct such an interval we (1) estimate the range within which a high proportion of sampling errors lie, and (2) add and subtract this range of error to and from our observed sample mean to form the interval which probably encloses the true mean.

This attack on the problem is analogous to the following homely situation. Encountering a man on the street, we may estimate his age to be 35 (point estimate). Upon being challenged, we reply: "Of course, this is only an estimate based on cues I recognize. However, from samples of men I have known, I feel reasonably confident that he is somewhere between 30 and 40 (interval estimate); and I feel more confident that he is between 25 and 45; and I'm absolutely certain that his age will be between 20 and 50." These intervals enjoy varying degrees of confidence, and for that reason are more realistically known as *confidence intervals.* The larger the interval, the more confidence we have that it includes the true age. Of course, the age of the gentleman is fixed. It is what it is, even though we are not completely informed. But as we shift the size of the interval, contracting and expanding it around the point estimate, we revise the degree of our confidence that the interval will enclose his true age. To be sure, the initial estimate of 35 years may have been brilliantly correct, but we shall never have the satisfaction of knowing it unless we ask him directly—which, under our assumption, we cannot do. So it is with sampling.

We already know how to draw a random sample; in addition, all we need to learn is how to construct the aforementioned confidence interval, the first step of which is to estimate the standard deviation of the sampling distribution of sample means. This standard deviation is called the *standard error.*

The Logic of the Standard Error of the Mean Our demonstration has revealed that the theoretical sampling distribution of means may be viewed as a normal curve (Figure 13.1.1), with the true mean at the center, as the grand mean of means. Statisticians express this truth more formally: The sampling distribution of the mean is approximately normal around the true mean, provided $n > 30$. Now, we do not know whether the mean of the sample which we happened to draw lies near the true mean, a sigma removed from it, or at any point on the base line; perhaps it is even three sigmas removed. But, if we have the value of the standard deviation of the sampling distribution, we could set up an interval for the true mean; hence, we must obtain an estimate of this value. We label this value a *standard error* (*SE*), since the deviations of all sample means from the true mean are actually sampling errors. A standard error is the standard deviation of a sampling distribution.

Our sample alone must furnish the information on which to base an estimate of the standard error, since it is all we have. Hence, we must explore the sample to uncover some clues to the magnitude of the sought-after standard error. Initially, we might guess that the size of the sample (n) and the degree of dispersion in the sample (*SD*) are related to the size of the standard error, according to the following reasoning.

First, as to sample size, we should expect larger samples to be more representative of the universe than smaller samples, and therefore we would expect larger samples to yield smaller sampling errors. In fact, when a sample contains every last item in a finite universe (a 100 percent sample), there can be no sampling error at all. Then every sample mean necessarily would be identical with the universe mean, and the standard error would be zero. At the other extreme, the smallest possible sample would contain only one case ($n = 1$). Then the distribution of sample means would be identical with that of individual universe values and would display the same amount of variation as the items in the universe itself. Hence, in that instance, the standard error would be equal to the standard deviation of the universe (σ). From this brief analysis, we may conclude that the standard error of the mean (1) will always be smaller than the standard deviation of the universe, since n will always be larger than 1; and (2) will decrease as sample size (n) increases, reaching zero when $n = 100$ percent of the universe.

Second, we intuitively expect the degree of scatter in the universe, as reflected by the degree of scatter in the sample, to affect the size of the sampling errors. Thus, if all values in the universe were alike, all sample means also would be alike, and there would be no sampling error whatever. The standard error of such a completely homogeneous universe would always be zero! Without sampling error, a sample of one girl would give us the average number of arms for all girls, since every girl has the same number of arms—a perfectly homogeneous universe; however, a sample of one girl would not give us the average height of all girls without error. There is a great variety of heights, which produces comparable variety within samples, and consequently variety among the sample means as well. Inverting this argument, we conjecture that a high degree of scatter in the sample is indicative of heterogeneity in the universe and therefore is also indicative of large sampling errors. Thus, we anticipate that the standard error for a given n will be larger when sampling a very heterogeneous mass of data than when sampling a relatively homogeneous mass.

Computation of the Standard Error Both of the foregoing insights are in accord with statistical theory and actually are explicit in the formula for the standard error of the mean, which contains n and σ as its basic terms. Without further proof, we give the formula:

$$\sigma_{\bar{X}} = \frac{\sigma}{\sqrt{n}} \tag{13.1.1}$$

where $\sigma_{\bar{X}}$ = standard error of the mean,

σ = standard deviation of the universe, and

n = sample size.

Of course, we do not have the standard deviation of the universe, as required by the formula; all we have is the standard deviation, s, of the sample.[1] We will therefore substitute it for the unknown parameter in the above formula:

$$s_{\bar{X}} = \frac{s}{\sqrt{n}}. \tag{13.1.2}$$

[1] Conventionally, Greek letters are used to represent *parameters* (for example, σ, μ); Latin letters (for example, s, $\bar{X}$) represent *sample statistics*, which are necessarily estimates. But this practice is not uniformly adhered to by writers in the field.

But the value of the substituted standard deviation of the sample is not identical with that of the universe, owing to sampling error; in fact, the former tends to be smaller than the latter because the scatter of values within the sample tends to be smaller than that in the universe. To match this understatement in the numerator (that is, to maintain the ratio in the fraction), we correspondingly reduce sample size (n) by one in the denominator:

$$s = \sqrt{\frac{\sum x^2}{n-1}}. \tag{13.1.3}$$

This formula may fittingly be substituted for the population sigma. The correction becomes trivial, of course, if sample size is very large. Nevertheless, it is conventionally included, as a matter of principle, even in large samples. Therefore, the working formula, with familiar symbols, will read as follows:

$$s_{\bar{X}} = \frac{\sqrt{\dfrac{\sum x^2}{n-1}}}{\sqrt{n}}$$

$$= \sqrt{\frac{\sum x^2}{n(n-1)}}. \tag{13.1.4}$$

This expression, then, yields the standard error of the mean which we have sought and from which we shall construct the interval estimate.

Construction of the Confidence Interval (Large Sample) Since the sampling distribution of the mean is approximately normal around the population mean, it follows that two out of three sample means (68.27 percent) will lie within one standard error of the true mean. Therefore, our particular sample mean will have 2 chances in 3 of falling within one standard error of the true mean. Hence, by adding and subtracting one standard error to and from our randomly selected sample mean, we have approximately 2 chances in 3 of enclosing the true mean. Similarly, 95 out of 100 sample means lie within 1.96 standard errors of the true mean; in consequence, if we attach 1.96 standard errors to either side of the sample mean, we have a 95 percent probability of enclosing the true mean. In general, it is possible to obtain any desired confidence interval by the simple technique of attaching the requisite multiple of the standard error to the observed sample mean. By thus making our interval estimate wide enough, we may be practically certain that the true mean has been contained within the interval, even though the observed sample mean is highly inaccurate.

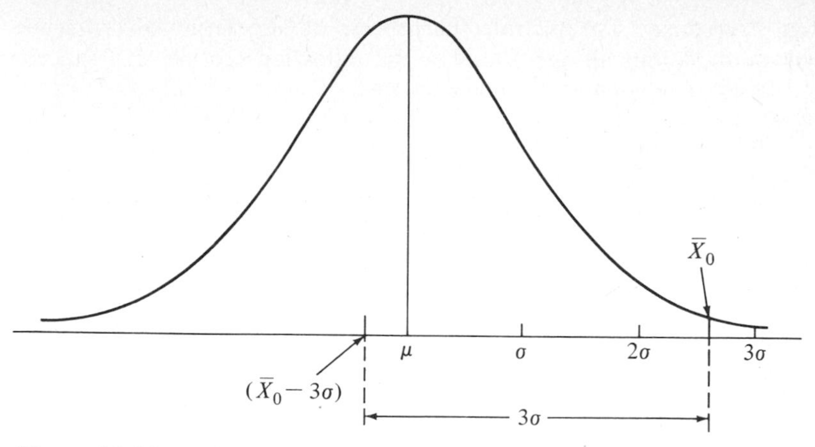

Figure 13.1.2 *Theoretical Sampling Distribution of Means, Observed Sample Mean,* $\bar{X}_0$

Let us suppose that we have drawn a "bad" sample, whose mean, $\bar{X}_0$, is located in the tail of the sampling distribution, far removed from the universe mean (μ) which it intends to represent (Figure 13.1.2). Clearly, in this instance, it would not have been sufficient to attach only 1 standard error to either side of the observed mean in order to catch the true mean; nor would it have been enough even to attach 2 standard errors, since the true mean is more than 2 standard errors distant from our illustrative sample mean. However, if we affix 3 standard errors to either side of the observed mean, we obtain an interval whose lower limit extends substantially beyond the true mean. Since practically all (99.74 percent) of the sample means lie within 3 standard errors of the universe mean, by adding and subtracting 3 standard errors to and from any randomly selected mean we become virtually certain that the resultant interval will trap, as it were, the target mean. Even the most pessimistic soul, who always fears the worst, may give himself a sense of relative security by attaching 3 standard errors to his observed sample mean.

Note that this method does not disclose the exact whereabouts of the true mean, as has been previously cautioned; it merely furnishes a stronger or weaker expectation that the true mean will be found within the specified interval estimate, derived from the observed sample. Wherever the true mean is, it is stationary. Since it is the interval estimate that either succeeds or fails in enclosing the parameter on a given sampling trial, we may properly speak of the probability of such an interval succeeding in encompassing the universe mean. Of course, once the sample has been drawn and the interval formed—once the trial is over—either it does or does not include the target mean between its calculated limits, although we will never know. Our confidence rating in that interval corresponds to its pretrial probability, which has been selected for our convenience and purpose.

Working Procedure To illustrate the process of calculating a confidence interval, we carry out all operations on the following sample of 30 suicide rates:

6.0	18.6	4.6	11.2	19.0
24.9	9.1	14.7	11.6	22.0
3.8	18.3	8.2	21.0	10.0
5.5	7.5	17.5	8.6	8.4
22.1	11.6	14.9	13.6	6.3
8.4	10.0	10.4	9.6	15.4

(1) We compute the sample mean:

$$\bar{X} = \frac{372.8}{30}$$

$$= 12.43.$$

(2) We compute the standard error of the sample mean:

$$s_{\bar{X}} = \sqrt{\frac{965.88}{(30)(29)}}$$

$$= \sqrt{1.11}$$

$$= 1.05.$$

(3) We multiply $s_{\bar{X}}$ by the number of sigma units to be added and subtracted for the agreed upon confidence interval—for example, 1.96 for the 95 percent interval:

$$1.96\ s_{\bar{X}} = 1.96(1.05)$$

$$= 2.06.$$

(4) We add and subtract 1.96 $s_{\bar{X}}$ to and from the observed sample mean, $\bar{X}$. Thus:

$$\bar{X} \pm 1.96\ s_{\bar{X}}$$

$$12.43 + 2.06 = 14.49;$$

$$12.43 - 2.06 = 10.37.$$

(5) Finally, we make the interpretation that the chances are 95 in 100 that the true mean will be found within the interval 10.37 to 14.49. We are reasonably sure that our interval contains the true mean, since only 5 intervals in 100 constructed in the same manner will fail to do so. However, if it doesn't, something has happened which would occur only 5 times in 100, a risk which we may be willing to face.

But how decide on the width of the confidence interval? Here one can offer only the most general guidance. A consideration of the consequences of the decision, the risk one will wish to run, and even the temperament of the person concerned will enter into the choice. In effect, a decision like this is comparable to any other of the numerous decisions we make in the face of life's uncertainties.

Demonstration of Experimental Confidence Intervals All statements of probability are based on the assumption of an infinite number of trials. While it is not possible to conduct such an infinity of trials, it is usually possible to conduct a fairly large number of experimental trials, the outcomes of which then serve to test the initial probability statement. To that end, we have calculated the 95 percent confidence limits for each of our 100 samples ($n = 30$) of suicide rates, to determine whether the resulting intervals actually would enclose the true mean approximately 95 percent of the time. Figure 13.1.3 portrays the locations of the 100 confidence intervals,

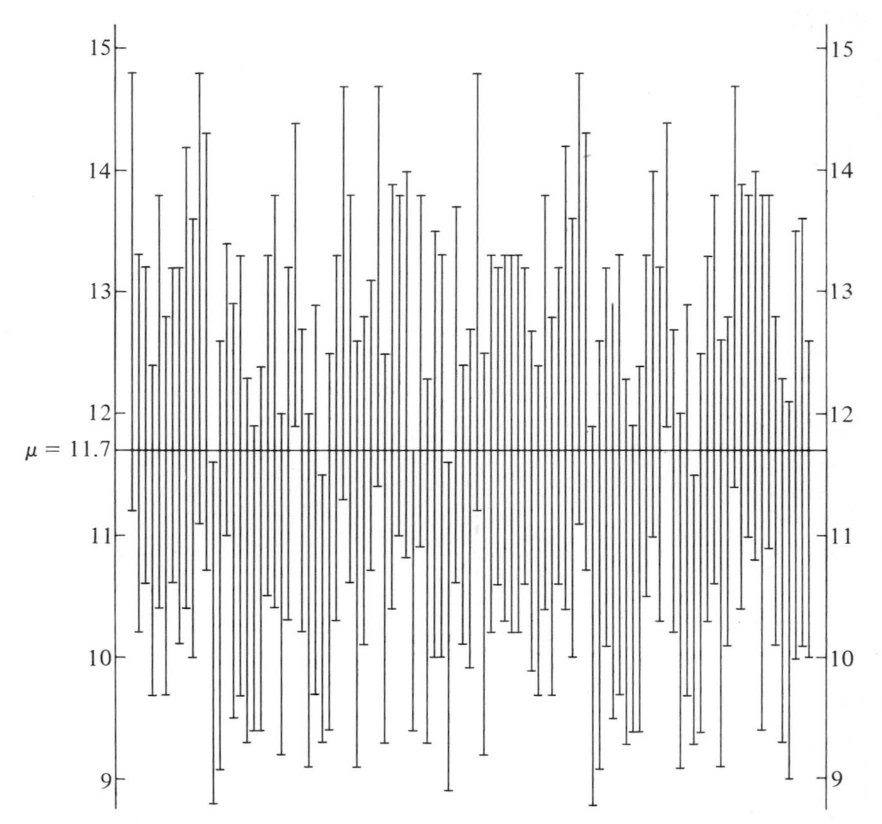

Figure 13.1.3 *100 Internal Estimates, 95 Percent Confidence, n = 30; Universe: Suicide Rates, 229 United States Standard Metropolitan Statistical Areas, 1970*

along with the known population mean of 11.7. It will be observed that exactly 95 out of 100 interval estimates do enclose the true mean, a result which, surprisingly enough, is identical with theoretical expectation. Here is tangible evidence that the confidence we place in a properly constructed interval estimate is amply justified.

Convenience of Large Samples Up to now, all probability statements have been based on the assumption that the sampling distribution of the mean is normal. It is reassuring that such a sampling distribution will always be normal when the universe itself is normal. A normal population produces normality in this sampling distribution, whatever the size of the sample.

However, it is particularly characteristic of sociological data that populations often are not normal. Size of families, income, and other social variables distinguish themselves from typical psychological data in that they are often severely skewed. Therefore, we may raise the question of whether such skewness disturbs the required normality of the sampling distribution. The answer is that it does. But at the same time, the effect of such skewness can be circumvented by appropriately increasing the size of the sample beyond the approximate minimum of 30. As sample size increases, the sampling distribution approaches normality. However, no general statement can be made about the rate at which the sampling distribution approaches normality, since this rate will vary according to the severity of the skew to be offset.

Table 13.1.6 *Population Distribution, 153 American Cities, 100,000 and Over, by Size of Population, 1970*

Size of Cities (in Thousands)	Frequency (f)	Percent
100–199	89	58.2
200–299	16	10.5
300–399	16	10.5
400–499	6	3.9
500–599	8	5.2
600–699	5	3.3
700–799	5	3.3
800–899	1	.7
900–999	1	.7
1,000 or more	6	3.9
Total	153	100.2

Source: U.S. Bureau of the Census, Census of Population, 1970. *Number of Inhabitants.* United States Summary PC(1)–A1. Table 28. Washington, D.C.: U.S. Government Printing Office. 1971.

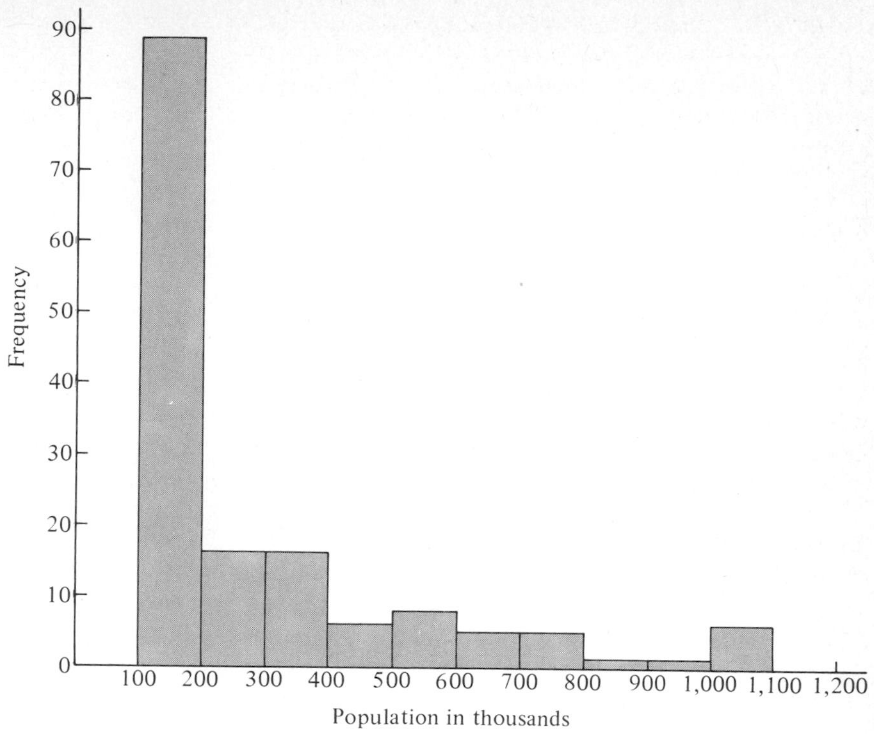

Figure 13.1.4 *Histogram, 153 American Cities by Size of Population*

But we may gain some insight into what might appear to be a rather remarkable phenomenon by examining the following experiment, in which we drew samples of size 20 and samples of size 40 from a compilation of 153 large American cities for size of population. The original distribution (Table 13.1.6) was highly skewed and consequently ideally suited for experimentally testing the effect of a skewed population on the sampling distribution of the mean. Figure 13.1.4 graphically portrays this highly skewed distribution.

The first set of 200 samples ($n = 20$) was drawn and the mean of each sample calculated and tabulated (Table 13.1.7). This experimental sampling distribution is still perceptibly skewed to the right, as is apparent in Figure 13.1.5, but not nearly to the same degree as the parent universe. This reduction in skewness is a result of the fundamental principle that the means are great levelers which necessarily cut down the individual extremes found in the universe. Therefore, from the very nature of the mean, it follows that a distribution of sample means never can be as irregular or as widely dispersed as are the individual values which make up the samples. Furthermore, the larger the sample, the more successfully will the dispersion be reduced. Thus, when sampling from a markedly skewed population, samples even as small as 20 will produce a sampling distribution displaying evidence of a strain toward normality, although obviously falling far short of that goal.

Table 13.1.7 *Distribution of 200 Sample Means,* n = *20*

Mean (in Thousands)	Frequency (f)	Percent
100–199	5	2.5
200–299	75	37.5
300–399	62	31.0
400–499	25	12.5
500–599	11	5.5
600–699	15	7.5
700–799	4	2.0
800–899	3	1.5
Total	200	100.0

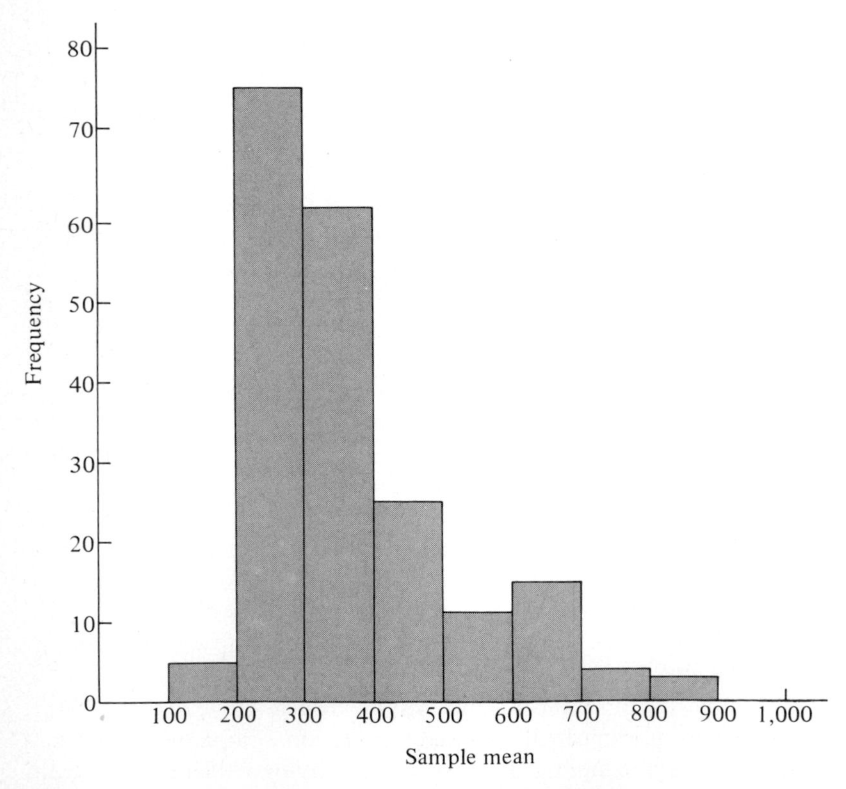

Figure 13.1.5 *Histogram, Distribution of 200 Sample Means,* n = *20*

Table 13.1.8 Distribution of 100 Sample Means, n = *40*

Mean (in Thousands)	Frequency (f)	Percent
100–199	0	0.0
200–299	23	23.0
300–399	41	41.0
400–499	19	19.0
500–599	15	15.0
600–699	2	2.0
700–799	0	0.0
800–899	0	0.0
Total	100	100.0

Therefore, samples of 40 should continue the trend toward normality, or at least toward symmetry. Such a distribution is tabulated in Table 13.1.8 and illustrated in Figure 13.1.6. Gone now is the long tail; instead, the beginnings of symmetry are clearly visible. Thus, sociologists, who are often called upon to deal with skewed populations, may draw considerable comfort from the important implications of this little experiment and may feel justified in applying normal probability calculations to their data by taking proper precautions.

The convenience of large samples consists in the assurance that the sampling distribution of the mean is approximately normal, an assumption which may be in doubt when samples are small. However, situations often do arise in which small samples are not only convenient, but may even be unavoidable. In such instances, modification of the foregoing procedures is required, although no new principles are involved.

Confidence Interval for a Percentage (Large Sample) Like the mean, the sample percentage also raises the issue of reliability, since sample percentages are equally subject to sampling errors. Thus, a given sample percentage (for example, 60 percent favoring military training) may be off a trifle or may deviate considerably from the true percentage. Similarly, 30 percent of a random selection of married women may be gainfully employed, but the percentage in the sampled universe may conceivably be as low as 20 or as high as 40. Fifty-one percent of the sample electorate may signify its intention of voting for the Republican nominee in the forthcoming election, but unless we can assess the accuracy of that estimate in some way, we cannot forecast the outcome of the election with any degree of confidence. Hence, in interpreting a sample percentage, again we must estimate the standard error to provide a confidence interval.

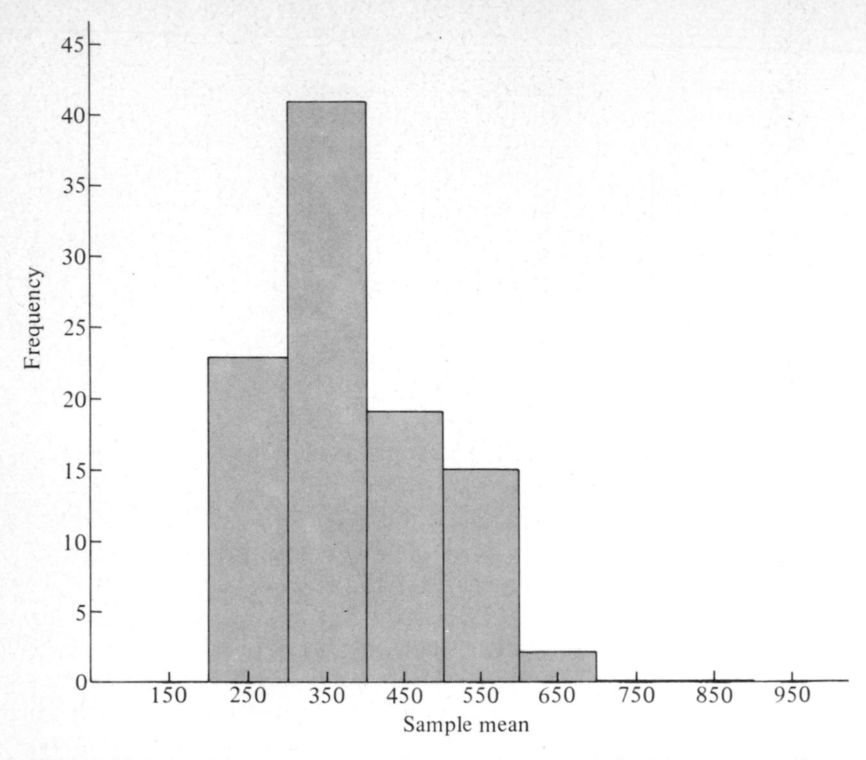

Figure 13.1.6. *Histogram, Distribution of 100 Sample Means,* n = *40.*

Essentially, the technique of constructing an interval estimate of a universe percentage is no different from that of the mean: we (1) fix the desired degree of confidence: (2) estimate the standard error; and (3) attach to the observed sample percentage the multiple of the standard error needed to attain the agreed upon confidence. The sampling theory on which this procedure rests is identical with that underlying the treatment of large sample means; namely, the distribution of all possible percentages (p) is approximately normal around the universe percentage, P, with standard error:

$$\sigma_p = \sqrt{\frac{PQ}{n}}, \tag{13.1.5}$$

where σ_p = standard error of percentage,

P = universe percentage, and

$Q = 100 - P$.

But since the parameters P and Q are unknown, of course, again we follow the convention of substituting the sample values p and q to obtain an estimate of the standard error:[1]

$$s_p = \sqrt{\frac{pq}{n}}. \qquad (13.1.6)$$

Computing an Interval Estimate of a Percentage Let us suppose that out of 900 randomly selected college students, 60 percent respond in favor of legalizing marijuana and 40 percent are unfavorable. To form a confidence interval, we must first estimate the standard error:

$$s_p = \sqrt{\frac{(60)(40)}{900}}$$
$$= \sqrt{2.667}$$
$$= 1.63.$$

By adding and subtracting this quantity to and from the sample percentage, we obtain the 68 percent confidence interval:

$$60\% \pm 1.63 \text{ percentage points}$$
$$58.37\% - 61.63\%.$$

The odds are 2 to 1 that the true percentage lies within this interval. If we should desire greater confidence—say, 95 percent—we would attach 1.96 standard errors:

$$60 \pm 1.96(1.63)$$
$$60 \pm 3.19$$
$$56.81\% - 63.19\%.$$

Since intervals constructed in this manner contain the true percentage 95 times out of 100, we may declare that we are confident at the 95 percent level that the true percentage is not less than 56.81, nor greater than 63.19. Since we assume that 5 percent will fail to include the true percentage, we thus take a 5 percent risk of being wrong.

The foregoing procedure is another instance of large sample technique in that it applies to samples having 30 or more cases. But the assumption of a normal sampling distribution hinges on the ratio, $P : Q$, as well as on sample

[1] As in the case of the mean, a better estimate of the standard error of p is provided by

$$\sqrt{\frac{pq}{n-1}}.$$

For a discussion of this point, see William G. Cochran, *Sampling Techniques*, New York: Wiley, 1953, 32–33.

Table 13.1.9 *Minimum* n *for Selected* p-*Values for Normal Approximation*

p	n
.5	30
.4	50
.3	80
.2	200
.1	600
.05	1,400

Adapted with permission from William G. Cochran, *Sampling Techniques*, p. 41. Copyright 1953. John Wiley & Sons, Inc. Reprinted with permission.

size. As the $P:Q$ balance departs more and more from 50 : 50, the sampling distribution becomes increasingly skewed.

As has already been pointed out in the study of the mean, in the case of percentages, the skew in the $P:Q$ balance communicates itself to the sampling distribution. Again, as with the mean, this skew can be circumvented only by a sharp increase in the size of n, so as to instill normality in the sampling distribution. Table 13.1.9 shows the sample sizes needed for varying sample percentages to permit the assumption of a normal sampling distribution.

Confidence Intervals for Other Parameters The general logic of interval estimates that take sampling error into account, illustrated above in confidence intervals for a mean and proportion, may be utilized to obtain confidence intervals for other parameters as well. Although the detailed computational procedures are not described in this book, the student should be aware of the possibility of constructing confidence intervals for the population variance, for various measures of association, and for the difference between two means or proportions. Like sample estimates of means and proportions, sample estimates of these parameters will vary from sample to sample, but nevertheless will cluster around their respective universe values in a patterned distribution. An estimate of the standard error may be made on the basis of sample information and this estimate may then be utilized in making an interval estimate of the parameter at any desired level of confidence. For some parameters, it should be noted, the procedures are more complicated than those discussed above for means and proportions. For example, the sampling distribution of the correlation coefficient is generally not a normal distribution, so usually a transformation must be made in order to utilize normal probability distributions in constructing a confidence interval for the coefficient of correlation. However, the general logic of constructing confidence intervals remains the same.

The Finite Population Multiplier In actual sociological investigations, we ordinarily encounter finite populations; yet in the foregoing instances we have calculated the standard error on the assumption of sampling from an infinite universe. But this is not quite so unrealistic as might appear on its face. Finite populations usually are large enough to be considered infinite, which is a practical convenience, since reliability procedures are somewhat less complex for infinite than for finite populations. For finite populations, the measurement of reliability must take into account both sample size (n) and the size of the universe (N), whereas for infinite populations, the computation of reliability must take into account only sample size, since the size of the universe is incalculable and hence cannot vary.

This principle is in agreement with common sense: a sizable fraction of a finite universe should be more representative than a negligible fraction thereof. A 100 percent sample would be most dependable, since in such cases there can be no sampling error at all. The standard error would be zero. Analogously, we expect a 50 percent sample to be more reliable than a 25 percent sample, which in turn we expect to yield more confidence than a 5 percent sample. In general, relatively large samples are more representative than small ones. The measurement of this improvement in reliability is effected by incorporating the *sampling ratio*, n/N, in the formula for the standard error. Thus, the standard error of the mean, on the assumption of a finite population, is:

$$\sigma_{\bar{X}} = \frac{\sigma}{\sqrt{n}}\sqrt{1 - \frac{n}{N}}. \tag{13.1.7}$$

The term, $1 - n/N$, or the proportion of the population not in the sample, is frequently referred to as the *finite population multiplier*,[1] because it measures the improvement in reliability attributable to the finite nature of the universe.

It is clear that the quantity, $1 - n/N$, will always be less than 1.00; hence, it will always produce a shrinkage in the standard error below that for the infinite population. When the sampling ratio is large, it will even lead to a substantial reduction in the size of the standard error. Thus, with a 25

[1] More commonly referred to as the *finite population corrector*. However, this term is somewhat misleading. The factor, $1 - n/N$, does not serve to "correct" a faulty result, rather it is an integral component in the standard error formula for finite populations. When the sampling ratio is small, this component may be sloughed off, whereupon the standard errors for finite and infinite populations become identical. If the standard error had been given for finite populations originally, then it would have been necessary to "correct" for the infinite population!

percent sample, such as that selected by the United States Bureau of the Census in 1960, employment of this term would lead to a reduction in the standard error as follows:

$$fpm = \sqrt{1 - .25}$$

$$= \sqrt{.75}$$

$$= .87.$$

Eighty-seven percent of the initial *SE* constitutes a 13 percent shrinkage. Hence, here it would be imperative to employ the multiplier; otherwise we would fail to do justice to the accuracy that actually resided in the sample data. On the other hand, when the sampling ratio is small, say less than 5 percent, the finite multiplier has little or no effect on the magnitude of the standard error and may be disregarded. Thus, in a sample of 100–200 cases it makes little difference whether the universe contains 10,000, or 1,000,000, or even 180,000,000 units. The fact that several hundred sample observations may be equally reliable, regardless of the size of the universe, has given birth to the paradox that sample reliability is affected only by the absolute number of cases and that the sampling ratio is of no consequence.

Since the sampling ratio is often relatively small, as in public opinion surveys, the finite population multiplier is correspondingly ignored, which explains its textual omission in some of the briefer discussions of sampling. However, it should be noted that whenever the universe is extremely small, as in a small community, the sampling ratio necessarily must be large; otherwise the number of cases in the sample would not be adequate to represent the sampled universe. Thus, the United States Bureau of the Census has seen fit to draw a 25 percent national sample to ensure adequate coverage in the very small enumeration areas. Of course, a much lower sampling ratio would be sufficient for national, state, and metropolitan coverage, but it would produce an insufficient number of sampling units from the numerous tiny subdistricts.

Even when the sampling ratio is large, other considerations may suggest that the finite population multiplier be ignored. Random errors of measurement as well as the random selection of sample cases lead to variation in sample estimates. Such random errors of measurement contribute less to the variation of sample means for large samples than for small samples, since positive and negative errors more nearly cancel each other in large samples. But a 100 percent sample will yield varied estimates of a parameter if the measure for each case is subject to random error of observation, even though the application of the finite population multiplier would suggest no variation with a 100 percent sample. Hence, the application of the finite population multiplier may be misleading when both random measurement errors and the sampling ratio are large.

QUESTIONS AND PROBLEMS

1. Define the following terms:
   - statistical inference
   - estimation
   - parameter
   - statistic
   - sampling error
   - sampling distribution
   - confidence interval
   - standard error of the mean
   - standard error of the percentage
   - interval estimate
   - point estimate
   - large sample
   - finite population multiplier

2. Using the table of random digits (Table VII, Appendix), draw two samples of 30 cases each from the list of 229 suicide rates (Table 13.1.1).
   (a) Compute the mean and the standard deviation of each sample.
   (b) Compute the sampling error of each sample mean on the basis of the known true mean.
   (c) As a class project, arrange the means from (a) above in a frequency table. Compare this distribution with the distribution of 201 rates. Comment on the difference.

3. (a) Estimate the standard error of the mean from each sample of 30 cases from the preceding exercise.
   (b) Explain why the estimated standard errors are not identical in value.
   (c) Establish 95 percent confidence intervals by adding to and subtracting from each corresponding sample mean 1.96 standard errors.
   (d) As a class project, determine what percentage of the 95 percent confidence intervals contain the true mean.

4. (a) For a given population, determine how sample size would have to be adjusted to reduce the standard error of the mean to:
   one-half its original size
   one-fourth its original size
   one-sixteenth its original size
   (Hint: Substitute in *SE* formula and solve.)
   (b) Explain how these results illustrate the statement that the reliability of the mean varies directly with sample size.

5. Explain the statement that the sampling distribution of the mean is the equivalent of the distribution of sampling errors.

6. A given sample has the following characteristics:

   $\bar{X} = 11$

   $s = 3$

   $n = 100.$

   If the population mean is known to be 12, what is the sampling error of the observed mean? What is the estimated standard error of the mean?

How often would you expect sampling error larger than $\pm.5$? (Hint: Find the required z-measure and refer to the table of normal areas.)

7. Explain in your own words why the sampling distribution of the percentage will be lacking in symmetry when n is small and $P:Q$ very unbalanced—say, 90 : 10. (Suggestion: Draw an appropriate graph with the percentage scale on the base line.)

8. Suppose that a complete set of attitude scores has a mean of 1,000 and a standard deviation of 200.
(a) If 25 scores are picked at random, what is the probability that their average score will be less than 950?
(b) Greater than 1,100?
(c) Between 900 and 1,100?
(Hint: Obtain the SE and z, and consult the table of normal areas.)

9. Discuss the statement: The wider the interval estimate, the greater the degree of reliability; the greater the degree of precision, the less reliable the interval estimate.

10. If available information suggests that the standard deviation of the distribution of ages in a population is approximately seven years, how large a sample will be required so we may be confident at the 95 percent level that the sample mean will miss the population mean by no more than one year?

11. Table 13.10 gives the proportion of survey respondents who reported to the police selected types of criminal activity of which the respondent was a victim. Before making any computations, indicate which of the proportions shown will have the smallest estimated standard error. The largest estimated standard error. Compute the estimated standard errors for each percentage and check your preliminary guesses.

Table 13.1.10 *Extent of Police Notification by the Victims of Selected Types of Crime*

Crime	Percent of Incidents in Which Police Are Notified	Number of Incidents
Robbery	65	31
Burglary	58	313
Larceny (over $50)	60	198
Larceny (under $50)	37	473
Vehicle theft	89	65
Fraud	26	82

Source: Philip H. Ennis, *Criminal Victimization in the United States: A Report of a National Survey* (A Report of a Research Study Submitted to the President's Commission on Law Enforcement and Administration of Justice). Table 22, page 42. Washington, D.C.: U.S. Government Printing Office, 1967.

2 DETERMINING APPROPRIATE SAMPLE SIZE

In any sample study, the inevitable question that arises early is the size of sample. Taking a sample in preference to the entire universe is an economy, of course. But there is no economy in a sample that is larger than necessary; and it is pound-foolish to be content with a sample that is too small to yield the requisite accuracy. The importance of this problem of sample size is much greater than one would deduce from the meager discussions available in elementary texts. And yet the question is bound to arise even in the simplest sampling study. The techniques appropriate to answering it are matters of fairly advanced statistics. However, there are some broad guidelines that do not require such advanced knowledge.

To the novice, it usually seems incredible how small a sample may produce dependable results. Literally, a few thousand interviews have successfully predicted the outcome of United States national elections. How is that observation to be reconciled with our common image of the diverse and unstable social world?

The most immediate explanation of this apparent incongruity is that even the social world may be viewed at times in simple dimensions. A political poll in a system comprising only two parties, and in which the voters are sampled according to well-established criteria, might very well yield highly reliable results—when there is no unforeseen intervention of novel circumstances. As a touchstone to our statistical judgment of sample size, we should visualize the supremely uncomplicated situation of a perfectly homogeneous universe which, of course, could be sampled by a single case. Such a sample cannot be spoiled by even the most arbitrary selection. Events of interest are rarely so homogeneous, however, and as the variability in the sampled universe increases, sample size must increase to maintain the same level of precision in parameter estimation.

Consequently, the question cannot be so simply put as: "What size sample?" In such an oversimplified form, the question can only be answered evasively: "It all depends." Upon what does it depend? It depends on the resources available, how the sample is to be drawn and analyzed, the anticipated loss of cases for final analysis, the characteristics of the population being sampled, and the precision required in the results. Our above discussion of confidence intervals has laid the foundation for an appreciation of how sample size affects precision and how different sample sizes may be required to reach the same level of precision for different populations. These factors, along with other considerations in determining sample size, are each discussed briefly below.

Resources The cost per case is an important budgetary issue in all research and will be a final determinant in the compromise between the optimum sample that we would like and the practical sample that we can afford. If resources are short and the cost per case is high, the investigator may have to be satisfied with a small sample and results that are suggestive rather than conclusive. If resources are more than ample, there remains the question of whether they are best invested in a large number of cases or in improved data quality for a smaller sample. Increasing the sample size beyond a certain point (which point depends on other factors discussed below) is wasteful of resources, even if they are more than ample, and commonly the better investment is in more complete data or data of higher quality.

Sampling Procedure For purposes of drawing inferences from sample to universe, a probability sample of limited size is preferable to a convenience sample of larger size. A large nonprobability sample *may* give reasonably close estimates of the parameters of interest, but more commonly such a sample will be biased and the estimates based on it will be distorted in unknown ways. Furthermore, confidence intervals may be constructed for parameters based on probability sample estimates (and other techniques of statistical inference may also be applied; see Chapters 14 and 15), but the theory of confidence intervals (and of statistical inference generally) does not apply to estimates based on nonprobability samples. Hence a probability sample of limited size will allow an investigator to estimate the precision of sample estimates, whereas even with a very large nonprobability sample, the investigator will have no knowledge of how much the sample estimates are likely to miss their mark. Although probability samples ordinarily increase the cost per case, a smaller probability sample permits more confident estimates of population parameters than a larger nonprobability sample.

By the judicious use of cluster and stratified sampling procedures, an investigator can often increase the precision of the sample estimates without increasing sample size. The use of complex sampling designs to increase the precision of estimates is beyond the scope of this book.

Analysis Planned A single sample of a human population is commonly analyzed in multiple ways to serve several different purposes. These purposes may require the subdivision of the total sample into parts for separate analysis, for example, into whites and blacks, males and females, urban and rural dwellers, and so forth. The purposes may also require the control of one or more variables by *elaboration* (See Chapter 8). If the analysis requires the subdivision of the sample into parts, the sample size must be sufficient to provide an adequate sample of each part to be separately analyzed, including the smallest of those parts. For example, since only about 10 percent of the American population is black, a random sample of Americans will yield about one black person for every nine whites. Thus a sample of 100 would be expected to yield about 90 whites and about 10 blacks. This may provide an adequate sample of the whites, but it will not yield a very useful sample of blacks. When subdivisions are planned, it may be wise to oversample those subdivisions that are least numerous in the population so that even the smallest grouping is represented by a sample of adequate size.

Although the sample size requirements for subdivision are intuitively evident, it may be less evident that other modes of statistical analysis also impose some minimum sample size requirements. The general principle is that the more variables incorporated simultaneously into the analysis, the larger the sample should be. For example, if one seeks only to estimate a population mean, a smaller sample will suffice than will be necessary if one wishes to make estimates for a long series of multiple regression coefficients.

Anticipated Loss of Cases Some cases originally included in a sample are typically lost by the time the analysis is done. Some survey respondents may not be traceable or may refuse to respond. A "mail-back" questionnaire typically results in less than one-half of the questionnaires being returned. Some questionnaires returned may not be usable. Observers may take days rather than hours to achieve an acceptable level of reliability, with the result of a diminished number of reliable observations for analysis. A serious loss of sample cases (for example, one-fourth or more) raises questions about the representativeness of the sample on which data are actually obtained. But even a relatively small loss of sample cases reduces sample size. In sampling human populations, it is wise to anticipate a loss of 10 to 20 percent of the cases in the original sample (and sometimes more!) and to plan the initial sample size accordingly.

Variation in the Population Sampled The standard error of the mean depends on sample size and on variation in the population sampled (see Formula 13.1.1). Standard errors for other parameter estimates are also affected by variation in the population sampled, although it is not always so evident in the formula as in the case of the standard error of the mean. The evident implication of Formula 13.1.1 is that to maintain a constant standard error of the mean, a smaller sample will suffice for a more homogeneous population than for a less homogeneous population. This should not entice one into assuming, however, that estimates of all parameters are similarly affected by population variation. Thus, for example, in estimating mean income at a given level of precision, a smaller sample will suffice for a population more homogeneous in income, but in estimating the regression of income on education at a given level of precision, it is not ordinarily the case that a smaller sample will suffice for a population more homogeneous in income. In fact, just the opposite would be expected, since the population variation that affects the precision of the estimate of a regression coefficient is not the variation in either the independent variable or the dependent variable, but the variation around the regression line. Such error variance ordinarily constitutes a higher proportion of the total when variation in either the independent or dependent variable is restricted. It is thus misleading to suggest that highly homogeneous populations make sampling less problematic in general. While this is true when our purpose is to estimate a mean, the very homogeneity that increases the precision in estimating the mean will diminish the precision in estimating regression coefficients.

The general principle holds that variation in the population sampled will affect precision and, hence, affect the sample size necessary to reach a given level of precision. The important rule in applying this general principle is to recognize that one must focus on that aspect of population variation that actually affects the precision of the estimate contemplated. This principle, when supplemented by empirical knowledge about the population to be sampled, can sometimes be useful in making judgments about the sample size required for one's purposes.

The Precision Required For some purposes, great precision may be useful while in other studies such precision would be superfluous. In the early stages of exploring a phenomenon, it may be useful to know if a correlation is positive or negative, whether or not one can give a very precise estimate of its magnitude. Such crude information as the mere direction of a correlation may be much less useful when the phenomenon has already been thoroughly explored. Pollsters strive to be quite precise in estimating the public's approval of the President because such judgments constitute a continuing series and considerable importance is attached to even minor fluctuations. On the other hand, in a poll of attitudes toward a certain voting referendum it may suffice simply to know that a majority of the voters is in favor or opposed. In most segments of contemporary social science, high precision (in the sense of very small sampling errors) is not absolutely necessary, and large samples are ordinarily utilized in social science research, not because of the demand for high precision, but because of the potential utility of analysis of the sample in subdivisions or parts. Each investigator must assess the precision required in the contemplated study, however, and plan a sample size that will be responsive to that need.

As a general guide it may be useful to know the sample sizes necessary to reach specified degrees of precision in the estimation of a population proportion or population mean. The width of the confidence interval at a given level of confidence is one way of measuring such precision. One-half of the width of the confidence interval has a meaning that is more intuitively grasped, since one-half the width of the confidence interval at the 95 percent level of confidence is the maximum by which the sample estimate would miss the population parameter in the most accurate 95 percent at the samples. Thus, if the width of the confidence interval for a proportion at the 95 percent level of confidence is .02, one-half the width is .01, and we may say that we are 95 percent confident that the sample estimate will miss the population proportion by no more than .01.

Utilizing one-half the confidence interval width at the 95 percent level of confidence as the measure of precision, we show in Table 13.2.1 the sample sizes necessary to reach selected levels of precision in the estimation of a

Table 13.2.1 *Sample Sizes Necessary to Reach Specified Degrees of Precision in the Estimation of a Population Proportion, Selected Assumptions about the Population*

Population Proportion	Maximum Error Allowable at the 95% Level of Confidence			
	.10	.05	.02	.01
.5	96	384	2400	9600
.6 or .4	92	368	2300	9200
.7 or .3	81	324	2025	8100
.8 or .2	[a]	244	1525	6100
.9 or .1	[a]	[a]	864	3456

[a] Sample size too small to justify use of the normal approximation.

population proportion for varying a priori assumptions about the magnitude of that proportion. Clearly, samples of several thousand cases are necessary if we require confidence at the 95 percent level that the sample estimate be within one percentage point of the population proportion. A few hundred cases will suffice if we require confidence at the 95 percent level that the estimate be within 5 percentage points of the population proportion, and slightly less than 100 cases will constitute an adequate sample if we increase this error allowance to 10 percentage points.

The sample sizes in Table 13.2.1 have been computed by the formula:

$$n = \frac{3.84\hat{p}\hat{q}}{e^2}, \tag{13.2.1}$$

where $\hat{p}$ = assumed value of population proportion,

$\hat{q} = 1 - \hat{p}$, and

e = maximum error allowable at the 95 percent level of confidence.

Analogous computations of sample sizes necessary to reach selected levels of precision in the estimation of a population mean may be made by the formula:

$$n = \frac{3.84\hat{\sigma}^2}{e^2} + 1, \tag{13.2.2}$$

where $\hat{\sigma}^2$ = assumed value of population variance, and

e = maximum error allowable at the 95 percent level of confidence.

QUESTIONS AND PROBLEMS

1. Approximately 30 percent of the students in a given university are graduate students. If a simple random sample of the entire student body is to be drawn, how large must that sample be to yield a subsample of approximately 100 graduate students?

2. A sample survey of households in a particular city must provide, among other items, a confidence interval for the proportion of gainfully employed males 65 years of age and over. What information might be utilized to estimate the proportion of households that would include at least one male 65 and over so that the total sample size could be set to provide an adequate subsample of males in this age grouping?

Testing a Statistical Hypothesis 14

1 THE GENERAL LOGIC OF HYPOTHESIS-TESTING

Hypothesis-Testing and Estimation Compared There are two general types of statistical inference: (1) *estimation*, which begins without any stated assumption about the value of the parameter and merely seeks to estimate descriptively what the value of that parameter is; and (2) *hypothesis-testing*, which begins with a hypothesis about the parameter and then uses the sample data to check the tenability of that statement.

In Chapter 13, we were concerned with problems of estimation. For example, we began with the observation of a sample mean, and from that observation we progressed to an interval estimate with a specified degree of confidence. We first drew our sample and then made our estimate of the population mean.

But in hypothesis-testing, we state our hypothesis about a parameter before we collect the sample data, which then is used to test our hypothesis. For example, we may hypothesize that, on the average, students from well-to-do and poor families, respectively, do not differ in their school marks. We begin with that supposition. To test it, we draw a sample of students from each population, compare the means of the two samples, and reach a decision whether, in the light of the observed differences, the hypothesis should be rejected. If we knew the mean for each population, the decision about whether these means were identical, as hypothesized, could be made with complete certainty. Any difference other than zero would refute the hypothesis. But the means we know are sample means, subject to sampling error.

Because of sampling error, two sample means may differ from each other even though the corresponding universe means are identical in value. The decision to be made in testing a statistical hypothesis is whether, after allowing for sampling error, the sample outcome casts doubt on the hypothesis.

Hypothesis-testing and estimation both rest on the patterned variation of sample statistics around a central value, that is, on the sampling distribution of a statistic. But hypothesis-testing and estimation entail quite different modes of reasoning. An interval estimate begins with the sample data and ends with a range of values which probably includes the population mean. In contrast, hypothesis-testing begins by assuming a population value and ends with a decision to reject or not to reject that assumption in the light of the sample data.

Null Hypotheses If the hypotheses tested by sample observations were merely whimsical guesses about population parameters, an elaborate procedure for deciding whether they should be rejected would hardly be worth the effort. To make hypothesis-testing more than a computational exercise, the hypothesis to be tested must have either theoretical or practical import. Some understanding of the process by which hypotheses are formulated is a prerequisite to the intelligent use and interpretation of procedures for testing them.

A hypothesis is a tentatively held supposition about what is, how things change, or how events are interconnected. Such a supposition may be drawn from substantive theory (for example, sociological theory), in which case we refer to it as a *substantive hypothesis*. On the other hand, our hypothesis may represent no more than a logical rival which must be discredited if the substantive hypothesis is to be accorded credibility. If the hypothesis we test runs counter to our substantive hypothesis, the rejection of that counterassertion will strengthen our belief in the substantive proposition.

Ideally, a substantive hypothesis is a logical implication of a theory that is more encompassing than that hypothesis alone. Hence, under ideal circumstances, the empirical support of a substantive hypothesis increases our confidence in the utility of the broader theory to help us understand specific events and to predict their occurrence. Before deciding that a particular empirical finding supports a substantive hypothesis (and, indirectly, the broader theory of which it is an implication), meticulous investigators will wish to be assured that plausible rival explanations are invalid. They will wish to assure themselves that their decision does not rest inadvertently on some flaw in procedure or some fluke in observation—the self-selection of subjects, the bias of measuring instruments, the intrusion of uncontrolled events, and so on. Of particular interest in the context of statistical inference is the possibility that the investigator will mistake a sampling error for a true difference between population values. For example, suppose a social psychologist wishes to test the substantive hypothesis that "punitive" supervision elicits more aggression than does "supportive" supervision. How much of a difference in the average aggression levels of punitively and supportively supervised sample groups must be observed to lend credibility to the hypothesis? Alternatively stated, how much of a departure from a difference of zero must be observed to cast reasonable doubt on the assumption that these types of supervision make no difference in aggression?

Clearly, we would risk an erroneous conclusion if we attached any importance to a sample difference that would occur frequently in repeated sampling when the true difference is zero. Likewise, we would risk a false conclusion if we dismissed as inconsequential a large difference which would occur only rarely in repeated sampling if the true difference is zero. These risks are inherent in drawing inferences from sample data. In interpreting the implications of such data for substantive hypotheses, a careful investigator will seek first to nullify propositions which could explain sample observations but which run counter to the substantive hypotheses.

The hypothesis to be nullified, or the *null hypothesis*, has undergone some shifts in meaning since it was launched by Sir Ronald Fisher in the 1920s. As a consequence, there is some diversity in the use of the term in standard textbooks, resulting in some confusion for readers. The beginning student should be aware of this confusion, which stems from the tendency for some writers to attach to the term a more restricted meaning than it was originally given. Originally intended to refer to any hypothesis subject to nullification by a sample statistic, the null hypothesis subsequently has become identified with the hypothesis that two or more populations are identical or that two or more variables are uncorrelated. Thus the term *null* frequently has been interpreted as synonymous with "no difference," "no effect," "no correlation," and so forth. It should be evident that the original meaning of "hypothesis to be nullified" includes the more restricted meaning, and no fundamental difference in the strategy of hypothesis-testing is implied by the adoption of one meaning rather than another. In this book null hypothesis is given the original broader meaning of "hypothesis to be tested."

Sociological theory seldom yields an exact substantive hypothesis as does, for example, biological theory with its prediction of genotypes distributed in the proportions 1/4, 1/2, and 1/4. However, sociological theory may suggest on a more modest level that one population has a larger mean than another, or a larger variance or kurtosis, without specifying the magnitude by which these values differ. Or our theory may suggest that two variables are positively correlated without specifying the degree of that correlation. In such cases, where the substantive hypothesis is loose rather than sharp, as mentioned above, we often set up a logical counter to our substantive proposition, which may be exactly stated and subjected to statistical test. Thus the hypothesis of no difference between population means contradicts the substantive hypothesis of some difference between population means, and the hypothesis of no correlation contradicts the hypothesis of some correlation. When the hypothesis to be tested stands in opposition to our substantive hypothesis, the rejection of that opposing hypothesis supports the substantive hypothesis. In such instances, the null hypothesis is formulated with the theoretical expectation that it will be nullified. Our theory is supported when we do nullify it; our theory is unsupported when we fail to nullify it.

In the following discussion, we symbolize the null hypothesis H_0. Alternatives to the null hypothesis, including the substantive hypothesis, if it is an alternative, will be symbolized by H_1, H_2, and so on, although we will rarely have occasion to specify more than a single alternative.

Testing the Null Hypothesis Having stated a null hypothesis, we proceed on the assumption that it is true, draw out its implications, and check to see if they are compatible with the observed sample outcome. The implications are in the form of probabilities for sets of sample values. These probabilities are conditional on the truth of H_0, and the same sample values will have different probabilities if the alternative (H_1) is true. The probabilities of sample outcomes under the null hypothesis are represented in the sampling distribution, the central tendency of which is determined by the null hypothesis. Knowing this sampling distribution, we may identify sample outcomes as relatively frequent or infrequent in occurrence. If a given sample value occurs frequently when the null hypothesis is true, observing it in the sample casts no doubt on that hypothesis. However, an important principle of our testing procedure is that we never can prove the null hypothesis true. The best possible evidence for the truth of the null hypothesis would be a sample value exactly equal to our hypothesized value. But even if we were to observe such a sample value, we still could not be positive that the null hypothesis was true. Such a value could very well be an instance of sampling error around H_1. From sample observations, the conclusions we draw about the truth of the null hypothesis are necessarily of a probabilistic nature.

If we never can prove the null hypothesis true, may we prove it definitely false? Here the statistical evidence may be more convincing. Knowing the sampling distribution under the null hypothesis, we can identify certain extreme sample outcomes that would be highly improbable if the null hypothesis were true. Upon observing any one of these extreme sample outcomes, we could conclude either (a) that the null hypothesis is true and we have observed a sample outcome that is highly improbable but possible, or (b) that the null hypothesis is false and some alternative hypothesis is more compatible with our sample value.

Given a sample outcome that would be very rare if the null hypothesis were true, quite reasonably we would choose the second of the above possible conclusions, that is, that the null hypothesis is false. But since highly improbable events do occur, there is always the possibility that we have made the wrong decision. The risk of making this kind of error may be made very small by the simple expedient of rejecting the null hypothesis only for those sample outcomes whose combined probabilities are very small. But unless we decide never to reject the null hypothesis at all, the risk of erroneously rejecting it can never be eliminated.

Example The process of reaching a decision about the tenability of the null hypothesis, outlined rather abstractly above, may be clarified by an illustration.

We wish to determine whether blue-collar workers of voting age are proportionately represented among registered voters. We know from census data that 60 percent of the males of voting age are blue-collar workers. From a sample of 100 male registered voters, we learn that 47 percent are blue-collar. Is this *sample* proportion lower than 60 percent by a magnitude sufficient to justify the conclusion that the proportion is also lower than 60 percent in the *population* of male registered voters? The null hypothesis to be tested is that the population proportion is equal to .60 and that the sample proportion differs from this only because of sampling variation. In symbols:

$$H_0 : P = .60$$

where P represents the proportion of blue-collar workers in the population of male registered voters. Since we anticipate that blue-collar workers will be underrepresented in the population of registered voters rather than being proportionately represented, as the null hypothesis asserts, we actually expect the proportion blue-collar among the voters to be less than .60. Hence we may state the alternative (that is, the substantive) hypothesis:

$$H_1 : P < .60.$$

If we knew the universe proportion, we could tell immediately, and with no chance of error, whether or not the null hypothesis is true. But the information available to us through the sample will be subject to sampling variation and, in deciding on the tenability of the null hypothesis, we must make due allowance for sampling error. Therefore, we refer to the sampling distribution we would obtain by drawing infinitely many samples of 100 cases from a population in which the null hypothesis is true. From our discussion in the previous chapter, we know that, for large samples, the sampling distribution of a proportion will be normal with mean equal to the population proportion and standard error equal to the population standard deviation divided by the square root of sample size. This latter ratio is conventionally symbolized:

$$\begin{aligned}\sigma_p &= \sqrt{\frac{PQ}{n}}\\ &= \sqrt{\frac{(.60)(.40)}{100}}\\ &= .049.\end{aligned}$$

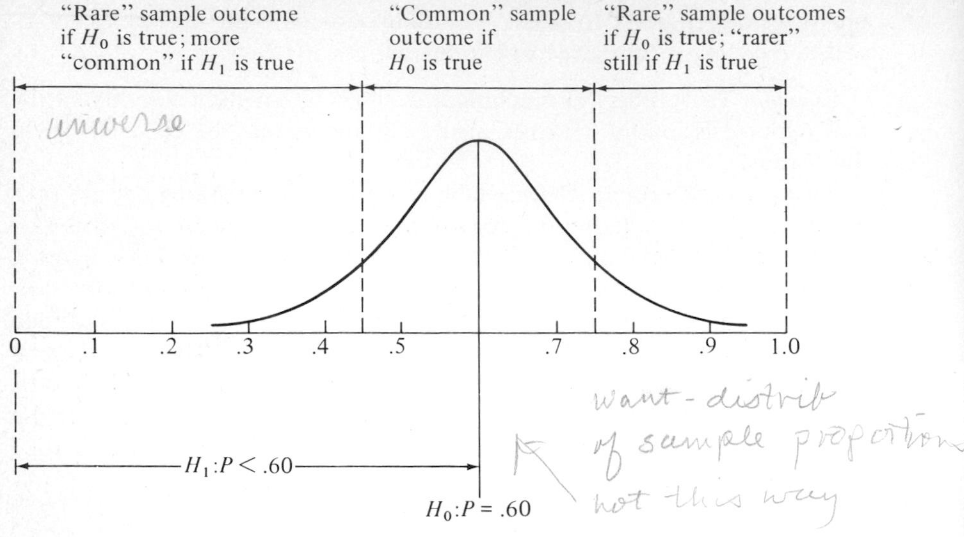

Figure 14.1.1 *Sampling Distribution of Sample Proportions for* n = *100,* $H_0 : P = .60$

Upon converting the sample proportion to a normal deviate—in this instance by subtracting .60 and dividing by .049—we may determine which sample outcomes will or will not lead to a rejection of the null hypothesis, with a specified risk of erroneously rejecting it.

Certain sample values in the "tails" of the sampling distribution (Figure 14.1.1) would rarely be observed if the null hypothesis were true. Those sample proportions in the left tail of the sampling distribution (that is, in the direction of H_1) would be rare if the null hypothesis were true but would be relatively frequent if H_1 were true. Thus, if we observe a sample value in the extreme lower tail, the data would suggest the rejection of H_0 in favor of H_1. Those extreme sample outcomes in the upper tail, if observed, also would cast doubt on H_0 but evidently would cast even more doubt on H_1, since they represent even more extreme deviations from H_1 than from H_0. We may decide in advance to reject H_0 if we observe a sample proportion at either extreme (two-tailed test), or to reject H_0 in favor of H_1 only if we observe a sample proportion in the extreme indicated by H_1 (one-tailed test). The decision to locate all rejection values in a single tail rather than to divide them equally between the two tails of the sampling distribution will depend primarily upon the investigator's confidence, whatever its grounds, in H_1 as descriptive of the true situation. In this illustration we will describe rejection values for both a one-tailed test and a two-tailed test.

Level of Significance But how rare must a sample outcome be if it is to cast doubt on the null hypothesis? There is no categorical answer to this question except that "it all depends." If the only point at issue were the risk we take of erroneously rejecting the null hypothesis, we would want to make that risk as small as possible. But by decreasing the risk of rejecting the null hypothesis when it is true, we increase the risk of failing to reject it when it is false! A reasonable resolution of this dilemma is to be sought, not in any misplaced faith in the invariable suitability of a fixed probability level, but by a careful assessment of the probable consequences of being wrong in either of these ways. If, as a consequence of rejecting the null hypothesis and thereby confirming our substantive hypothesis, a new series of more intensive studies will be launched to clarify and extend the line of reasoning on which the study was originally based, perhaps the investigator can afford a high risk of erroneous rejection. If the wrong decision is made, the subsequent studies will correct it. But if the false rejection of the null hypothesis will lead to action that will not in itself provide a corrective for a wrong decision, an investigator may insist that the null hypothesis be rejected only if there is an exceedingly small probability of doing so erroneously. Although hypothesis-testing procedures are useful in measuring the risk of being wrong, they cannot advise the investigator about how much risk should be taken.

The probability of rejecting the null hypothesis when it is true is called the *level of significance* and symbolized α (alpha). We may adopt whatever level seems appropriate to the purposes and probable consequences of the investigation. With this freedom, different investigators viewing the same data may, of course, set different levels of significance and come to different conclusions about whether or not to reject the null hypothesis. But there are reasonable conventions that govern our choice of level of significance. Ideally, we would choose whichever level is most appropriate for a given test, but this ideal is practically impossible to attain because of the difficulty of determining the most appropriate level for each application of a test of significance. In the absence of any compelling reason to do otherwise, we find it easier to follow precedent in setting significance levels than to work out extended justifications for each selection. When the null hypothesis opposes the substantive hypothesis, it is common practice to set the level of significance at .05 or .01. These conventional levels are arbitrary, of course and they may be disregarded when there are clear reasons for taking a greater or lesser risk.

Zone of Rejection The *zone of rejection* consists of those sample values, located in one or both tails of the sampling distribution, whose combined probability under the null hypothesis is equal to α. Since we reject the null hypothesis if the observed sample value falls in the rejection zone, it is evident that the probability of rejecting the null hypothesis, if it is true, is precisely equal to the level of significance. With a two-tailed test, we actually have two rejection zones, one in each tail, with the level of significance usually divided equally between them.

To locate the zone of rejection for a two-tailed test, we divide α by 2 and, referring to a table of normal probabilities, find the z-value corresponding to that probability. Thus with $\alpha = .05$, we find the z-value that will mark off the most extreme .025 of the area in each tail. Recall that the conventional table of normal probabilities is so constructed that we locate $.475 = .500 - .025$ in the body of the table and then find the corresponding z-value, which is 1.96. Thus the zone of rejection at the .05 level of significance, two-tailed test, consists of all points more than 1.96 standard errors away from the mean in either direction. The student may wish to test his or her understanding of this procedure by verifying that the zone of rejection at the .01 level of significance, two-tailed test, consists of all points more than 2.58 standard errors away from the mean in either direction.

To locate the point marking off the zone of rejection for a one-tailed test with the level of significance set at .05, we find the z-value that will mark off the most extreme .05 of the area in a single tail. Therefore we locate $.45 = (.50 - .05)$ in the body of the table and then find the corresponding z-value, which is 1.65. Thus the zone of rejection at the .05 level of significance, one-tailed test, consists of all points more than 1.65 standard errors away from the mean in the direction of H_1. The z-value for the level of significance at .01, one-tailed test, is 2.33, as the student may wish to verify by referring to the table of normal probabilities.

Having located the zone of rejection, we now draw our sample, compute the sample proportion, and determine whether it lies in the zone of rejection. We locate the sample proportion in the sampling distribution by finding its deviation from H_0 in standard error units. Thus, we compute:

$$z = \frac{p - P}{\sqrt{\dfrac{PQ}{n}}}$$

where $p =$ the observed sample proportion,
$P =$ the population proportion hypothesized by H_0,
$Q = 1 - P$, and
$n =$ sample size.

If the resulting z-value falls in the zone of rejection, we reject the null hypothesis at the specified level of significance. If it does not fall in the zone of rejection, we conclude that, at this level of significance, the sample data do not cast doubt on the null hypothesis.

If we had decided upon a one-tailed test at the .05 level of significance for our test of $H_0: P = .60$, against the alternative hypothesis $H_1 : P < .60$, the zone of rejection would consist of all sample proportions that yield a z-value less than -1.65 (that is, larger in absolute value than 1.65, but in the negative direction). If our sample of 100 yields 47 percent blue-collar workers, we compute:

$$z = \frac{.47 - .60}{\sqrt{\frac{(.6)(.4)}{100}}}$$

$$= -\frac{.13}{.049}$$

$$= -2.65.$$

Since the z-value for our observed sample proportion lies in the zone of rejection (that is, beyond -1.65) the null hypothesis is rejected at our previously selected level of significance.

The above null hypothesis could have been rejected at a still lower level, since 2.65 is some distance from 1.65. We may determine the probability under the null hypothesis of obtaining a sample proportion smaller than .47 for a sample of 100 by utilizing the table of normal probabilities. The obtained z-value of -2.65 indicates that samples as extreme or more so would occur approximately 4 times in 1,000 if the null hypothesis were true. If another investigator had selected a smaller α than ours, the decision to reject the null hypothesis with a one-tailed test would still be made unless they insisted on making α smaller than .004. Even a confirmed skeptic might take comfort in such a small probability of error in rejecting the null hypothesis, and therefore might concur in the conclusion that these data support the alternative substantive hypothesis that blue-collar workers are underrepresented among male registered voters. Of course, the goal toward which the entire procedure was directed was a decision about the substantive hypothesis, and when this decision is made, sociological theory rather than statistical theory becomes the guide for assessing its importance and its implications.

In the illustration above, several features common to all tests of the null hypothesis can be identified:

(1) In all statistical hypothesis testing, the null hypothesis is a proposition about some population or populations, which proposition can be proven neither true nor false because available data are always subject to sampling variation.

(2) A test of the null hypothesis requires knowledge of the sampling distribution of some test statistic. Knowledge of sampling distributions is derived mathematically, but it will be helpful to think of the sampling distribution as that distribution which would be generated by drawing infinitely many samples of size n from a population in which the null hypothesis is true.

(3) To give the null hypothesis an opportunity to be rejected, it is necessary to run some risk of falsely rejecting it. This risk level, set by the investigator, is called the level of significance. The level of significance and the decision to locate it in either one tail or both tails determine which values of the test statistic fall in the zone of rejection.

(4) The decision to reject or not to reject the null hypothesis is based on the location of the observed test statistic in the sampling distribution. If it is located in the zone of rejection, as determined by the level of significance, the hypothesis is rejected at that level. Otherwise, the appropriate conclusion is that the data do not cast doubt on the null hypothesis.

(5) The probability of failing to reject the null hypothesis when false (sometimes referred to as a Type II risk), cannot be set at an arbitrary level selected by the investigator and ordinarily is not known. This probability depends upon which of many possible alternative hypotheses is true, the nature of the test being used, the size of the sample, and the level of significance. By making the level of significance smaller, the probability of this second kind of error is increased when the null hypothesis is false. By increasing the sample size, the probability of this second kind of error is made smaller.

QUESTIONS AND PROBLEMS

1. Define the following terms:

hypothesis-testing	one-tailed test
null hypothesis	level of significance
alternative hypothesis	zone of rejection
substantive hypothesis	test statistic
two-tailed test	

2. A candidate claimed the support of 60 percent of the electorate. In a sample of 1,000 registered voters, 55 percent declared for that candidate. Test the credibility of the candidate's claim.

3. Census sources describe a small city as being composed of 9,000 adult males and 11,000 adult females. In a sample survey of 400 adults residing in this city, 228 respondents were females. Does this proportion of females in the sample cast doubt on the assumption that the sampling was random with respect to sex?

2 TESTING THE HYPOTHESIS OF NO DIFFERENCE BETWEEN MEANS

In this section we consider three procedures for testing the hypothesis that the means of two populations are identical. The first may be used for very large independent samples whether or not the populations from which the samples are drawn are normally distributed. The second test is not limited to large samples but requires that the populations be distributed approximately normally. The third test is designed for the special problem of testing the hypothesis of no difference between the means of two distributions in which every element in one distribution has a matched counterpart in the other, and where the sampling unit is a pair of elements.

Large Sample Test for Difference between Means To center our discussion around a concrete problem, we consider the mean achievement scores of two large samples of black students in the twelfth grade (Table 14.2.1). One sample consists of blacks in schools where the majority of students are white. The other sample consists of blacks in schools where the majority of students are black. We pose the question whether the achievement scores differ in the two populations represented by these samples. If the difference between the sample means is not due to sampling variation, we may then speculate about the reasons for the difference. But if a difference between sample means of this magnitude would be a probable result of sampling variation, such speculation would be ill-advised.

The null hypothesis to be tested asserts that black students in predominantly black schools do not differ in mathematical achievement test scores from black students in predominantly white schools. The alternative or substantive hypothesis is that black students with predominantly white classmates will have higher achievement scores. In symbols:

$$H_0 : \mu_1 - \mu_2 = 0$$

$$H_1 : \mu_1 - \mu_2 > 0$$

where μ_1 is the mean score for black students with predominantly white classmates and μ_2 is the mean score for black students with predominantly black classmates.

Table 14.2.1 *Mathematical Achievement Test Scores of Black Students in the Twelfth Grade by Racial Composition of Senior Class, Metropolitan Areas of the Northeastern United States, 1965*

Sample	Number	Mean Achievement Test Score	Standard Deviation of Test Scores
Black students with majority of classmates black	1,689	40.9	9.7
Black students with majority of classmates white	2,545	42.9	12.2

Source: Adapted from James S. Coleman et al., *Equality of Educational Opportunity*. pp. 332–333. U.S. Department of Health, Education and Welfare, Office of Education. Washington, D.C.: U.S. Government Printing Office. 1966.

Inspection of the data reveals a difference between sample means of two score units in the expected direction. We wish to determine the probability of a difference as large or larger if the null hypothesis of equal population means is true.

In Chapter 13 we considered the sampling distribution of sample means. Now we must consider the sampling distribution of *differences* between two sample means. To define and illustrate this concept, we may imagine a sampling experiment in which we alternately sample from two well-defined universes, selecting a large random sample of size n_1 from one population, and a large sample of size n_2 from the other. By replacing samples, we may continue to draw samples in this manner indefinitely. For each pair of independently selected samples we compute the difference between the two sample means, amassing in this way as many differences as there are pairs of samples. We now pose the problem: How will these hypothetical differences between paired means distribute themselves, and what will their mean and standard deviation (standard error of the difference) be? According to sampling theory, an infinite supply of such differences will have a normal distribution, with mean equal to the true difference between the universe means $(\mu_1 - \mu_2)$, and standard error equal to the square root of the sum of the respective variances of the means. The standard error of the difference, expressed in symbols, is as follows:

$$\sigma_{\bar{X}_1 - \bar{X}_2} = \sqrt{\frac{\sigma_1^2}{n_1} + \frac{\sigma_2^2}{n_2}}, \tag{14.2.1}$$

where σ_1^2 and σ_2^2 represent the population variances of the respective populations. It is clear from this formula that the standard error of the difference between sample means, like the standard error of the mean itself, reflects both the degree of variation within the sampled populations and the respective sample sizes. The smaller the population variances and the larger the sample n's, the smaller will be the standard error of the difference.

Subtracting the difference between population means from the observed difference between sample means and dividing by the standard error converts the observed difference into a normal deviate:

$$z = \frac{(\bar{X}_1 - \bar{X}_2) - (\mu_1 - \mu_2)}{\sqrt{\frac{\sigma_1^2}{n_1} + \frac{\sigma_2^2}{n_2}}}. \tag{14.2.2}$$

Under the null hypothesis that $\mu_1 - \mu_2 = 0$, this statistic becomes:

$$z = \frac{\bar{X}_1 - \bar{X}_2}{\sqrt{\frac{\sigma_1^2}{n_1} + \frac{\sigma_2^2}{n_2}}} = \frac{\bar{X}_1 - \bar{X}_2}{\sigma_{\bar{X}_1 - \bar{X}_2}}. \tag{14.2.3}$$

But this ratio cannot be calculated from our sample information, because the variances of the two populations, σ_1^2 and σ_2^2, are unknown. However, from the sample data we may compute estimates of these population variances, s_1^2 and s_2^2. Substituting these estimates in Formula 14.2.3, we obtain:

$$z = \frac{\bar{X}_1 - \bar{X}_2}{\sqrt{\dfrac{s_1^2}{n_1} + \dfrac{s_2^2}{n_2}}} = \frac{\bar{X}_1 - \bar{X}_2}{s_{\bar{X}_1 - \bar{X}_2}}. \quad (14.2.4)$$

If the null hypothesis is true, the statistic computed by Formula 14.2.4 for large samples will be approximately normally distributed with mean zero and standard deviation 1.0, that is, it will be distributed as a normal deviate (z). Therefore, this statistic may be located in the normal probability distribution to determine the probability of our observed difference or one more extreme, if the null hypothesis is true.

The statistic obtained by Formula 14.2.4 is sometimes called the *critical ratio*, since its magnitude is critical in determining whether we judge the observed difference to be statistically significant.

To illustrate the calculation of such a critical ratio, we proceed with the test of the null hypothesis, using the data of Table 14.2.1. We select .05 as the level of significance. Since H_1 specifies direction, a one-tailed test is required with the zone of rejection in the right tail of the distribution toward H_1. We find the z-value corresponding to our significance level, α. This will be a point in the right tail beyond which 5 percent of the differences between sample means will lie if the null hypothesis is true. Reading from the table, we find that $\alpha = .05$ corresponds to $z = 1.65$. Thus, critical ratios of 1.65 or larger constitute the zone of rejection for a one-tailed test at the .05 level of significance, as illustrated in Figure 14.2.1.

Applying Formula 14.2.4 to the data of Table 14.2.1 gives:

$$z = \frac{42.9 - 40.9}{\sqrt{\dfrac{148.84}{2545} + \dfrac{94.09}{1689}}} = \frac{2.0}{\sqrt{.058 + .056}} = \frac{2.0}{\sqrt{.114}} = \frac{2.0}{.338} = 5.9.$$

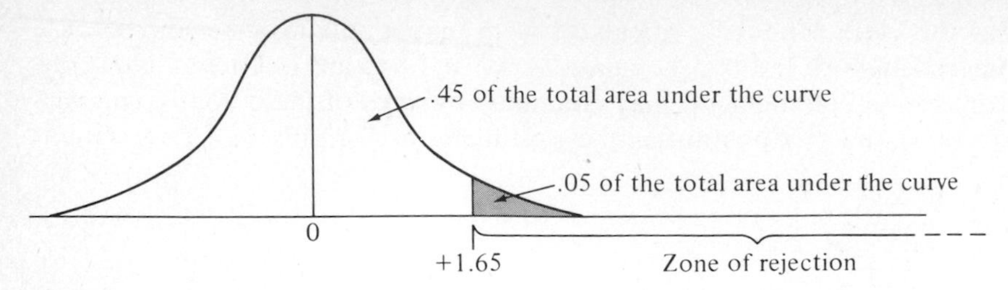

Figure 14.2.1 *Zone of Rejection in a Normal Sampling Distribution* (z) *for a One-Tailed Test, Level of Significance = .05, Upper Tail Critical*

Since the observed critical ratio lies in the zone of rejection, we conclude that the data have cast doubt on the null hypothesis of no difference between population means. We therefore reject the null hypothesis in favor of H_1.

We might wish an estimate of the precise probability of obtaining a critical ratio as large or larger if the null hypothesis is true. For a critical ratio as large as the one observed in this instance, Table I (Appendix) will provide only an estimate, because the observed ratio is larger than any shown in the table. Since the probability of getting a critical ratio of 4.0 or larger is approximately 6 in 100,000, the probability of obtaining a critical ratio of 5.9 or larger must be considerably less. Therefore, we would reject the null hypothesis at a much lower level of significance than .05; if the null hypothesis were true, this observed difference between sample means truly would be a one-in-a-million event.

These data, then, indicate that the null hypothesis of no difference between means should be rejected in favor of the substantive hypothesis that black students in predominantly black schools have lower mathematical achievement test scores than black students in predominantly white schools. Several additional points should be made in connection with this conclusion:

(1) The magnitude of the critical ratio is not necessarily indicative of the magnitude of the difference between means. In this case, with very large samples, the critical ratio is large, but the means are not far apart. Although the rejection of the null hypothesis gives assurance that the population means differ, this difference, when compared to the total variation of such scores from all sources, is relatively small. This is indicated by the proportion of the total variance in mathematical achievement test scores accounted for by the racial composition of the class; a correlation ratio (relative reduction in prediction error; see Chapter 9) may be utilized for this purpose. Computing the correlation ratio for the data of Table 14.2.1, we get .007, indicating that less than 1 percent of the total variation in test scores is associated with the racial composition of the class, as measured by a 0, 1 dummy variable. If our goal were to account for most or all of the variance in test scores, we have taken only a very minor step by focusing on racial composition. But if our goal is to determine whether or not black students in

the two types of schools differ in mean test scores, we have made a decision that they do differ with high confidence.

(2) Rejection of the null hypothesis does not imply that the observed difference is the true difference. Recognizing that the observed difference will vary from sample to sample, we may wish to construct a confidence interval for the difference between population means, utilizing the techniques discussed in Chapter 13. A 95 percent confidence interval for the difference may be constructed by adding and subtracting 1.96 standard errors, $(\bar{X}_1 - \bar{X}_2) \pm 1.96\, s_{\bar{X}_1 - \bar{X}_2}$. For the data of Table 14.2.1, the 95 percent confidence interval for the difference between population means is 1.34 – 2.66. We assume that intervals constructed in this way would include the true difference 95 times in 100.

(3) A statistically significant difference does not necessarily indicate a cause-and-effect relationship; that is, the conclusion that the population means differ does not necessarily imply that transferring students from one type of school to the other would have an effect on their test scores. The difference *may* be a result of class composition, but it also may be an effect of the selection of students into the two types of schools or an effect of some other characteristic of schools that is associated with the proportion of white students. In rejecting the null hypothesis, we conclude only that the observed difference would be unlikely if the null hypothesis were true and hence that sampling variation alone probably is not responsible for the observed difference. In this instance, there remain logically plausible rival explanations for an observed difference other than the assumption that a predominantly black class has the effect of reducing the test scores. These other alternatives also would have to be discredited before concluding that an effect has been observed.

(4) The statistical inference to be drawn from a test of the null hypothesis is limited to the population sampled; the extension of this conclusion to some larger target population should not be made on the basis of a rejection of the null hypothesis alone. The data of Table 14.2.1 pertain to black students in the twelfth grade in metropolitan schools in the northeastern United States in 1965, and we reject the null hypothesis for that sampled population. This rejection does not necessarily imply that the same null hypothesis would be rejected for twelfth-grade students in another region, for sixth-grade students in the same region, or for twelfth-grade students in the same region in nonmetropolitan schools. If we generalize to populations other than the population sampled, we do so on the basis of the assumption that those other populations are similar in all crucial respects to the sampled population.

It should be evident that an assessment and interpretation of the substantive meaning of the difference between means must rest on considerations not encompassed in the theory of statistical inference. But had we not rejected the null hypothesis of no difference for these data, no search for a substantive interpretation of the difference would have been initiated; a difference that would occur commonly as a result of sampling variation around a true difference of zero requires no further interpretation.

The Difference between Two Proportions The procedure for testing the null hypothesis of no difference between two population proportions or percentages is identical in fundamentals with that set forth for two means. The student will recall from Chapter 5 that proportions and percentages *are* means, and it should therefore not be surprising that the same general procedure is applicable. Only a minor modification in the computing formula for z given above is required to describe the test statistic for the difference between proportions.

The standard error of the difference between proportions depends on the standard errors of each of the proportions being compared. When testing the null hypothesis of no difference between population proportions, the estimated standard error of the difference may be based on a single population proportion since the proportions in the two populations are identical by hypothesis. The estimate of that single population proportion, p_w, is the weighted mean of the two sample proportions, and it is given by:

$$p_w = \frac{n_1 p_1 + n_2 p_2}{n_1 + n_2}. \tag{14.2.5}$$

The standard error of the difference between proportions is then:

$$\begin{aligned} \sigma_{p_1 - p_2} &= \sqrt{\frac{p_w q_w}{n_1} + \frac{p_w q_w}{n_2}} \\ &= \sqrt{p_w q_w \left(\frac{1}{n_1} + \frac{1}{n_2}\right)} \\ &= \sqrt{p_w q_w \left(\frac{n_1 + n_2}{n_1 n_2}\right)}. \end{aligned} \tag{14.2.6}$$

With this estimate of the standard error, Formula 14.2.4 becomes:

$$z = \frac{p_1 - p_2}{\sqrt{p_w q_w \left(\dfrac{n_1 + n_2}{n_1 n_2}\right)}}. \tag{14.2.7}$$

To illustrate the computations, we consider the data of Table 14.2.2 showing the percentage distribution of responses to an equal opportunity question among persons in two samples. Our interest here centers on the question of whether the percentage answering "No" is sufficiently different among junior high school and senior high school student samples to conclude that there is a difference in the populations these samples represent. We proceed to test the null hypothesis with a two-tailed test at the .01 level of significance. The critical value of z is therefore 2.58. If the standard deviate for our observed difference between proportions is greater than 2.58, or less

Table 14.2.2 *Opinions on Equal Opportunity among Junior High School and Senior High School Students in Baltimore*

Responses to the question: "Do all kids in America have the same chance to grow up and get the good things in life?"		
	Junior High School Students	Senior High School Students
Yes	21%	20%
No	63	70
Don't Know[a]	15	10
N	518	335

[a] In this case, "Don't Know" responses are assumed to have substantive meaning instead of indicating a procedural defect (see Chapter 3), and it is therefore appropriate to compute percentages on the basis of the total frequency instead of on the basis of those responding "Yes" or "No."

Source: Roberta G. Simmons and Morris Rosenberg, "Functions of Children's Perceptions of the Stratification System," *American Sociological Review* 36 (April, 1971): 235–249. Adapted from Table 2, p. 240. Reprinted with permission.

than -2.58, we will reject the null hypothesis. Otherwise, we will conclude that the observed difference between proportions is not of sufficient magnitude to cast doubt on the null hypothesis at this level of significance.

The estimated common population proportion is given by:

$$p_w = \frac{518(.63) + 335(.70)}{518 + 335}$$

$$= \frac{560.8}{853}$$

$$= .66.$$

To test the null hypothesis, we compute:

$$z = \frac{.70 - .63}{\sqrt{(.66)(.34)\left(\frac{518 + 335}{518 \times 335}\right)}}$$

$$= \frac{.07}{\sqrt{(.2244)(.0049)}}$$

$$= \frac{.07}{\sqrt{.0011}}$$

$$= \frac{.07}{.033}$$

$$= 2.12.$$

Since the normal deviate for the observed difference between proportions does not fall in the zone of rejection for the .01 level of significance, we cannot reject the null hypothesis at that level. Although we would have rejected the null hypothesis at the .05 level of significance (critical value = 1.96), having selected the .01 level of significance in advance we conclude that these data fail to cast strong doubt on the hypothesis that junior and senior high school students are equally skeptical about equal opportunity.

We reiterate that the procedure set forth above for testing the hypothesis that two population percentages or proportions are identical is applicable only to large samples. Only then will the theoretical sampling distribution of differences be normal, as is assumed in determining the probabilities. Moreover, the samples must be extra large insofar as the sample percentages deviate markedly from 50 percent. Therefore, it is necessary to scrutinize each application to ensure that the sample data meet the conditions presupposed in the application of this test, a rule which holds for every other statistical application as well. An alternate but equivalent test for the difference between two proportions, called the chi-square test, will be discussed in Chapter 15.

Small Sample Test for Difference between Means With small samples we must undertake a slightly different procedure. We continue to express our observed difference between sample means as a multiple of the estimated standard error, but we modify our formula for estimating that standard error. This ratio is symmetrically distributed and becomes increasingly normal as the sample sizes increase. It is known as "Student's t" distribution, so named because the papers in which it was originally presented did not give the author's name but were simply signed "Student." [1] To test the hypothesis of no difference between population means (small samples) we set up the ratio:

$$t = \frac{\bar{X}_1 - \bar{X}_2}{\sqrt{\dfrac{n_1 s_1^2 + n_2 s_2^2}{n_1 + n_2 - 2}}\sqrt{\dfrac{n_1 + n_2}{n_1 n_2}}}. \tag{14.2.8}$$

Although it may not be evident immediately, Formula 14.2.8 is very similar to Formula 14.2.4 for large samples. In shifting from Formula 14.2.3 to Formula 14.2.4, we simply substituted sample estimates, s_1^2 and s_2^2, for their

[1] The author did not remain anonymous forever. He was William Sealy Gosset (1876–1947), whose firm relaxed its restriction against research publications by employees and allowed him to publish anonymously the paper on the t distribution in 1908.

respective population variances. In moving from Formula 14.2.3, via 14.2.4, to 14.2.8, we assume $\sigma_1^2 = \sigma_2^2 = \sigma^2$, and thus Formula 14.2.3 becomes:

$$z = \frac{\bar{X}_1 - \bar{X}_2}{\sqrt{\sigma^2\left(\frac{1}{n_1} + \frac{1}{n_2}\right)}}$$

$$= \frac{\bar{X}_1 - \bar{X}_2}{\sigma\sqrt{\frac{n_1 + n_2}{n_1 n_2}}}. \tag{14.2.9}$$

To estimate the common population standard deviation, (σ), we pool sums of squares from the respective samples and divide by degrees of freedom:

$$s = \sqrt{\frac{n_1 s_1^2 + n_2 s_2^2}{n_1 + n_2 - 2}}. \tag{14.2.10}$$

By substituting this value as an estimate for σ in Formula 14.2.9, we obtain Formula 14.2.8.

Student's t distribution actually is a family of distributions; there is not one distribution, but many. They are alike in their essential properties and share a common formula. They differ according to the degrees of freedom possessed by the sample estimate. Since that number is affected by sample size, we say that the shape of the t distribution varies by sample size. In order to know which specific distribution is to be utilized in finding the probability of a given t value, the appropriate degrees of freedom must be determined. For testing the hypothesis of no difference between means by Formula 14.2.8, the degrees of freedom may be found by calculating $n_1 + n_2 - 2$.

Whatever the degrees of freedom, Student's t distribution is symmetrical and has the general form of a bell-shaped curve, similar in appearance to the normal distribution. For very small degrees of freedom, however, the t distribution is notably flatter (*platykurtic*) than a normal distribution. This difference is shown in Figure 14.2.2, comparing the normal distribution with Student's t distribution with four degrees of freedom. As we increase the degrees of freedom, the distribution of t becomes more nearly identical to the distribution of z. Hence, for large samples, the t and z distributions are practically identical, and for infinite degrees of freedom, they are identical.

Critical values of t for selected probability levels are shown in the Appendix (Table V) for selected degrees of freedom. The student may note that, with infinite degrees of freedom (last row of table), the critical values of t are the same as the critical values of z with corresponding probability levels. For any given probability level, as degrees of freedom become larger (running down any column), the critical values approach more and more closely the critical value of z; even with degrees of freedom as small as 30, the critical values of t are only slightly larger than those of z.

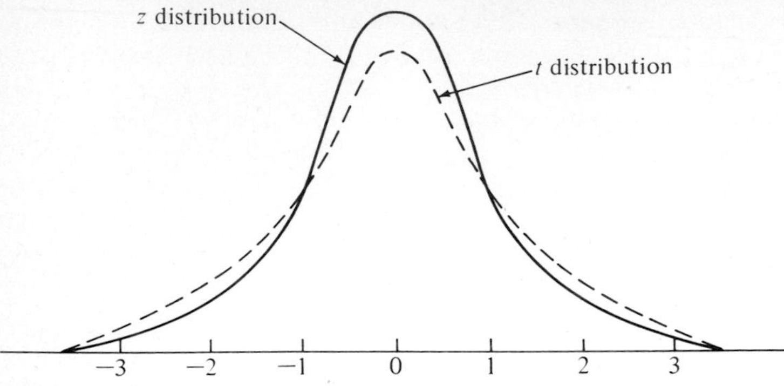

Figure 14.2.2 *Distribution of* z *and Distribution of* t *with 4 Degrees of Freedom*

To illustrate the application of the *t*-test and the use of the table of critical values of *t*, we turn now to a concrete illustration. In Table 14.2.3 we show the mean ratings on quality of performance of subjects in an experiment designed to determine the effect of contrasting types of competition. Standard deviations and sample sizes have been given hypothetical values selected for ease of manipulation. Subjects were randomly assigned to the two experimental conditions and subsequently divided into teams of four members each. The instructions to the "Group Competition" teams specified that the grade for all members of a team would be the same, and that the teams would be competing against each other; the grade for each team would depend upon the rank of its collective performance relative to other teams. The instructions to the "Individual Competition" teams specified that, although the other members of their team would be their "lab partners," grades would be assigned individually and individuals would be competing with one another; the grade for a given individual would depend upon the rank of his or her performance relative to all class members.

Table 14.2.3 *Means and Standard Deviations of Quality of Performance Ratings for Students Performing under Different Competitive Conditions*

	Competitive Conditions	
	Individual Competition	Group Competition
Mean	6.3	5.5
Standard Deviation	2.1	1.9
n	16	16

Source: James W. Julian and Franklyn A. Perry, "Cooperation Contrasted with Intra-Group and Inter-Group Competition," *Sociometry* 30 (March, 1967), pp. 79–90. Abbreviated Table 2, p. 85. Reprinted with permission. Standard deviations and *ns* are hypothetical for illustrative purposes only.

Lacking a clear theoretical rationale for anticipating which of the two groupings would have the higher quality of performance, we formulate the substantive hypothesis of some difference between means without specifying the direction of that difference. We then test the rival null hypothesis of no difference between means. In symbols, this is:

$$H_0 : \mu_1 - \mu_2 = 0,$$

$$H_1 : \mu_1 - \mu_2 \neq 0.$$

The nondirectional character of the alternative hypothesis, H_1, dictates a two-tailed test.

In this instance, the two sets of observed performance scores have *not* been obtained by a series of random draws from two separate populations of cases differentiated by the competitive conditions under which they perform. The cases observed are the only cases that have been exposed to these experimental conditions. Since the observed means do not represent randomly drawn samples from two populations of cases, why is it necessary to test the null hypothesis of no difference between means and thereby come to a decision on the null hypothesis? To clarify the relevance of a test of the null hypothesis under such circumstances, we imagine that we have a single set of $n_1 + n_2$ cases, each case having a score not yet measured. We randomly select n_1 cases to constitute the first sample, allowing the remaining n_2 cases to constitute the second sample. Then, for each sample we measure the scores and compute the means. By repeating this set of operations many times, we may obtain

$$C_{n_1}^{n_1+n_2} = \frac{(n_1 + n_2)!}{n_1!\,n_2!}$$

different pairs of samples and the same number of differences between sample means. As a consequence of random assignment, the differences between sample means would vary from one pair of samples to the other in a manner analogous to the variation of randomly drawn samples. Between assignment and measurement, if nothing intervenes to change differentially the scores in the two samples, the mean of this sampling distribution of differences would be zero, since all pairs of samples would be samples from the same population. On the other hand, suppose something (for example, an experimental treatment) does intervene between assignment and measurement which changes the scores in the two samples differentially. Then the observed difference between means will be affected, not only by random assignment, but also by the differential change in scores as a result of intervening events. Thus the question to be answered by testing the null hypothesis in this situation is this: Is the observed difference between means such that it would be a commonly occurring difference in an endless series of random assignments, or is the observed difference of such magnitude that some differential change in scores between assignment and measurement is indicated? If the two experimental conditions have identical effects, the expected means would be identical and differences between sample means would be distributed as expected under the null hypothesis, varying around zero as a result of random assignment. If the two experimental conditions have different effects, differences between sample means would vary around some nonzero difference. Differences resulting from random assignment thus would be increased by a constant amount—the differential effect—and large differences would become more likely than if the null hypothesis were true. Hence, rejection of the null hypothesis under these circumstances leads to the conclusion that the experimental conditions have different effects.

The mathematical derivation of the t distribution requires that the parent populations (that is, the populations from which the samples are randomly drawn) be normally distributed. In this instance, which is typical of many actual applications of the t-test for the difference between means, the conceptual populations are only approximately normal. Relatively minor departures from normality in the population distributions have been found to have only trivial effects on the t distribution, provided the total number of cases is no smaller than about 25 and provided the cases are about evenly divided between the two samples. If either of these conditions is not met, or if the departure from normality in the population distributions is very great, the use of the t-test is ill-advised.

We now proceed to determine a critical value of t for testing the null hypothesis for the data of Table 14.2.3. Since $n_1 + n_2 = 32$, the degrees of freedom are $n_1 + n_2 - 2 = 30$, and we will find the critical value in that row of the t-table (Appendix Table III). We have noted above that a two-tailed test is required and we select .05 as the level of significance. Therefore, the critical value of t, from the table, is 2.042. For 30 degrees of freedom, the probability of obtaining an observed t as large or larger (disregarding sign) than 2.042 is 5 in 100, if the null hypothesis is true. The corresponding critical value of z is 1.96; rejection of the null hypothesis utilizing t with 30 degrees of freedom therefore requires a slightly larger observed value of t than would be required if we had very large samples.

Using Formula 14.2.8, we now compute t and ascertain whether our obtained result lies inside or outside the zone(s) of rejection.

$$t = \frac{(6.3 - 5.5) - 0}{\sqrt{\dfrac{(16 \cdot 4.41) + (16 \cdot 3.61)}{30}}\sqrt{\dfrac{32}{256}}}$$

$$= \frac{.8}{\sqrt{\dfrac{128.32}{240}}}$$

$$= \frac{.8}{\sqrt{.53}}$$

$$= \frac{.8}{.728}$$

$$= 1.1.$$

Since the obtained t value does not fall in the predetermined zone of rejection, these data do not lead us to reject the null hypothesis. We conclude that our findings do not cast doubt on the hypothesis of no difference.

Several features of this decision and of the t-test should be emphasized:

(1) In failing to reject the null hypothesis, we have not demonstrated that it is true. The data of Table 14.2.3 do not demonstrate that the two experimental conditions have identical effects; rather, they fail to cast serious doubt on the assumption of identical effects. If a larger sample or controls for additional variables produced smaller estimates of the standard error of the difference, the test would be more sensitive to the existence of an effect, if there is one, and under such other circumstances, we *might* reject the null hypothesis. To avoid a premature conclusion, then, it is appropriate to conclude simply that the data fail to support the substantive hypothesis of an effect rather than concluding that the data *demonstrate* that no effect occurs.

(2) The conclusion following from a rejection or nonrejection of the null hypothesis is expressed appropriately in terms of "effects" only if alternative hypotheses have been ruled out in the study design. In the illustrative problem above, the subjects were treated identically except for the difference in competitive conditions; hence no difference in the conditions under which they performed, except for the competitive conditions, could contribute to the observed difference. Furthermore, subjects were assigned randomly to the two groupings so that they would be alike except for random differences; the two samples did not differ initially because of self-selection into preferred competitive conditions or some other assortative process. In discussing the difference between test scores for students in segregated and integrated schools, rejection of the null hypothesis did not necessarily imply an effect of segregation (although the rejection of the null hypothesis was not inconsistent with that assumption) because the study design did not include the random assignment of students to schools. Any number of processes by which students sort themselves or are sorted into schools may have been responsible for the observed difference in scores. But in the experimental illustration of this section, rejection of the null hypothesis would have been interpreted appropriately as indicating an effect and not simply a difference. Such assortative processes as otherwise might have occurred to produce an initial difference were not allowed to operate, because subjects had been assigned randomly to the two conditions. Of course, the applicability of the t-test for small samples is not limited to studies involving random assignment of conceptual populations; this test also may be used to test the hypothesis of no difference between the means of "real" populations when samples are small.

(3) The critical values of Student's t distribution have been computed on the basis of three assumptions. First, it is assumed that the sampling distribution of the differences between sample means (that is, the numerator of t) is normal with mean zero. Second, it is assumed that the standard error of the

difference (the denominator of t) also has a known, albeit not normal, sampling distribution. For small samples, these two assumptions will be true only if the populations from which the samples are drawn are distributed approximately normally. Third, it is assumed that the population variances are equal. Departure from any of these assumptions in any particular application will render the probabilities inaccurate to some degree. If the sample variances are strikingly different—if the variance of one sample is twice as large as the variance of the other, for example—a difference between *population* variances is suggested, and another test would be appropriate. If the sample distributions are nonnormal—if they are bimodal or very heavily skewed, for example—the assumption of normal populations would be questionable and one of the nonparametric tests discussed in a subsequent section may be utilized.

The *t*-Test for Matched or Correlated Samples Both tests discussed above are appropriate for *independent* random samples. Now we consider a test of the hypothesis of no difference between means that is appropriate for correlated random samples. If we wish to compare husbands and wives, we may draw a sample of husbands only; wives then enter the sample by virtue of their marriage to husbands in the sample rather than by random selection. By sampling husbands, we obtain two interdependent samples; a sample of husbands and a sample of wives. If we drew independent samples of husbands and wives, the inclusion of a given husband in the sample would not necessarily imply the inclusion of his wife in the other sample. Our comparison would be between married men and women, not between husbands and wives. Two samples are said to be dependent if the inclusion of a given element in one sample determines or influences the inclusion of an element in the second sample. The t-test discussed above is less sensitive to the existence of a true difference between the means of two populations represented by correlated samples than is an alternative t-test, discussed below, that uses a different estimate of the standard error of the difference.

Suppose we wish to test the null hypothesis of no difference between the mean number of children desired by husbands and wives. Also, suppose that we have available a sample of married couples, hence dependent samples of husbands and of wives. Then we may take advantage of whatever correlation there is between the number of children desired by each husband and wife in estimating the sampling variation of the difference between means. The consequences of such a correlation are illustrated in the hypothetical data of Table 14.2.4, in which there is a perfect correlation (although not perfect agreement) between the number of offspring desired by husbands and their wives. The variation among the differences between each husband and wife is reduced to zero, although evidently there is variation among husbands and among wives. Given correlated samples, it is perhaps more convenient to conceive of a single population of differences and the mean of that population than it is to conceive of two populations and the difference between

Table 14.2.4 Dependent Samples: Number of Offspring Desired by Husbands and Their Wives

Married Couple	Number of Offspring Desired: Wife (X_1)	Husband (X_2)	Difference $(X_1 - X_2) = 0$
A	3	2	1
B	5	4	1
C	1	0	1
D	2	1	1
E	4	3	1
Mean	3	2	1

Difference between means = $\bar{X}_1 - \bar{X}_2 = 3 - 2 = 1$
Mean of the differences = $\bar{D} = 1$
Standard deviations of X_1 and $X_2 = s_{X_1} = s_{X_2} = \sqrt{2}$
Standard deviation of the differences = $s_D = 0$

those population means. As illustrated in Table 14.2.4, the difference between sample means is equal to the mean of the differences. Thus the two models will not differ in the observed difference but in the procedure for estimating the standard error of that difference.

The sampling distribution of a single sample mean (which includes, of course, the mean of the difference for matched samples) has been discussed in relation to estimation procedures in Chapter 13. In that discussion, large samples were assumed so that the z statistic could be utilized. Here, we lift that restriction since our test statistic distributes as t, regardless of sample size (Formula 14.2.11).

Under the null hypothesis of a mean difference of zero, the ratio of that mean difference to its standard error will have a t distribution with $n - 1$ degrees of freedom:

$$t = \frac{\bar{D}}{s_D/\sqrt{n-1}}$$

$$= \frac{\bar{D}}{s_{\bar{D}}} \qquad (14.2.11)$$

where $\bar{D}$ = the mean of the pair-by-pair differences,

s_D = standard deviation of pair-by-pair differences,

n = number of pairs.

At least three kinds of correlated samples may be encountered in sociological data: (1) cases matched in pairs by design as a control for variables not of immediate interest; (2) cases occurring in "natural" pairs, as in the illustration of married couples; and (3) measures on the same case at different times. We illustrate the application of the *t*-test for correlated samples with the third of these three types.

Assume that a panel study of consensus (on conceptions of women's roles) for a sample of student living groups yields the data of Table 14.2.5. The consensus of each of 10 groups has been determined at the beginning and at the end of the school year; interest centers in whether mean consensus has increased during this period as a consequence of social interaction within the groups. The substantive hypothesis is that the mean of the change scores $(X_2 - X_1)$ will be positive. The null hypothesis of a mean change of zero is tested to determine whether the observed mean difference would be common as a sampling error around a true difference of zero. In symbols:

$H_0 : \mu_D = 0,$

$H_1 : \mu_D > 0.$

Table 14.2.5 *Consensus among Student Living Groups on Conception of Women's Rules, September and June*

Living Group	Consensus Scores		Change $= X_2 - X_1 = D$ (June–September)
	September X_1	June X_2	
a	25	26	1
b	39	45	6
c	42	45	3
d	56	52	−4
e	67	78	11
f	68	75	7
g	72	86	14
h	77	77	0
i	81	76	−5
j	85	87	2

$\bar{D} = 3.5$

$s_D = 5.8$

$$s_{\bar{D}} = \frac{5.8}{\sqrt{9}} = 1.95$$

Having specified a direction in the alternative hypothesis, a one-tailed test is indicated.

To test the null hypothesis, we utilize formula 14.2.11, substituting in it the appropriate values from Table 14.2.5:

$$t = \frac{3.50}{1.95}$$

$$= 1.80.$$

Since we have 10 differences in the sample, we have 9 degrees of freedom for the t-test. Referring to Appendix Table V, we find that, with 9 degrees of freedom, the observed t must exceed 1.833 to reject the null hypothesis at the .05 level of significance, one-tailed test. Our observed t is smaller than this critical t; therefore we conclude that, at this level of significance, these data do not provide convincing evidence that consensus has increased during the course of the year. Because the sample is small, making the test relatively insensitive to the existence of a true change if there is one, and because the observed t value falls only barely short of the critical value, the finding might be interpreted as suggesting the need for an expanded study with an increase in sample size.

QUESTIONS AND PROBLEMS

1. Define the following terms:
   critical ratio
   sampling distribution of the difference between means
   standard error of the difference between means
   Student's t distribution
   degrees of freedom
   random assignment
   correlated samples
   independent samples

2. Each member in a series of laboratory groups was allowed to propose, by secret ballot, a portion of group "earnings" to be allocated to their respective group leaders. A randomly selected 10 of these groups had been "successful" by design, while the other 10 had been "unsuccessful" by design, although the amount of earnings available for each group to allocate was identical for all groups. Interest centered in the effect of group success on the amount allocated to the group leader. Hypothetical results are shown in Table 14.2.6. Use the t-test to determine whether the null hypothesis of no difference in mean reward should be rejected at the .05 level of significance.

Table 14.2.6 *Reward Allocated to Group Leader in "Successful" and "Unsuccessful" Laboratory Groups*

Mean Reward (in Cents) Suggested for the Group Leader by Four Other Members of the Group			
Successful Groups		Unsuccessful Groups	
Group Number	Reward	Group Number	Reward
1	137	11	102
2	94	12	81
3	165	13	106
4	107	14	92
5	120	15	116
6	131	16	73
7	124	17	98
8	111	18	105
9	117	19	119
10	128	20	122

4. In a survey of heads of households, respondents were asked whether they expected to move to a new address within the next five years. Those who answered "Yes, definitely" or "Yes, probably" were combined; they numbered 283 and had a mean age of 38.2 years with a standard deviation of 8.1. Those who answered "No, definitely not," "No, probably not," or "Don't know" were combined; these respondents numbered 277 and had a mean age of 41.5 years with a standard deviation of 7.6. With these data should the investigator conclude, as was originally proposed, that in the population sampled, those who move are younger than those who do not? (Use .05 level of significance.)

5. From a pool of potential experimental subjects, 12 pairs matched on GPA were selected. One subject in each pair then was randomly assigned to an experimental treatment, the other subject being assigned to the control. Hypothetical results are shown in Table 14.2.7. Use the t-test for correlated samples and the .05 level of significance to test the hypothesis of no experimental effect.

6. In a sample survey study of public perceptions of the Watts riot (Los Angeles, 1965) the investigators were interested in the correlates of perception of the riot as social protest. They found that among the 278 respondents who indicated a positive attitude toward civil rights, 59 percent perceived the riot as social protest, as compared to 37 percent of the 257 respondents who indicated a negative attitude toward civil rights. Test the hypothesis of no relationship between civil rights attitude and perception of the Watts riot as social protest in the population from

Table 14.2.7 *Scores for Matched Pairs Assigned to Experimental and Control Groups*

Matched Pair Number	Scores	
	Experimental Pair Member	Control Pair Member
1	13	5
2	7	3
3	10	9
4	12	8
5	12	6
6	12	3
7	10	5
8	5	1
9	8	1
10	9	4
11	9	2
12	13	7

which these investigators drew their sample (that is, test the null hypothesis that the proportion perceiving the riot as social protest does not differ among those with positive and negative attitudes toward civil rights). Use the .01 level of significance.[1]

3 DISTRIBUTION-FREE TESTS FOR A DIFFERENCE IN LOCATION

Although the above tests for a difference between population means are useful for many hypothesis-testing problems, the probabilities yielded by their application will be valid only to the degree that specified conditions are met: Testing the hypothesis of no difference between means requires an interval level of measurement and, for small samples, normally distributed populations. When available clues indicate radical departures from these conditions, we have recourse to tests which utilize only the ordinal features of the data and which do not specify the form of the parent distributions. Because no assumptions about the form of the parent distribution are required, they are called *distribution-free tests.* And because such comparisons do not entail the estimation of parameters, they are also called *nonparametric tests.*

[1] See Vincent Jeffries, Ralph Turner, and Richard T. Morris, "The Public Perception of the Watts Riot as Social Protest," *American Sociological Review* 36 (June, 1971): 443–451, Table 3 at p. 447.

Like the tests for a difference between means, the nonparametric tests described in this section are designed to detect a difference in the *location* of two distributions along a continuum, if such a difference exists. With interval measures, the location of each distribution may be represented by its mean, and testing for a difference between means thus provides a test for a difference in location. The nonparametric tests discussed in this section utilize clues to a difference in location other than a difference in central tendency. If interval measures are lacking, the location of each distribution and the difference between them cannot be in terms of such quantities as the mean or median, range or quartile range. But if cases from two populations can be ordered relative to each other along the same continuum, clues to a *difference* in location will appear, even though the two distributions cannot be precisely located along their common scale. If the members of one group are known typically to exceed members of the other group in income, we readily sense that their respective income distributions are differently located along the continuum of money income, even though the actual incomes of individuals remain unknown. Paradoxically, then, we may discover a difference in location of two groups even though we do not know the location of either.

The Mann–Whitney Test Imagine two population distributions which are identical in every respect. To make this image more concrete, imagine such identical distributions of morale scores (interval measures) for men and women. We now draw a sample of cases from each population. Since the two populations are assumed identical, we would expect the values in the two samples to have very similar magnitudes. One way of measuring the degree to which the two samples are similar in their values is to compare each value in one sample with every value in the other, determining for each such comparison which value is larger. If, as anticipated, the two samples are composed of similar values, we would expect that in approximately one-half of such pairs, the man's score would be larger than the woman's score, whereas in the other one-half of the pairs, the woman's score would exceed the man's. The pairing and comparison of values from two very small hypothetical samples is illustrated in Table 14.3.1.

With n_1 cases in the first sample and n_2 cases in the second, there are $n_1 \times n_2$ possible pairs of cases composed of one case from each sample. Thus, in Table 14.3.1, with four cases in each sample, there are $4 \times 4 = 16$ possible pairs, all of which are shown in the table. Assuming no pair is composed of tied scores, in each of these pairs the score from the first sample either exceeds the score from the second, or vice versa. We count the number of pairs in which the score from the first exceeds the score from the second and designate that number by the symbol U. We also count the number of pairs in which the score from the second sample exceeds the score from the first (or subtract U from $n_1 \times n_2$) and designate that number by the symbol U'. The sum of U and U' must be equal to the total number of pairs, or $n_1 n_2$. If the populations from which the samples are drawn are identical in every

Table 14.3.1 *All Possible Pairs of Elements from Two Samples and a Comparison of Their Magnitudes*

Sample of Women's Score: 9, 16, 35, 63
Sample of Men's Score: 12, 20, 28, 71

Pair Number	Score		Woman's Score Larger than Man's	Man's Score Larger than Woman's
	Woman	Man		
1	9	12		×
2	9	20		×
3	9	28		×
4	9	71		×
5	16	12	×	
6	16	20		×
7	16	28		×
8	16	71		×
9	35	12	×	
10	35	20	×	
11	35	28	×	
12	35	71		×
13	63	12	×	
14	63	20	×	
15	63	28	×	
16	63	71		×

Total number of pairs in which woman's score exceeds man's: $7 = U$
Total number of pairs in which man's score exceeds woman's: $9 = U'$

respect, we expect U and U' to be approximately equal to $\frac{1}{2}(n_1 n_2)$. However, the statistics U and U', like other sample statistics, are subject to sampling variation, and therefore we must allow for sampling error in the observed values.

From the sampling distribution of U under the hypothesis that the populations differ in no respects, we may determine the probability of each possible value of U; then we may designate certain values of U, far removed from the expected value of $n_1 n_2/2$, which would cast doubt on the assumption that the populations are identical. On the other hand, if the populations were not identical but were differently located, we would expect this difference in location typically to be reflected in the sample Us. Then we would expect the scores in one sample to be larger than the scores in the other, and expect U to deviate markedly from $n_1 n_2/2$. (U' also would deviate from $n_1 n_2/2$ but in the opposite direction.) Thus, observed values of U that would be improbable under the hypothesis of no difference between population distributions would be more common under the alternative hypothesis of a difference in location.

To utilize the statistics U and U' in testing the null hypothesis, their exact sampling distribution must be determined. Such computations have been made and critical values of these statistics have been tabled for reference purposes (Appendix Table V). The tables are constructed so that the smaller of the two statistics serves as the test statistic; either U or U', *whichever is smaller*, is compared to the critical value in the table for the appropriate sample sizes and level of significance. If the observed value is as small or smaller than the critical value shown in the table, the hypothesis is rejected at that level of significance.

Except for extremely small samples, the computation of U and U' by listing all possible pairs, as in Table 14.3.1, would be very tedious. A more convenient computing procedure is:

(1) Rank the values from both samples in a single series. Thus, for the data in Table 14.3.1, we assign ranks as follows:
Ranks for sample of women's scores: 1, 3, 6, 7.
Ranks for sample of men's scores: 2, 4, 5, 8.

(2) Sum the ranks for each sample separately.
Sum of ranks for women's scores: $1 + 3 + 6 + 7 = 17 = R_1$.
Sum of ranks for men's scores: $2 + 4 + 5 + 8 = 19 = R_2$.
As a check,

$$R_1 + R_2 = \frac{n(n+1)}{2},$$

$$17 + 19 = \frac{8 \times 9}{2} = 36.$$

(3) Compute U:

$$U = n_1 n_2 + \frac{n_2(n_2+1)}{2} - R_2. \tag{14.3.1}$$

For the data of Table 14.3.1:

$$\begin{aligned} U &= 16 + \frac{4 \cdot 5}{2} - 19 \\ &= 26 - 19 \\ &= 7. \end{aligned}$$

(4) Compute U':

$$\begin{aligned} U' &= n_1 n_2 - U \\ &= 16 - 7 \\ &= 9. \end{aligned} \tag{14.3.2}$$

(5) Select the smaller of the two values, U or U', for comparison with the critical values in the appropriate table. (The sample sizes in Table 14.3.1 are so small that no critical values are given in Appendix Table V.)

The hypothetical data of Table 14.3.2 may be used to illustrate the application of the Mann–Whitney test and the use of Appendix Table V. These data give the results of a study of supervisors and work crews, randomly drawn from a population of supervisors in a large industrial plant. On the basis of their responses to a set of questions, each supervisor has been classified as "employee oriented" or as "company oriented." From company records, an absentee rate has been computed for each crew, with the result shown in the table. The question to be answered is whether these data cast doubt on the assumption of identically distributed rates for each type of

Table 14.3.2 *Absentee Rates for Work Crews with "Company Oriented" and "Employee Oriented" Supervisors*

Crews with "Company Oriented" Supervisors ($n = 10$)		Crews with "Employee Oriented" Supervisors ($n = 14$)	
Absentee Rate (X_c)	Rank	Absentee Rate (X_e)	Rank
		.7	1
		1.2	2
		1.3	3
1.5	4		
		1.6	5.5
		1.6	5.5
		1.8	7
		1.9	8
		2.0	9
		2.1	10
2.7	11		
		2.8	12
3.2	13		
		3.3	14
3.5	15		
3.6	16		
3.7	17		
		3.9	18
		4.1	19
4.4	20		
		5.2	21
6.3	22		
7.8	23		
8.0	24		
Sum of Ranks	$165 = R_1$		$135 = R_2$

supervisor. We assume that the company records indicate that the distribution of absentee rates for such crews is not normal but is heavily skewed to the right, making the application of the t-test inadvisable. However, the Mann–Whitney test is applicable regardless of what form the population distribution may take.

On the assumption that employee oriented supervisors will have a positive effect on employee morale and, as a consequence, will reduce absenteeism, we formulate the substantive hypothesis that the absentee rates for crews under employee oriented supervisors will tend to be lower than the rates for crews under company oriented supervisors. Therefore the rival null hypothesis of identical distributions of rates is tested. In symbols:

$$H_0: Pr(X_e < X_c) = .5,$$

$$H_1: Pr(X_e < X_c) > .5,$$

where X_e = absentee rate under employee oriented supervisors,

X_c = absentee rate under company oriented supervisors.

The null hypothesis may be read as follows: The probability is one-half that, among all possible pairs in the population sampled, the rate for crews under an employee oriented supervisor (X_e) is smaller than the rate for crews under a company oriented supervisor (X_c). The alternative hypothesis states that this probability exceeds 0.5. A one-tailed test is required, and we select .05 as the level of significance.

With two samples of size 10 and 14, we could form 140 pairs, compare the magnitudes of rates in each pair, and compute U and U' by counting. To avoid such listing we apply Formulas 14.3.1 and 14.3.2:

$$\begin{aligned} U &= 140 + \frac{14 \times 15}{2} - 135 \\ &= 245 - 135 \\ &= 110. \\ U' &= n_1 n_2 - U \\ &= 140 - 110 \\ &= 30. \end{aligned}$$

The smaller of the two values, $U' = 30$, serves as the test statistic. Referring to the table for the level of significance at .05, direction predicted, with $n_1 = 10$ and $n_2 = 14$, the critical value is found to be 41. Since we reject the null hypothesis when the smaller of the two observed statistics, U or U', is less than the critical value, these data, with $U' = 30$, lead to the rejection of the hypothesis at this level of significance.

Appendix Table V is designed for small samples only; in the illustration above, if the number of cases in either sample had exceeded 20, the table would not have provided the critical value. If both samples are larger than 20, the following statistic, which bypasses the computation of U and U' but which is based on the same line of reasoning, has an approximately normal distribution:

$$z = \frac{R_1 - R_2 - (n_1 - n_2)(n_1 + n_2 + 1)/2}{\sqrt{n_1 n_2(n + n_2 + 1)/3}}. \tag{14.3.3}$$

Critical values for this statistic may be determined by referring to the table of normal probabilities.

The Kolmogorov–Smirnov (K–S) Test This easily applied test may be used to test for a difference in location when two samples are large and there are many tied ranks between samples. Under these same circumstances, the Mann–Whitney test is tedious and complex in application. The K–S test is based on the sampling distribution of the maximum difference between the cumulative frequency distributions of the two samples, under the hypothesis that the populations from which the samples come are identical in every respect. When this hypothesis is true, random samples from the two populations are expected to have similar cumulative distributions so that the largest difference between them would typically be small, but subject to sampling variation. A large maximum difference, which would rarely occur in samples from identically distributed populations, would be more common in samples from populations that differ from each other. The hypothesis of no difference between population distributions is rejected when the maximum difference between cumulative frequency distributions is so large that it would be rare if the hypothesis were true.

The data of Table 14.3.3 will be used to illustrate the test of the null

Table 14.3.3 *Number of Correct Definitions of Criminal Argot (Slang) Terms for Boys and Girls, 16–19 Years of Age*

Correct Definitions	Boys			Girls			Difference
	f	*c/f*	*c/p*	*f*	*c/f*	*c/p*	
None	25	25	.40	36	36	.43	.43 − .40 = .03
1–2	9	34	.54	21	57	.68	.68 − .54 = .14
3–5	8	42	.67	17	74	.88	.88 − .67 = .21
6 or more	21	63	1.00	10	84	1.00	
Total	63			84			

Source: Adapted from Paul Lerman, "Argot, Symbolic Deviance and Subcultural Delinquency," *American Sociological Review* 32 (April, 1967), Table 1, p. 216. Reprinted with permission.

hypothesis with the K–S test. Because the frequency distribution of the number of correctly defined criminal argot terms among boys suggests either bimodality or a clear skew to the right, and because the distribution among girls is also skewed to the right, the application of the t-test for a difference between means is inadvisable. If the number of ordered categories (that is, the class intervals of number of correct definitions) had been less than four, the K–S test also would have been inapplicable. However, the test may be applied in this instance, and the existence of unequal and open intervals and the apparently nonnormal population distribution presents no difficulty. In the sample, boys correctly define more argot terms, on the average, than do girls. The question to be answered by the K–S test is whether or not the cumulative distributions for boys and girls are sufficiently different to reject the hypothesis of identical population distributions for boys and girls.

On the basis of the frequency distributions, we obtain the cumulative distribution of proportions for each sample. The differences between the cumulative proportions for the two samples are computed, and the largest of these differences is found to be .21.

Assuming that both samples are larger than 40, we may test the two-tailed hypothesis of identical distributions against the critical values given in Table 14.3.4. At the .05 level of significance, with $n_1 = 63$ and $n_2 = 84$, as in the illustrative data on knowledge of criminal argot, the maximum difference between cumulative proportion distributions must exceed:

$$1.36\sqrt{\frac{63 + 84}{63 \cdot 84}} = 1.36\sqrt{.0278}$$

$$= 1.36(.167)$$

$$= .227.$$

Since the observed maximum difference is less than the critical value, the null hypothesis cannot be rejected at the .05 level of significance.

Table 14.3.4 *Critical Values of* D *for the Two-Tailed Kolmogorov–Smirnov Test*

Level of Significance	Critical D
.10	$1.22\sqrt{\frac{n_1 + n_2}{n_1 n_2}}$
.05	$1.36\sqrt{\frac{n_1 + n_2}{n_1 n_2}}$
.01	$1.63\sqrt{\frac{n_1 + n_2}{n_1 n_2}}$

The Relative Power of Nonparametric Tests Tests of the null hypothesis sometimes are compared with each other in their sensitivity to the falsity of the null hypothesis. The probability of rejecting the null hypothesis when it is false is termed the *power* of the test, and this probability depends upon the alternative true hypothesis, sample size, the level of significance, and the choice of a one- or two-tailed test. When all these factors remain constant, tests may be compared in terms of their power. Under circumstances where the t-test and the Mann–Whitney test are both appropriate, the Mann–Whitney test is found to be somewhat less sensitive to the falsity of the null hypothesis, that is, it has lower power, than the t-test. The same is true for the K–S test. The difference in power between the t-test and the nonparametric tests discussed here is relatively slight—an increase of approximately 6 percent in sample size is usually sufficient to offset the lower power of the nonparametric test. Because of the difference in power, the t-test would be preferred whenever the necessary assumptions are met. If the necessary assumptions for the t-test are not met, however, the nonparametric tests are preferable since, under those circumstances, the probabilities derived from the use of the t-test would be less accurate.

QUESTIONS AND PROBLEMS

1. Define the following terms:
   nonparametric test
   distribution-free test
   sum of ranks
   maximum D
   power of a test

2. A sample of youth gangs observed by detached workers were ranked on their cohesion by a set of judges who were given only that information presumed to be relevant to cohesion. On the basis of other information not available to the judges, the gangs were classified as "heavily delinquent" and "mildly delinquent." The rankings (Rank 1 to the most cohesive, Rank 2 to the next most cohesive, and so on) of the gangs, classified by degree of involvement in delinquency, were as follows:
   (a) Heavily delinquent: 3, 7, 9, 10, 11, 14, 16, 17, 18, 19, 21.
   (b) Mildly delinquent: 1, 2, 4, 5, 6, 8, 12, 13, 15, 20.
   Do these data cast doubt on the hypothesis of no difference in cohesion between heavily delinquent and mildly delinquent gangs at the .05 level of significance?

3. Table 14.3.5 represents an attempt to measure the impact of the Kinsey findings on tolerance toward sexual practices.
   Use the Kolmogorov–Smirnov test and the .05 level of significance to test the hypothesis that exposure to the Kinsey data has no impact on tolerance toward sexual practices.

Table 14.3.5 *Impact of Kinsey Findings, Sex Tolerance before and after Exposure, 1952*

Groups	Point Change in Direction of Tolerance				Total
	0	1	2	3 & over	
Experimental (with Kinsey data)	32	27	21	24	104
Control (without Kinsey data)	52	21	21	12	106
					210

Source: C. Kirkpatrick, S. Stryker, and P. Buell, "Attitudes Toward Male Sex Behavior," *American Sociological Review* 17 (1952), abbreviated Table 3, p. 584. Reprinted with permission.

4 TESTING HYPOTHESES ABOUT CORRELATION AND REGRESSION COEFFICIENTS

The magnitude of the product-moment correlation is as subject to sampling variation as any other sample statistic. Thus the correlation between talkativeness and popularity may be zero in the population, but the observed correlation coefficients in a series of random samples will vary around zero due to sampling variation, and for small samples the observed correlations will vary widely around a population coefficient of zero. The single sample available to a particular investigator may thus yield a correlation coefficient that differs from zero only because of sampling variation. As a rule, therefore, before rendering an elaborate substantive interpretation of a sample r, we should reckon with the possibility that there is actually no linear correlation at all.

In principle, testing the null hypothesis that an observed r differs from a population correlation of specified magnitude only because of sampling variation does not differ from the other tests against the null hypothesis discussed above: We assume the truth of the null hypothesis and then test that assumption by calculating the probability of obtaining an r, in a sample of the given size, that differs from the hypothesized value as much or more as the observed r. We may once again employ the normal probability distribution, because the sampling distribution of r is approximately normal around a population correlation of 0, with standard error, σ_r, estimated by the following formula:

$$\sigma_r = \frac{1 - r^2}{\sqrt{n - 2}}. \tag{14.4.1}$$

The t value for an observed r based on a sample of size n is thus:

$$t = \frac{(r - \rho)\sqrt{n-2}}{1 - r^2}. \tag{14.4.2}$$

To test the null hypothesis that the correlation in the population is zero (H_0: $\rho = 0$), we assume $\rho = 0$, in which case the sample rs will be distributed normally around zero, and t for the observed r is given by:

$$t = \frac{r\sqrt{n-2}}{1 - r^2}. \tag{14.4.3}$$

We may also have occasion to test the null hypothesis that the partial regression coefficient, β (beta), for any one of the predictor variables in a multiple regression equation differs from zero only because of sampling variation. This null hypothesis is taken as the equivalent of the hypothesis that in the population the predictor in question adds nothing to the variance explained by the other predictors, and consequently that the appearance of an *addition* to the explained variance in the sample is the result of sampling variation alone. Hence the test statistic is based on the observed difference between the variance explained with and without the predictor in question. We do not present here the derivation of the test statistic, which may be generally described as follows:

$$t = \sqrt{\frac{\text{Explained variation added by } X_i}{\frac{1}{\text{d.f.}}\ \text{Error variation}}}. \tag{14.4.4}$$

A convenient computational formula relies on the proportion of variation explained (R^2), and may be written as follows:

$$\begin{aligned} t_{n-k-1} &= \sqrt{\frac{(n-k-1)(R^2 \sum y^2 - R_1^2 \sum y^2)}{(1 - R^2)\sum y^2}} \\ &= \sqrt{n-k-1}\sqrt{\frac{R^2 - R_1^2}{1 - R^2}} \end{aligned} \tag{14.4.5}$$

where n = the number of cases;

k = the number of predictors (including the predictor in question);

R^2 = the square of the multiple correlation coefficient based on all predictors;

R_1^2 = the square of the multiple correlation coefficient based on all predictors except X_i;

t_{n-k-1} = indicates that the sampling distribution of the statistic (under the null hypothesis that the regression coefficient for X_i is zero in the population) is the t distribution with $n - k - 1$ degrees of freedom.

We illustrate the application of this test against the null hypothesis by reference to the data of Table 10.3.2, showing the zero-order correlations between Occupational Attainment (Y), Parental SES (X_1), and High School Grades (X_2) for 435 black American males. As indicated in the text accompanying Table 10.3.2, $b_{Y1.2} = .155$, $b_{Y2.1} = 4.12$, and $R^2 = .034$.

We test the null hypothesis that Parental SES (X_1) makes no contribution to explained variance beyond that achieved by High School Grades (X_2)—that is, $H_0\colon \beta_{y1.2} = 0$—against the alternative—$H_1\colon \beta_{y1.2} > 0$, at the .05 level of significance. The directionally specific alternative indicates that a one-tailed test is appropriate. We note that in this instance, with only two predictors, the variance explained *without* X_1 is given by $r_{y2}^2 = (.134)^2 = .018$. Hence

$$
\begin{aligned}
t &= \sqrt{435 - 2 - 1}\sqrt{\frac{.035 - .018}{1 - .035}} \\
&= \sqrt{432}\sqrt{\frac{.017}{.965}} \\
&= 20.8\sqrt{.018} \\
&= 20.8(.134) \\
&= 2.79.
\end{aligned}
$$

The critical level of t with 432 degrees of freedom for a one-tailed test at the .05 level of significance is the corresponding critical level of z: 1.645. Since the t value for the observed regression coefficient is in excess of this critical value, the null hypothesis that $\beta_{Y1.2} = 0$ may be rejected. As an exercise, the student may wish to confirm that the null hypothesis that $\beta_{Y2.1} = 0$ may also be rejected at the .05 level of significance ($t = 2.79$).

In the special case of a single predictor (or bivariate regression), $R^2 = r^2$ and all explained variation is achieved by the single predictor; hence the explained variation "added" by that predictor is r^2. Thus in the special case of bivariate regression, the t-test for the null hypothesis that $\beta = 0$ (Formula 14.4.5) becomes

$$
\begin{aligned}
t &= \sqrt{n - 1 - 1}\sqrt{\frac{r^2 - 0}{1 - r^2}} \\
&= \frac{r\sqrt{n - 2}}{1 - r^2}.
\end{aligned}
$$

The student will note that this equation is identical to the test statistic for the null hypothesis that $\rho = 0$ (Formula 14.4.3). This is correct because the population correlation is zero if and only if the regression coefficient in the population is zero.

The standard error of an unstandardized regression coefficient is frequently useful, especially in setting confidence intervals for such coefficients, and standard errors for regression coefficients are typically included in the print-out of computer programs for regression analysis. Such a standard error is, of course, the estimated standard deviation of the normal sampling distribution of the regression coefficient. Although such standard errors are, in this sense, conceptually similar to the standard errors previously discussed, the computation of the standard error of a partial regression coefficient is beyond the scope of this book. Interested readers with a facility in matrix algebra are referred to Johnston (1972) for an authoritative treatment of this topic.

QUESTIONS AND PROBLEMS

1. In a study of 16 social welfare organizations, the correlation betweeen the degree of general participation in organizational decision making and the rate of program change was found to be .49. Assuming the 16 organizations constitute a random sample of welfare organizations, should the null hypothesis of no correlation be rejected at the .05 level of significance?[1]

2. The partial regression coefficients for occupational attainment on parental SES and high school grades for white and for black males are provided in Chapter 10, Section 3. Test the null hypothesis that these regression coefficients are zero in the populations from which each sample was drawn.

SELECTED REFERENCES

Fisher, Ronald A.
1960 The Design of Experiments. Edinburgh: Oliver and Boyd. Chapter 1.

Johnston, J. J.
1972 Econometric Methods. 2nd Edition. New York: McGraw-Hill.

Kish, Leslie
1959 "Some statistical problems in research design." American Sociological Review 24 (June): 328–338.

[1] Source: Jerald Hage and Michael Aiken, "Program Change and Organizational Properties: A Comparative Analysis," *American Journal of Sociology* 72 (March 1967): 509.

McGinnis, Robert
1959 "Randomization and inference in sociological research." American Sociological Review 23 (August): 408–414.

Selvin, Hannan C.
1957 "A critique of tests of significance in survey research." American Sociological Review 22 (October): 519–527.

Siegel, Sidney
1956 Non-parametric Statistics for the Social Sciences. New York: McGraw-Hill.

Hypothesis-Testing 15

The procedure for testing the hypothesis that two populations have the same mean is all right as far as it goes, but it does not go far enough. This procedure will not suffice if we wish to test the hypothesis that *three* populations have the same mean; it will not do to test the hypothesis that *four* populations have the same mean. To test these hypotheses, and extensions of them, we require a more general method, the *analysis of variance*. This method will permit us to compare any number (k) of samples at the same time and to decide whether they came from populations identical in their means. Thus, we would have recourse to the analysis of variance to test the hypothesis that Baptists, Methodists, and Presbyterians have the same average annual family income. We would also call upon the analysis of variance to decide whether differences between the averages of these denominations are the same from one region to another. For instance, we might suppose that Presbyterians rank first in the North and Baptists in the South, and that Methodists rank second in both regions. Such speculation will force itself upon us whenever measures have been classified by two or more characteristics (factors), rather than by one alone. The analysis of variance is a remarkably versatile technique for drawing inferences about the differences between populations and the factors producing those differences. In this respect, it is closely similar to multiple regression; its very simplest forms are presented in Sections 1 and 2 of this chapter.

Just as the analysis of variance permits us to compare the means of any number of populations, *contingency table analysis* (CTA) permits us to compare any number of populations in their proportions (or ratios). It stands in relation to attributes as the analysis of variance stands in relation to interval measures. Thus, we would use *CTA* to determine whether the proportion favoring equal job chances for blacks is the same for grade school, high school, and college graduates. If respondents had also been classified by region, we could apply *CTA* to investigate the effect of education on that opinion after making allowances for the effect of region, the effect of region after making allowances for education, and the effect of region on the effect of education, and vice-versa. Regarding this latter possibility, we might suppose, as a working hypothesis, that education makes less difference in the South than in the North. The introduction of yet another variable, like religion, would probably enhance the sociological significance of our investigation and permit us to assess the effect of education after making allowances for both religion and region, for example. Contingency table analysis is designed to handle such multiple classifications. In its versatility it is strictly parallel to the analysis of variance. Its simplest forms are presented in Sections 3 and 4 of this chapter.

1 ONE-WAY ANALYSIS OF VARIANCE

The procedure for handling three or more sample means differs in detail but not in principle from the procedure for comparing two means at a time. We (1) state the hypothesis to be tested, (2) fix the risk of rejecting the hypothesis when true, (3) draw the samples, (4) calculate the test-statistic, (5) locate the obtained test-statistic in the sampling distribution, and (6) decide for or against the null hypothesis.

The hypothesis to be tested is that k $(k \geq 2)$ populations have the same mean. In symbols:

$$H_0\colon \mu_1 = \mu_2 = \cdots = \mu_k\,.$$

When we are dealing with three sample means, our hypothesis reduces to

$$H_0\colon \mu_1 = \mu_2 = \mu_3.$$

To simplify our presentation, we will limit ourselves to the case of three populations.

As noted in previous discussions, our significance level will depend on the practical consequences of rejecting a hypothesis when true, or committing a so-called *Type I Error*. If the consequences are very serious, we will want to take only a small risk of rejecting a true hypothesis; on the other hand, if the consequences are of no gravity, and perhaps are no more than academic, we will be willing to risk a much larger Type I Error.

In testing the hypothesis that three populations have the same mean, our procedure is to select cases randomly from each of these three populations. It is not necessary that samples be equal in size, nor do they have to be large; but cases must be randomly drawn both within and between samples. The theory behind the analysis of variance rests on the assumption that measures are mutually independent; the fulfillment of this assumption is assured by random sampling.

Next is the calculation of the test-statistic. This is the F-ratio, named in honor of R. A. Fisher, the British statistician who derived it. It reflects the magnitude of differences between sample means relative to the magnitude of differences among individual measures within samples. Since relatively large differences between sample means occur infrequently when the null hypothesis is true, such differences weaken our confidence in that hypothesis; on the other hand, relatively small differences will strengthen our confidence in the hypothesis, since they will occur frequently. This generalization will be expressed more precisely in the ensuing discussion.

Total Sum of Squares and Component Sums of Squares In unfolding the logic of the F-ratio, we take as our point of departure the connection between the total sum of squares and its component sums of squares. To demonstrate this connection, we operate on three samples of three measures each for a total of nine measures. These measures are shown in Table 15.1.1. Since measures in this table are classified in one way, we speak of it as a one-way table. It is pertinent that measures within the rows of a one-way table may be arranged in any order without affecting in any way the results of the analysis of variance.

In this demonstration, our first step is to calculate the sum of the squared deviations from the mean of all measures combined—the so-called *total sum of squares*. In symbols,

$$TSS = \sum_{1}^{k} \sum_{1}^{n_i} (Y_{ij} - \bar{Y}_{.})^2 \qquad (15.1.1)$$

where TSS is an abbreviation for total sum of squares; k = number of samples; n_i = number of cases in ith sample; Y_{ij} is jth measure in ith sample; $\bar{Y}_{.}$ = overall mean. In our problem, the mean of all nine measures is 8.

Table 15.1.1 *Given Measures*

		n_i	Sum	Mean
A_1	7, 8, 9	3	24	8
A_2	7, 8, 9	3	24	8
A_3	7, 8, 9	3	24	8

Subtracting 8 from all measures, squaring these deviations, and summing gives

$$
\begin{aligned}
TSS &= (7\text{–}8)^2 + (8\text{–}8)^2 + (9\text{–}8)^2 \\
&\quad + (7\text{–}8)^2 + (8\text{–}8)^2 + (9\text{–}8)^2 \\
&\quad + (7\text{–}8)^2 + (8\text{–}8)^2 + (9\text{–}8)^2 \\
&= 6.
\end{aligned}
$$

This is the sum of the squared deviations around the mean for all three samples combined.

Second we get the sum of the squared deviations between each measure and the mean of its own sample—the so-called *within-groups sum of squares*. It may be written as the sum of as many sums as there are samples (three in our case):

$$WSS = \sum_{1}^{k} \sum_{1}^{n_i} (Y_{ij} - \bar{Y}_i) \qquad (15.1.2)$$

where WSS is an abbreviation for "within-groups sum of squares," and $\bar{Y}_i$= mean of ith sample. Carrying out the required operations on each sample separately, we get:

A_1	A_2	A_3
$(7\text{–}8)^2 = 1$	$(8\text{–}8)^2 = 0$	$(8\text{–}8)^2 = 0$
$(8\text{–}8)^2 = 0$	$(7\text{–}8)^2 = 1$	$(9\text{–}8)^2 = 1$
$(9\text{–}8)^2 = 1$	$(9\text{–}8)^2 = 1$	$(7\text{–}8)^2 = 1$
2	2	2

Summing these within-sample sums of squares gives

$$
\begin{aligned}
WSS &= 2 + 2 + 2 \\
&= 6.
\end{aligned}
$$

Continuing with our demonstration, we now calculate the sum of the weighted squared deviations between the respective sample means and the overall mean, with the number of the cases in each sample (n_i) serving as weights. This sum has come to be known as the *between-groups sum of squares*. In symbols:

$$BSS = \sum_{1}^{k} n_i(\bar{Y}_i - \bar{Y}_{.})^2 \qquad (15.1.3)$$

Table 15.1.2 Once-Adjusted Measures

		n_i	Sum	Mean
A_1	7, 8, 9	3	24	8
A_2	10, 9, 11	3	30	10
A_3	12, 13, 11	3	36	12

where *BSS* is an abbreviation for "between-groups sum of squares." Subtracting the overall mean of 8 from the mean of the first sample gives 0; applying the same procedure to the second sample also gives 0, as does applying that procedure to the third sample. Because each deviation between sample mean and overall mean is 0, the sum of the weighted squared deviations between sample means and the overall mean is 0. In short $BSS = 0$.

We conclude from the foregoing computations that in the absence of differences between sample means, the total sum of squares (*TSS*) is equal to the within-groups sum of squares (*WSS*). Furthermore, since the between-groups sum of squares (*BSS*) is equal to zero, the ratio of *BSS* to *WSS*, which reflects the strength of *BSS* relative to *WSS*, is also equal to zero. Although this ratio is not our test-statistic, it is, as we shall presently see, an important element in the *F*-ratio, which is our test-statistic.

Continuing the demonstration: Let us now disturb the equality between the sample means of Table 15.1.1 by adding a constant value of 2 to each value in the second row and a constant of 4 to each value in the third row, leaving unchanged the values in the first row. These once-adjusted measures are presented in Table 15.1.2. Upon analyzing these measures, we will find that the within-groups sum of squares is unchanged, and that the increase in the total sum of squares is exactly (and necessarily) equal to the increase in the between-groups sum of squares.

To confirm these claims we first calculate *WSS*:

A_1	A_2	A_3
$(7-8)^2 = 1$	$(10-10)^2 = 0$	$(12-12)^2 = 0$
$(8-8)^2 = 0$	$(9-10)^2 = 1$	$(13-12)^2 = 1$
$(9-8)^2 = 1$	$(11-10)^2 = 1$	$(11-12)^2 = 1$
2	2	2

Adding up within-group sums of squares gives $WSS = 6$, as in the case of the given, or unadjusted measures.

Next we calculate the between-groups sum of squares according to Formula 15.1.3:

Group	Mean	Squared Deviation	Weighted Squared Deviation
1	8	$(8-10)^2 = 4$	$3(4) = 12$
2	10	$(10-10)^2 = 0$	$3(0) = 0$
3	12	$(12-10)^2 = 4$	$3(4) = 12$
			24

Adding up the weighted squared deviations gives $BSS = 24$. To verify that the inflation in TSS is due solely to the differences between sample means, we independently calculate TSS and subtract WSS from that sum to get BSS. Taking the overall mean of 10 out of the once-adjusted measures, squaring, and summing gives $TSS = 30$. Then

$$
\begin{aligned}
BSS &= TSS - WSS \\
&= 30 - 6 \\
&= 24,
\end{aligned}
$$

confirming our point that the enlargement of TSS from 6 to 30 is exactly equal to the increase in BSS from 0 to 24. And we might now say that the differences between sample means account for 80 percent of the total sum of squares. The remainder of 20 percent is attributable to the variation within samples.

Instead of focusing on the contribution of differences among means to the total sum of squares, as in the foregoing analysis, we might have focused advantageously on the total sum of squares as the sum of the within-groups sum of squares and the between-groups sum of squares. In symbols,

$$TSS = WSS + BSS. \tag{15.1.4}$$

This identity is practically important because, given any two of these sums, we may get the third by addition or subtraction: given TSS and BSS, we may obtain WSS by subtraction; given BSS and WSS, we may get TSS by addition. In getting component sums of squares, the usual procedure is to get TSS and BSS independently, and then subtract the latter from the former to get WSS.

Twice-Adjusted Measures Continuing with our analysis, let us now inflate the variation of measures within samples by adding 1 to the largest value and subtracting 1 from the smallest value in each sample. These twice-adjusted measures are displayed in Table 15.1.3. Since sample means are unchanged by this adjustment, we anticipate that the change in the total sum of squares will be equal to the change in the within-groups sum of squares. Upon calculating the total sum of squares, we get 48, an increase of $48 - 30 = 18$; upon calculating the within-groups sum of squares, we obtain

Table 15.1.3 *Twice-Adjusted Measures*

		n_i	Sum	Mean
A_1	6, 8, 10	3	24	8
A_2	10, 8, 12	3	30	10
A_3	12, 14, 10	3	36	12

24, an increase of 24 − 6 = 18. Since *TSS* and *WSS* both increase by 18, and since *BSS* is unaltered, we conclude that the greater total variation is due solely to the greater variation of measures within samples.

Although the absolute magnitude of *BSS* is unchanged, its relative contribution to *TSS* has declined, as reflected by the ratio of *BSS* to *WSS*. For the once-adjusted measures this ratio is 4 to 1; for the twice-adjusted measures, it is 1 to 1. For the once-adjusted measures, *BSS* represents 80 percent of the total sum of squares; for the twice-adjusted measures, it represents 50 percent of the total. Our conclusion is that constant differences between sample means will contribute a little or a great deal to the total variation, depending on the heterogeneity of measures within samples.

The *F*-Ratio While the ratio of *BSS* to *WSS* is simple both to obtain and to interpret, it is practically impossible to use as a test of the hypothesis of identical population means. To convert that ratio into a useable test-statistic, we divide it by the ratio of $k-1$ to $N-k$ (or multiply it by the ratio of $N-k$ to $k-1$). This ratio of ratios (or product of ratios) is the *F*-ratio:

$$F = \frac{BSS}{WSS} \bigg/ \frac{k-1}{N-k} = \left(\frac{BSS}{WSS}\right) \times \left(\frac{N-k}{k-1}\right). \tag{15.1.5}$$

Depending on the calculated value of *F* in a given problem, and our chosen level of significance, we will either reject the null hypothesis at that level of significance, or retain it.

The *F*-Table In making our decision to retain or reject the null hypothesis, we refer to a table of *F*-ratios, each corresponding to a given combination of $k-1 = v_1$ and $N-k = v_2$, that will be exceeded a specified proportion of times when the null hypothesis of identical means is true. Here are a few *F*-ratios (see Table IV, Appendix) that will be exceeded 5 percent of the time when the null hypothesis of identical means is true:

$N{-}k = v_2$	$k{-}1 = v_1$			
	1	2	3	4
3	10.13	9.55	9.28	9.12
4	7.71	6.94	6.59	6.39
5	6.61	5.79	5.41	5.19
6	5.99	5.14	4.76	4.53
8	5.32	4.46	4.07	3.84
10	4.96	4.10	3.71	3.48
12	4.75	3.88	3.49	3.26
14	4.60	3.74	3.34	3.11
16	4.49	3.63	3.24	3.01
18	4.41	3.55	3.16	2.93
20	4.35	3.49	3.10	2.87
25	4.24	3.38	2.99	2.76
40	4.08	3.23	2.84	2.61

Scanning this sample of entries we find, for instance, that F will exceed 3.24 five times in 100 with $k{-}1 = 3$ and $N{-}k = 16$; with $k{-}1 = 1$ and $N{-}k = 40$, it will exceed 4.08 five times in 100. If our calculated value of F is larger than the tabulated value of F at our chosen level of significance, we reject the null hypothesis of identical means; otherwise we retain it, at least tentatively. By way of example, consider the F-ratio for Table 15.1.3.

$$F = \tfrac{24}{24} \times \tfrac{6}{2} = 3.$$

We ask: Does this F-ratio of 3 require rejection of the null hypothesis at the 5 percent significance level? Examining the above table, we find that for $k{-}1 = 2$ and $N{-}k = 6$, F will exceed 5.14 five times in 100 when the null hypothesis is true. Since our calculated value of F is smaller than the significance value, we do not decide against the null hypothesis.

The *F*-Distribution The application of Formula 15.1.5 to a one-way table will always yield an F-ratio, but that formula yields little or no insight into the statistical principle on which the use of the F-ratio as a test-statistic rests. This principle pertains to the sampling distribution of ratios of independent estimates of the variance of a normal distribution, given that the estimate in the numerator has v_1 degrees of freedom and the estimate in the denominator has v_2 degrees of freedom. This is the *F-distribution.* Depending on the values of v_1 and v_2, it will be more or less skewed to the right. The F-distribution is thus, like the distribution of t, a whole family of distributions. Its mean value of $v_2/(v_2 - 2)$ approaches 1.00 as v_2 increases, and we therefore say that the expected value of F is approximately 1.

An analysis of variance, by definition, yields independent estimates of the variance of a normal population. One estimate of that variance is obtained by dividing the within-groups sum of squares by its degrees of freedom:

$$MS(W) = \frac{WSS}{N-k} \tag{15.1.6}$$

where MS stands for mean square and the parenthesis gives the source of that mean square.

A second estimate of the postulated common variance is obtained by dividing the between-groups sum of squares by its degrees of freedom:

$$MS(B) = \frac{BSS}{k-1}. \tag{15.1.7}$$

The within-groups mean square gives an unbiased estimate of the common variance, regardless of the truth of the hypothesis of identical means. The between-groups mean square will give a consistently inflated estimate when the null hypothesis is false, that is, when population means actually differ. For that reason the ratio of $MS(B)$, based on $k-1 = v_1$ degrees of freedom, to $MS(W)$, based on $N-k = v_2$ degrees of freedom, will tend to exceed its expected value of $v_2/(v_2 - 2)$ when the null hypothesis is false, and thereby increase the likelihood of its being rejected.

In analyzing the variance of Table 15.1.3, we decided against rejecting the null hypothesis because our calculated F-value of 3 was smaller than the 5 percent significance value of 5.14. Our new view of F as the ratio of independent estimates of the same variance puts that decision in a statistically more meaningful light. In getting the F-ratio for Table 15.1.3 by Formula 15.1.5, we implicitly calculated

$$MS(W) = \tfrac{24}{6} = 4$$

and

$$MS(B) = \tfrac{24}{2} = 12,$$

and the ratio between them,

$$F = \frac{MS(B)}{MS(W)} = \frac{12}{4} = 3.$$

The sampling distribution of ratios formed in this way permits us to test the null hypothesis of identical means, provided of course that sampling assumptions have been met, or have not been severely violated.

Unequal n_i Although samples were equal in size in the foregoing demonstration, such equality in the n_i is not a requirement of the method. The method is perfectly general, and applies to any set of n_i, whether equal or not. (Equal n_i may be regarded as that special set in which all differences between the n_i are zero.) However, the user will need to keep in mind that in getting *BSS* different weights are attached to the squared differences between sampled means and the overall mean. To illustrate this point, we analyze the variance of three samples of 2, 3, and 4 cases, respectively.

		n_i	Sum	Mean
A_1	6, 8, 10	3	24	8.00
A_2	10, 9	2	19	9.50
A_3	12, 14, 10, 11	4	47	11.75
		9	90	10.00

$$TSS = 42.00$$

$$BSS = 3(8.00–10.00)^2 + 2(9.50–10.00)^2 + 4(11.75–10.00)^2$$
$$= 24.75.$$

$$WSS = TSS - BSS$$
$$= 42.00 - 24.75$$
$$= 17.25.$$

Analysis of Variance

Source	*SS*	*df*	*MS*	*F*
Total	42.00	8		
Within	17.25	6	2.890	
Between	24.75	2	12.375	4.28

Computing ***BSS*** To get the sum of the squared deviations around the overall mean, we apply the standard computing formula:

$$TSS = \sum_1^k \sum_1^{n_i} Y_{ij}^2 - \frac{(\sum_1^k \sum_1^{n_i} Y_{ij})^2}{N}. \qquad (15.1.8)$$

We apply the same formula to get the sum of the squared deviations around the mean of any sample:

$$\sum_1^{n_i} (Y_{ij} - \bar{Y}_i)^2 = \sum_1^{n_i} Y_{ij}^2 - \frac{(\sum_1^{n_i} Y_{ij})^2}{n_i}. \qquad (15.1.9)$$

Is there a formula of comparable convenience for computing the sum of the weighted squared deviations between sample means and the overall mean? The student will correctly surmise that there is such a formula, otherwise we would not have raised this possibility. It is:

$$BSS = \sum_{1}^{k} \frac{(\sum Y_{ij})^2}{n_i} - \frac{(\sum_1^k \sum_1^{n_i} Y_{ij})^2}{N}. \tag{15.1.10}$$

Inspecting Formula 15.1.10, we see that it obviates the need to calculate deviations between sample means and the overall mean. We need only to manipulate sample sums and the sum of those sums. From a practical standpoint, two features of Formula 15.1.10 are noteworthy: (1) this formula has the same correction term as the total sum of squares, namely,

$$C = \frac{(\sum_1^k \sum_1^{n_i} Y_{ij})^2}{N};$$

and (2) with equal n_i, we need not divide each squared sample sum by n_i before summing those sums, rather we may sum the squared sums and then divide by the common n_i. In symbols:

$$\sum_{1}^{k} \frac{(\sum_1^{n_i} Y_{ij})^2}{n_i} = \frac{\sum_1^k (\sum_1^{n_i} Y_{ij})^2}{n_i}$$

where the n_i are equal.

To illustrate Formula 15.1.10 we apply it to the measures of Table 15.1.3. Carrying out prescribed operations, we get $BSS = 24$, in agreement with the direct method. Here is the detail:

$$\begin{aligned} BSS &= \frac{(24)^2}{3} + \frac{(30)^2}{3} + \frac{(36)^2}{3} - \frac{(90)^2}{9} \\ &= \frac{(24)^2 + (30)^2 + (36)^2}{3} - \frac{(90)^2}{9} \\ &= \frac{576 + 900 + 1{,}296}{3} - \frac{(90)^2}{9} \\ &= 924 - 900 \\ &= 24. \end{aligned}$$

Although the computer has lessened the practical utility of Formula 15.1.10, it nevertheless will be handy for the student who does exercises by hand, either of necessity or for the sake of a better understanding of the method.

Analysis of Variance Assumptions With measures sorted into two or more groups, as in Table 15.1.1, it is always possible to partition the total sum of squares into its component sums. From these component sums it is always possible to calculate an F-ratio by Formula 15.1.5. However, that F-ratio will not be subject to statistical interpretation unless the conditions under which the F-distribution will realize itself have been met. Since we may have lost sight of these assumptions in the preceding discussion, we repeat them here for the sake of emphasis and summary.

(1) We assume *interval measures* (constant scale units)—this perhaps goes without saying. Although it is possible to calculate an F-ratio for ordinal measures, and even for 0,1 dummy variates, the interpretation to be placed on such ratios will usually be uncertain at best. Such applications should be foregone in the absence of compelling justification to the contrary.

(2) We assume *random sampling* from *normal populations* with a *common variance*. The requirement of random sampling is inflexible, but the assumption of normal populations with a common variance is subject to some relaxation, owing to the so-called robustness of the F-ratio. This means that the F-distribution will maintain itself without serious distortion even though the sampled populations are less than normal and their variances not exactly equal. However, it is impossible to give a rule for finding the point at which the F-table loses its validity for purposes of assessing statistical significance. It is possible only to state that the F-table may have validity in a given application notwithstanding that assumptions have not been strictly met.

(3) The calculation of F will generally depend on whether our hypothesis pertains only to the classes in the table (fixed effects), or whether it pertains to infinitely many classes (random effects), of which the k classes in the table are a sample.

The one-way analysis of variance represents an exception to this generalization. In this special case, the composition of F is the same, whether our hypothesis covers all classes in the sampled universe or whether it is limited to those classes appearing in the table. Nevertheless, even in this case, it is important to distinguish between hypotheses, since they constitute different claims. To maintain this distinction, the restricted hypothesis for fixed effects is sometimes put in the form:

$$\frac{1}{k-1}\sum_{1}^{k}(\mu_i - \mu)^2 = 0 \qquad (15.1.11)$$

where k = number of classes in table; and the hypothesis for random effects is put in the form

$$\sigma^2_{(\mu_i - \mu)} = 0 \qquad (i = 1, 2, \ldots, \infty). \qquad (15.1.12)$$

Both represent claims that population means are identical; in the first case (Formula 15.1.11), the claim holds only for the k classes in the one-way table; in the second case (Formula 15.1.12), the claim holds for all classes in the sampled universe of classes.

QUESTIONS AND PROBLEMS

1. Define the following terms:
   one-way analysis of variance
   total sum of squares
   between-groups sum of squares
   within-groups sum of squares
   F-ratio
   significance level (α)
   fixed effects

2. Calculate

$$F = \frac{(BSS)(N-k)}{(WSS)(k-1)}$$

   for the following one-way table:

B_1	B_2	B_3
10.1	13.6	9.7
11.1	10.8	10.9
13.9	11.5	11.2

3. Calculate

$$F = \frac{(BSS)(N-k)}{(WSS)(k-1)}$$

   for the following one-way table:

B_1	B_2
10.1	13.6
11.1	10.8
13.9	11.5
9.7	10.9
	11.2

   Calculate t^2 for the same table and compare with F.

4. An investigator wishes to test H_0 that $\mu_1 = \mu_2 = \cdots = \mu_k$ by the variance ratio:

$$F = \frac{BSS}{k-1} \bigg/ \frac{WSS}{N-k}.$$

Find the critical value of F for the following combinations of k and α (significance level), given that $N = 40$:

F	α	k
——	.05	5
——	.01	10
——	.05	2
——	.05	4
——	.01	3
——	.01	8

5. Calculate

$$F = \frac{BSS}{k-1} \bigg/ \frac{WSS}{N-k}$$

given the following entries:

Source	SS	df	MS	F
Total		105		
Between		5	19	
Within	500			

2 TWO-WAY ANALYSIS OF VARIANCE

Preliminary Comment A two-way classification differs from a one-way classification in that each value (Y) is classified in two ways instead of one—by A and B, rather than by A alone. Since two-way groupings may be regarded as classes within classes, we sometimes refer to them as subclasses, or subsamples. With values arranged in subclasses, it is possible to divide the total sum of squares into four parts based on the differences (1) among values within subclasses; (2) among row means; (3) among column means; and (4) among differences among subsample means within columns or rows. From these results we may test the hypothesis that the means of the populations in rows are identical, the hypothesis that the means of the populations in columns are identical, the hypothesis that differences among subsample means within columns (or rows) are constant, or the equivalent hypothesis of no interaction between A and B.

For example, if measures have been classified by ethnic background and occupation, we may test the hypothesis that the means of the ethnic groups do not differ, the hypothesis that the means of the occupational groups do not differ, the hypothesis that differences among ethnic groups are unaffected by occupation, or equivalently, that differences among occupations are unaffected by ethnic background. The latter is the hypothesis of no interaction between occupation and ethnic background.

In doing a two-way analysis of variance we routinely get as many F-ratios as there are different hypotheses—usually three. When a given hypothesis is true, sample ratios will conform to the F-distribution (as tabulated); when that hypothesis is false, sample ratios will diverge from the path of that distribution in contradiction to the null hypothesis.

As with the one-way table, it is instructive to investigate the effect of changing differences between row and column means on the total sum of squares, and also the effect of changing differences between differences among subclass means on that sum. This demonstration may provide further insight into the total sum of squares as the sum of its component sums of squares, and thereby into the differences among the F-ratios themselves.

Let us start with nine subsamples whose means and variances are equal (Table 15.2.1). To analyze the variance of these nine subsamples, we compute first the sum of the squared deviations around the overall mean. Squaring the given measures and summing, and subtracting the correction factor gives:

$$TSS = 1{,}746 - 1{,}728$$
$$= 18.$$

Next, we sum the sums of squared deviations within subsamples to get the within-groups sum of squares. Since, in our example, the sum of the squared deviations within each subsample is identical, we need only to obtain the

Table 15.2.1 *Given Measures*

	B_1	B_2	B_3
A_1	7	7	7
	8	8	8
	9	9	9
A_2	7	7	7
	8	8	8
	9	9	9
A_3	7	7	7
	8	8	8
	9	9	9

sum of the squared deviations in any one of them and multiply that result by the number of subsamples. Carrying out the required operations gives

$$\begin{aligned}
WSS &= (rc) \sum_{1}^{3} (Y_{ijk} - \bar{Y}_{ij})^2 \\
&= (3 \times 3)[(7-8)^2 + (8-8)^2 + (9-8)^2] \\
&= (9)(1 + 0 + 1) \\
&= 18
\end{aligned}$$

where r = number of rows, c = number of columns.

Lastly, we calculate the between-groups sum of squares by Formula 15.1.10. Application of this formula yields

$$\begin{aligned}
BSS &= \frac{9(576)}{3} - 1{,}728 \\
&= 1{,}728 - 1{,}728 \\
&= 0.
\end{aligned}$$

This answer, $BSS = 0$ was a foregone conclusion, because when there are no differences among subsample means, $TSS = WSS$.

Once-Adjusted Measures Let us now deliberately disturb the equality among row means while leaving undisturbed the equality of sums of squared deviations within subsamples. Of the many ways in which this might be accomplished we subtract 1 from each measure in the first row (A_1) and add that quantity to each measure in the third row (A_3). These once-revised measures are displayed in Table 15.2.2.

Because measures within subsamples were adjusted by the same constant (including 0 in the second row [A_2]), $WSS = 18$ remains unchanged.

Table 15.2.2 *Once-Adjusted Measures*

	B_1	B_2	B_3
A_1	6	6	6
	7	7	7
	8	8	8
A_2	7	7	7
	8	8	8
	9	9	9
A_3	8	8	8
	9	9	9
	10	10	10

However, the total sum of squared deviations of 36 is larger by 18, reflecting the greater variation of these once-revised measures around the overall mean of 8. Since column means are unchanged, as is the internal variation within subsamples, it is logical to attribute this excess of 18 to the differences between the row means.

To confirm this conclusion, we calculate both the sum of the squared deviations between the subsample means and the grand mean, and the sum of the weighted squared deviations between row means and the grand mean (*RSS*). Carrying out operations, we get:

$$BSS = 1{,}746 - 1{,}728$$
$$= 18.$$
$$RSS = 9(7-8)^2 + 9(8-8)^2 + 9(9-8)^2$$
$$= 18.$$

These operations verify that the increase in *BSS* from 0 to 18 is exactly (and necessarily) equal to the increase in *RSS* from 0 to 18.

Twice-Adjusted Measures Our next step is to disturb the equality among column means, leaving undisturbed the differences between row means, as well as the differences among measures within subsamples. To do this, we will add the same constant to measures in a given column but different constants to measures in different columns. Adhering to this rule, we reduce each value in the first column (B_1) by 1, and increase each value in the third column (B_3) by 1, leaving intact the values in the second column (B_2). These twice-revised measures are displayed in Table 15.2.3. Our interest lies in the effect of this adjustment on the between-groups sum of squares, since the sum of the squared deviations within subsamples will remain unchanged.

Table 15.2.3 *Twice-Adjusted Measures*

	B_1	B_2	B_3
A_1	5	6	7
	6	7	8
	7	8	9
A_2	6	7	8
	7	8	9
	8	9	10
A_3	7	8	9
	8	9	10
	9	10	11

Calculating the between-groups sum of squares (by Formula 15.1.10) for Table 15.2.3 gives

$$BSS = 1{,}764 - 1{,}728$$
$$= 36.$$

Calculating the sum of the weighted squared deviations between column means and the grand mean gives

$$CSS = 9(7 - 8)^2 + 9(8 - 8)^2 + 9(9 - 8)^2$$
$$= 9 + 0 + 9$$
$$= 18.$$

Subtracting *CSS* from *BSS* gives 18, which is the sum of the squared deviations based on the differences between the row means of the once-adjusted measures.

Putting results together, we conclude that the differences between column means has inflated the total sum of squares by 18, and that this total now has three nonzero components where in the case of the unadjusted measures it had one, and in the case of the once-revised measures it had two. The total variation of Table 15.2.3 reflects differences among measures within subsamples, differences between row means, and differences between column means.

Thrice-Adjusted Measures To complete our demonstration, we now establish differences among differences between subsample means within rows (or columns). However, before making this adjustment let us focus on the pattern of differences between subsample means within rows (or columns) for the twice-adjusted measures of Table 15.2.3. Within the first row (A_1), the subsample means are 6, 7, and 8, respectively; within the second row (A_2), the means are 7, 8, and 9; within the third row (A_3), the cell means are 8, 9, and 10, respectively. It is pertinent that, within each row, differences between means are identical. In the first row the mean of the second column (B_2) exceeds the mean of the first column (B_1) by 1, and the mean of the third column (B_3) exceeds the mean of the second column (B_2) by 1. This pattern of differences prevails in each row. The absence of interaction is reflected by the absence of differences among such differences; conversely the presence of differences among such differences is evidence for the hypothesis of interaction.

We now proceed to disturb the equality among differences of differences, leaving differences between row and column means, and between measures within subsamples unchanged. Of the many ways in which this might be accomplished, we subtract 2 from each measure in A_1B_1 and A_3B_3 and add that quantity to each measure in A_1B_3 and A_3B_1. With these revisions, we have the thrice-adjusted measures for Table 15.2.4.

Table 15.2.4 *Thrice-Adjusted Measures*

	B_1	B_2	B_3
A_1	3	6	9
	4	7	10
	5	8	11
A_2	6	7	8
	7	8	9
	8	9	10
A_3	9	8	7
	10	9	8
	11	10	9

Calculating sums of squares for these measures, we obtain:

$TSS = 102;$

$WSS = 18;$

$BSS = 84.$

Comparing these sums with corresponding sums from Table 15.2.3, we find that BSS is larger by 48. By disturbing the equality of differences between means within rows (or columns), we have produced an inflation in the between-groups sum of squares.

We could have calculated the magnitude of this inflation directly by means of the following formula:

$$ISS = \sum \sum n_{ij}[(\bar{Y}_{ij} + \bar{Y}_{..}) - (\bar{Y}_{i.} + \bar{Y}_{.j})]^2 \tag{15.2.1}$$

where ISS is an abbreviation for *Interaction Sum of Squares*. Applying this formula to Table 15.2.4, we get:

Row	Column	$(\bar{Y}_{ij} + \bar{Y}_{..}) - (\bar{Y}_{i.} + \bar{Y}_{.j}) = d_{ij}$	d_{ij}^2	$n_{ij}d_{ij}^2$
1	1	$(4 + 8) - (7 + 7) = -2$	4	$(3)(4) = 12$
1	2	$(7 + 8) - (7 + 8) = 0$	0	$(3)(0) = 0$
1	3	$(10 + 8) - (7 + 9) = 2$	4	$(3)(4) = 12$
2	1	$(7 + 8) - (8 + 7) = 0$	0	$(3)(0) = 0$
2	2	$(8 + 8) - (8 + 8) = 0$	0	$(3)(0) = 0$
2	3	$(9 + 8) - (8 + 9) = 0$	0	$(3)(0) = 0$
3	1	$(10 + 8) - (9 + 7) = 2$	4	$(3)(4) = 12$
3	2	$(9 + 8) - (9 + 8) = 0$	0	$(3)(0) = 0$
3	3	$(8 + 8) - (9 + 9) = -2$	4	$(3)(4) = 12$
				$ISS = 48$

The sum of the squared deviations, calculated according to Formula 15.2.1, has come to be known as the interaction sum of squares since an interaction between the main classifications will contribute to its magnitude. A statistical interaction between A and B may be regarded as a failure of a pattern of differences among subclass means within a given row (or column) to reproduce the pattern of differences among column (or row) means, rows (or columns) disregarded. This failure is a demonstration that the effect of one factor is conditioned by the effect of the other. We speak of such conditional effects as *interaction effects.*

Testing Hypotheses Let us now view the entries in Table 15.2.4 in a more realistic and practical light. Let us regard these entries as sample measures whose variance is to be analyzed and whose F-ratios are to be used for hypothesis testing, as in an actual sociological study. The hypotheses to be tested are these: (1) that row means are identical, (2) that column means are identical, (3) that differences between cell means within rows (or columns) are equal to corresponding differences between column (or row) means. In symbols:

(1) $\mu_{1.} = \mu_{2.} = \mu_{3.}$;

(2) $\mu_{.1} = \mu_{.2} = \mu_{.3}$;

(3) $(\mu_{ij} + \mu_{..}) = (\mu_{i.} + \mu_{.j})$.

If we decide against Hypotheses 1 and 2, we conclude that main effects (A and B) are significant; if we decide against Hypothesis 3, we conclude that interaction (A × B) is significant. Main effects may be significantly present when interaction is absent; and interaction may be significantly present when one or both main effects are absent.

In doing a two-way analysis of variance we first partition the total sum of squares into the within-subgroups sum of squares and the between-subgroups sum of squares, exactly as in a one-way analysis of variance. These sums of squares will appear in the top half of the analysis of variance summary table (Table 15.2.5). Second, we partition the between-groups sum of squares into its respective component sums of squares: the between-rows sum of squares, the between-columns sum of squares, and the interaction sum of squares. These sums of squares will appear in the bottom half of the summary table (Table 15.2.5).

Having calculated the component sums of squares, we next divide each by its degrees of freedom for estimates of the common variance (excluding the between-groups sum of squares, since its mean square is not required for hypothesis-testing). Since component sums of squares have as many degrees of freedom as TSS, and since TSS has N–1 *df*, we must split up that many *df* among the component sums of squares. Of the total, N–rc go to the within-groups sum of squares (WSS), and rc–1 go to the between-groups sum of

Table 15.2.5 *Analysis of Variance of Table 15.2.4*

Source	SS	df	MS	F
Total	102	26		
Within	18	18	1.00	
Between	84	8		
Rows	18	2	9.00	9.00
Columns	18	2	9.00	9.00
Interaction	48	4	12.00	12.00

squares (BSS). Of the $rc-1$ going to BSS, $r-1$ go to the between-rows sum of squares, $c-1$ to the between-columns sum of squares, and the remainder—$(r-1)(c-1)$—go to interaction. After degrees of freedom have been determined by means of these counting formulas, they are recorded in the third column of the summary table (Table 15.2.5) alongside the sums of squares themselves.

Getting mean squares amounts to dividing the second column of the summary table by the third. These mean squares are recorded in the next to the last, or the fourth, column of the summary table. $MS(W)$ provides an unbiased estimate of the common variance whether main effects or interaction are present or not. $MS(R)$ gives an unbiased estimate of that variance when row effects are absent; otherwise, it will tend to be too large. Similarly, $MS(C)$ gives an unbiased estimate of the common variance when column effects are absent, otherwise it will be excessive. And $MS(I)$ gives the unbiased estimate of the common variance in the absence of interaction, otherwise it will be inflated.

With mean squares in hand, we are ready to set up the F-ratios or test-statistics. These ratios are shown in the last column of the summary table (Table 15.2.5). The ratio of $MS(R)$ to $MS(W)$ provides a test of the hypothesis that row means are identical; the ratio of $MS(C)$ to $MS(W)$ provides a test of the hypothesis that column means are identical; and the ratio of $MS(I)$ to $MS(W)$ provides a test of the hypothesis that the effect of one factor is not contingent upon the effect of the other and vice versa. In deciding for or against a given hypothesis we match our calculated F-ratio against a tabulated significance value which will be exceeded a specified proportion of times when the null hypothesis is true. If we are willing to run a 5 percent risk of rejecting a true hypothesis, we refer to that set of values that will be exceeded 5 percent of the time when the null hypothesis is true. If we are willing to take no more than a 1 percent chance, we refer to those values that will be exceeded 1 percent of the time when the null hypothesis is true.

Whether a calculated F-ratio is for or against its corresponding hypothesis will depend on our chosen significance level. Let us match our calculated

F-values against the 5 percent significance values, as if we were willing to take a 5 percent chance of rejecting a true hypothesis. Turning to a table of 5 percent significance values (Appendix, Table IV), we find:

v_1	v_2	$F_{.95}$
2	18	3.55
4	18	2.93

Since our calculated F-ratios of 9.00 and 9.00 for row means and column means, respectively, exceed the 5 percent significance level, we would decide in favor of rejecting the hypothesis of identical means. Similarly, since the F-ratio of 12.00 for interaction is larger than its significance value of 2.93, we would decide against the hypothesis of no interaction. Our general conclusion would be that main effects and interaction are significant; we would interpret this to mean that the magnitude of a given measure depends not only on factors A and B separately but also on the interaction between them. By the technique of two-way analysis of variance, we have been able to test three hypotheses about the composition of our sample measures.

Case of $n_{ij} = 1$ When we have only a single measure in each subgroup, it is impossible to get the within-groups sum of squares; hence in this limiting case, $TSS = BSS$. Under these special circumstances our analysis will be limited to partitioning BSS into its component parts: RSS, CSS, and ISS. For the analysis to proceed further, we must make an additional assumption. On the assumption of no interaction we may test for the significance of main effects.

To illustrate this procedure, consider a set of nine measures, each classified by A and B, with no two measures in the same subgroup. These measures are shown in Table 15.2.6. We get the total sum of squares in the

Table 15.2.6

	B_1	B_2	B_3	Sum	Mean
A_1	3	9	9	21	7
A_2	5	3	4	12	4
A_3	4	6	20	30	10
Sum	12	18	33	63	
Mean	4	6	11		7

usual way, subtracting the correction factor from the sum of the squared values. Thus,

$$TSS = 673 - 441$$
$$= 232.$$

Since WSS necessarily equals zero, $BSS = TSS$.

Upon decomposing BSS into its constituent sums, we get

$$CSS = 519 - 441$$
$$= 78$$

and

$$RSS = 495 - 441$$
$$= 54.$$

Adding 54 and 78 we get 132, which falls short of the total sum of squares by 100. Obviously, some residual variation remains which can be ascribed neither to differences among row means nor to differences among column means.

To explore the composition of this remainder of 100 in limited fashion, let us determine the values expected in the event of no differences among row means and column means, respectively, with the overall mean unchanged. We need only to adjust values so that differences among marginal means are eliminated. Where before we created differences to discern the effect of that adjustment, we now eliminate differences in order to discern the effect of that reverse adjustment.

To eliminate differences among column means, we adjust each individual measure by an amount equal to the deviation of its column mean from the overall mean. We lower values when the column mean is larger than the overall mean; we raise them when the column mean is smaller. Thus, we raise values in the first and second columns by 3 and 1, respectively; we lower values in the third column by 4. After this adjustment, we have the once-adjusted values shown in Table 15.2.7.

Table 15.2.7

	B_1	B_2	B_3	Sum	Mean
A_1	6	10	5	21	7
A_2	8	4	0	12	4
A_3	7	7	16	30	10
$\bar{Y}_{.j} - \bar{Y}_{..}$	−3	−1	4	63	
Mean	7	7	7		7

Table 15.2.8

	B_1	B_2	B_3	$\bar{Y}_{i.} - \bar{Y}_{..}$	Mean
A_1	6	10	5	0	7
A_2	11	7	3	−3	7
A_3	4	4	13	3	7
Mean	7	7	7		7

Since the deviations of the observed column means sum to zero, row means are unaffected by this adjustment, as are the initial differences among them.

To equalize row means, we apply the deviations of row means from the overall mean to our once-adjusted values, an operation that will not disturb the newly established equality of column means since the net increment within each column is necessarily zero. Again, we lower values where row means are in excess of the overall mean; we raise values where row means are below the overall mean. By this adjustment, we obtain a set of measures which still vary among themselves internally but whose row and column totals are now equal.

Upon calculating the sum of squares for this set, we find it to be 100, the result previously obtained by subtracting row and column sums of squares from the between-groups sum of squares, *BSS*. Although this remainder may be obtained readily by deducting the quantity (*RSS* + *CSS*) from *BSS*, the student should apply the foregoing technique to easy numerical examples in order to heighten his or her appreciation of its essential nature.

After calculating the sums of squares and recording them in the summary table (Table 15.2.9), we determine degrees of freedom by means of counting formulas. The total sum of squares has $N-1 = rc-1$ degrees of freedom. Of this total, $r-1$ go to rows, $c-1$ go to columns, and $(r-1)(c-1)$ go to interaction. Dividing sums of squares by degrees of freedom gives the mean squares shown in the fourth column of the summary table (excluding the total sum of squares since its mean square is not required for hypothesis-testing). On the

Table 15.2.9

Source	*SS*	*df*	*MS*	*F*
RSS	54	2	27	1.08
CSS	78	2	39	1.56
ISS	100	4	25	

assumption that interaction is zero, the ratio of $MS(R)$ to $MS(I)$ will have an F-distribution; similarly, on the assumption of no interaction, the ratio of $MS(C)$ to $MS(I)$ will have an F-distribution. These ratios are shown in the last column of the summary table. If the assumption of no interaction is false, then $MS(I)$ will be too large and the F-ratios will be too small. In that event, we will tend to favor a hypothesis that is actually false, that is, we will tend to overlook differences between population means that actually exist.

Despite this hazard, the case of $n_{ij} = 1$ is not infrequent in sociological research. When we have two or more measures on the same set of individuals and are interested both in differences between individuals and in differences between repeated measures, then we may wish to use the foregoing procedure for testing main effects.

Review of Assumptions The use of the F-ratio as a test of significance requires that certain assumptions be met. These are: (1) *random sampling* from (2) *normal populations* with (3) *equal variances.* Our demonstration (Table 15.2.5) was based on the assumption that samples were randomly drawn from normally distributed populations with a common variance. The assumption that sampled populations be normal and have a common variance is subject to relaxation, owing to the aforementioned robustness of the F-ratio; in each case the user must decide whether it is safe to use the F-ratio as a significance test in the face of ungrounded assumptions. However, the requirement that cases be randomly selected is subject to no relaxation. Use of the F-distribution absolutely requires that cases be randomly selected or randomly assigned.

The ratio of $MS(R)$ to $MS(W)$ and the ratio of $MS(C)$ to $MS(W)$ (as calculated in Table 15.2.5) are valid for significance testing provided that interest is restricted to the classes appearing in the table. In other words, these formulas hold for fixed effects; with random effects other formulas are required.

Similarly, the formulas in Table 15.2.5 hold only for equal n_{ij}. Unequal n_{ij} call for different procedures. When ratios between cell frequencies are proportional to ratios between marginal frequencies, the modifications are slight; however, when that proportionality does not hold, then what would appear to be a radically different approach must be used. These modifications are beyond the scope of this book.

QUESTIONS AND PROBLEMS

1. Define the following terms:
   two-way analysis of variance
   interaction effect
   main effect
   fixed effects

2. Do a two-way analysis of variance of the following table:

	B_1	B_2	B_3
A_1	10.1	13.6	9.7
	11.1	10.8	10.9
	13.9	11.5	11.2
A_2	9.1	13.3	10.7
	16.2	14.1	8.4
	10.1	14.0	10.1

3. Given six (6) subsamples of five measures each,

	B_1	B_2	B_3
	3	3	11
	6	12	20
A_1	9	10	16
	10	9	19
	2	10	16
	0	11	11
	8	2	14
A_2	1	4	14
	4	4	19
	7	5	14

calculate:

$$F = \frac{(RSS)(N-rc)}{(WSS)(r-1)}$$

$$F = \frac{(CSS)(N-rc)}{(WSS)(c-1)}$$

$$F = \frac{(ISS)(N-rc)}{(WSS)(r-1)(c-1)}$$

Interpret these F-ratios.

3 A × B CONTINGENCY TABLE

We have previously learned (p. 420) that for large samples the statistic

$$z = \frac{p_1 - p_2}{s_{(p_1 - p_2)}}$$

has an approximately normal distribution with mean equal to zero and standard deviation equal to 1. Therefore, to test the hypothesis that

$P_1 = P_2$, we calculate z from our sample data and compare that calculated value with our preselected significance value. To reject the null hypothesis at the 5 percent level, our z-value must exceed $|1.96|$; to reject at the 1 percent level, our z-value must exceed $|2.58|$; and so on.

This procedure will suffice for comparing two populations but not for comparing three or more populations. Just as we turned from t to F in order to compare three or more population means, here we turn from z to χ^2 in order to test the hypothesis that three or more populations have the same proportion of cases in a given category. The general procedure for testing a statistical hypothesis is unchanged: (1) state the hypothesis to be tested, (2) set the significance level, (3) select sample cases, (4) calculate the test statistic, and (5) make a decision for or against the stated hypothesis. As we shall see, large values of χ^2 are against the hypothesis of no differences between population proportions, while small values of χ^2 tend to favor that hypothesis, other things being equal.

To be a little more concrete, let us compare the responses of whites by years of schooling to the following question on job opportunities for blacks: "Do you think blacks should have as good a chance as white people to get any kind of job or do you think that white people should have the first chance at any job?" Responses are presented in Table 15.3.1. In the table, acceptance of the first alternative is coded "Yes," rejection of the first alternative in favor of the second is coded "No." Judging from the pattern of proportions, responses to the question are correlated with educational background. The proportion favoring equal job opportunities for blacks increases as the level of schooling increases; persons with the most education have the most favorable attitude, persons with the least education have the

Table 15.3.1 *Proportion of White Respondents Favoring Equal Job Chances for Blacks, by Years of Schooling*

Equal Chances	No High School	High School	Some College	Total
Yes[a]	.46	.63	.75	.59
No[b]	.54	.37	.25	.41
Total	1.00 (124)	1.00 (165)	1.00 (65)	1.00 (354)

[a] Code for accepting alternative that "Blacks should have as good a chance as white people to get any kind of job. ..."
[b] Code for rejecting alternative above in favor of alternative that "White people should have the first chance at any kind of job."
Source: Abbreviated and adapted from Table 8 in Davis, James A. "Hierarchical Models for Significance Tests in Multivariate Contingency Tables: An Exegesis of Goodman's Recent Papers," in H. L. Costner, ed., *Sociological Methodology 1973–1974*. San Francisco: Jossey Bass. 1974. Reprinted with permission.

least favorable attitudes toward blacks. Education appears to make a difference in an individual's answer to the question.

However, we cannot automatically rule out the hypothesis that population proportions are identical, since large differences will arise in the course of random sampling. To investigate that possibility, we calculate χ^2 and refer that value to the χ^2-distribution. Just as the F-ratio gives us the probability of larger differences between sample means when the hypothesis of populations identical in means is true, the χ^2 statistic gives us the probability of larger differences between sample proportions when the hypothesis that populations are identical in their proportions is true. In our problem, the hypothesis to be tested is:

$$H_0 : P_{11} = P_{12} = P_{13}$$

where the first number in the subscript indicates the row of the 2×3 contingency table, and the second entry gives the column.

Make-up of χ^2 The computation of χ^2, as with F, depends on the hypothesis to be tested. In testing the hypothesis that population proportions are identical, the computation starts with the observed cell frequencies (rather than proportions). These are displayed in Table 15.3.2. Since responses are classified in two ways and years of schooling are classified in three, our contingency table has two rows and three columns, therefore we speak of it as a 2×3 contingency table. It is necessary that we count not only the number of cases in each cell, but also the number in each column, the number in each row, and the grand total (N) of all cases. These various totals, given in Table 15.3.2, are required for subsequent calculations.

Having set up the observed frequencies, we next calculate the frequencies expected under the null hypothesis that population proportions are equal. These expected frequencies must meet two conditions: (1) the proportion of cases in each row must be the same in every column, and (2) their marginal totals must equal the observed marginals. To meet these criteria, we need

Table 15.3.2 *Number Favoring Equal Job Chances for Blacks, by Years of Schooling*

Equal Chances	No High School	High School	Some College	Total
Yes	57	104	49	210
No	67	61	16	144
Total	124	165	65	354

Source: Table 15.3.1.

Table 15.3.3 *Expected Numbers* ($\hat{n}_{ij}$) *Favoring Equal Job Chances for Blacks, by Years of Schooling*

Equal Chances	No High School	High School	Some College	Total
Yes	73.56	97.88	38.56	210.00
No	50.44	67.12	26.44	144.00
Total	124.00	165.00	65.00	354.00

Source: Table 15.3.2.

only to multiply column totals by the proportion of cases in each row for all cases combined. In symbols:

$$\hat{n}_{ij} = \frac{n_{i.}}{N}(n_{.j}) \tag{15.3.1}$$

where $\hat{n}_{ij}$ is the expected frequency for the cell standing at the intersection of the ith row and jth column. Applying this formula to get the expected frequencies for the first row, we multiply column totals by .59; applying .41 to the same column totals gives the expected frequencies for the second row. These results are presented in Table 15.3.3.

Making frequencies within columns proportional to row totals is equivalent to finding the expected number of cases in each cell on the assumption that the A and B classifications are statistically independent. We have previously learned (Chapter 11) that the probability of independent events occurring together is the product of their individual probabilities. If, referring to Table 15.3.2, we take $n_{i.}/N$ as the probability of A_i, and $n_{.j}/N$ as the probability of B_j, then the probability of their occurring together is $n_{i.}n_{.j}/N^2$; and the expected frequency of the event A_i and B_j on N trials would be

$$\hat{n}_{ij} = Pr(A_i\,B_j)N$$

$$= \frac{n_{i.}n_{.j}}{N}. \tag{15.3.2}$$

But this is simply the total for the jth column multiplied by the proportion of cases in the ith row for all columns combined, as given by Formula 15.3.1. Making frequencies within columns proportional to row totals is thus equivalent to finding the expected number of cases in each cell on the assumption that the A and B classifications are statistically independent. But however the expected frequencies are construed, they may always be easily calculated according to Formula 15.3.2.

Table 15.3.4 *Deviations between Observed Numbers* (n_{ij}) *and Expected Numbers* ($\hat{n}_{ij}$)

Equal Chances	No High School	High School	Some College	Total
Yes	−16.56	6.12	10.44	0.00
No	+16.56	−6.12	−10.44	0.00
Total	0.00	0.00	0.00	0.00

Source: Tables 15.3.2 and 15.3.3.

Once calculated, expected frequencies are subtracted from observed frequencies for as many deviations as there are cells in the contingency table. These deviations are shown in Table 15.3.4. The requirement that expected frequencies have the same marginal totals as the observed frequencies causes the deviations in each row, and hence causes all the deviations collectively to sum to zero. What this means practically is that the sum of these deviations cannot be used to gauge how well the observed frequencies fit the expected frequencies. And so, as our next step, we do the obvious thing: We square the deviations. These squared deviations are shown in Table 15.3.5.

At this juncture, it would seem natural to divide each squared deviation by N before summing, as we did in getting the variance. However, before summing, in this case we divide each squared deviation by its expected *frequency* before summing. In symbols:

$$\chi^2 = \sum_{1}^{i} \sum_{1}^{j} \frac{(n_{ij} - \hat{n}_{ij})^2}{\hat{n}_{ij}}. \tag{15.3.3}$$

Applying this formula to the numbers in our problem, we get $\chi^2 = 17.06$ (calculations shown in Table 15.3.6). This is the test statistic that permits us to decide for or against our hypothesis that population proportions are equal.

Table 15.3.5 *Squared Deviations between Observed and Expected Numbers*

Equal Chances	No High School	High School	Some College
Yes	274.23	37.45	108.99
No	274.23	37.45	108.99

Source: Table 15.3.4.

Table 15.3.6 *Sum of Squared Deviations between Observed and Expected Numbers Divided by Expected Numbers*

Equal Chances	No High School	High School	Some College	Total
Yes	3.73	.38	2.83	6.94
No	5.44	.56	4.12	10.12
				$\chi^2 = 17.06$

Source: Tables 15.3.5 and 15.3.3.

In reaching this decision we ask "How often will we obtain χ^2 as large or larger than 17.06 under the null hypothesis?" To answer this question, we locate our calculated χ^2 in the χ^2-distribution. But we cannot do that without first establishing the number of degrees of freedom on which our calculated χ^2 rests.

Degrees of Freedom If we force three values to sum to 10, then we may write two of the values arbitrarily, but the third will be determined by the requirement that the three values sum to 10. If we arbitrarily choose 2 and 3 for our first two values then the third value will necessarily be 5, in order to satisfy the condition that the sum of the three values equal 10. Because only two of the three values are free to vary, we state that the set of three has two degrees of freedom (*df*).

Our problem is to find the number of expected frequencies that are free to vary and still satisfy the condition that the expected marginals be equal to the observed marginals. We may arrive at that number by the following method: If we arbitrarily write $\hat{n}_{11}$, then $\hat{n}_{21}$ is fixed by the condition that the sum of the expected frequencies in the first column equal the sum of the observed frequencies in that column. If, as our second step, we arbitrarily write $\hat{n}_{12}$, then $\hat{n}_{22}$ is determined by the requirement that expected frequencies in the second column be equal to the observed frequencies in that column. Furthermore, having selected $\hat{n}_{12}$ and $\hat{n}_{21}$ for values in the first row, the remaining frequency in that row is determined by the requirement that $\hat{n}_{11} + \hat{n}_{12} + \hat{n}_{13} = \Sigma n_{1j} = n_{1.}$. By the same token we are no longer free to write $\hat{n}_{23}$, since $\hat{n}_{21} + \hat{n}_{22} + \hat{n}_{23}$ must equal $\Sigma n_{2j} = n_{2.}$. Apparently, in our problem we can write only two of the expected frequencies freely and still satisfy the conditions imposed on the entire set; and our conclusion is that this set of expected frequencies, and the χ^2 based on them, has two degrees of freedom.

This tedious procedure of writing frequencies arbitrarily until we have exhausted our degrees of freedom is obviated by what are essentially counting formulas. In testing the hypothesis that the proportion of cases in each category (row) of A is the same in each population (column) of B, the

Table 15.3.7 *Table of χ^2-Values, by Selected Probability Values*

df	.70	.50	.30	.10	.05	.02	.01
1	.15	.46	1.07	2.71	3.84	5.41	6.64
2	.71	1.39	2.41	4.60	5.99	7.82	9.21
3	1.42	2.37	3.67	6.25	7.82	9.84	11.34
4	2.20	3.36	4.88	7.78	9.49	11.67	13.28
5	3.00	4.35	6.06	9.24	11.07	13.39	15.09
10	7.27	9.34	11.78	15.99	18.31	21.16	23.21
15	11.72	14.34	17.32	22.31	24.00	28.26	30.58
20	16.27	19.34	22.78	28.41	31.41	35.02	37.57
25	20.87	24.34	28.17	34.38	37.65	41.57	44.31
30	25.51	29.34	33.53	40.26	43.77	47.96	50.89

expected frequencies ($\hat{n}_{ij}$), and the χ^2 derived from them, will have $(a-1)(b-1)$ degrees of freedom, where a = number of rows in the A × B contingency table, and b = number of columns. When there are only two categories (rows) of A, as in our problem, this formula reduces to $df = b-1$. As a rule: When the hypothesis to be tested is that proportions in a given category are identical, the χ^2 test-statistic will have as many degrees of freedom as there are columns (populations) less one.

Having established that our $\chi^2 = 17.06$ has 2 degrees of freedom, we are ready to refer it to a χ^2 table. Although there is some variation, a χ^2 table ordinarily gives selected centiles of χ^2 for all degrees of freedom, 1 through 30. A few such entries are given in Table 15.3.7. The first row gives the centile values of χ^2 for one degree of freedom; the fifth row gives centile values of χ^2 for five degrees of freedom; and so on. With thirty degrees of freedom or more, one may use the z-table (table of normal probabilities) because of the close correspondence between z and $\sqrt{\chi^2}$ on $df > 30$.

We may use Table 15.3.7 to assess the significance of $\chi^2 = 17.06$. Our interest will be limited to the entries in the second row, since our χ^2 has two degrees of freedom. Scanning the second row, we find that $\chi^2 = 1.39$ will be exceeded fifty percent of the time when the null hypothesis is true; moving to the right, we see that $\chi^2 = 5.99$ will be exceeded five percent of the time when the null hypothesis is true. If we had set $\alpha = .05$, our decision would have been to reject the null hypothesis, since our calculated value is larger than $\chi^2_{.95} = 5.99$. We would have rejected the null hypothesis at the 1 percent level for that matter, since $\chi^2 = 17.06$ is larger than $\chi^2_{.99} = 9.21$. Note: the χ^2-table offers no guidance in choosing the significance level; it merely indicates whether our obtained value of χ^2 will be exceeded a specified proportion of sampling trials under the null hypothesis. It is a mere reference table, although a very useful one.

Assumption of χ^2-Test In testing the hypothesis that populations are identical in their proportions, we assume that cases are randomly selected both between and within populations. The χ^2-test requires that observations be mutually independent and random sampling is the technique for meeting that requirement. With makeshift and make-do samples—which are common in social research—statistical inference in the strict sense of the term is out of the question, although there is nothing to stop the investigator from speculating about the representativeness of the materials. Nevertheless, when the object is to assess the risk of rejecting a true hypothesis (and the risk of accepting a false one), random selection is an absolute necessity.

It is possible to calculate χ^2 for any contingency table; however, the χ^2-table is valid as a reference only for relatively large expected frequencies. How large is large? Although there is no categorical answer to this question, authorities agree that the χ^2-table may be used provided that no expected frequency is smaller than five. Hence, the rule is that the χ^2-test requires all *expected frequencies* to be larger than five. When expected frequencies are smaller, the χ^2-table should be set aside in favor of more exact probabilities.

QUESTIONS AND PROBLEMS

1. Define the following terms:
   contingency table analysis
   A × B contingency table
   hypothesis of equal proportions
   χ^2-test

2. An investigator wishes to test H_0 that $P_1 = P_2 = \cdots P_k$ by the χ^2 statistic:

$$\chi^2 = \sum_1^2 \sum_1^k \frac{(n_{ij} - \hat{n}_{ij})^2}{n_{ij}}.$$

   Find the critical values of χ^2 for the following combinations of α and *df*.

χ^2	α	*df*
———	.05	6
———	.10	6
———	.01	4
———	.02	2
———	.001	5
———	.20	2

3. Calculate χ^2 for Table 15.3.8. Is this result for or against the hypothesis that religious denominations are identical in their occupational distributions?

Table 15.3.8 *Lawyers, Professors, and Engineers by Religious Affiliations, 1967*

Religion	Lawyers	Professors	Engineers
Protestant	108	31	145
Catholic	54	33	31
Jewish	41	13	0
Other or none	4	22	8

Source: Adapted from Harold L. Wilensky and Jack Ladinsky. "From Religious Community to Occupational Group: Structural Assimilation Among Professors, Lawyers, and Engineers," *American Sociological Review* 32 (August, 1967), Table 1, p. 544. Reprinted with permission.

4. Calculate χ^2 for Table 15.3.9. Does this result favor the idea that irregular employment promotes revolutionary attitudes? Explain.

Table 15.3.9 *Pre-Revolutionary Employment Status and Attitude toward the Cuban Revolution among Cuban Workers, 1962*

Months Worked Per Year Before Revolution	Attitude		
	Favorable	Indecisive	Hostile
6 or less	54	6	3
7–9	14	2	3
10 or more	65	14	26

Source: Adapted from Maurice Zeitlin, *Revolutionary Politics and the Cuban Working Class*. Princeton: Princeton University Press, 1967, pp. 55. Copyright © 1967 by Princeton University Press. Reprinted by permission of Princeton University Press.

5. Calculate χ^2 for Table 15.3.10. Is this calculation consistent with the idea that age affects one's opinion about ending life? Explain.

Table 15.3.10 *Opinion Concerning Euthanasia by Age, Percentage Distribution, United States, April, 1975*

"Do you think a person has a moral right to end his or her life under these conditions: When this person has a disease that is incurable?"				
Age	Yes	No	No opinion	Total[a]
18–29	54	40	6	100
30–49	40	52	8	100
50 and over	29	63	8	100

[a] Sample sizes for age groups as listed from young to old are 445, 500, 555.

Source: *The Gallup Opinion Index*, Report No. 122, August, 1975, p. 22. Reprinted with permission.

4 A × B × C CONTINGENCY TABLE

Preliminary Remark An A × B × C contingency table differs from an A × B table in that cases are classified in three ways instead of two. For example, if, in addition to classifying cases by, say, response (Yes = A_1 or No = A_2) and by years of schooling (Grade = B_1, High = B_2, College = B_3), we classify them by region (South = C_1 or North = C_2), we get an A × B × C contingency table. An example is supplied by Table 15.4.1.

This table permits us to investigate the effect not only of education (B) on the attitude of whites towards jobs for blacks (A), but also the effect of region (C) on the attitudes of whites, and the effect of region on the effect of education on that attitude. In a manner analogous to the two-way analysis of variance of interval measures, we test the hypothesis that (1) educational groupings are identical in their proportions within regions; (2) regional groupings are identical in their proportions within educational classes; and (3) the degree of association between response and educational background does not differ from one region to another (or identically that the degree of association between region and response does not differ from one educational grouping to another). Hypothesis 1 claims that the relation between A and B will vanish with C constant; Hypothesis 2 claims that the relation between A and C will vanish with B constant; Hypothesis 3 claims that the degree of association between A and B is the same for all categories of C. Although the last hypothesis is sometimes referred to as the hypothesis of no interaction, its meaning differs from the meaning of its counterpart in the analysis of variance.

We note parenthetically the similarity between A × B × C contingency table analysis and the process of elaboration, as given in Chapter 8. Our object in elaborating a total relation is to determine whether the relation between A and B vanishes with C constant (explanation), or whether the

Table 15.4.1 *Response by Education by Region*

	South				North			
Response	Grade School	High School	Some College	T	Grade School	High School	Some College	Total
Yes	11	19	9	39	46	85	40	171
No	25	19	7	51	42	42	9	93
Total	36	38	16	90	88	127	49	264

Source: Table 15.3.1.

association between A and B differs from one level of C to another (specification).

Procedure The procedure for testing each of the above hypotheses adheres to the common rule of calculating the deviations between observed frequencies and expected frequencies under a given hypothesis, squaring these deviations, adjusting these squared deviations on their own expected frequencies, and summing these squared deviations for χ^2. Generally speaking, we determine whether our observed frequencies fit those given by the model.

In testing Hypotheses 1 and 2, we may calculate expected frequencies in accordance with Formula 15.3.2, but in testing Hypothesis 3, we get the expected frequencies by *iteration.* This means that trial frequencies are repeatedly adjusted by the same operation until they satisfy, in addition to the marginal constraints, the requirement that the association between A and B be the same at each level of C. Except for iteration, which is arduous to carry out by hand, hypothesis-testing with an A × B × C table, as an arithmetic operation, is no different from hypothesis-testing with an A × B contingency table.

Hypothesis 1 The hypothesis to be tested is that educational classes are identical in their proportions within regions. In symbols:

$$H_0 : P_{11k} = P_{12k} = P_{13k} \qquad (k = 1, 2)$$

where P_{11k} = proportion giving first response (yes) in first educational grouping (grade school) in kth region; and so forth. This hypothesis requires that we set up a separate 2 × 3 table for each region, each identical in form to Table 15.4.1. For each 2 × 3 table, we calculate χ^2, subject to the restriction that expected marginals equal observed marginals. The sum of these separate χ^2s is our test-statistic.

Since each component χ^2 in this sum has two degrees of freedom, the sum itself has four degrees of freedom. The restrictions on the degrees of freedom may be conveniently displayed in tabular form. Table 15.4.2 gives the row and column marginals for each 2 × 3 table of expected frequencies. Scanning entries we find that, given A_1 and C_1, the expected frequencies for B_1, B_2, and B_3 must sum to 39. Continuing, we find that the expected frequencies for A_1 and A_2, given B_3 and C_2, must sum to 49. The other entries in this table may similarly be construed as constraints on the expected frequencies, or degrees of freedom.

The calculation of χ^2 for Hypothesis 1 is given in Table 15.4.3. In this table, n_{ijk} stands for the number of persons giving the ith response (Yes or No) in the jth educational grouping (Grade School, High School, or College) in the kth region (North or South). The expected frequencies (the

Table 15.4.2 *Marginal Constraints in Testing Hypothesis 1*

	C_1	C_2	
A_1	39	171	
A_2	51	93	
	B_1	B_2	B_3
C_1	36	38	16
C_2	88	127	49

Source: Table 15.4.1.

$\hat{n}_{ijk}$) are calculated according to the following formula, which, except for a change in notation, is the same as Formula 15.3.2:

$$\hat{n}_{ijk} = \frac{n_{i.k}n_{.jk}}{N_{..k}}. \tag{15.4.1}$$

With expected frequencies in hand we proceed step-wise to χ^2: To obtain χ^2, we (1) get deviations between observed and expected frequencies; (2) square these deviations; (3) express these squares as multiples of expected

Table 15.4.3 *Testing Hypothesis that Educational Groupings Are Identical within Regions*

ijk	n_{ijk}	$\hat{n}_{ijk}$	$n_{ijk} - \hat{n}_{ijk}$	$(n_{ijk} - \hat{n}_{ijk})^2$	$(n_{ijk} - \hat{n}_{ijk})^2/\hat{n}_{ijk}$
111	11	15.60	−4.60	21.1600	1.3564
121	19	16.47	2.53	6.4009	.3886
131	9	6.93	2.07	4.2849	.6183
211	25	20.40	4.60	21.1600	1.0372
221	19	21.53	−2.53	6.4009	.2973
231	7	9.07	−2.07	4.2849	.4724
					4.1702
112	46	57.00	−11.00	121.0000	2.1228
122	85	82.26	2.74	7.5076	.0913
132	40	31.74	8.26	68.2276	2.1496
212	42	31.00	11.00	121.0000	3.9032
222	42	44.74	− 2.74	7.5076	.1678
232	9	17.26	− 8.26	68.2276	3.9529
					12.3876
					$\chi^2 = 16.5578$

frequencies; and (4) sum these multiples for χ^2. Carrying out these calculations, we get $\chi^2 = 4.1702 + 12.3876 = 16.5578$ on four degrees of freedom. Since we would obtain larger χ^2s fewer than 1 time in 100 when the null hypothesis of identical proportions within regions is true (see Table 15.3.7), we would probably reject Hypothesis 1. Our substantive conclusion would be that the effect of education on the opinions of whites toward job chances for blacks is not spurious, since it maintains itself after adjusting for the effect of region.

Hypothesis 2 The hypothesis to be tested is that regions are identical in their proportions within educational classes. In symbols:

$$H_0: P_{1j1} = P_{1j2} \qquad (j = 1, 2, 3).$$

In running this test, we set up a separate 2 × 2 table for each educational grouping, and for each of these tables, we calculate χ^2, subject to the restriction that expected marginals equal the observed marginals in that table. These 2 × 2 contingency tables comprise Table 15.4.4. Our test-statistic is the sum of these individual χ^2s. Since each of the component χ^2s has one degree of freedom, their sum will have three degrees of freedom.

As with Hypothesis 1, the restrictions on the expected frequencies may be conveniently displayed in tabular form. Table 15.4.5 gives the row marginals for each 2 × 2 table of expected frequencies; column marginals are given in the bottom half of Table 15.4.2. Scanning Table 15.4.5 we see that, given A_1 and B_1, the expected frequencies for C_1 and C_2 must add up to 57; returning to Table 15.4.2, we find that, given B_2 and C_2, the expected frequencies must sum to 127.

The calculation of χ^2 for Hypothesis 2 is given in Table 15.4.6. Expected frequencies are calculated in accordance with the following formula (which is essentially the same as Formula (15.4.1):

$$\hat{n}_{ijk} = \frac{(n_{ij.})(n_{.jk})}{N_{.j.}}. \qquad (15.4.2)$$

Table 15.4.4 *Response by Region by Education*

	Grade School			High School			Some College		
Response	South	North	Total	South	North	Total	South	North	Total
Yes	11	46	57	19	85	104	9	40	49
No	25	42	67	19	42	61	7	9	16
Total	36	88	124	38	127	165	16	49	65

Source: Table 15.4.1.

Table 15.4.5 A × B Marginal Constraints in Testing Hypothesis 2

	B_1	B_2	B_3
A_1	57	104	49
A_2	67	61	16

Source: Table 15.4.1.

The sums of the ratios of squared deviations to expected frequencies for the three tables are 4.8542, 3.5951, and 4.1828; summing these component χ^2s gives us our test statistic of 12.6321 with three degrees of freedom. Since we expect to obtain larger χ^2s fewer than 1 time in 100 when the null hypothesis is true, we would probably reject Hypothesis 2 that regions are identical in their proportions within educational groupings. Our substantive conclusion would be that the influence of region cannot be ascribed to the concealed effect of education, since the regional effect persists with the effect of education statistically removed.

Table 15.4.6 Testing Hypothesis that Regions are Identical within Educational Classes by χ^2

ijk	n_{ijk}	$\hat{n}_{ijk}$	$n_{ijk} - \hat{n}_{ijk}$	$(n_{ijk} - \hat{n}_{ijk})^2$	$(n_{ijk} - \hat{n}_{ijk})^2/\hat{n}_{ijk}$
111	11	16.55	−5.55	30.8025	1.8612
112	46	40.45	5.55	30.8025	.7615
211	25	19.45	5.55	30.8025	1.5837
212	42	47.55	−5.55	30.8025	.6478
					4.8542
121	19	23.95	−4.95	24.5025	1.0231
122	85	80.05	4.95	24.5025	.3061
221	19	14.05	4.95	24.5025	1.7440
222	42	46.95	−4.95	24.5025	.5219
					3.5951
131	9	12.06	−3.06	9.3636	.7764
132	40	36.94	3.06	9.3636	.2535
231	7	3.94	3.06	9.3636	2.3765
232	9	12.06	−3.06	9.3636	.7764
					4.1828
					$\chi^2 = 12.6321$

Hypothesis 3 This hypothesis claims that the association between response (A) and region (C) does not differ by educational class (B). It is identical to the hypothesis that the association between response (A) and education (B) does not differ by region (C), since the truth of the former requires the truth of the latter, and vice versa. But stated more precisely, the hypothesis to be tested is that the pattern of joint frequencies in 2 × 2 contingency tables showing response and region do not differ from one educational grouping to another; or equivalently that 2 × 2 tables showing response and education (educational categories two at a time) do not differ from one region to the other. In symbols:

$$H_3\colon \frac{n_{111}n_{212}}{n_{112}n_{211}} = \frac{n_{121}n_{222}}{n_{122}n_{221}} = \frac{n_{131}n_{232}}{n_{132}n_{231}}$$

where n_{111} = number coded "Yes" in subpopulation B_1C_1; n_{212} = number coded "No" in subpopulation B_1C_2; and so on.

Degrees of Freedom In testing H_3, we force the expected frequencies to add up to three sets of marginal totals instead of two as in the case of Hypothesis 1 or Hypothesis 2. To clarify this extension of restrictions, let us return to Table 15.4.1 and Hypothesis 1. In testing this hypothesis we forced the marginal frequencies to add up to the marginals shown in Table 15.4.2. Now, if instead of analyzing the relationship between A and B for C_1 and C_2 separately, we form a single table with C_2 added as a second layer to C_1, then we have in effect a two-layer table (that is, a 2 × 3 × 2 table) with three sets of marginals: AB, AC, and BC. The hypothesis that populations are identical in degree of association requires that the expected frequencies reproduce all three sets of marginals. For example, B_1, B_2, and B_3 must sum to 171, given A_1 and C_2; C_1 and C_2 must sum to 61, given A_2 and B_2; A_1 and A_2 must sum to 88, given B_1 and C_2.

A little exploration will reveal that with this additional constraint, we are free to write only two cell entries in the two-layer table (the 2 × 3 × 2 table), the remaining ten entries being determined by the restrictions we have imposed upon the expected frequencies. What this means practically is that χ^2 for H_3 will have two degrees of freedom. The number of degrees of freedom for Hypothesis 3, given any A × B × C contingency table, can always be readily calculated by the following formula:

$$df = (a-1)(b-1)(c-1)$$

where a is the number of categories of A, b is the number of categories of B, and c is the number of categories of C. In our case,

$$df = (2-1)(3-1)(2-1)$$
$$= 2,$$

in agreement with our previous answer.

Iterating for Expected Frequencies As previously noted, the calculation of χ^2 for Hypothesis 3 is no different in essentials from the calculation of χ^2 for Hypotheses 1 and 2, except that the expected frequencies must be derived by a process of iteration. This means that arbitrary trial values are repeatedly adjusted until they conform to the model of no differences between populations in degree of association. When the working limit of improvement is reached, the iterative process is discontinued. Since the same operations are repeated in each cycle, it will suffice for present purposes to go through the operations in the initial cycle alone. Unless an investigator has access to a computer, a complete iteration takes forever. For that reason an iteration is either done electronically or not done at all. The object of the demonstration here is to merely clarify the student's concept of the process of arriving at expected frequencies by iteration.

To begin iteration, we require the three-variable table of frequencies, and the two-variable tables of collapsed frequencies—these appear as Steps (1), (2), (3), and (4) in Table 15.4.7. The following step-wise operations comprise the first cycle of iteration; they are repeated in every subsequent cycle:

(1) Choose any of the two-variable tables, say (2), and divide the frequencies of that table equally between the levels of B for (5).

(2) Add the frequencies of (5) across categories C to get (6).

(3) Divide the frequencies of (3) by the corresponding frequencies of (6) to get the weights of (7).

(4) Multiply the A_iB_j frequencies of (5) by the A_iB_j weights of (7) for the adjusted frequencies of (8).

(5) Add the frequencies of (8) over categories of A to get (9).

(6) Divide the frequencies of (4) by the frequencies of (9) to obtain the weights of (10).

(7) Multiply the B_iC_j entries of (8) by the B_iC_j weights of (10) to get (11).

Step number (11) in Table 15.4.7 marks the end of the first round of iteration and/or the beginning of the second round. If we were to continue into the second cycle, we would collapse the frequencies of (11) to get the required two-variable tables. Beginning with any one of these two-variable tables, say AC, we would split its frequencies evenly among categories of B; and then add these divided frequencies across levels of B to get an AB table of frequencies. We would continue in this manner through the second cycle. Upon completing the second cycle, we would proceed as necessary to reach the agreed upon working limit of improvement. Four cycles were required to obtain the expected frequencies of Table 15.4.8.

With the expected frequencies given in Table 15.4.8, we are ready to proceed with the calculation of χ^2, which is also shown in Table 15.4.8. Comparing our obtained χ^2 with $\chi^2_{.95}$ on two degrees of freedom, we find that it is greatly deficient, and we would decide against rejecting the hypothesis of no differences in degree of association between opinion and region from one educational grouping to another. Apparently the impact of region

Table 15.4.7 *Step-wise Routine for Getting Expected Frequencies under Hypothesis 3*

1

	B_1		B_2		B_3	
	C_1	C_2	C_1	C_2	C_1	C_2
A_1	11	46	19	85	9	40
A_2	25	42	19	42	7	9

2

	C_1	C_2
A_1	39	171
A_2	51	93

3

	B_1	B_2	B_3
A_1	57	104	49
A_2	67	61	16

4

	B_1	B_2	B_3
C_1	36	38	16
C_2	88	127	49

5

	B_1		B_2		B_3	
	C_1	C_2	C_1	C_2	C_1	C_2
A_1	13	57	13	57	13	57
A_2	17	31	17	31	17	31

6

	B_1	B_2	B_3
A_1	70	70	70
A_2	48	48	48

7

	B_1	B_2	B_3
A_1	.8143	1.4857	.7000
A_2	1.3958	1.2708	.3333

8

	B_1		B_2		B_3	
	C_1	C_2	C_1	C_2	C_1	C_2
A_1	10.5859	46.4151	19.3141	84.6849	9.1000	39.9000
A_2	23.7286	43.2698	21.6036	39.3948	5.6661	10.3323

9

	B_1	B_2	B_3
C_1	34.3145	40.9177	14.7661
C_2	89.6849	124.0797	50.2323

(*continued*)

Table 15.4.7 (*continued*)

		B_1	B_2	B_3
10	C_1	1.0491	.9287	1.0836
	C_2	.9812	1.0235	.9755

		B_1		B_2		B_3	
		C_1	C_2	C_1	C_2	C_1	C_2
11	A_1	11.1057	45.5425	17.9370	86.6750	9.8608	38.9224
	A_2	24.8937	42.4563	20.0633	40.3205	6.1398	10.0792

on opinion is equally intense from one educational level to another. Or we could reverse the roles of region and education and say that the impact of education on opinion is equally intense from one region to the other.

Perspective on Data Analysis If we are required to determine what part of the total variation in a set of interval measures can be attributed to differences between group means, we have recourse to the analysis of variance. We might, for example, employ that method to determine what part of the

Table 15.4.8 *Testing Hypothesis that Degree of Association between A and C Is the Same at Each Level of B*

ijk	n_{ijk}	$\hat{n}_{ijk}$	$n_{ijk} - \hat{n}_{ijk}$	$(n_{ijk} - \hat{n}_{ijk})^2$	$(n_{ijk} - \hat{n}_{ijk})^2/\hat{n}_{ijk}$
111	11	11.23	−0.23	.0529	.0047
112	46	45.77	+0.23	.0529	.0012
211	25	24.77	+0.23	.0529	.0021
212	42	42.23	−0.23	.0529	.0013
121	19	17.82	1.18	1.3924	.0781
122	85	86.18	−1.18	1.3924	.0162
221	19	20.18	−1.18	1.3924	.0690
222	42	40.82	1.18	1.3924	.0341
131	9	9.95	−0.95	.9025	.0907
132	40	39.05	0.95	.9025	.0231
231	7	6.05	0.95	.9025	.1492
232	9	9.95	−0.95	.9025	.0907
					$\chi^2 = 0.5604$ $\chi^2_{.95}$ on 2 $df = 5.991$

variation of the suicide rates in Table 4.1.4 is due to differences between regional means. The analysis of variance may thus be viewed as a technique for measuring the statistical (but not necessarily causal) effect of a given factor on a given variable (for example, the effect of region on the suicide rate). Leaving aside its practical utility in sociological research, we consider it worthy of study because of the mentality it represents. With its emphasis on contrasts between groups free of confounding factors, it fosters a habit of planning sequentially so that the effect of a given factor on a given variable may be isolated and measured, and spurious ascriptions of causality may be avoided. Even though its utility in social research is limited at the moment, due to sampling and measurement conditions that cannot be readily met, an examination of its essentials is worthwhile because of the analytical approach they exemplify. As a research model, the analysis of variance may be valuable to the sociologist even though its use, particularly in regards to significance testing, cannot always be justified.

The same remarks apply to contingency table analysis, with the qualification that opportunities for its application in sociology appear to be relatively more numerous. This is so because so many of the variables of sociology take the form of unordered sets, for example, sex, race, nationality, political party, religion, occupation, marital status, and so forth. Disentangling relations among such variables is the special province of contingency table analysis. We are likely to utilize this method whenever our job is to unravel or ransack the relations between two or more qualitative variables.

As with the F-ratio, significance testing by χ^2 requires that certain assumptions be met. However, even when these assumptions are not met and significance testing is passed up, we can justify application of contingency table analysis because it throws light on connections that might be overlooked or misunderstood if frequency tables were cursorily evaluated, or "eye-balled." Students wishing to pursue this topic should consult the references at the end of this chapter.

QUESTIONS AND PROBLEMS

1. Define the following terms:
   $A \times B \times C$ contingency table
   hypothesis of uniform association

2. Collapse the frequencies of Table 15.4.9 to set up an $A \times B \times C$ table: knowledge (A) by read newspapers (B) by listen to radio (C). Test H that there are no differences in knowledge between listeners within categories of readers by the appropriate χ^2. Test H that degree of association between knowledge and reading is uniform with categories of listeners by appropriate χ^2.

Table 15.4.9 *Cross-Classification of Individuals According to Five Dichotomized Variables*[a]

Knowledge	Radio				No radio			
	Solid reading		No solid reading		Solid reading		No solid reading	
	Good	Poor	Good	Poor	Good	Poor	Good	Poor
Newspapers								
Lectures	23	8	8	4	27	18	7	6
No lectures	102	67	35	59	201	177	75	156
No newspapers								
Lectures	1	3	4	3	3	8	2	10
No lectures	16	16	13	50	67	83	84	393

[a] The five variables classify individuals according to whether they (1) read newspapers, (2) listen to radio, (3) read books and magazines, (4) attend lectures, and (5) whether their knowledge of cancer is good or poor. Source: Leo A. Goodman, "The Multivariate Analysis of Qualitative Data: Interactions Among Multiple Classifications." *Journal of the American Statistical Association* 65 (March, 1970): 226 256. Table 1, page 227. Reprinted with permission.

SELECTED REFERENCES

Blalock, Hubert M., Jr.

1970 Social Statistics. Second edition. New York: McGraw-Hill. Chapters 13–15.

Burke, Peter J. and Austin T. Turk

1975 "Factors affecting postarrest dispositions: A model for analysis." Social Problems 22: 313–331.

Davis, James, A.

1974 "Hierarchical models for significance tests in multivariate contingency tables: An exegesis of Goodman's recent papers," in H. L. Costner (ed.), Sociological Methodology 1973–1974. San Francisco: Jossey Bass.

Fisher, Ronald A.

1958 Statistical Methods for Research Workers. Thirteenth edition. Edinburgh: Oliver and Boyd.

Goodman, Leo

1969 "How to ransack social mobility tables and other kinds of cross-classification tables." American Journal of Sociology 75: 1–39.

1970 "The multivariate analysis of qualitative data: Interactions among multiple classifications." Journal of the American Statistical Association 65: 225–256.

1972 "A general model for the analysis of surveys." American Journal of Sociology 77: 1035–1086.

Snedecor, G. W. and W. G. Cochran

1967 Statistical Methods. Sixth edition. Ames, Iowa: Iowa State University Press.

Appendix

Table I *Areas of the Normal Curve*

$\frac{x}{\sigma}$	.00	.01	.02	.03	.04	.05	.06	.07	.08	.09
0.0	.0000	.0040	.0080	.0120	.0159	.0199	.0239	.0279	.0319	.0359
0.1	.0398	.0438	.0478	.0517	.0557	.0596	.0636	.0675	.0714	.0753
0.2	.0793	.0832	.0871	.0910	.0948	.0987	.1026	.1064	.1103	.1141
0.3	.1179	.1217	.1255	.1293	.1331	.1368	.1406	.1443	.1480	.1517
0.4	.1554	.1591	.1628	.1664	.1700	.1736	.1772	.1808	.1844	.1879
0.5	.1915	.1950	.1985	.2019	.2054	.2088	.2123	.2157	.2190	.2224
0.6	.2257	.2291	.2324	.2357	.2389	.2422	.2454	.2486	.2518	.2549
0.7	.2580	.2612	.2642	.2673	.2704	.2734	.2764	.2794	.2823	.2852
0.8	.2881	.2910	.2939	.2967	.2995	.3023	.3051	.3078	.3106	.3133
0.9	.3159	.3186	.3212	.3238	.3264	.3289	.3315	.3340	.3365	.3389
1.0	.3413	.3438	.3461	.3485	.3508	.3531	.3554	.3577	.3599	.3621
1.1	.3643	.3665	.3686	.3718	.3729	.3749	.3770	.3790	.3810	.3830
1.2	.3849	.3869	.3888	.3907	.3925	.3944	.3962	.3980	.3997	.4015
1.3	.4032	.4049	.4066	.4083	.4099	.4115	.4131	.4147	.4162	.4177
1.4	.4192	.4207	.4222	.4236	.4251	.4265	.4279	.4292	.4306	.4319
1.5	.4332	.4345	.4357	.4370	.4382	.4394	.4406	.4418	.4430	.4441
1.6	.4452	.4463	.4474	.4485	.4495	.4505	.4515	.4525	.4535	.4545
1.7	.4554	.4564	.4573	.4582	.4591	.4599	.4608	.4616	.4625	.4633
1.8	.4641	.4649	.4656	.4664	.4671	.4678	.4686	.4693	.4699	.4706
1.9	.4713	.4719	.4726	.4732	.4738	.4744	.4750	.4758	.4762	.4767
2.0	.4772	.4778	.4783	.4788	.4793	.4798	.4803	.4808	.4812	.4817
2.1	.4821	.4826	.4830	.4834	.4838	.4842	.4846	.4850	.4854	.4857
2.2	.4861	.4865	.4868	.4871	.4875	.4878	.4881	.4884	.4887	.4890
2.3	.4893	.4896	.4898	.4901	.4904	.4906	.4909	.4911	.4913	.4916
2.4	.4918	.4920	.4922	.4925	.4927	.4929	.4931	.4932	.4934	.4936
2.5	.4938	.4940	.4941	.4943	.4945	.4946	.4948	.4949	.4951	.4952
2.6	.4953	.4955	.4956	.4957	.4959	.4960	.4961	.4962	.4963	.4964
2.7	.4965	.4966	.4967	.4968	.4969	.4970	.4971	.4972	.4973	.4974
2.8	.4974	.4975	.4976	.4977	.4977	.4978	.4979	.4980	.4980	.4981
2.9	.4981	.4982	.4983	.4984	.4984	.4984	.4985	.4985	.4986	.4986
3.0	.49865	.4987	.4987	.4988	.4988	.4988	.4989	.4989	.4989	.4990
3.1	.49903	.4991	.4991	.4991	.4992	.4992	.4992	.4992	.4993	.4993
4.0	.49997									

Source: Table 10, "Areas of the Normal Curve," in *Tables for Statisticians.* 2nd edition, by Herbert Arkin and Raymond R. Colton, Harper & Row, 1963, Barnes & Noble Division.

Table II *Ordinates of the Normal Curve*

$\frac{x}{\sigma}$	.00	.01	.02	.03	.04	.05	.06	.07	.08	.09
0.0	1.00000	.99995	.99980	.99955	.99920	.99875	.99820	.99755	.99685	.99596
0.1	.99501	.99396	.99283	.99158	.99025	.98881	.98728	.98565	.98393	.98211
0.2	.98020	.97819	.97609	.97390	.97161	.96923	.96676	.96420	.96156	.95882
0.3	.95600	.95309	.95010	.94702	.94387	.94055	.93723	.93382	.93024	.92677
0.4	.92312	.91799	.91558	.91169	.90774	.90371	.89961	.89543	.89119	.88688
0.5	.88250	.87805	.87353	.86896	.86432	.85962	.85488	.85006	.84519	.84060
0.6	.83527	.83023	.82514	.82010	.81481	.80957	.80429	.79896	.79459	.78817
0.7	.78270	.77721	.77167	.76610	.76048	.75484	.74916	.74342	.73769	.73193
0.8	.72615	.72033	.71448	.70861	.70272	.69681	.69087	.68493	.67896	.67298
0.9	.66689	.66097	.65494	.64891	.64287	.63683	.63077	.62472	.61865	.61259
1.0	.60653	.60047	.59440	.58834	.58228	.57623	.57017	.56414	.55810	.55209
1.1	.54607	.54007	.53409	.52812	.52214	.51620	.51027	.50437	.49848	.49260
1.2	.48675	.48092	.47511	.46933	.46357	.45793	.45212	.44644	.44078	.43516
1.3	.42956	.42399	.41845	.41294	.40747	.40202	.39661	.39123	.38569	.38058
1.4	.37531	.37007	.36487	.35971	.35459	.34950	.34445	.33944	.33447	.32954
1.5	.32465	.31980	.31500	.31023	.30550	.30082	.29618	.29158	.28702	.28251
1.6	.27804	.27361	.26923	.26489	.26059	.25634	.25213	.24797	.24385	.23978
1.7	.23575	.23176	.22782	.22392	.22008	.21627	.21251	.20879	.20511	.20148
1.8	.19790	.19436	.19086	.18741	.18400	.18064	.17732	.17404	.17081	.16762
1.9	.16448	.16137	.15831	.15530	.15232	.14939	.14650	.14364	.14083	.13806
2.0	.13534	.13265	.13000	.12740	.12483	.12230	.11981	.11737	.11496	.11259
2.1	.11025	.10795	.10570	.10347	.10129	.09914	.09702	.09495	.09290	.09090
2.2	.08892	.08698	.08507	.08320	.08136	.07956	.07778	.07604	.07433	.07265
2.3	.07100	.06939	.06780	.06624	.06471	.06321	.06174	.06029	.05888	.05750
2.4	.05614	.05481	.05350	.05222	.05096	.04973	.04852	.04737	.04618	.04505
2.5	.04394	.04285	.04179	.04074	.03972	.03873	.03775	.03680	.03586	.03494
2.6	.03405	.03317	.03232	.03148	.03066	.02986	.02908	.02831	.02757	.02684
2.7	.02612	.02542	.02474	.02408	.02343	.02280	.02218	.02157	.02098	.02040
2.8	.01984	.01929	.01876	.01823	.01772	.01723	.01674	.01627	.01581	.01536
2.9	.01492	.01449	.01408	.01367	.01328	.01288	.01252	.01215	.01179	.01145
3.0	.01111									
4.0	.00034									

Source: Table 11, "Ordinates of the Normal Curve," in *Tables for Statisticians*, 2nd edition, by Herbert Arkin and Raymond R. Colton, Harper & Row, 1963, Barnes & Noble Division.

Table III *Distribution of* t

df	Level of significance for one-tailed test					
	.10	.05	.025	.01	.005	.0005
	Level of significance for two-tailed test					
	.20	.10	.05	.02	.01	.001
1	3.078	6.314	12.706	31.821	63.657	636.619
2	1.886	2.920	4.303	6.965	9.925	31.598
3	1.638	2.353	3.182	4.541	5.841	12.941
4	1.533	2.132	2.776	3.747	4.604	8.610
5	1.476	2.015	2.571	3.365	4.032	6.859
6	1.440	1.943	2.447	3.143	3.707	5.959
7	1.415	1.895	2.365	2.998	3.499	5.405
8	1.397	1.860	2.306	2.896	3.355	5.041
9	1.383	1.833	2.262	2.821	3.250	4.781
10	1.372	1.812	2.228	2.764	3.169	4.587
11	1.363	1.796	2.201	2.718	3.106	4.437
12	1.356	1.782	2.179	2.681	3.055	4.318
13	1.350	1.771	2.160	2.650	3.012	4.221
14	1.345	1.761	2.145	2.624	2.977	4.140
15	1.341	1.753	2.131	2.602	2.947	4.073
16	1.337	1.746	2.120	2.583	2.921	4.015
17	1.333	1.740	2.110	2.567	2.898	3.965
18	1.330	1.734	2.101	2.552	2.878	3.922
19	1.328	1.729	2.093	2.539	2.861	3.883
20	1.325	1.725	2.086	2.528	2.845	3.850
21	1.323	1.721	2.080	2.518	2.831	3.819
22	1.321	1.717	2.074	2.508	2.819	3.792
23	1.319	1.714	2.069	2.500	2.807	3.767
24	1.318	1.711	2.064	2.492	2.797	3.745
25	1.316	1.708	2.060	2.485	2.787	3.725
26	1.315	1.706	2.056	2.479	2.779	3.707
27	1.314	1.703	2.052	2.473	2.771	3.690
28	1.313	1.701	2.048	2.467	2.763	3.674
29	1.311	1.699	2.045	2.462	2.756	3.659
30	1.310	1.697	2.042	2.457	2.750	3.646
40	1.303	1.684	2.021	2.423	2.704	3.551
60	1.296	1.671	2.000	2.390	2.660	3.460
120	1.289	1.658	1.980	2.358	2.617	3.373
∞	1.282	1.645	1.960	2.326	2.576	3.291

Source: Table III is taken from Table III of Fisher and Yates: *Statistical Tables for Biological, Agricultural and Medical Research,* published by Longman Group, Ltd., London (previously published by Oliver and Boyd, Edinburgh), and used by permission of the authors and publishers.

Table IV-a *Variance Ratio* (F-*Ratio*) 5% *Points*

ν_2	ν_1 = 1	2	3	4	5	6	8	12	24	∞
1	161.4	199.5	215.7	224.6	230.2	234.0	238.9	243.9	249.0	254.3
2	18.51	19.00	19.16	19.25	19.30	19.33	19.37	19.41	19.45	19.50
3	10.13	9.55	9.28	9.12	9.01	8.94	8.84	8.74	8.64	8.53
4	7.71	6.94	6.59	6.39	6.26	6.16	6.04	5.91	5.77	5.63
5	6.61	5.79	5.41	5.19	5.05	4.95	4.82	4.68	4.53	4.36
6	5.99	5.14	4.76	4.53	4.39	4.28	4.15	4.00	3.84	3.67
7	5.59	4.74	4.35	4.12	3.97	3.87	3.73	3.57	3.41	3.23
8	5.32	4.46	4.07	3.84	3.69	3.58	3.44	3.28	3.12	2.93
9	5.12	4.26	3.86	3.63	3.48	3.37	3.23	3.07	2.90	2.71
10	4.96	4.10	3.71	3.48	3.33	3.22	3.07	2.91	2.74	2.54
11	4.84	3.98	3.59	3.36	3.20	3.09	2.95	2.79	2.61	2.40
12	4.75	3.88	3.49	3.26	3.11	3.00	2.85	2.69	2.50	2.30
13	4.67	3.80	3.41	3.18	3.02	2.92	2.77	2.60	2.42	2.21
14	4.60	3.74	3.34	3.11	2.96	2.85	2.70	2.53	2.35	2.13
15	4.54	3.68	3.29	3.06	2.90	2.79	2.64	2.48	2.29	2.07
16	4.49	3.63	3.24	3.01	2.85	2.74	2.59	2.42	2.24	2.01
17	4.45	3.59	3.20	2.96	2.81	2.70	2.55	2.38	2.19	1.96
18	4.41	3.55	3.16	2.93	2.77	2.66	2.51	2.34	2.15	1.92
19	4.38	3.52	3.13	2.90	2.74	2.63	2.48	2.31	2.11	1.88
20	4.35	3.49	3.10	2.87	2.71	2.60	2.45	2.28	2.08	1.84
21	4.32	3.47	3.07	2.84	2.68	2.57	2.42	2.25	2.05	1.81
22	4.30	3.44	3.05	2.82	2.66	2.55	2.40	2.23	2.03	1.78
23	4.28	3.42	3.03	2.80	2.64	2.53	2.38	2.20	2.00	1.76
24	4.26	3.40	3.01	2.78	2.62	2.51	2.36	2.18	1.98	1.73
25	4.24	3.38	2.99	2.76	2.60	2.49	2.34	2.16	1.96	1.71
26	4.22	3.37	2.98	2.74	2.59	2.47	2.32	2.15	1.95	1.69
27	4.21	3.35	2.96	2.73	2.57	2.46	2.30	2.13	1.93	1.67
28	4.20	3.34	2.95	2.71	2.56	2.44	2.29	2.12	1.91	1.65
29	4.18	3.33	2.93	2.70	2.54	2.43	2.28	2.10	1.90	1.64
30	4.17	3.32	2.92	2.69	2.53	2.42	2.27	2.09	1.89	1.62
40	4.08	3.23	2.84	2.61	2.45	2.34	2.18	2.00	1.79	1.51
60	4.00	3.15	2.76	2.52	2.37	2.25	2.10	1.92	1.70	1.39
120	3.92	3.07	2.68	2.45	2.29	2.17	2.02	1.83	1.61	1.25
∞	3.84	2.99	2.60	2.37	2.21	2.09	1.94	1.75	1.52	1.00

ν_1 = degrees of freedom in numerator; ν_2 = degrees of freedom in denominator. Lower 5% points are found by interchange of ν_1 and ν_2; i.e., ν_1 must always correspond with the greater mean square.

Source: Table IV is taken from Table V of Fisher and Yates: *Statistical Tables for Biological, Agricultural and Medical Research*, published by Longman Group, Ltd., London (previously published by Oliver and Boyd, Edinburgh), and used by permission of the authors and publishers.

Table IV-b *Variance Ratio* (F-*Ratio*) *1 % Points*

ν_2	ν_1 = 1	2	3	4	5	6	8	12	24	∞
1	4052	4999	5403	5625	5764	5859	5981	6106	6234	6366
2	98.49	99.01	99.17	99.25	99.30	99.33	99.36	99.42	99.46	99.50
3	34.12	30.81	29.46	28.71	28.24	27.91	27.49	27.05	26.60	26.12
4	21.20	18.00	16.69	15.98	15.52	15.21	14.80	14.37	13.93	13.46
5	16.26	13.27	12.06	11.39	10.97	10.67	10.27	9.89	9.47	9.02
6	13.74	10.92	9.78	9.15	8.75	8.47	8.10	7.72	7.31	6.88
7	12.25	9.55	8.45	7.85	7.46	7.19	6.84	6.47	6.07	5.65
8	11.26	8.65	7.59	7.01	6.63	6.37	6.03	5.67	5.28	4.86
9	10.56	8.02	6.99	6.42	6.06	5.80	5.47	5.11	4.73	4.31
10	10.04	7.56	6.55	5.99	5.64	5.39	5.06	4.71	4.33	3.91
11	9.65	7.20	6.22	5.67	5.32	5.07	4.74	4.40	4.02	3.60
12	9.33	6.93	5.95	5.41	5.06	4.82	4.50	4.16	3.78	3.36
13	9.07	6.70	5.74	5.20	4.86	4.62	4.30	3.96	3.59	3.16
14	8.86	6.51	5.56	5.03	4.69	4.46	4.14	3.80	3.43	3.00
15	8.68	6.36	5.42	4.89	4.56	4.32	4.00	3.67	3.29	2.87
16	8.53	6.23	5.29	4.77	4.44	4.20	3.89	3.55	3.18	2.75
17	8.40	6.11	5.18	4.67	4.34	4.10	3.79	3.45	3.08	2.65
18	8.28	6.01	5.09	4.58	4.25	4.01	3.71	3.37	3.00	2.57
19	8.18	5.93	5.01	4.50	4.17	3.94	3.63	3.30	2.92	2.49
20	8.10	5.85	4.94	4.43	4.10	3.87	3.56	3.23	2.86	2.42
21	8.02	5.78	4.87	4.37	4.04	3.81	3.51	3.17	2.80	2.36
22	7.94	5.72	4.82	4.31	3.99	3.76	3.45	3.12	2.75	2.31
23	7.88	5.66	4.76	4.26	3.94	3.71	3.41	3.07	2.70	2.26
24	7.82	5.61	4.72	4.22	3.90	3.67	3.36	3.03	2.66	2.21
25	7.77	5.57	4.68	4.18	3.86	3.63	3.32	2.99	2.62	2.17
26	7.72	5.53	4.64	4.14	3.82	3.59	3.29	2.96	2.58	2.13
27	7.68	5.49	4.60	4.11	3.78	3.56	3.26	2.93	2.55	2.10
28	7.64	5.45	4.57	4.07	3.75	3.53	3.23	2.90	2.52	2.06
29	7.60	5.42	4.54	4.04	3.73	3.50	3.20	2.87	2.49	2.03
30	7.56	5.39	4.51	4.02	3.70	3.47	3.17	2.84	2.47	2.01
40	7.31	5.18	4.31	3.83	3.51	3.29	2.99	2.66	2.29	1.80
60	7.08	4.98	4.13	3.65	3.34	3.12	2.82	2.50	2.12	1.60
120	6.85	4.79	3.95	3.48	3.17	2.96	2.66	2.34	1.95	1.38
∞	6.64	4.60	3.78	3.32	3.02	2.80	2.51	2.18	1.79	1.00

ν_1 = degrees of freedom in numerator; ν_2 = degrees of freedom in denominator. Lower 1% points are found by interchange of ν_1 and ν_2; i.e., ν_1 must always correspond with the greater mean square.

Source: Table IV is taken from Table V of Fisher and Yates: *Statistical Tables for Biological, Agricultural and Medical Research*, published by Longman Group, Ltd., London (previously published by Oliver and Boyd, Edinburgh), and used by permission of the authors and publishers.

Table V-a *Critical Values of* U *in the Mann-Whitney Test at* $\alpha = .05$ *with Direction Predicted or at* $\alpha = .10$ *with Direction Not Predicted*

N_2 / N_1	9	10	11	12	13	14	15	16	17	18	19	20
1											0	0
2	1	1	1	2	2	2	3	3	3	4	4	4
3	3	4	5	5	6	7	7	8	9	9	10	11
4	6	7	8	9	10	11	12	14	15	16	17	18
5	9	11	12	13	15	16	18	19	20	22	23	25
6	12	14	16	17	19	21	23	25	26	28	30	32
7	15	17	19	21	24	26	28	30	33	35	37	39
8	18	20	23	26	28	31	33	36	39	41	44	47
9	21	24	27	30	33	36	39	42	45	48	51	54
10	24	27	31	34	37	41	44	48	51	55	58	62
11	27	31	34	38	42	46	50	54	57	61	65	69
12	30	34	38	42	47	51	55	60	64	68	72	77
13	33	37	42	47	51	56	61	65	70	75	80	84
14	36	41	46	51	56	61	66	71	77	82	87	92
15	39	44	50	55	61	66	72	77	83	88	94	100
16	42	48	54	60	65	71	77	83	89	95	101	107
17	45	51	57	64	70	77	83	89	96	102	109	115
18	48	55	61	68	75	82	88	95	102	109	116	123
19	51	58	65	72	80	87	94	101	109	116	123	130
20	54	62	69	77	84	92	100	107	115	123	130	138

Source: D. Auble, "Extended Tables for the Mann-Whitney Statistic," *Bulletin of the Institute of Educational Research* **1**, 1953, Tables 1, 3, 5, and 7 (as adapted in S. Siegel, *Non-parametric Statistics for the Behavioral Sciences*, McGraw-Hill, 1956, table K). Reprinted with permission.

Table V-b *Critical Values of* U *in the Mann-Whitney Test at* $\alpha = .025$ *with Direction Predicted or at* $\alpha = .05$ *with Direction Not Predicted*

N_1 \ N_2	9	10	11	12	13	14	15	16	17	18	19	20
1												
2	0	0	0	1	1	1	1	1	2	2	2	2
3	2	3	3	4	4	5	5	6	6	7	7	8
4	4	5	6	7	8	9	10	11	11	12	13	13
5	7	8	9	11	12	13	14	15	17	18	19	20
6	10	11	13	14	16	17	19	21	22	24	25	27
7	12	14	16	18	20	22	24	26	28	30	32	34
8	15	17	19	22	24	26	29	31	34	36	38	41
9	17	20	23	26	28	31	34	37	39	42	45	48
10	20	23	26	29	33	36	39	42	45	48	52	55
11	23	26	30	33	37	40	44	47	51	55	58	62
12	26	29	33	37	41	45	49	53	57	61	65	69
13	28	33	37	41	45	50	54	59	63	67	72	76
14	31	36	40	45	50	55	59	64	67	74	78	83
15	34	39	44	49	54	59	64	70	75	80	85	90
16	37	42	47	53	59	64	70	75	81	86	92	98
17	39	45	51	57	63	67	75	81	87	93	99	105
18	42	48	55	61	67	74	80	86	93	99	106	112
19	45	52	58	65	72	78	85	92	99	106	113	119
20	48	55	62	69	76	83	90	98	105	112	119	127

Source: D. Auble, "Extended Tables for the Mann-Whitney Statistic," *Bulletin of the Institute of Educational Research* **1**, 1953, Tables 1, 3, 5, and 7 (as adapted in S. Siegel, *Non-parametric Statistics for the Behavioral Sciences*, McGraw-Hill, 1956, table K). Reprinted with permission.

Table V-c *Critical Values of* U *in the Mann-Whitney Test at* $\alpha = .01$ *with Direction Predicted or at* $\alpha = .02$ *with Direction Not Predicted*

N_2 / N_1	9	10	11	12	13	14	15	16	17	18	19	20
1												
2					0	0	0	0	0	0	1	1
3	1	1	1	2	2	2	3	3	4	4	4	5
4	3	3	4	5	5	6	7	7	8	9	9	10
5	5	6	7	8	9	10	11	12	13	14	15	16
6	7	8	9	11	12	13	15	16	18	19	20	22
7	9	11	12	14	16	17	19	21	23	24	26	28
8	11	13	15	17	20	22	24	26	28	30	32	34
9	14	16	18	21	23	26	28	31	33	36	38	40
10	16	19	22	24	27	30	33	36	38	41	44	47
11	18	22	25	28	31	34	37	41	44	47	50	53
12	21	24	28	31	35	38	42	46	49	53	56	60
13	23	27	31	35	39	43	47	51	55	59	63	67
14	26	30	34	38	43	47	51	56	60	65	69	73
15	28	33	37	42	47	51	56	61	66	70	75	80
16	31	36	41	46	51	56	61	66	71	76	82	87
17	33	38	44	49	55	60	66	71	77	82	88	93
18	36	41	47	53	59	65	70	76	82	88	94	100
19	38	44	50	56	63	69	75	82	88	94	101	107
20	40	47	53	60	67	73	80	87	93	100	107	114

Source: D. Auble, "Extended Tables for the Mann-Whitney Statistic," *Bulletin of the Institute of Educational Research* **1**, 1953, Tables 1, 3, 5, and 7 (as adapted in S. Siegel, *Non-parametric Statistics for the Behavioral Sciences*, McGraw-Hill, 1956, table K). Reprinted with permission.

Table VI *Values of* χ^2

df	*P* = .99	.98	.95	.90	.80	.70	.50
1	.000157	.000628	.00393	.0158	.0642	.148	.455
2	.0201	.0404	.103	.211	.446	.713	1.386
3	.115	.185	.352	.584	1.005	1.424	2.366
4	.297	.429	.711	1.064	1.649	2.195	3.357
5	.554	.752	1.145	1.610	2.343	3.000	4.351
6	.872	1.134	1.635	2.204	3.070	3.828	5.348
7	1.239	1.564	2.167	2.833	3.822	4.671	6.346
8	1.646	2.032	2.733	3.490	4.594	5.527	7.344
9	2.088	2.532	3.325	4.168	5.380	6.393	8.343
10	2.558	3.059	3.940	4.865	6.179	7.267	9.342
11	3.053	3.609	4.575	5.578	6.989	8.148	10.341
12	3.571	4.178	5.226	6.304	7.807	9.034	11.340
13	4.107	4.765	5.892	7.042	8.634	9.926	12.340
14	4.660	5.368	6.571	7.790	9.467	10.821	13.339
15	5.229	5.985	7.261	8.547	10.307	11.721	14.339
16	5.812	6.614	7.962	9.312	11.152	12.624	15.338
17	6.408	7.255	8.672	10.085	12.002	13.531	16.338
18	7.015	7.906	9.390	10.865	12.857	14.440	17.338
19	7.633	8.567	10.117	11.651	13.716	15.352	18.338
20	8.260	9.237	10.851	12.443	14.578	16.266	19.337
21	8.897	9.915	11.591	13.240	15.445	17.182	20.337
22	9.542	10.600	12.338	14.041	16.314	18.101	21.337
23	10.196	11.293	13.091	14.848	17.187	19.021	22.337
24	10.856	11.992	13.848	15.659	18.062	19.943	23.337
25	11.524	12.697	14.611	16.473	18.940	20.867	24.337
26	12.198	13.409	15.379	17.292	19.820	21.792	25.336
27	12.879	14.125	16.151	18.114	20.703	22.719	26.336
28	13.565	14.847	16.928	18.939	21.588	23.647	27.336
29	14.256	15.574	17.708	19.768	22.475	24.577	28.336
30	14.953	16.306	18.493	20.599	23.364	25.508	29.336

Table VI *Values of* χ^2 *(Continued)*

df	$P = .30$	.20	.10	.05	.02	.01	.001
1	1.074	1.642	2.706	3.841	5.412	6.635	10.827
2	2.408	3.219	4.605	5.991	7.824	9.210	13.815
3	3.665	4.642	6.251	7.815	9.837	11.345	16.268
4	4.878	5.989	7.779	9.488	11.668	13.277	18.465
5	6.064	7.289	9.236	11.070	13.388	15.086	20.517
6	7.231	8.558	10.645	12.592	15.033	16.812	22.457
7	8.383	9.803	12.017	14.067	16.622	18.475	24.322
8	9.524	11.030	13.362	15.507	18.168	20.090	26.125
9	10.656	12.242	14.684	16.919	19.679	21.666	27.877
10	11.781	13.442	15.987	18.307	21.161	23.209	29.588
11	12.899	14.631	17.275	19.675	22.618	24.725	31.264
12	14.011	15.812	18.549	21.026	24.054	26.217	32.909
13	15.119	16.985	19.812	22.362	25.472	27.688	34.528
14	16.222	18.151	21.064	23.685	26.873	29.141	36.123
15	17.322	19.311	22.307	24.996	28.259	30.578	37.697
16	18.418	20.465	23.542	26.296	29.633	32.000	39.252
17	19.511	21.615	24.769	27.587	30.995	33.409	40.790
18	20.601	22.760	25.989	28.869	32.346	34.805	42.312
19	21.689	23.900	27.204	30.144	33.687	36.191	43.820
20	22.775	25.038	28.412	31.410	35.020	37.566	45.315
21	23.858	26.171	29.615	32.671	36.343	38.932	46.797
22	24.939	27.301	30.813	33.924	37.659	40.289	48.268
23	26.018	28.429	32.007	35.172	38.968	41.638	49.728
24	27.096	29.553	33.196	36.415	40.270	42.980	51.179
25	28.172	30.675	34.382	37.652	41.566	44.314	52.620
26	29.246	31.795	35.563	38.885	42.856	45.642	54.052
27	30.319	32.912	36.741	40.113	44.140	46.963	55.476
28	31.391	34.027	37.916	41.337	45.419	48.278	56.893
29	32.461	35.139	39.087	42.557	46.693	49.588	58.302
30	33.530	36.250	40.256	43.773	47.962	50.892	59.703

For larger values of *df*, the expression $\sqrt{2\chi^2} - \sqrt{2df - 1}$ may be used as a normal deviate with unit variance, remembering that the probability of χ^2 corresponds with that of a single tail of the normal curve.

Source: Table VI is taken from Table IV of Fisher and Yates: *Statistical Tables for Biological, Agricultural and Medical Research*, published by Longman Group, Ltd., London (previously published by Oliver and Boyd, Edinburgh), and used by permission of the authors and publishers.

Table VII *Five Thousand Random Digits*

	50–54	55–59	60–64	65–69	70–74	75–79	80–84	85–89	90–94	95–99
00	59391	58030	52098	82718	87024	82848	04190	96574	90464	29065
01	99567	76364	77204	04615	27062	96621	43918	01896	83991	51141
02	10363	97518	51400	25670	98342	61891	27101	37855	06235	33316
03	86859	19558	64432	16706	99612	59798	32803	67708	15297	28612
04	11258	24591	36863	55368	31721	94335	34936	02566	80972	08188
05	95068	88628	35911	14530	33020	80428	39936	31855	34334	64865
06	54463	47237	73800	91017	36239	71824	83671	39892	60518	37092
07	16874	62677	57412	13215	31389	62233	80827	73917	82802	84420
08	92494	63157	76593	91316	03505	72389	96363	52887	01087	66091
09	15669	56689	35682	40844	53256	81872	35213	09840	34471	74441
10	99116	75486	84989	23476	52967	67104	39495	39100	17217	74073
11	15696	10703	65178	90637	63110	17622	53988	71087	84148	11670
12	97720	15369	51269	69620	03388	13699	33423	67453	43269	56720
13	11666	13841	71681	98000	35979	39719	81899	07449	47985	46967
14	71628	73130	78783	75691	41632	09847	61547	18707	85489	69944
15	40501	51089	99943	91843	41995	88931	73631	69361	05375	15417
16	22518	55576	98215	82068	10798	86211	36584	67466	69373	40054
17	75112	30485	62173	02132	14878	92879	22281	16783	86352	00077
18	80327	02671	98191	84342	90813	49268	95441	15496	20168	09271
19	60251	45548	02146	05597	48228	81366	34598	72856	66762	17002
20	57430	82270	10421	05540	43648	75888	66049	21511	47676	33444
21	73528	39559	34434	88596	54076	71693	43132	14414	79949	85193
22	25991	65959	70769	64721	86413	33475	42740	06175	82758	66248
23	78388	16638	09134	59880	63806	48472	39318	35434	24057	74739
24	12477	09965	96657	57994	59439	76330	24596	77515	09577	91871
25	83266	32883	42451	15579	38155	29793	40914	65990	16255	17777
26	76970	80876	10237	39515	79152	74798	39357	09054	73579	92359
27	37074	65198	44785	68624	98336	84481	97610	78735	46703	98265
28	83712	06514	30101	78295	54656	85417	43189	60048	72781	72606
29	20287	56862	69727	94443	64936	08366	27227	05158	50326	59566
30	74261	32592	86538	27041	65172	85532	07571	80609	39285	65340
31	64081	49863	08478	96001	18888	14810	70545	89755	59064	07210
32	05617	75818	47750	67814	29575	10526	66192	44464	27058	40467
33	26793	74951	95466	74307	13330	42664	85515	20632	05497	33625
34	65988	72850	48737	54719	52056	01596	03845	35067	03134	70322

Table VII *Five Thousand Random Digits (Continued)*

	50–54	55–59	60–64	65–69	70–74	75–79	80–84	85–89	90–94	95–99
35	27366	42271	44300	73399	21105	03280	73457	43093	05192	48657
36	56760	10909	98147	34736	33863	95256	12731	66598	50771	83665
37	72880	43338	93643	58904	59543	23943	11231	83268	65938	81581
38	77888	38100	03062	58103	47961	83841	25878	23746	55903	44115
39	28440	07819	21580	51459	47971	29882	13990	29226	23608	15873
40	63525	94441	77033	12147	51054	49955	58312	76923	96071	05813
41	47606	93410	16359	89033	89696	47231	64498	31776	05383	39902
42	52669	45030	96279	14709	52372	87832	02735	50803	72744	88208
43	16738	60159	07425	62369	07515	82721	37875	71153	21315	00132
44	59348	11695	45751	15865	74739	05572	32688	20271	65128	14551
45	12900	71775	29845	60774	94924	21810	38636	33717	67598	82521
46	75086	23537	49939	33595	13484	97588	28617	17979	70749	35234
47	99495	51434	29181	09993	38190	42553	68922	52125	91077	40197
48	26075	31671	45386	36583	93459	48599	52022	41330	60651	91321
49	13636	93596	23377	51133	95126	61496	42474	45141	46660	42338
50	32847	31282	03345	89593	69214	70381	78285	20054	91018	16742
51	16916	00041	30236	55023	14253	76582	12092	86533	92426	37655
52	66176	34047	21005	27137	03191	48970	64625	22394	39622	79085
53	46299	13335	12180	16861	38043	59292	62675	63631	37020	78195
54	22847	47839	45385	23289	47526	54098	45683	55849	51575	64689
55	41851	54160	92320	69936	34803	92479	33399	71160	64777	83378
56	28444	59497	91586	95917	68553	28639	96455	34174	11130	91994
57	47520	62378	98855	83174	13088	16561	68559	26679	06238	51254
58	34978	63271	13142	82681	05271	08822	06490	44984	49307	62717
59	37404	80416	69035	92980	49486	74378	75610	74976	70056	15478
60	32400	65482	52099	53676	74648	94148	65095	69597	52771	71551
61	89262	86332	51718	70663	11623	29834	79820	73002	84886	03591
62	86866	09127	98021	03871	27789	58444	44832	36505	40672	30180
63	90814	14833	08759	74645	05046	94056	99094	65091	32663	73040
64	19192	82756	20553	58446	55376	88914	75096	26119	83898	43816
65	77585	52593	56612	95766	10019	29531	73064	20953	53523	58136
66	23757	16364	05096	03192	62386	45389	85332	18877	55710	96459
67	45989	96257	23850	26216	23309	21526	07425	50254	19455	29315
68	92970	94243	07316	41467	64837	52406	25225	51553	31220	14032
69	74346	59596	40088	98176	17896	86900	20249	77753	19099	48885
70	87646	41309	27636	45153	29988	94770	07255	70908	05340	99751
71	50099	71038	45146	06146	55211	99429	43169	66259	97786	59180
72	10127	46900	64984	75348	04115	33624	68774	60013	35515	62556
73	67995	81977	18984	64091	02785	27762	42529	97144	80407	64524
74	26304	80217	84934	82657	69291	35397	98714	35104	08187	48109

Table VII *Five Thousand Random Digits (Continued)*

	50–54	55–59	60–64	65–69	70–74	75–79	80–84	85–89	90–94	95–99
75	81994	41070	56642	64091	31229	02595	13513	45148	78722	30144
76	59537	34662	79631	89403	65212	09975	06118	86197	58208	16162
77	51228	10937	62396	81460	47331	91403	95007	06047	16846	64809
78	31089	37995	29577	07828	42272	54016	21950	86192	99046	84864
79	38207	97938	93459	75174	79460	55436	57206	87644	21296	43395
80	88666	31142	09474	89712	63153	62333	42212	06140	42594	43671
81	53365	56134	67582	92557	89520	33452	05134	70628	27612	33738
82	89807	74530	38004	90102	11693	90257	05500	79920	62700	43325
83	18682	81038	85662	90915	91631	22223	91588	80774	07716	12548
84	63571	32579	63942	25371	09234	94592	98475	76884	37635	33608
85	68927	56492	67799	95398	77642	54913	91853	08424	81450	76229
86	56401	63186	39389	88798	31356	89235	97036	32341	33292	73757
87	24333	95603	02359	72942	46287	95382	08452	62862	97869	71775
88	17025	84202	95199	62272	06366	16175	97577	99304	41587	03686
89	02804	08253	52133	20224	68034	50865	57868	22343	55111	03607
90	08298	03879	20995	19850	73090	13191	18963	82244	78479	99121
91	59883	01785	82403	96062	03785	03488	12970	64896	38336	30030
92	46982	06682	62864	91837	74021	89094	39952	64158	79614	78235
93	31121	47266	07661	02051	67599	24471	69843	83696	71402	76287
94	97867	56641	63416	17577	30161	87320	37752	73276	48969	41915
95	57364	86746	08415	14621	49430	22311	15836	72492	49372	44103
96	09559	26263	69511	28064	75999	44540	13337	10918	79846	54809
97	53873	55571	00608	42661	91332	63956	74087	59008	47493	99581
98	35531	19162	86406	05299	77511	24311	57257	22826	77555	05941
99	28229	88629	25695	94932	30721	16197	78742	34974	97528	45447

Source: Reprinted by permission from *Statistical Methods* by George W. Snedecor and William G. Cochran, sixth edition © 1967 by Iowa State University Press, Ames, Iowa.

Table VIII *Squares and Square Roots of the Numbers from 1 to 1,000*

Number	*Square*	*Square Root*	*Number*	*Square*	*Square Root*
1	1	1.000	41	16 81	6.403
2	4	1.414	42	17 64	6.481
3	9	1.732	43	18 49	6.557
4	16	2.000	44	19 36	6.633
5	25	2.236	45	20 25	6.708
6	36	2.449	46	21 16	6.782
7	49	2.646	47	22 09	6.856
8	64	2.828	48	23 04	6.928
9	81	3.000	49	24 01	7.000
10	1 00	3.162	50	25 00	7.071
11	1 21	3.317	51	26 01	7.141
12	1 44	3.464	52	27 04	7.211
13	1 69	3.606	53	28 09	7.280
14	1 96	3.742	54	29 16	7.348
15	2 25	3.873	55	30 25	7.416
16	2 56	4.000	56	31 36	7.483
17	2 89	4.123	57	32 49	7.550
18	3 24	4.243	58	33 64	7.616
19	3 61	4.359	59	34 81	7.681
20	4 00	4.472	60	36 00	7.746
21	4 41	4.583	61	37 21	7.810
22	4 84	4.690	62	38 44	7.874
23	5 29	4.796	63	39 69	7.937
24	5 76	4.899	64	40 96	8.000
25	6 25	5.000	65	42 25	8.062
26	6 76	5.099	66	43 56	8.124
27	7 29	5.196	67	44 89	8.185
28	7 84	5.292	68	46 24	8.246
29	8 41	5.385	69	47 61	8.307
30	9 00	5.477	70	49 00	8.367
31	9 61	5.568	71	50 41	8.426
32	10 24	5.657	72	51 84	8.485
33	10 89	5.745	73	53 29	8.544
34	11 56	5.831	74	54 76	8.602
35	12 25	5.916	75	56 25	8.660
36	12 96	6.000	76	57 76	8.718
37	13 69	6.083	77	59 29	8.775
38	14 44	6.164	78	60 84	8.832
39	15 21	6.245	79	62 41	8.888
40	16 00	6.325	80	64 00	8.944

Table VIII *Squares and Square Roots of the Numbers from 1 to 1,000 (Continued)*

Number	*Square*	*Square Root*	*Number*	*Square*	*Square Root*
81	65 61	9.000	121	1 46 41	11.000
82	67 24	9.055	122	1 48 84	11.045
83	68 89	9.110	123	1 51 29	11.091
84	70 56	9.165	124	1 53 76	11.136
85	72 25	9.220	125	1 56 25	11.180
86	73 96	9.274	126	1 58 76	11.225
87	75 69	9.327	127	1 61 29	11.269
88	77 44	9.381	128	1 63 84	11.314
89	79 21	9.434	129	1 66 41	11.358
90	81 00	9.487	130	1 69 00	11.402
91	82 81	9.539	131	1 71 61	11.446
92	84 64	9.592	132	1 74 24	11.489
93	86 49	9.644	133	1 76 89	11.533
94	88 36	9.695	134	1 79 56	11.576
95	90 25	9.747	135	1 82 25	11.619
96	92 16	9.798	136	1 84 96	11.662
97	94 09	9.849	137	1 87 69	11.705
98	96 04	9.899	138	1 90 44	11.747
99	98 01	9.950	139	1 93 21	11.790
100	1 00 00	10.000	140	1 96 00	11.832
101	1 02 01	10.050	141	1 98 81	11.874
102	1 04 04	10.100	142	2 01 64	11.916
103	1 06 09	10.149	143	2 04 49	11.958
104	1 08 16	10.198	144	2 07 36	12.000
105	1 10 25	10.247	145	2 10 25	12.042
106	1 12 36	10.296	146	2 13 16	12.083
107	1 14 49	10.344	147	2 16 09	12.124
108	1 16 64	10.392	148	2 19 04	12.166
109	1 18 81	10.440	149	2 22 01	12.207
110	1 21 00	10.488	150	2 25 00	12.247
111	1 23 21	10.536	151	2 28 01	12.288
112	1 25 44	10.583	152	2 31 04	12.329
113	1 27 69	10.630	153	2 34 09	12.369
114	1 29 96	10.677	154	2 37 16	12.410
115	1 32 25	10.724	155	2 40 25	12.450
116	1 34 56	10.770	156	2 43 36	12.490
117	1 36 89	10.817	157	2 46 49	12.530
118	1 39 24	10.863	158	2 49 64	12.570
119	1 41 61	10.909	159	2 52 81	12.610
120	1 44 00	10.954	160	2 56 00	12.649

Table VIII *Squares and Square Roots of the Numbers from 1 to 1,000 (Continued)*

Number	*Square*	*Square Root*	*Number*	*Square*	*Square Root*
161	2 59 21	12.689	201	4 04 01	14.177
162	2 62 44	12.728	202	4 08 04	14.213
163	2 65 69	12.767	203	4 12 09	14.248
164	2 68 96	12.806	204	4 16 16	14.283
165	2 72 25	12.845	205	4 20 25	14.318
166	2 75 56	12.884	206	4 24 36	14.353
167	2 78 89	12.923	207	4 28 49	14.387
168	2 82 24	12.961	208	4 32 64	14.422
169	2 85 61	13.000	209	4 36 81	14.457
170	2 89 00	13.038	210	4 41 00	14.491
171	2 92 41	13.077	211	4 45 21	14.526
172	2 95 84	13.115	212	4 49 44	14.560
173	2 99 29	13.153	213	4 53 69	14.595
174	3 02 76	13.191	214	4 57 96	14.629
175	3 06 25	13.229	215	4 62 25	14.663
176	3 09 76	13.266	216	4 66 56	14.697
177	3 13 29	13.304	217	4 70 89	14.731
178	3 16 84	13.342	218	4 75 24	14.765
179	3 20 41	13.379	219	4 79 61	14.799
180	3 24 00	13.416	220	4 84 00	14.832
181	3 27 61	13.454	221	4 88 41	14.866
182	3 31 24	13.491	222	4 92 84	14.900
183	3 34 89	13.528	223	4 97 29	14.933
184	3 38 56	13.565	224	5 01 76	14.967
185	3 42 25	13.601	225	5 06 25	15.000
186	3 45 96	13.638	226	5 10 76	15.033
187	3 49 69	13.675	227	5 15 29	15.067
188	3 53 44	13.711	228	5 19 84	15.100
189	3 57 21	13.748	229	5 24 41	15.133
190	3 61 00	13.784	230	5 29 00	15.166
191	3 64 81	13.820	231	5 33 61	15.199
192	3 68 64	13.856	232	5 38 24	15.232
193	3 72 49	13.892	233	5 42 89	15.264
194	3 76 36	13.928	234	5 47 56	15.297
195	3 80 25	13.964	235	5 52 25	15.330
196	3 84 16	14.000	236	5 56 96	15.362
197	3 88 09	14.036	237	5 61 69	15.395
198	3 92 04	14.071	238	5 66 44	15.427
199	3 96 01	14.107	239	5 71 21	15.460
200	4 00 00	14.142	240	5 76 00	15.492

Table VIII *Squares and Square Roots of the Numbers from 1 to 1,000 (Continued)*

Number	*Square*	*Square Root*	*Number*	*Square*	*Square Root*
241	5 80 81	15.524	281	7 89 61	16.763
242	5 85 64	15.556	282	7 95 24	16.793
243	5 90 49	15.588	283	8 00 89	16.823
244	5 95 36	15.620	284	8 06 56	16.852
245	6 00 25	15.652	285	8 12 25	16.882
246	6 05 16	15.684	286	8 17 96	16.912
247	6 10 09	15.716	287	8 23 69	16.941
248	6 15 04	15.748	288	8 29 44	16.971
249	6 20 01	15.780	289	8 35 21	17.000
250	6 25 00	15.811	290	8 41 00	17.029
251	6 30 01	15.843	291	8 46 81	17.059
252	6 35 04	15.875	292	8 52 64	17.088
253	6 40 09	15.906	293	8 58 49	17.117
254	6 45 16	15.937	294	8 64 36	17.146
255	6 50 25	15.969	295	8 70 25	17.176
256	6 55 36	16.000	296	8 76 16	17.205
257	6 60 49	16.031	297	8 82 09	17.234
258	6 65 64	16.062	298	8 88 04	17.263
259	6 70 81	16.093	299	8 94 01	17.292
260	6 76 00	16.125	300	9 00 00	17.321
261	6 81 21	16.155	301	9 06 01	17.349
262	6 86 44	16.186	302	9 12 04	17.378
263	6 91 69	16.217	303	9 18 09	17.407
264	6 96 96	16.248	304	9 24 16	17.436
265	7 02 25	16.279	305	9 30 25	17.464
266	7 07 56	16.310	306	9 36 36	17.493
267	7 12 89	16.340	307	9 42 49	17.521
268	7 18 24	16.371	308	9 48 64	17.550
269	3 23 61	16.401	309	9 54 81	17.578
270	7 29 00	16.432	310	9 61 00	17.607
271	7 34 41	16.462	311	9 67 21	17.635
272	7 39 84	16.492	312	9 73 44	17.664
273	7 45 29	16.523	313	9 79 69	17.692
274	7 50 76	16.553	314	9 85 96	17.720
275	7 56 25	16.583	315	9 92 25	17.748
276	7 61 76	16.613	316	9 98 56	17.776
277	7 67 29	16.643	317	10 04 89	17.804
278	7 72 84	16.673	318	10 11 24	17.833
279	7 78 41	16.703	319	10 17 61	17.861
280	7 84 00	16.733	320	10 24 00	17.889

Table VIII *Squares and Square Roots of the Numbers from 1 to 1,000 (Continued)*

Number	Square	Square Root	Number	Square	Square Root
321	10 30 41	17.916	361	13 03 21	19.000
322	10 36 84	17.944	362	13 10 44	19.026
323	10 43 29	17.972	363	13 17 69	19.053
324	10 49 76	18.000	364	13 24 96	19.079
325	10 56 25	18.028	365	13 32 25	19.105
326	10 62 76	18.055	366	13 39 56	19.131
327	10 69 29	18.083	367	13 46 89	19.157
328	10 75 84	18.111	368	13 54 24	19.183
329	10 82 41	18.138	369	13 61 61	19.209
330	10 89 00	18.166	370	13 69 00	19.235
331	10 95 61	18.193	371	13 76 41	19.261
332	11 02 24	18.221	372	13 83 84	19.287
333	11 08 89	18.248	373	13 91 29	19.313
334	11 15 56	18.276	374	13 98 76	19.339
335	11 22 25	18.303	375	14 06 25	19.363
336	11 28 96	18.330	376	14 13 76	19.391
337	11 35 69	18.358	377	14 21 29	19.416
338	11 42 44	18.385	378	14 28 84	19.442
339	11 49 21	18.412	379	14 36 41	19.468
340	11 56 00	18.439	380	14 44 00	19.494
341	11 62 81	18.466	381	14 51 61	19.519
342	11 69 64	18.493	382	14 59 24	19.545
343	11 76 49	18.520	383	14 66 89	19.570
344	11 83 36	18.547	384	14 74 56	19.596
345	11 90 25	18.574	385	14 82 25	19.621
346	11 97 16	18.601	386	14 89 96	19.647
347	12 04 09	18.628	387	14 97 69	19.672
348	12 11 04	18.655	388	15 05 44	19.698
349	12 18 01	18.682	389	15 13 21	19.723
350	12 25 00	18.708	390	15 21 00	19.748
351	12 32 01	18.735	391	15 28 81	19.774
352	12 39 04	18.762	392	15 36 64	19.799
353	12 46 09	18.788	393	15 44 49	19.824
354	12 53 16	18.815	394	15 52 36	19.849
355	12 60 25	18.841	395	15 60 25	19.875
356	12 67 36	18.868	396	15 68 16	19.900
357	12 74 49	18.894	397	15 76 09	19.925
358	12 81 64	18.921	398	15 84 04	19.950
359	12 88 81	18.947	399	15 92 01	19.975
360	12 96 00	18.974	400	16 00 00	20.000

Table VIII *Squares and Square Roots of the Numbers from 1 to 1,000 (Continued)*

Number	*Square*	*Square Root*	*Number*	*Square*	*Square Root*
401	16 08 01	20.025	441	19 44 81	21.000
402	16 16 04	20.050	442	19 53 64	21.024
403	16 24 09	20.075	443	19 62 49	21.048
404	16 32 16	20.100	444	19 71 36	21.071
405	16 40 25	20.125	445	19 80 25	21.095
406	16 48 36	20.149	446	19 89 16	21.119
407	16 56 49	20.174	447	19 98 09	21.142
408	16 64 64	20.199	448	20 07 04	21.166
409	16 72 81	20.224	449	20 16 01	21.190
410	16 81 00	20.248	450	20 25 00	21.213
411	16 89 21	20.273	451	20 34 01	21.237
412	16 97 44	20.298	452	20 43 04	21.260
413	17 05 69	20.322	453	20 52 09	21.284
414	17 13 96	20.347	454	20 61 16	21.307
415	17 22 25	20.372	455	20 70 25	21.331
416	17 30 56	20.396	456	20 79 36	21.354
417	17 38 89	20.421	457	20 88 49	21.378
418	17 47 24	20.445	458	20 97 64	21.401
419	17 55 61	20.469	459	21 06 81	21.424
420	17 64 00	20.494	460	21 16 00	21.448
421	17 72 41	20.518	461	21 25 21	21.471
422	17 80 84	20.543	462	21 34 44	21.494
423	17 89 29	20.567	463	21 43 69	21.517
424	17 97 76	20.591	464	21 52 96	21.541
425	18 06 25	20.616	465	21 62 25	21.564
426	18 14 76	20.640	466	21 71 56	21.587
427	18 23 29	20.664	467	21 80 89	21.610
428	18 31 84	20.688	468	21 90 24	21.633
429	18 40 41	20.712	469	21 99 61	21.656
430	18 49 00	20.736	470	22 09 00	21.679
431	18 57 61	20.761	471	22 18 41	21.703
432	18 66 24	20.785	472	22 27 84	21.726
433	18 74 89	20.809	473	22 37 29	21.749
434	18 83 56	20.833	474	22 46 76	21.772
435	18 92 25	20.857	475	22 56 25	21.794
436	19 00 96	20.881	476	22 65 76	21.817
437	19 09 69	20.905	477	22 75 29	21.840
438	19 18 44	20.928	478	22 84 84	21.863
439	19 27 21	20.952	479	22 94 41	21.886
440	19 36 00	20.976	480	23 04 00	21.909

Table VIII *Squares and Square Roots of the Numbers from 1 to 1,000 (Continued)*

Number	*Square*	*Square Root*	*Number*	*Square*	*Square Root*
481	23 13 61	21.932	521	27 14 41	22.825
482	23 23 24	21.954	522	27 24 84	22.847
483	23 32 89	21.977	523	27 35 29	22.869
484	23 42 56	22.000	524	27 45 76	22.891
485	23 52 25	22.023	525	27 56 25	22.913
486	23 61 96	22.045	526	27 66 76	22.935
487	23 71 69	22.068	527	27 77 29	22.956
488	23 81 44	22.091	528	27 87 84	22.978
489	23 91 21	22.113	529	27 98 41	23.000
490	24 01 00	22.136	530	28 09 00	23.022
491	24 10 81	22.159	531	28 19 61	23.043
492	24 20 64	22.181	532	28 30 24	23.065
493	24 30 49	22.204	533	28 40 89	23.087
494	24 40 36	22.226	534	28 51 56	23.108
495	24 50 25	22.249	535	28 62 25	23.130
496	24 60 16	22.271	536	28 72 96	23.152
497	24 70 09	22.293	537	28 83 69	23.173
498	24 80 04	22.316	538	28 94 44	23.195
499	24 90 01	22.338	539	29 05 21	23.216
500	25 00 00	22.361	540	29 16 00	23.238
501	25 10 01	22.383	541	29 26 81	23.259
502	25 20 04	22.405	542	29 37 64	23.281
503	25 30 09	22.428	543	29 48 49	23.302
504	25 40 16	22.450	544	29 59 36	23.324
505	25 50 25	22.472	545	29 70 25	23.345
506	25 60 36	22.494	546	29 81 16	23.367
507	25 70 49	22.517	547	29 92 09	23.388
508	25 80 64	22.539	548	30 03 04	23.409
509	25 90 81	22.561	549	30 14 01	23.431
510	26 01 00	22.583	550	30 25 00	23.452
511	26 11 21	22.605	551	30 36 01	23.473
512	26 21 44	22.627	552	30 47 04	23.495
513	26 31 69	22.650	553	30 58 09	23.516
514	26 41 96	22.672	554	30 69 16	23.537
515	26 52 25	22.694	555	30 80 25	23.558
516	26 62 56	22.716	556	30 91 36	23.580
577	26 72 89	22.738	557	31 02 49	23.601
518	26 83 24	22.760	558	31 13 64	23.622
519	26 93 61	22.782	559	31 24 81	23.643
520	27 04 00	22.804	560	31 36 00	23.664

Table VIII *Squares and Square Roots of the Numbers from 1 to 1,000 (Continued)*

Number	*Square*	*Square Root*	*Number*	*Square*	*Square Root*
561	31 47 21	23.685	601	36 12 01	24.515
562	31 58 44	23.707	602	36 24 04	24.536
563	31 69 69	23.728	603	36 36 09	24.556
564	31 80 96	23.749	604	36 48 16	24.576
565	31 92 25	23.770	605	36 60 25	24.597
566	32 03 56	23.791	606	36 72 36	24.617
567	32 14 89	23.812	607	36 84 49	24.637
568	32 26 24	23.833	608	36 96 64	24.658
569	32 37 61	23.854	609	37 08 81	24.678
570	32 49 00	23.875	610	37 21 00	24.698
571	32 60 41	23.896	611	37 33 21	24.718
572	32 71 84	23.917	612	37 45 44	24.739
573	32 83 29	23.937	613	37 57 69	24.759
574	32 94 76	23.958	614	37 69 96	24.779
575	33 06 25	23.979	615	37 82 25	24.799
576	33 17 76	24.000	616	37 94 56	24.819
577	33 29 29	24.021	617	38 06 89	24.839
578	33 40 84	24.042	618	38 19 24	24.860
579	33 52 41	24.062	619	38 31 61	24.880
580	33 64 00	24.083	620	38 44 00	24.900
581	33 75 61	24.104	621	38 56 41	24.920
582	33 87 24	24.125	622	38 68 84	24.940
583	33 98 89	24.145	623	38 81 29	24.960
584	34 10 56	24.166	624	38 93 76	24.980
585	34 22 25	24.187	625	39 06 25	25.000
586	34 33 96	24.207	626	39 18 76	25.020
587	34 45 69	24.228	627	39 31 29	25.040
588	34 57 44	24.249	628	39 43 84	25.060
589	34 69 21	24.269	629	39 56 41	25.080
590	34 81 00	24.290	630	39 69 00	25.100
591	34 92 81	24.310	631	39 81 61	25.120
592	35 04 64	24.331	632	39 94 24	25.140
593	35 16 49	24.352	633	40 06 89	25.159
594	35 28 36	24.372	634	40 19 56	25.179
595	35 40 25	24.393	635	40 32 25	25.199
596	35 52 16	24.413	636	40 44 96	25.219
597	35 64 09	24.434	637	40 57 69	25.239
598	35 76 04	24.454	638	40 70 44	25.259
599	35 88 01	24.474	639	40 83 21	25.278
600	36 00 00	24.495	640	40 96 00	25.298

Table VIII *Squares and Square Roots of the Numbers from 1 to 1,000 (Continued)*

Number	*Square*	*Square Root*	*Number*	*Square*	*Square Root*
641	41 08 81	25.318	681	46 37 61	26.096
642	41 21 64	25.338	682	46 51 24	26.115
643	41 34 49	25.357	683	46 64 89	26.134
644	41 47 36	25.377	684	46 78 56	26.153
645	41 60 25	25.397	685	46 92 25	26.173
646	41 73 16	25.417	686	47 05 96	26.192
647	41 86 09	25.436	687	47 19 69	26.211
648	41 99 04	25.456	688	47 33 44	26.230
649	42 12 01	25.475	689	47 47 21	26.249
650	42 25 00	25.495	690	47 61 00	26.268
651	42 38 01	25.515	691	47 74 81	26.287
652	42 51 04	25.534	692	47 88 64	26.306
653	42 64 09	25.554	693	48 02 49	26.325
654	42 77 16	25.573	694	48 16 36	26.344
655	42 90 25	25.593	695	48 30 25	26.363
656	43 03 36	25.612	696	48 44 16	26.382
657	43 16 49	25.632	697	48 58 09	26.401
658	43 29 64	25.652	698	48 72 04	26.420
659	43 42 81	25.671	699	48 86 01	26.439
660	43 56 00	25.690	700	49 00 00	26.458
661	43 69 21	25.710	701	49 14 01	26.476
662	43 82 44	25.729	702	49 28 04	26.495
663	43 95 69	25.749	703	49 42 09	26.514
664	44 08 96	25.768	704	49 56 16	26.533
665	44 22 25	25.788	705	49 70 25	26.552
666	44 35 56	25.807	706	49 84 36	26.571
667	44 48 89	25.826	707	49 98 49	26.589
668	44 62 24	25.846	708	50 12 64	26.608
669	44 75 61	25.865	709	50 26 81	26.627
670	44 89 00	25.884	710	50 41 00	26.646
671	45 02 41	25.904	711	50 55 21	26.665
672	45 15 84	25.923	712	50 69 44	26.683
673	45 29 29	25.942	713	50 83 69	26.702
674	45 42 76	25.962	714	50 97 96	26.721
675	45 56 25	25.981	715	51 12 25	26.739
676	45 69 76	26.000	716	51 26 56	26.758
677	45 83 29	26.019	717	51 40 89	26.777
678	45 96 84	26.038	718	51 55 24	26.796
679	46 10 41	26.058	719	51 69 61	26.814
680	46 24 00	26.077	720	51 84 00	26.833

Table VIII *Squares and Square Roots of the Numbers from 1 to 1,000 (Continued)*

Number	*Square*	*Square Root*	*Number*	*Square*	*Square Root*
721	51 98 41	26.851	761	57 91 21	27.586
722	52 12 84	26.870	762	58 06 44	27.604
723	52 27 29	26.889	763	58 21 69	27.622
724	52 41 76	26.907	764	58 36 96	27.641
725	52 56 25	26.926	765	58 52 25	27.659
726	52 70 76	26.944	766	58 67 56	27.677
727	52 85 29	26.963	767	58 82 89	27.695
728	52 99 84	26.981	768	58 98 24	27.713
729	53 14 41	27.000	769	59 13 61	27.731
730	53 29 00	27.019	770	59 29 00	27.749
731	53 43 61	27.037	771	59 44 41	27.767
732	53 58 24	27.055	772	59 59 84	27.785
733	53 72 89	27.074	773	59 75 29	27.803
734	53 87 56	27.092	774	59 90 76	27.821
735	54 02 25	27.111	775	60 06 25	27.839
736	54 16 96	27.129	776	60 21 76	27.857
737	54 31 69	27.148	777	60 37 29	27.875
738	54 46 44	27.166	778	60 52 84	27.893
739	54 61 21	27.185	779	60 68 41	27.911
740	54 76 00	27.203	780	60 84 00	27.928
741	54 90 81	27.221	781	60 99 61	27.946
742	55 05 64	27.240	782	61 15 24	27.964
743	55 20 49	27.258	783	61 30 89	27.982
744	55 35 36	27.276	784	61 46 56	28.000
745	55 50 25	27.295	785	61 62 25	28.018
746	55 65 16	27.313	786	61 77 96	28.036
747	55 80 09	27.331	787	61 93 69	28.054
748	55 95 04	27.350	788	62 09 44	28.071
749	56 10 01	27.368	789	62 25 21	28.089
750	56 25 00	27.386	790	62 41 00	28.107
751	56 40 01	27.404	791	62 56 81	28.125
752	56 55 04	27.423	792	62 72 64	28.142
753	56 70 09	27.441	793	62 88 49	28.160
754	56 85 16	27.459	794	63 04 36	28.178
755	57 00 25	27.477	795	63 20 25	28.196
756	57 15 36	27.495	796	63 36 16	28.213
757	57 30 49	27.514	797	63 52 09	28.231
758	57 45 64	27.532	798	63 68 04	28.249
759	57 60 81	27.550	799	63 84 01	28.267
760	57 76 00	27.568	800	64 00 00	28.284

Table VIII *Squares and Square Roots of the Numbers from 1 to 1,000 (Continued)*

Number	*Square*	*Square Root*	*Number*	*Square*	*Square Root*
801	64 16 01	28.302	841	70 72 81	29.000
802	64 32 04	28.320	842	70 89 64	29.017
803	64 48 09	28.337	843	71 06 49	29.034
804	64 64 16	28.355	844	71 23 36	29.052
805	64 80 25	28.373	845	71 40 25	29.069
806	64 96 36	28.390	846	71 57 16	29.086
807	65 12 49	28.408	847	71 74 09	29.103
808	65 28 64	28.425	848	71 91 04	29.120
809	65 44 81	28.443	849	72 08 01	29.138
810	65 61 00	28.460	850	72 25 00	29.155
811	65 77 21	28.478	851	72 42 01	29.172
812	65 93 44	28.496	852	72 59 04	29.189
813	66 09 69	28.513	853	72 76 09	29.206
814	66 25 96	28.531	854	72 93 16	29.223
815	66 42 25	28.548	855	73 10 25	29.240
816	66 58 56	28.566	856	73 27 36	29.257
817	66 74 89	28.583	857	73 44 49	29.275
818	66 91 24	28.601	858	73 61 64	29.292
819	67 07 61	28.618	859	73 78 81	29.309
820	67 24 00	28.636	860	73 96 00	29.326
821	67 40 41	28.653	861	74 13 21	29.343
822	67 56 84	28.671	862	74 30 44	29.360
823	67 73 29	28.688	863	74 47 69	29.377
824	67 89 76	28.705	864	74 64 96	29.394
825	68 06 25	28.723	865	74 82 25	29.411
826	68 22 76	28.740	866	74 99 56	29.428
827	68 39 29	28.758	867	75 16 89	29.445
828	68 55 84	28.775	868	75 34 24	29.462
829	68 72 41	28.792	869	75 51 61	29.479
830	68 89 00	28.810	870	75 69 00	29.496
831	69 05 61	28.827	871	75 86 41	29.513
832	69 22 24	28.844	872	76 03 84	29.530
833	69 38 89	28.862	873	76 21 29	29.547
834	69 55 56	28.879	874	76 38 76	29.563
835	69 72 25	28.896	875	76 56 25	29.580
836	69 88 96	28.914	876	76 73 76	29.597
837	70 05 69	28.931	877	76 91 29	29.614
838	70 22 44	28.948	878	77 08 84	29.631
839	70 39 21	28.965	879	77 26 41	29.648
840	70 56 00	28.983	880	77 44 00	29.665

Table VIII *Squares and Square Roots of the Numbers from 1 to 1,000 (Continued)*

Number	*Square*	*Square Root*	*Number*	*Square*	*Square Root*
881	77 61 61	29.682	921	84 82 41	30.348
882	77 79 24	29.698	922	85 00 84	30.364
883	77 96 89	29.715	923	85 19 29	30.381
884	78 14 56	29.732	924	85 37 76	30.397
885	78 32 25	29.749	925	85 56 25	30.414
886	78 49 96	29.766	926	85 74 76	30.430
887	78 67 69	29.783	927	85 93 29	30.447
888	78 85 44	29.799	928	86 11 84	30.463
889	79 03 21	29.816	929	86 30 41	30.480
890	79 21 00	29.833	930	86 49 00	30.496
891	79 38 81	29.850	931	86 67 61	30.512
892	79 56 64	29.866	932	86 86 24	30.529
893	79 74 49	29.883	933	87 04 89	30.545
894	79 92 36	29.900	934	87 23 56	30.561
895	80 10 25	29.916	935	87 42 25	30.578
896	80 28 16	29.933	936	87 60 96	30.594
897	80 46 09	29.950	937	87 79 69	30.610
898	80 64 04	29.967	938	87 98 44	30.627
899	80 82 01	29.983	939	88 17 21	30.643
900	81 00 00	30.000	940	88 36 00	30.659
901	81 80 01	30.017	941	88 54 81	30.676
902	81 36 04	30.033	942	88 73 64	30.692
903	81 54 09	30.050	943	88 92 49	30.708
904	81 72 16	30.067	944	89 11 36	30.725
905	81 90 25	30.083	945	89 30 25	30.741
906	82 08 36	30.100	946	89 49 16	30.757
907	82 26 49	30.116	947	89 68 09	30.773
908	82 44 64	30.133	948	89 87 04	30.790
909	82 62 81	30.150	949	90 06 01	30.806
910	82 81 00	30.166	950	90 25 00	30.822
911	82 99 21	30.183	951	90 44 01	30.838
912	83 17 44	30.199	952	90 63 04	30.854
913	83 35 69	30.216	953	90 82 09	30.871
914	83 53 96	30.232	954	91 01 16	30.887
915	83 72 25	30.249	955	91 20 25	30.903
916	83 90 56	30.265	956	91 39 36	30.919
917	84 08 89	30.282	957	91 58 49	30.935
918	84 27 24	30.299	958	91 77 64	30.952
919	84 45 61	30.315	959	91 96 81	30.968
920	84 64 00	30.332	960	92 16 00	30.984

Table VIII *Squares and Square Roots of the Numbers from 1 to 1,000 (Continued)*

Number	*Square*	*Square Root*	*Number*	*Square*	*Square Root*
961	92 35 21	31.000	981	96 23 61	31.321
962	92 54 44	31.016	982	96 43 24	31.337
963	92 73 69	31.032	983	96 62 89	31.353
964	92 92 96	31.048	984	96 82 56	31.369
965	93 12 25	31.064	985	97 02 25	31.385
966	93 31 56	31.081	986	97 21 96	31.401
967	93 50 89	31.097	987	97 41 69	31.417
968	93 70 24	31.113	988	97 61 44	31.432
969	93 89 61	31.129	989	97 81 21	31.448
970	94 09 00	31.145	990	98 01 00	31.464
971	94 28 41	31.161	991	98 20 81	31.480
972	94 47 84	31.177	992	98 40 64	31.496
973	94 67 29	31.193	993	98 60 49	31.512
974	94 86 76	31.209	994	98 80 36	31.528
975	95 06 25	31.225	995	99 00 25	31.544
976	95 25 76	31.241	996	99 20 16	31.559
977	95 45 29	31.257	997	99 40 09	31.575
978	95 64 84	31.273	998	99 60 04	31.591
979	95 84 41	31.289	999	99 80 01	31.607
980	96 04 00	31.305	1000	100 00 00	31.623

Table IX Populations, Suicides per 100,000, Divorces per 100,000 for 229 Standard Metropolitan Statistical Areas, United States. 1970

SMSA	Population (in Thousands)	Suicide Rate (per 100,000)	Divorce Rate (per 100,000 14 Years and Over)
Abilene, Texas	114	7.9	24.9
Akron, Ohio	679	11.6	30.4
Albany, Georgia	90	3.3	24.1
Albany, New York	722	7.3	15.4
Albuquerque, New Mexico	316	20.6	34.8
Allentown, Pennsylvania	544	13.6	20.9
Altoona, Pennsylvania	135	14.0	18.9
Amarillo, Texas	144	13.2	33.1
Anaheim, California	1,420	16.1	34.1
Anderson, Indiana	138	6.5	40.4
Ann Arbor, Michigan	234	10.7	23.7
Appleton, Wisconsin	277	7.9	18.0
Asheville, North Carolina	145	11.7	24.0
Atlanta, Georgia	1,390	14.7	33.4
Atlantic City, New Jersey	175	7.4	24.7
Augusta, Georgia	253	13.0	26.4
Austin, Texas	296	9.1	26.8
Bakersfield, California	329	14.6	36.1
Baltimore, Maryland	2,071	10.7	25.8
Baton Rouge, Louisiana	285	11.6	18.5
Bay City, Michigan	117	18.7	23.8
Beaumont, Texas	316	8.9	31.4
Billings, Montana	87	9.2	32.6
Biloxi, Mississippi	135	7.4	32.1
Binghamton, New York	303	7.6	16.9
Birmingham, Alabama	739	9.2	30.7
Bloomington, Illinois	104	6.7	19.9
Boise City, Idaho	112	17.8	37.4
Boston, Massachusetts	2,754	9.4	20.6
Bridgeport, Connecticut	390	24.9	21.2
Brockton, Massachusetts	190	11.6	25.2
Brownsville, Texas	140	5.0	19.1
Bryan, Texas	58	5.2	15.7
Buffalo, New York	1,349	9.3	17.2
Canton, Ohio	372	9.7	30.0
Cedar Rapids, Iowa	163	10.4	29.8
Champaign-Urbana, Illinois	163	8.6	17.2
Charleston, South Carolina	304	9.5	16.6
Charleston, West Virginia	230	11.3	25.5
Charlotte, North Carolina	409	12.0	18.8
Chattanooga, Tennessee	306	7.8	32.3

Table IX *Populations, Suicides per 100,000, Divorces per 100,000 for 229 Standard Metropolitan Statistical Areas, United States, 1970 (Continued)*

SMSA	Population (in Thousands)	Suicide Rate (per 100,000)	Divorce Rate (per 100,000 14 Years and Over)
Chicago, Illinois	6,975	8.4	29.1
Cincinnati, Ohio	1,385	11.6	28.9
Cleveland, Ohio	2,064	13.1	30.6
Colorado Springs, Colorado	236	11.9	26.0
Columbia, Missouri	81	9.9	18.4
Columbia, South Carolina	323	7.7	18.7
Columbus, Georgia	239	7.1	25.3
Columbus, Ohio	916	13.2	34.0
Corpus Christi, Texas	285	14.0	25.7
Dallas, Texas	1,556	11.8	39.9
Davenport, Iowa	363	9.7	31.7
Dayton, Ohio	850	11.1	34.8
Decatur, Illinois	125	12.8	29.7
Denver, Colorado	1,228	21.8	37.3
Des Moines, Iowa	286	12.2	34.5
Detroit, Michigan	4,200	11.3	32.6
Dubuque, Iowa	91	7.7	15.6
Duluth, Minnesota	265	17.0	26.2
Durham, North Carolina	190	7.4	17.2
El Paso, Texas	359	10.0	24.4
Erie, Pennsylvania	264	7.2	21.1
Eugene, Oregon	213	14.1	32.6
Evansville, Indiana	233	9.5	33.8
Fall River, Massachusetts	150	20.0	19.0
Fargo, North Dakota	120	6.7	17.8
Fayetteville, North Carolina	212	7.5	17.5
Flint, Michigan	497	9.1	31.4
Fort Lauderdale, Florida	620	21.0	32.7
Fort Smith, Arkansas	160	6.9	35.0
Fort Wayne, Indiana	280	6.4	28.8
Fort Worth, Texas	762	12.7	37.7
Fresno, California	413	16.0	31.8
Gadsden, Alabama	94	10.6	26.6
Gainesville, Florida	105	14.3	23.3
Galveston, Texas	170	19.4	40.8
Gary, Indiana	633	8.4	32.5
Grand Rapids, Michigan	539	8.7	24.3
Great Falls, Montana	82	13.5	37.2
Green Bay, Wisconsin	158	8.8	17.7
Greensboro, North Carolina	604	12.3	20.2
Greensville, South Carolina	300	11.0	18.9

Table IX *Populations, Suicides per 100,000, Divorces per 100,000 for 229 Standard Metropolitan Statistical Areas, United States, 1970 (Continued)*

SMSA	Population (in Thousands)	Suicide Rate (per 100,000)	Divorce Rate (per 100,000 14 Years and Over)
Hamilton, Ohio	226	12.8	27.1
Harrisburg, Pennsylvania	411	9.0	21.3
Hartford, Connecticut	664	12.1	22.1
Honolulu, Hawaii,	629	9.5	28.7
Houston, Texas	1,985	14.3	38.7
Huntington, West Virginia	254	8.7	30.4
Huntsville, Alabama	228	8.8	22.7
Indianapolis, Indiana	1,110	11.5	36.9
Jackson, Michigan	143	11.2	42.0
Jackson, Mississippi	259	8.5	22.3
Jacksonville, Florida	529	12.9	37.3
Jersey City, New Jersey	609	3.8	18.1
Johnstown, Pennsylvania	263	8.0	14.5
Kalamazoo, Michigan	202	12.4	25.1
Kansas City, Missouri	1,254	11.5	36.7
Kenosha, Wisconsin	118	11.0	25.0
Knoxville, Tennessee	400	12.7	27.6
La Crosse, Wisconsin	80	21.1	20.2
Lafayette, Louisiana	110	13.7	13.3
Lafayette, Indiana	109	7.3	21.1
Lake Charles, Louisiana	145	6.9	19.8
Lancaster, Pennsylvania	320	8.8	19.6
Lansing, Michigan	378	10.0	24.0
Laredo, Texas	73	2.7	14.6
Las Vegas, Nevada	273	22.0	68.8
Lawton, Oklahoma	108	4.6	28.0
Lexington, Kentucky	174	16.1	34.1
Lima, Ohio	171	9.9	26.2
Lincoln, Nebraska	168	7.7	26.7
Little Rock Arkansas	323	9.6	36.6
Lorain, Ohio	257	12.8	24.4
Los Angeles, California	7,036	20.9	52.8
Louisville, Kentucky	827	12.5	32.5
Lubbock, Texas	179	10.6	24.3
Lynchburg, Virginia	123	9.7	22.8
Macon, Georgia	206	14.5	27.5
Madison, Wisconsin	290	8.6	20.6
Manchester, New Hampshire	108	17.5	24.3
Mansfield, Ohio	130	11.5	31.3
McAllen, Texas	182	7.2	13.2
Memphis, Tennessee	770	11.8	27.1

Table IX *Populations, Suicides per 100,000, Divorces per 100,000 for 229 Standard Metropolitan Statistical Areas, United States, 1970 (Continued)*

SMSA	Population (in Thousands)	Suicide Rate (per 100,000)	Divorce Rate (per 100,000 14 Years and Over)
Miami, Florida	1,268	19.0	41.2
Midland, Texas	65	12.2	27.7
Milwaukee, Wisconsin	1,404	12.0	28.3
Minneapolis, Minnesota	1,814	9.4	26.7
Mobile, Alabama	377	8.2	27.6
Modesto, California	195	12.3	38.1
Monroe, Louisiana	115	8.7	21.7
Montgomery, Alabama	201	11.4	28.6
Muncie, Indiana	129	18.6	33.3
Muskegon, Michigan	157	7.6	27.6
Nashville, Tennessee	541	16.1	32.0
New Haven, Connecticut	356	15.2	19.1
New Orleans, Louisiana	1,046	9.0	26.7
New York, New York	11,572	6.3	17.2
Newark, New Jersey	1,857	5.5	16.7
Newport News, Virginia	292	8.9	24.0
Norfolk, Virginia	681	9.6	23.9
Odessa, Texas	92	6.5	33.7
Ogden, Utah	126	12.7	32.3
Oklahoma City, Oklahoma	641	10.0	41.8
Omaha, Nebraska	540	10.7	30.6
Orlando, Florida	428	11.7	30.5
Owensboro, Kentucky	79	7.5	25.4
Oxnard, California	376	12.8	30.9
Paterson, New Jersey	1,359	6.3	14.5
Pensacola, Florida	243	10.3	27.9
Peoria, Illinois	342	9.6	29.3
Petersburg, Virginia	129	10.1	22.0
Philadelphia, Pennsylvania	4,818	9.8	20.1
Phoenix, Arizona	968	14.0	35.5
Pine Bluff, Arkansas	85	10.5	26.5
Pittsburgh, Pennsylvania	2,401	11.9	20.7
Pittsfield, Massachusetts	80	20.1	20.9
Portland, Maine	142	16.9	32.7
Portland, Oregon	1,009	15.1	45.6
Providence, Rhode Island	913	6.6	20.6
Provo, Utah	138	9.4	15.2
Pueblo, Colorado	118	9.3	36.6
Racine, Wisconsin	171	10.5	25.5
Raleigh, North Carolina	228	11.4	19.4
Reading, Pennsylvania	296	17.2	29.8

***Table IX** Populations, Suicides per 100,000, Divorces per 100,000 for 229 Standard Metropolitan Statistical Areas, United States, 1970 (Continued)*

SMSA	Population (in Thousands)	Suicide Rate (per 100,000)	Divorce Rate (per 100,000 14 Years and Over)
Reno, Nevada	121	24.8	73.6
Richmond, Virginia	518	14.1	28.1
Roanoke, Virginia	181	15.4	29.1
Rochester, Minnesota	84	25.0	17.5
Rochester, New York	883	12.0	17.1
Rockford, Illinois	272	14.0	32.2
Sacramento, California	801	16.0	37.8
Saginaw, Michigan	220	7.3	24.6
St. Joseph, Missouri	87	18.4	38.9
St. Louis, Missouri	2,363	9.8	28.8
Salem, Oregon	187	16.1	35.0
Salinas, California	250	11.2	33.8
Salt Lake City, Utah	558	16.5	30.8
San Angelo, Texas	71	22.5	28.5
San Antonio, Texas	864	10.5	29.5
San Bernardino, California	1,140	14.9	36.8
San Diego, California	1,358	17.9	37.6
San Francisco, California	3,110	22.1	47.2
San Jose, California	1,065	15.1	35.6
Santa Barbara, California	264	12.9	33.8
Santa Rosa, California	205	22.5	35.3
Savannah, Georgia	188	8.5	30.5
Scranton, Pennsylvania	234	9.4	16.1
Seattle, Washington	1,422	16.9	44.7
Sherman, Texas	83	6.0	30.0
Shreveport, Louisiana	294	11.6	25.6
Sioux City, Iowa	116	15.5	28.6
Sioux Falls, South Dakota	95	7.4	24.1
South Bend, Indiana	280	10.7	24.6
Spokane, Washington	287	16.7	39.9
Springfield, Illinois	161	13.0	29.5
Springfield, Missouri	153	11.8	32.3
Springfield, Ohio	157	10.2	32.4
Springfield, Massachusetts	530	11.7	24.7
Steubenville, Ohio	166	8.5	26.7
Stockton, California	290	16.2	39.7
Syracuse, New York	637	9.9	16.8
Tacoma, Washington	411	13.6	37.5
Tallahassee, Florida	103	11.6	25.5
Tampa, Florida	1,013	16.4	35.7
Terre Haute, Indiana	175	12.0	34.8

Table IX *Populations; Suicides per 100,000, Divorces per 100,000 for 229 Standard Metropolitan Statistical Areas, United States, 1970 (Continued)*

SMSA	Population (in Thousands)	Suicide Rate (per 100,000)	Divorce Rate (per 100,000 14 Years and Over)
Texarkana, Arkansas	101	10.9	35.8
Toledo, Ohio	693	15.2	31.6
Topeka, Kansas	155	12.2	35.9
Trenton, New Jersey	304	8.9	19.0
Tucson, Arizona	352	15.6	31.9
Tulsa, Oklahoma	477	9.4	42.8
Tuscaloosa, Alabama	116	5.2	27.7
Tyler, Texas	97	11.3	29.2
Utica, New York	341	10.0	15.6
Vallejo, California	249	12.8	40.7
Vineland, New Jersey	121	6.6	22.9
Waco, Texas	148	10.8	34.6
Washington, D.C.	2,861	9.8	25.2
Waterloo, Iowa	133	12.8	23.9
West Palm Beach, Florida	349	16.3	32.5
Wheeling, West Virginia	183	10.9	26.3
Wichita, Kansas	389	9.2	34.6
Wichita Falls, Texas	126	10.3	31.4
Wilkes-Barre, Pennsylvania	342	6.7	15.2
Wilmington, Delaware	499	13.2	24.3
Wilmington, North Carolina	107	11.2	20.2
Worcester, Massachusetts	344	18.3	21.7
York, Pennsylvania	330	12.7	22.9
Youngstown, Ohio	536	9.9	26.8

Sources: For suicide rates—U.S. Department of Health, Education and Welfare, *Vital Statistics of the United States: 1970*, Vol. II, Mortality, Part B. Washington, D.C.: U.S. Government Printing Office. 1974. For SMSA's—U.S. Bureau of the Census, *County and City Data Book, 1972* (*A Statistical Abstract Supplement*). Table 3. Washington, D.C.: U.S. Government Printing Office. 1973. For divorce rates—U.S. Bureau of the Census, "Marital Status and Household Relationship by Race and Sex for Areas and Place: 1970," *Census of Population: 1970 General Population Characteristics*. Washington, D.C.: U.S. Government Printing Office. 1974.

Index of Names

Index of Subjects

BCDEFGHIJ–H–7987